Mr Carbon Ato

And The Theory Of

Carbon Entromorphology

Mark Andrew Janes

Publishing details

First edition published in January 2012

ISBN 978-1-907140-53-2

Images from Mark Andrew Janes & Istockphoto.com (under licence) and
GimpSavy.com public domain images (Sun, Earth and spaceman)

Periodic table data taken from the Royal Society of Chemistry's site
http://www.rsc.org/education/teachers/learnnet/ptdata/table/index.htm

IMPORTANT STATEMENT ABOUT OFFENSIVE CONTENT

This book contains text and some imagery associated with sexual reproduction utilised in a scientific context, for scientific demonstration. It also contains a section which contains swear words and other offensive medical images which may not be suitable for children.

Proof read and edited by Martin Williams 2011.

CONTENTS

PART THREE

The 'blind physicist'

PART FOUR

The 'quantum gender' model

PART FIVE

Neurological genetics through an extension to classical genetics

PART SIX

Analytical hypotheses in support of carbon entromorphology

PART SEVEN

Frustrating progress – a letter

PART EIGHT

The fractional dimensions of natural living organisation

PART NINE

Extending evolution by natural selection

(Neurological genetics continued)

PART TEN

The physical consequence of amplified atomic properties

'Carbon electromorphology'

- Neurochromosome electromorphology - the biological effects of quantum gravity, emerging structures with electromagnetic origins and characteristic morphologies, pages 224 to 231.
- Neurochromosome extended cell cycle - an extension to cell size and function and its consequence for the cell cycle, pages 232 to 240.
- Neurochromosome bio electric structures - conservation of electromagnetic atomic logic in amplified entromorphic atoms, pages 240 to 242.

PART ELEVEN

Conservation of atomic logic

- Neurochromosome entrochiraloctets - the basic carbon based fractionally dimensional template for living anatomy and physiology, pages 243 to 245.
- Neurochromosome quantum taxonomy - living classification of fractionally dimensional systems based on carbon logic, pages 245 to 249.
- Neurochromosome metabolic organelles - quantum energy balancing through the living valance shell and conductance band by ECASSE, pages 249 to 253.

PART TWELVE

Thermodynamics and morality

Acknowledgments

Dedicated to my mother Margaret and father Richard and Pip (my little electron), and the memory of Ellie and the rest of my wonderful family for their support. And especially to my niece Emma for her enthusiasm, ideas, help and support as my personal assistant. A huge thank you to my brother Stephen, for his help in all areas of my work, including his excellent advice regarding publication and graphic design. To Martin Williams for proof reading and editing.

Thanks also to Andy at Apex Macclesfield for his excellent help with graphics. A huge thanks to the profession of physics for doing the really really hard work, I am constantly in amazement at their staggering achievements in creating quantum theory in the first place. I have just simply extended its use and my only criticism concerns its current lack of application in biology. Thanks to the old group at prontoprint Macclesfield and especially to Ryan for his incredible help with graphic design.

To Libb Thims and all the people associated with the 'Encyclopaedia of Human Thermodynamics EoHT'–Libb has made incredible contributions to the development of human thermodynamics.

Also to my colleagues and friends at AstraZeneca for their incredible support over the years.

About the author

Mark Andrew Janes has a BSc (Hons) degree in Biology and is a full member of the 'Society of Biology MSB' and a 'Chartered Biologist CBiol'. Having worked in the pharmaceutical industry for over fifteen years; working in analytical chemistry for over ten years, two years in cancer research, and then for ICI Pharmaceuticals in biotechnology working on all aspects of chemical and biochemical engineering and aseptic processing and fermentation on 'ricin immunotoxins'.

The past four years have been devoted to developing 'carbon entromorphology' and the concepts associated with it, battling with a highly conservative scientific world especially in Great Britain. Carbon entromorphology is the culmination of twenty years of scientific study, investigation and development. Carbon entromorphology is also a modern art form and forms the basis of a way of life; its author has developed the ideas on the website www.mrcarbonatom.com.

There is also a vast array of lectures available through the website homepage, covering all aspects of the subject and the gritty scientific citations associated with the theory. The website has extensive audio lectures supporting the published material and a vast number of broad references and citations.

Carbon entromorphology is also documented thoroughly through the 'Encyclopaedia of Human Thermodynamics' which can be found on the website at www.EoHT.com.

Mark Andrew Janes in 'South Park' Macclesfield in autumn of 2007.

Preface

Identification of a major problem regarding resolution across science that segregates knowledge.

When one tries to consider the true extent of the conscious experience which is referred to broadly as life, we find ourselves in a cornucopia of extraordinary complexity. Life even at its simplest level is ludicrously complex where even a simple protozoan, a single celled organism, invisible to the naked eye produces obscene levels of variation and functional complexity. The profession of biology is all too aware of this complexity and with it the limitations of physical evaluation and modelling more commonly associated with the physical sciences; through sound numerical measurement and mathematical modelling. Living organisms therefore cannot be considered against conventional measurement, and mathematical treatment and investigation. The trillions of atoms which make the protozoan are impossible to measure and as such initial state conditions are typically impossible to establish with any physical meaning. But unlike the atom which is illusive and incomprehensible and subject to 'none visual properties' (where only mathematical models make sense of that level of natural scale) the protozoan can be visualised under the microscope. Its physical shape, colour, texture and internal complexity can be measured through observation and the biologist capitalises on this with glorious visual evidence of the immense variation in life. The culmination of this ability is never better illustrated than through the great works of Robert Hooke, the 'Micrographia' 1665 where Hooke is quoted as saying : -

'The science of Nature.... should return to the plainness and soundness of observations.'

Compare this to the approach of one of the great fathers of modern atomic theory and quantum mechanics Niels Bohr who in the early part of the 20th century suggested that when it came to atoms :-

'Words (language) can only be used as poetry.'

Richard Feynman one of the grand masters of theoretical physics and QED (Quantum Electrodynamics) used a visual system to represent how atomic particles interact, and was resultantly ridiculed by his peers for suggesting that any aspect of atomic physics could be visualised under any circumstances. To a physicist, especially those concerned with the atom and it's illusive mysteries where atoms appear to be in more than one place, and produce 'counter-intuitiveness' this seems like a none scientific approach. This is due to their ability to form hard dense natural structures such as rock, although each atom within the rock is found to be more or less empty space. It is not easy to convince an atomic physicist that visible observation of everyday structures is an observation of atoms and their logic, where a physicist is constantly reaching for their calculator. As a result biology and physics remain elusive to each other and hold little understanding of each other's limitations in measurement. As a result science is anything but integrated and any notion of bringing the two together in one system of logic seems ludicrously futile. Physics is mathematically observable and biology is at best statistically and probabilistically observable, until science bridges the gap then the conscious nature of existence will remain a distant fantasy.

There is a 'resolution problem' in science which produces segregation of the natural levels of scale. Physicists have nothing to do with biologists because biology is a largely none mathematical system using words and pictures, and vice versa where a biologist has nothing to do with physics because it is a mathematical system, even though both groups are studying the nature of matter. This produces duplication of effort and knowledge such as nuclear physics (physics) and genetic engineering (biology), theoretically considered by this author as the same logic on different levels of scale.

Carbon entromorphology is a new broad theory that acts to organise and explain the nature of living conscious awareness and the mind, by unification of knowledge. It therefore acts to define life by spanning and integrating human knowledge (from the arts to the sciences), now defined by this theory as the basis of the 'human neurological genome' (all human knowledge but expressed by the logic of genetics). In carbon entromorphology life is described as a 'soulatrophic pathway' from the word the 'soul' meaning the origin of living things. It runs all the way back to the 'Big Bang' and is defined by specific discrete natural levels such as atomic, microbial, somatic (radial), cognitive (bilateral) and spiritual (familial). Mathematically the pathway is described as a 'fractional dimension D'; a fractal geometry. Each natural level of scale is self-symmetrical to the other levels and is driven by iterative (an unbroken ancestry) cell cycling and is made up of the fraction of the lower levels and acts as a fraction of the levels above it. Building up levels of organisation to produce an integrated complete organism (the diagrams in part 1on page 1 will simplify this process for the reader).

This author's motivation to establish these new theories is founded in his everyday experience as three people in one body due to a mood disorder; the passionate deep-seated nature of my mind is utterly prolific and my hyperactive mania drives my creativity. I call it focussed mania because it's highly productive but under some physiological control. The daily toil of a man and his extraordinary and deeply analytical mind and mental instability has precipitated this theory due to the fact that I originally started thinking of my moods as 'energy levels' which is an expression associated with quantum mechanics.

Was I identifying the quantum mechanics of my own life? I suspect this is what has happened here. Depression as a dominant particle state and mania as a dominant waveform, in other words atomic physics was all around me, the very essence of me. Fundamentally I have attempted to explained how my mind works using a cross fertilisation of genetics and nuclear physics as a basis and as a result, I have also discovered life on the level of the humble atom and on the periodic table.

I realised some time ago that the brain could be hypothesised to be a huge super coiled condensed chromosome, a neurological chromosome. That's why it looks the way it does. On the level of a classical cell, DNA super coils into classical chromosomes; which have a very similar organisation? I realised that a book could also be hypothesised to be a type of chromosome; this article you are reading is also hypothesised as a nucleonic chromosome, so is a CD, a DVD, also the humble videotape, any recording information device. The very words you are reading are organised symmetrically i.e. in a binary information linear helix in direct comparison to the DNA in this author's body. The words are the genes (memories), neurological genes; the sentences form basic genomes and build up to super genomes such as the entire subject of physics. The black and white binary nature of the words are hypothesised as a powerful illustration of the four forces of nature, and specifically the strong interaction and its gluonic binding acting to hold words and therefore language together with its pionic links forming sentences and quark like letters. An extraordinary awakening for humanity to the hidden truth around us every day; the nuclear physics of language and information systems. Again this is further explained and supporting evidence is presented throughout the book.

I am constantly aware of the four forces in my every waking moment, another reason why my theory stands strongly. When one reads a gene in a cell the chromosome is opened up, in the same way you open a book in the same way you opened up this article. The chromosome has a centromere; your book has its binding in the same way, a DVD, CD ROM, an LP record all have centromere holes in the middle. They are all chromosomic and all store information through super coiled information systems, think of following the groove on a CD or record, could it be based on the helix in DNA? More agreement and continuity between fundamental logic and these new theories. When you read, as you are doing right now your eyes go from left to right then right to left to read the next line down. You are in fact forming a helix because information and genes are the very essence of the four-force model of reality and they interact as an open and closed linear super coil and the nature of quantum information is driven by expression through electroweak logic. You read the words and interpret them into actions (proteinascious forms) this is done through your electroweak interface namely your body, your hands open the book your eyes read the detail, your cell also reads and interprets your DNA genes into proteins by the same logic. To myself quantum mechanics and nuclear physics is all around me (contrary to current thinking in physics), it allows me to make sense of all life by seeing fermionic and bosonic logic and the four forces of nature everywhere I look.

The thermodynamic driver for these theories is called ECASSE which stands for 'Entromorphic Carbon Amplification by Solar Stimulated Emission' the same type of logic as a LASER only for carbon based particles with significant rest mass, partially sharing similar energy levels (flock of birds). In essence life is like a coherent beam of particles in closely related energy states like a LASER beam, which is a process called 'stimulated bosonic emission'. A flock of birds is an ECASSE beam, Manchester United football club operates with in-phase coherence >60,000 stadium supporters per game and is therefore an ECASSE beam. By reading this article you are demonstrating nuclear and field amplification as some of the terms you are reading are deposited to memory (condensed matter and memories), this is a non locality effect (which will be explained later on) in your mind, the entromorphic genome has been passed through a four-force logic interface with its surroundings. To reiterate the effect is theorized by LASER logic a process called 'Bose condensate'. In LASER beams, one photon stimulate release of another, one becomes two. In ECASSE one classical cell divides into two through mitosis and undergoes mitosis twice in meiosis, again stimulated emission where the energy has come from phototropism (the Sun).

The energy for this ECASSE beam comes from the Sun, the Sun pumps 'free energy' photons into the Earth's open system, into a gain medium known as the atmosphere and the oceans where the photons concentrate and become amplified by Bose condensate logic. The same process comparative logic but an intermediate boson as life contains lots of mass, the beam still has the distribution conserved from carbon atomic physics, which makes it a partial fermion and a partial boson (for associated particle statistics), the particle wave duality of this in-phase coherent particle is fermionically and bosonically conserved. Neurological genomics included are models of law and order as more evidence of the strong interaction in amplified carbon cells. Atomic nuclei, DNA nuclei, neurological nuclei, technological nuclei are all-nucleonic and act as amplified regions of atomic logic, the expression 'stored information ROM' is a nucleonic strong interaction. **As you read this book these theories will become clearer, so please stay with it if you feel a little lost at the present time!**

We also find a bit of shock when we examine the evolution of science because science and religion display complementarities in this new science, which seems unlikely. Where science accepts uncertainty in measurement and religion accepts that all life follows evolutionary patterns, also science must accept a goal for evolution to avoid death and aiming to live almost indefinitely. My work predicts breathtaking and beautiful consequences to this new and powerful spirituality as life enters the final stage of evolution the emergence of artificial life and divine stability in a heavenly energy rich future. The following images will help to clarify these theories which are difficult to understand because they are very broad in scope, as a result there are a great many images used in this book to try to help the reader understand often complicated scientific ideas, and support the theory with very clear photographic evidence. Thank you for participating and I hope this book will be enlightening and thought provoking.

'May the events of your day bring you ever closer to quantum stability through phototropism!'

Best wishes and thank you for your interest,

Mark Andrew Janes

Mark Andrew Janes as 'Mr Carbon Atom' in my parents garden in Macclesfield in autumn of 2009.

The basic fundamental principles of fractal geometry

Carbon entromorphology is a scientific approach to modelling all the natural levels of scale in living organisms using the powerful logic of 'fractal geometry' – the true geometry of natural order. In nature humans, in particular, are used to a very wide perception of natural scale. For the most broadly educated individual, this ranges from the tiny world of atoms and their challengingly small components (99.9999999999999% empty space) to the unimaginable vastness of the universe where insignificant dots in the night sky are actually galaxies millions of light years across.

These extremes are defined as the boundaries of modern physics but these boundaries are constantly being pushed further into the abyss of knowledge. Between these extremes is a vast array of other levels of natural detail and, in particular, those associated with life in its current limited definition. Life in particular modern humans are a unique case because they are self-aware of all the natural levels as they are composed of all those levels. From particle wave systems such as atoms to molecules, to cells, to tissues and organs to complete multicellular organisms and on to multi organism groups and so on, up to the enormous world of geological structures and the universe.

Consciousness exists within this spectrum of natural scale as a shifting focus, which ancient civilisations refer to as the soul. Science may refer to it as a point of consciousness but the fascinating reality of our understanding of what we are is that we cannot define it as an absolute particle point neither can we refer to it as a none specific wave, we are at best a 'living duality'.

Conventional geometry is called Euclidean geometry, a world of perfect shapes such as circles, lines and whole number dimensions. This perfect image of reality is very powerful but lacks the true reality of the way our existence is built up from the tiny world through to the massive world with seamless connectivity (accretion building processes; building blocks accumulate to form larger structures).

Fractal geometry is different from Euclidean geometry as it has the correct natural dichotomy of natural scale. In Euclidean geometry a line is still a line on any level of magnification, in fractal geometry a line when amplified more and more presents an infinite level of complexity. A shape such as a circle in fractal geometry has a fixed area but an infinite circumference as more and more detail appears through amplification. Another fascinating and powerful aspect of fractal geometry is 'self symmetry'. The original geometric seed known as the 'generating element' appears complete as part of the increasing detail in a fractal geometry at regular intervals. In the case of the circular shape the entire element appears over and over again as we amplify the generating lines.

Fractal geometry produces this very natural logic because of one other critical mathematical property called 'iteration'. Iteration is a mathematical process where a mathematical formula is applied time and time again. The resultant solution of the mathematical function f(x) is pumped back into the equation conserving the original generating functional element which may be a number input from a geometrical shape.

The greatest testament to the unimaginable mathematical beauty of fractal geometry has originated from the work of Henri Poincare and Arthur Cayley 100 years before the great Benôit Mandelbrot produced the Madelbrot set. This set of numbers was generated using the simplest of mathematical equations, but applying the power of iteration, to produce outrageously complex and mind bendingly beautiful shapes and order, infinitely complex but heralding the power of fractals to describe the true nature of Nature.

The iterative formula is: -

Biological growth involves iterative conserved cell cycling!

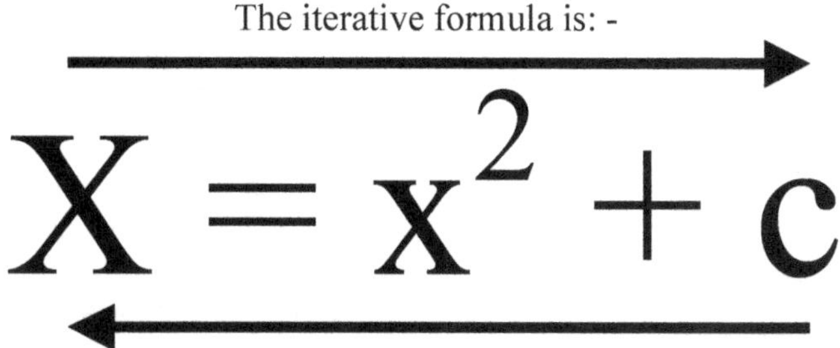

Firstly a value is chosen for c such as 0.5 which fixes the function and starting with x = 0, the first round of the function is 0.5 which becomes x for the second round, and so on cycling the numbers through the function over and over. It is important to re-emphasize that because of the mathematical cycling the original number generating element is always present, hence it appears in any geometrical representation as 'self symmetry'. This conserved approach is very important when considering living organisms and the logic of carbon entromorphology where carbon is the generating element for life.

Mandelbrot had to use more complexity than this using his equation: -

The cell cycling in nature is a demonstration of iterative determinism.

$$Z = z^2 + c$$

Where z and c are two dimensional complex numbers (a complex number has a real and imaginary component) producing a two dimensional set and the breathtaking beauty of the final set.

A simpler example of the fractal process is the Sierpiński gasket, and the Koch snowflake. Fractals have successfully been used to model geological landscapes, plants such as ferns where the overall structure is comprised of smaller copies (self symmetrical generating elements). Fractal geometry has been widely hypothesised to be the true logic of nature and through its use in carbon entromorphology (where carbon and even more fundamentally hydrogen are the self symmetrical generating elements) producing all naturally living structures.

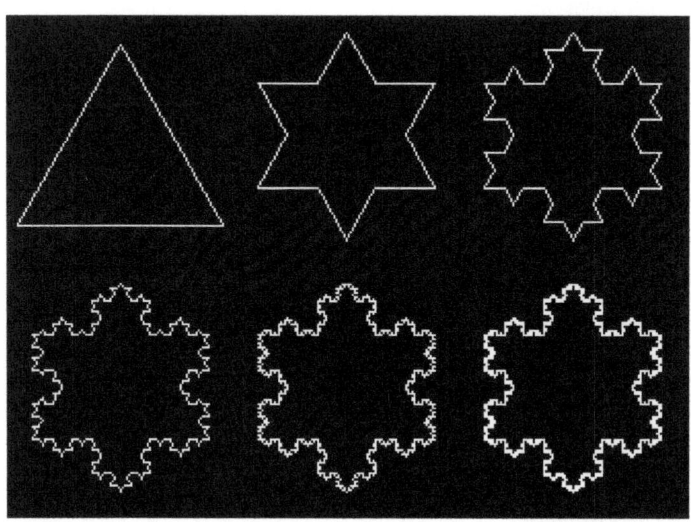

The Koch snowflake.

The increasing detail through each cycle can clearly be seen.

The equilateral triangle is the generating element; in the same way carbon (and even more fundamentally hydrogen) is the generating element in all life.

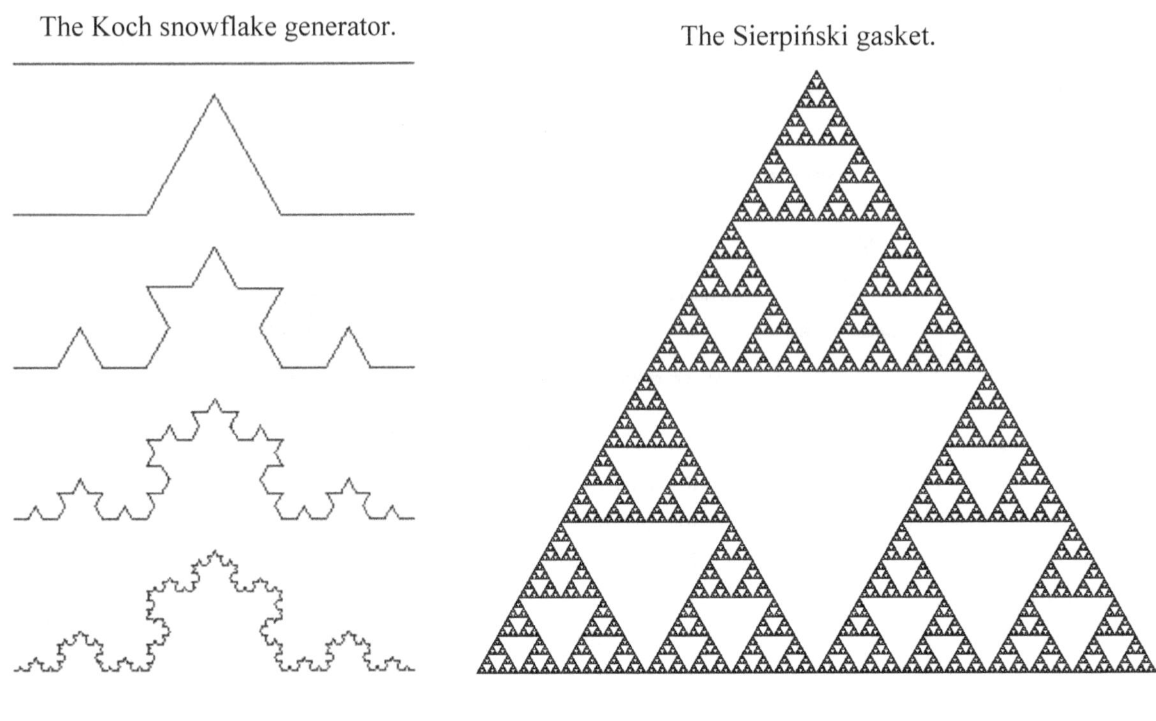

The Koch snowflake generator.

The Sierpiński gasket.

'Self symmetry' on smaller levels.

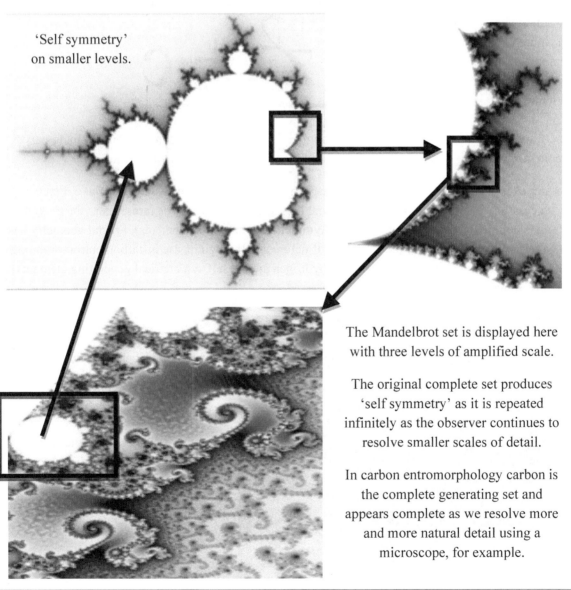

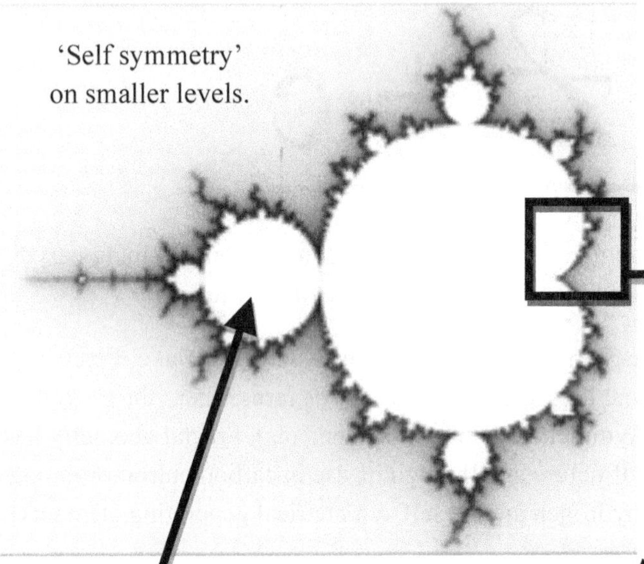

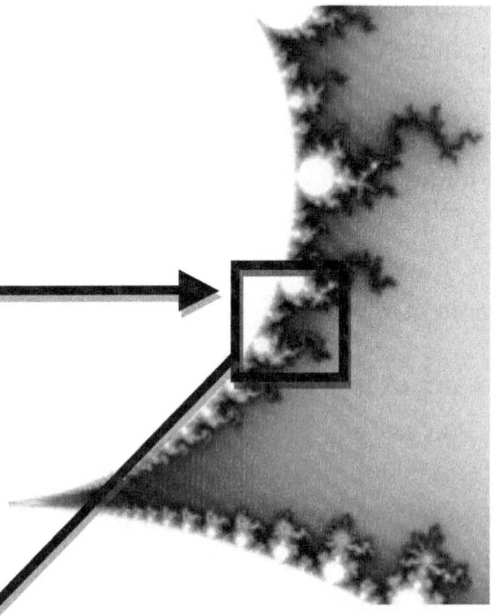

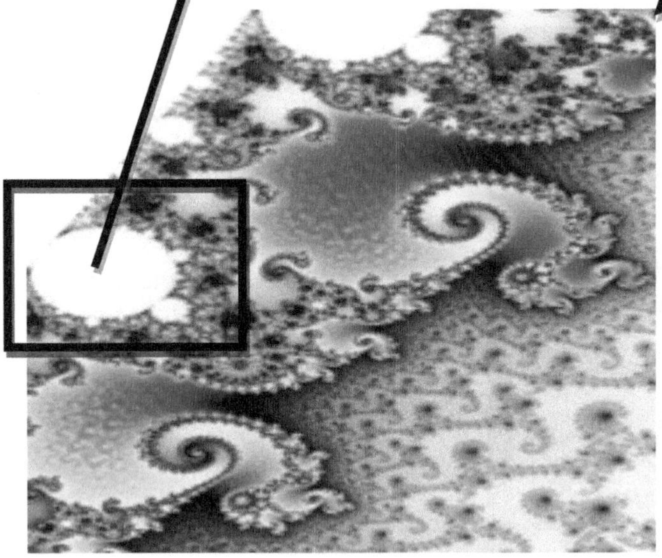

The Mandelbrot set is displayed here with three levels of amplified scale.

The original complete set produces 'self symmetry' as it is repeated infinitely as the observer continues to resolve smaller scales of detail.

In carbon entromorphology carbon is the complete generating set and appears complete as we resolve more and more natural detail using a microscope, for example.

'Self symmetry' in plants is extensively demonstrated.

Fractional dimensions

Felix Hausdorff investigated fractal logistics in a very innovative and practical mathematical way.

This point is particularly important when considering the fractal nature of carbon entromorphology. As mentioned earlier on in this section on fractals, a living conscious being is integrated into stacks of naturally occurring scale which define 'the observer', and covers the smallest and largest observable and measurable structures in the universe. Scale was Hausdorff's genius and his evaluation can be thought of through some simple steps.

1. A line (1 dimension in Euclidean geometry) scaled (amplified) up by a factor of 3 so the line is $3 = 3^1$ the index (power) is a whole number giving a dimension of 1.

2. A solid square (2 dimensions in Euclidean geometry) scaled (amplified) up by a factor 3 so the area is $9 = 3^2$ the index (power) is a whole number giving a dimension of 2.

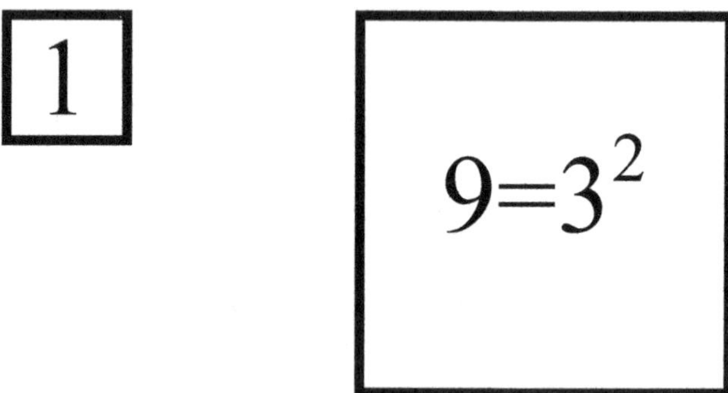

3. A solid cube (3 dimensions in Euclidean geometry) scaled (amplified) up by a factor 3 so the line is $27 = 3^3$ the index (power) is a whole number giving a dimension of 3.

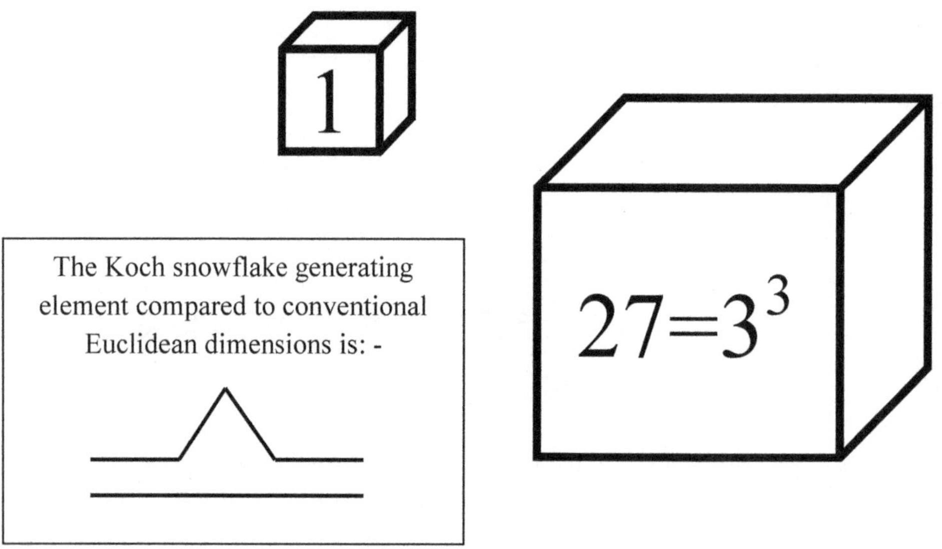

This is the expected logic for a line, square and cube. If the Koch curve is scaled up (amplified) by 3 then it becomes 4 times longer as the extra detail is present and the scale is resultantly 'fractional'.

The Hausdoff dimension takes into account this phenomenon by producing a mathematical method for presenting the integrated iterative logic of fractional dimension. Where: -

$$4 = 3^D$$

$$D = \frac{\log 4}{\log 3} = 1.262$$

With the fractal dimension D the value is typically more than the original ordinary Euclidian dimension. In this way we can consider all natural orders of scale as being fractional dimensions, such as the sub atomic fraction, the atomic fraction, the molecular fraction, the cellular fraction and so on and so forth. Each dimension seamlessly translates two integrated dimensions interfaced with each other.

Fractals have already been successfully used to model marine growth rates for somatic (radial) organisms of the most basic multi-cellularity, such as corals and sponges. This evidence already supports the use of fractal logic as the absolutely defining logic for modelling life.

Brain activity, population dynamics and many more aspects are opening up to the enormous power of fractals as the true geometry of nature. Each level of scale defined next in this book exists by its iterative link to the levels it is fractionally defined by and acts as a fractional dimension of any levels above it.

It is also true to postulate that the true value for fractional dimensions is 'e' the Naperian logarithm and 'π' the circular constant. In both quantum mechanics and the majority of biological science these constants appear everywhere and reflect natural dimensional structures such as field logic and population growth. The link to fractals is based on the growth rate in ALL living organisms which is based exclusively on 'e'. It is the absolute model of mitosis and meiosis, the geometric progression which produces all Earth based life; exponential growth.

It makes perfect sense that a structure composed of smaller building blocks should, when bonded into larger compositions, still retain the conserved logic of its building blocks. This is very important in biology because living organisms are built up from smaller naturally occurring building blocks like atoms, the process known as 'accretion'.

The academic spectrum and the collapse of scientific collaboration.

The segregation of the academic model of human knowledge is displayed below for the scientific components. It should be noted that 'the arts' and 'languages' can also be differentiated in this way to produce a universal academic spectrum. All aspects of scientific endeavour should be applicable to all academic investigation including 'the arts'. For example the works of William Shakespeare can be investigated using nuclear physics and quantum mechanics, in other words 'everything is physics'. This fits with the statements made by the father of modern atomic theory Ernest Rutherford where he stated '*everything is either science or stamp collecting*'; although this author feels that even stamp collecting is science. There is a 'resolution problem' across the academic spectrum.

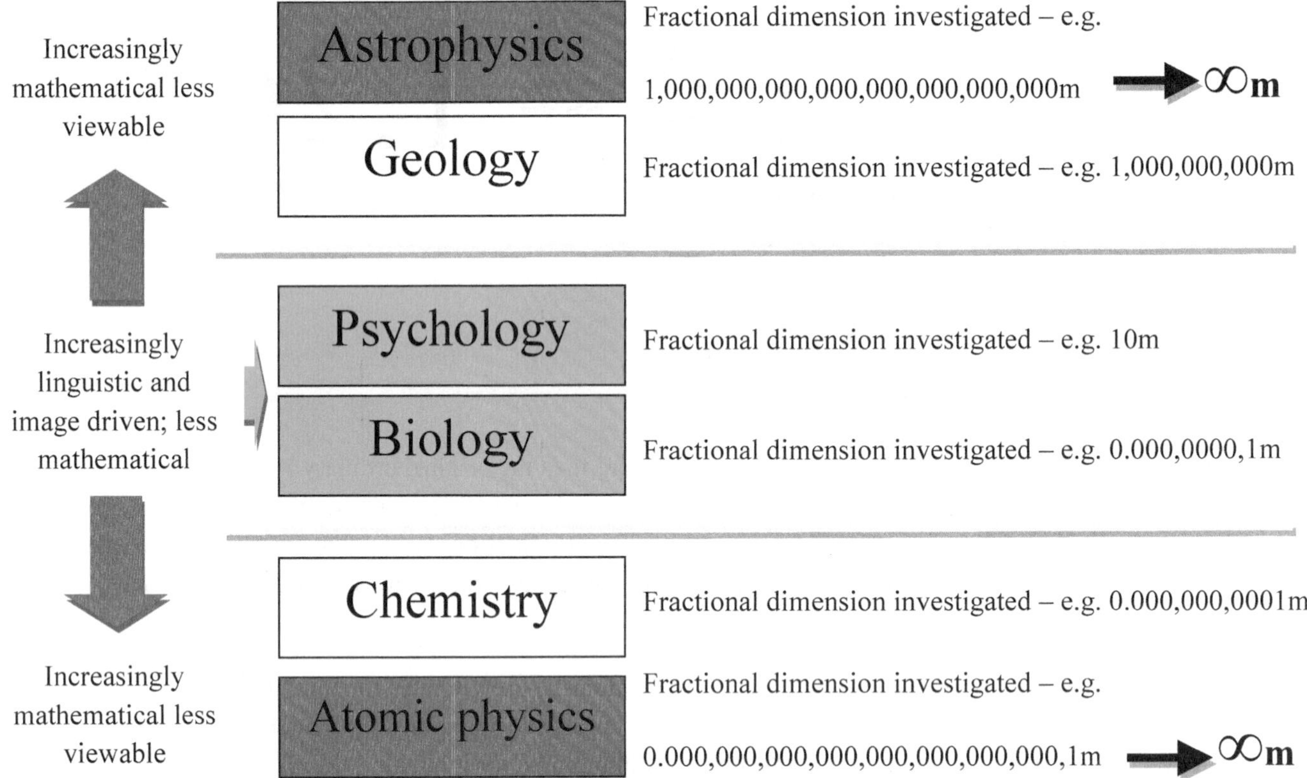

The image of the academic spectrum is displayed here against a value for a typical 'single' level of natural scale referred to as the 'fractional dimension D'. These values illustrate how the physics profession concern themselves with the absolute extremes of matter, the largest but often most distant objects through astrophysics, and the smallest particles in atomic physics. Atoms like distant galaxies are very difficult to resolve in a visual way especially atoms and because of this fact, these particle objects can be treated as simple particles (spatial points) which can be modelled against mathematical methods. Because of the mathematical axiom physicists make little or no attempt to contemplate life in their investigations and simple summarise them by two words 'the observer'. As we draw the model into the centre we find the professions most associated with life, biological science.

This profession deals with a broad selection of particle sizes but uses visual investigation and language to do science because living particles such as cells and their components are viewable. As a result, biologists don't need strict mathematical tools and often only use statistical methods, which produce a vague and inconclusive result. Biologists typically avoid mathematics and have no interest in physics especially in its pure form. Since this spectrum is based on a difference in the investigative tools associated with the methods of measurement across science has precipitated a complete breakdown in cross fertilisation of ideas across science. Carbon entromorphology aims to correct this segregation by compromise. Below is the basic model of a modern human (animal model) based on carbon entromorphology, where the self symmetry (carbon properties such as shape and physical potential) is self evident as we explore three of the fractal levels, based on carbon as the 'generating element'. Atomic, cellular and organismal self symmetry – based on carbon generating elements, levels of natural scale.

The 'Big Bang' singularity 'element, group and period zero' on the periodic table (13.7 billion years ago).

Nucleogenesis the formation of the elements (>5 billion years ago).

(1 billion years ago).

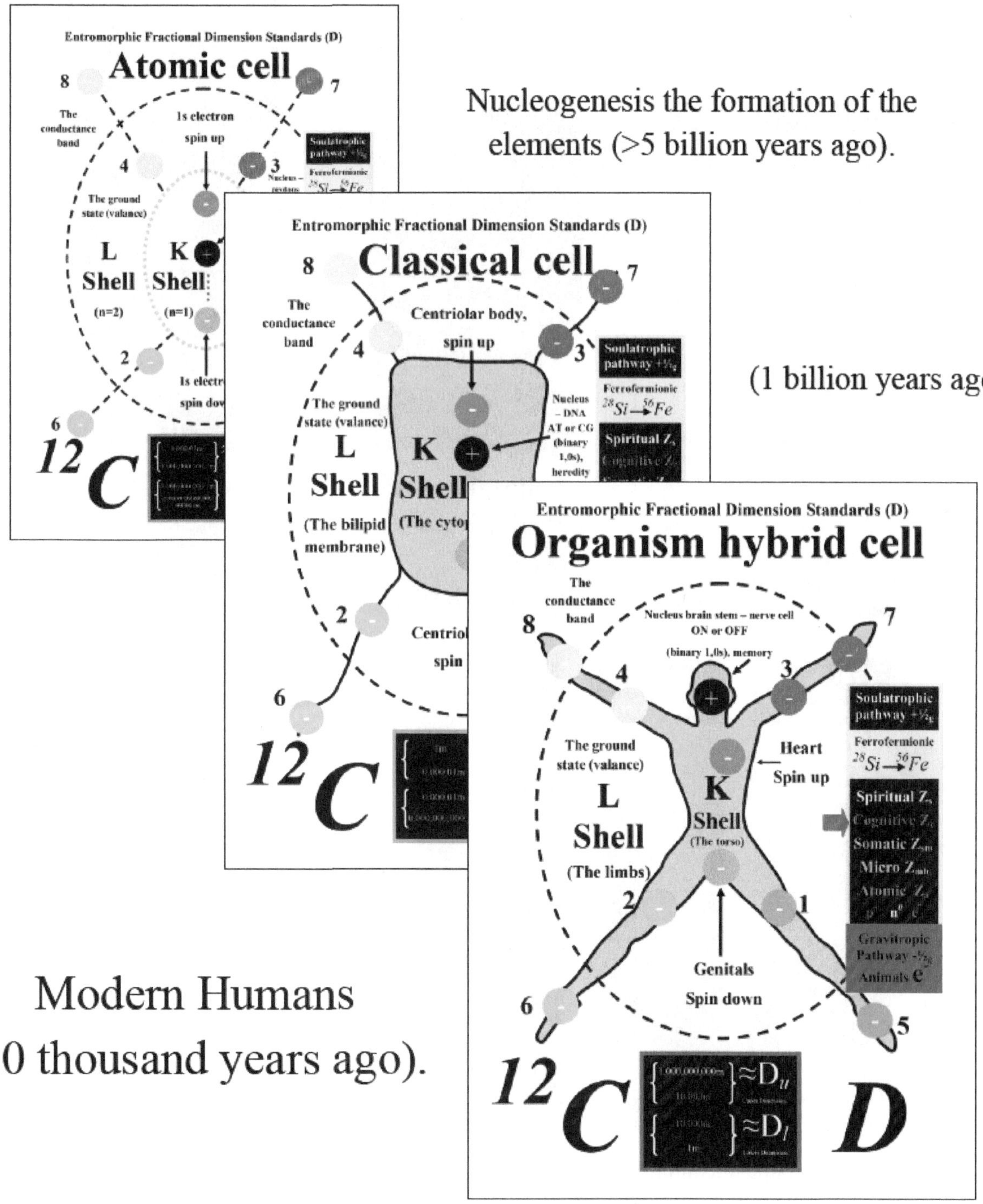

Modern Humans (20 thousand years ago).

The fractional dimensions and soulatrophic pathways of life

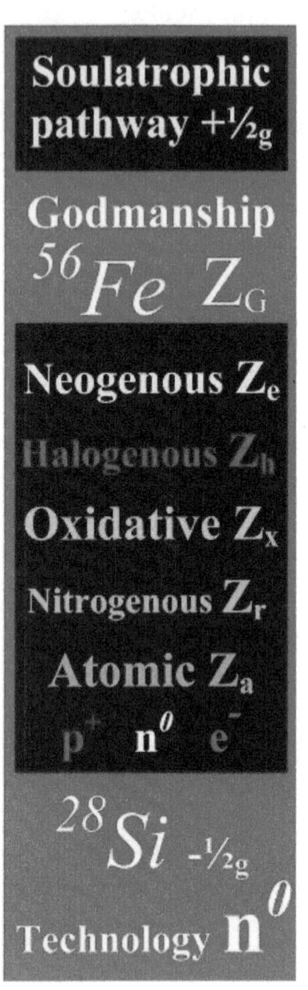

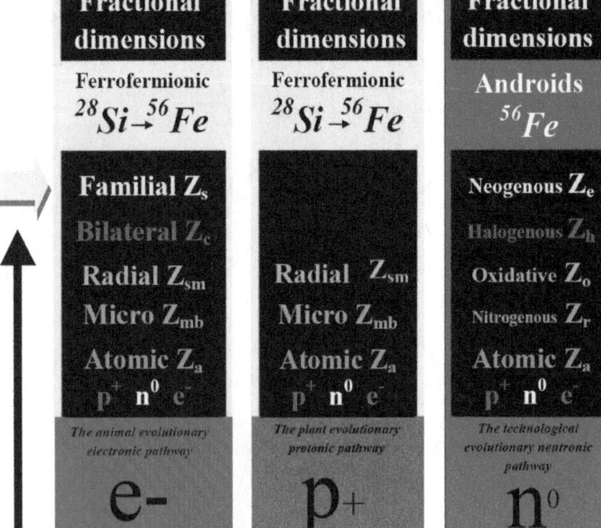

The soulatrophic models above are historically used in carbon entromorphology to describe the fractal scale systems in all living organisms.

A more scientifically appropriate system pictured as 'fractional dimensions' here has been produced although either model is appropriate.

They are used throughout this book where a small arrow is placed next to the 'highest' level appropriate to the living organism in focus.

For example modern humans have all three pathways complete, where as a lizard has a maximum dimension of bilateral and no familial level.

The arrow points to the maximum fractional dimension (soulatrophic energy level), for the organism of focus. All the lower levels are also filled but the convention simply uses one arrow to establish evolutionary hierarchy.

D is represented in carbon entromorphology by a representative value of natural scale

Units are (SI) metres

Soulatrophic pathway $+\frac{1}{2}_g$

Ferrofermionic

$^{28}Si \rightarrow {}^{56}Fe$

∞ (13.7billion years)

Spiritual Z_s	1,000,000,000 m	$\approx$D Fractional Dimension
Cognitive Z_c	10,000 m	
Somatic Z_{sm}	1 m	
Micro Z_{mb}	0.000,01 m	
Atomic Z_a	0.000,000,000,1 m	
p^+ n^0 e^-	0.000,000,000,000,000,000,000,1 m	

Gravitropic Pathway $-\frac{1}{2}_g$

Animals e^-

D is represented in carbon entromorphology by a representative value of natural scale

Fractional dimensions

Units are (SI) metres

Ferrofermionic $^{28}Si \rightarrow ^{56}Fe$

∞ (13.7billion years)

Familial Z_s	1,000,000,000 m
Bilateral Z_c	10,000 m
Radial Z_{sm}	1 m
Micro Z_{mb}	0.000,01 m
Atomic Z_a	0.000,000,000,1 m
p^+ n^0 e^-	0.000,000,000,000,000,000,000,1 m

{1,000,000,000 m; 10,000 m} ≈D
Fractional Dimension

The animal evolutionary electronic pathway

e-

(Big Bang)

The fractal generating element is carbon

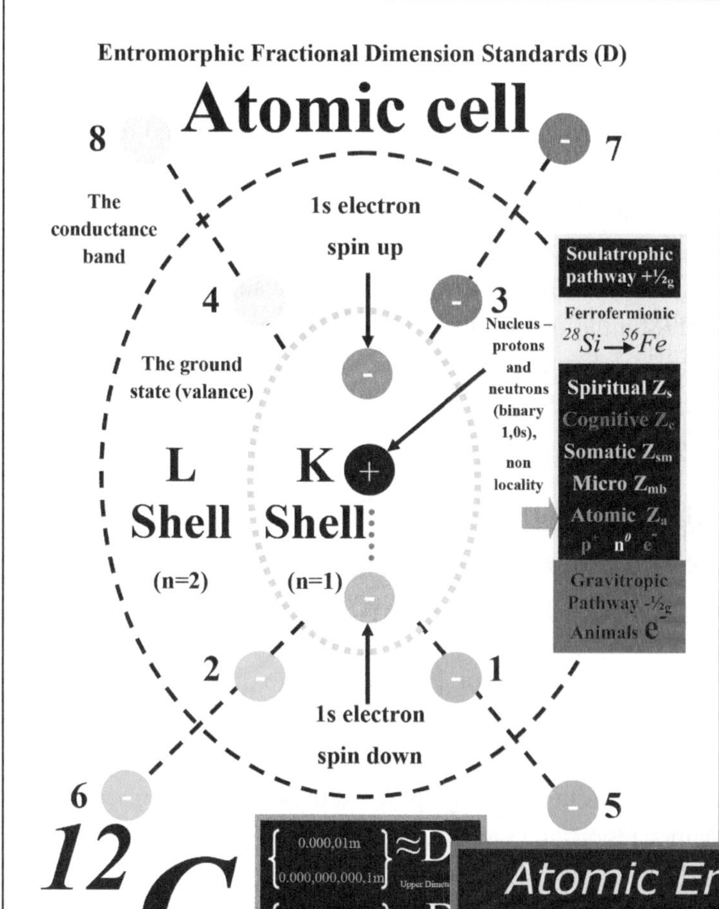

The atomic cell model on the left of this page is one of the newest models of carbon used throughout this book. It is simpler than the one below.

It is the generating element for all life on Earth. Carbon however is fractionally defined by hydrogen, but the story sits effectively with a carbon base.

The atomic cell model on the right of this page is one of the oldest models of carbon used throughout this book.

It is a more technically detailed model containing physical properties such as spin and sub shells.

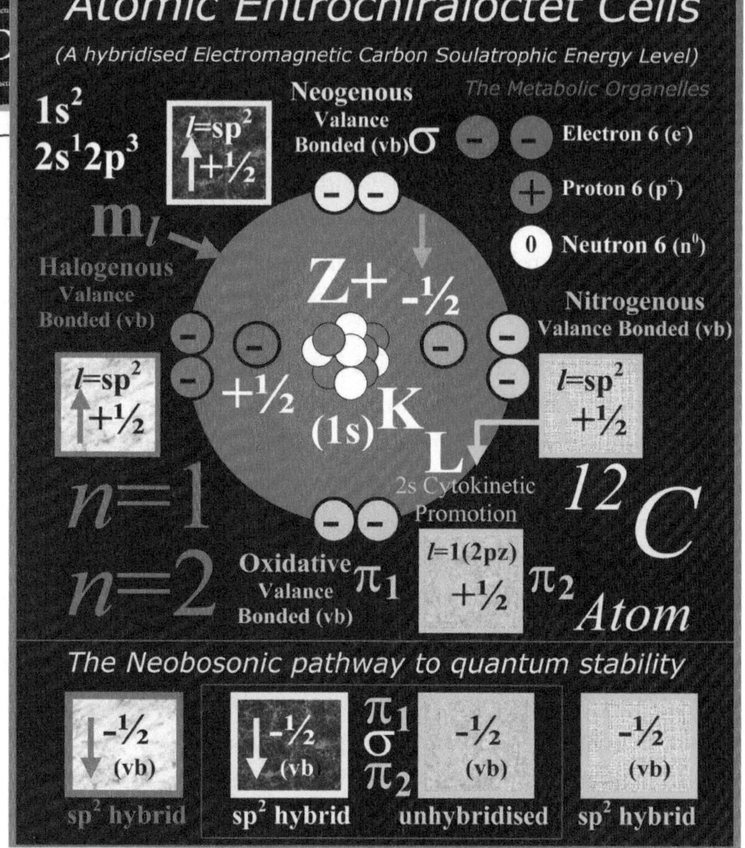

Proof and evidence in support of the theory of carbon entromorphology.

This book has been designed to present the very broad theory of carbon entromorphology such that the reader can prepare themselves with the basic tools of fractal geometry and its relationship to nature. The preface preparation leads into a profound fundamental illustration of the basic fractional dimensions found in a modern human 'animal model'. The demonstration allows the reader to identify the phenomena in their own bodies and lives, and focuses their attention on the need to find fault with the hypothesis presented. From this point a simple story is used to re-emphasize the natural levels of scale in nature and the way forces change their equilibrial determinism as we traverse the smallest to the largest scales of organisation in the universe. The book then presents a technical paper which may be a demanding read for most people although it does provide a strong scientific angle to allow for further elaboration and suggests a fundamental limitation in science called the 'resolution problem'. The analytical hypotheses are presented to force carbon entromorphology to be scrutinised against its claims and provides evidence to support them. This leads to the central dogma of carbon entromorphology which is described as follows: -

Life is centrally composed of thermodynamically filtered carbon atom accretion reactions; therefore life has the properties of carbon atoms, seamlessly on all observable levels of natural scale.

This runs in the face of current thinking in physics where atomic logic appears to defy belief and seems subtly and sublimely invisible to our everyday experiences. This is contrary to Einstein's argument that science is simply a refinement of every day experience. Evidence in carbon entromorphology is mainly provided in this book through photographic example. This is high level and very thorough evidence of 'a picture paints a thousand words'. It is also important to realise that for every example presented there are quite literally billions of others which also agree with the logical interpretation.

Numerical and mathematical evidence is also provided and a powerful mathematical calculation is presented in the 'blind physicist' (pages 91 and 92). This calculation provides excellent agreement between carbon entromorphology through nuclear and field amplification and a powerful universal constant of nature. There is also further mathematical elaboration about mathematics being the 'language of Nature' and how a reductionist model of mathematics can reveal the true nature of reality using Euler's formula and some other interpretations of its relationship to physics.

Many of the classification theories in carbon entromorphology are produced because of the central dogma. In other words if the theory is correct then ANY living situation should be organised against carbon atomic logic and can be modelled and classified against such atomic properties. An example is the 'theory of quantum taxonomy' (page 245 to 249) where carbon properties on a GLOBAL scale allow science to classify organisms against the particle anatomy in carbon and its physical potential. For example a worm is classed as a nuclear organism, a modern human classed as a valance bonded neogenous particle. Where gravity and electromagnetic links connect seamlessly.

Often the viewpoint of classification is being forced on the reader through the series of supportive statements, often without mathematical proof. This is a typical situation in biological science and should be considered as limited by physical scientists, biological entities are too complex to treat with precise mathematics, and it's simply unavoidable. This however, should not compromise development as long as the central dogma is true then science should be able work in this way.

Science check

Do the observations correlate with the theoretical model?

YES ✓ ***NO***

PART ONE

An introduction to natural living order

Step 1- Getting started by following 6 basic steps; develop an understanding of natural living order.

Step 2 - Look at pages 1 to 10, understand that each fractional level fits together as a composite.

Step 3 - Identify the different levels of natural order displayed for a human (animal model) subject that collectively forms a living organism such as you; look for conserved symmetry in the structures.

Step 4 - Consider life to be hypothesised by this theory. To be constructed out of discrete fractional dimensions (also known as soulatrophic energy levels, atomic, micro, somatic or radial, cognitive or bilateral and spiritual or familial) detailed by the following points 1 to 5 and describing a pathway of thermodynamically accumulated accretion amplified properties.

Postulate 1 - The basic argument of the theory of carbon entromorphology. That the images and their properties are true and are a representative integrated model of life as a pathway of solar amplified discrete fractional dimensions (soulatrophicity meaning the origin of life the 'Big Bang'). Displaying conserved properties from hydrogen and carbon (but also nitrogen and oxygen making 96.3% of the components of this human life form and to a lesser extent 22 further trace elements). This model is a thorough and complete picture of living organisms from the smallest observable scale to the largest seamlessly over billions of years of integrated conservation (Thims, 2010).

Step 5 - Realise that all life on Earth is a staggering 13.7 billion years old. That life, including yourself, started at the 'Big Bang' and began this process of growth (through thermodynamic accretion reactions) and amplification of their properties through carbon centralised and catalysed bonding.

Postulate 2 - Understand that life is a continuous pathway propagated by thermodynamically stable ancestors, where interactions are conserved and filtered through a process known as heredity through evolution by natural selection. This link to the past allows small scale atomic properties to undergo conserved amplification (because of links or bonds in carbon but in conjunction with nitrogen and oxygen and also hydrogen) into larger structures, such as the jump from atoms to the integrated organisms they form through a cellular fractional dimensional level.

Step 6 - Consider and identify your nuclear regions such as your brain (neurological cells), your DNA and your atomic nuclei, and their symmetrical properties. Consider and identify your field regions such as your limbs and torso, your cellular cytoplasm and membrane, your atomic K shell and L shell.

Point 1 –We are made of three types of sub atomic particles the electron the proton and neutron.

This reduced level of organisation hypothesises basic sexual dimorphic gender in living beings (as defined by carbon entromorphology) and shows high levels of conservation of these particles properties on higher levels of natural scale, observation suggests many levels of conserved morphology.

***Point 2* -** The simplest atom is hydrogen and appears conserved in the larger structures it creates such as atoms, molecules and cells; hydrogen is made of just one proton and electron (polarised).

***Point 3* -** We are made of 26 atoms (Thims, 2010) but carbons p block properties (and nitrogen and oxygen and hydrogen compose 96.3% of all living beings) are the essential central nucleonic atoms hence 'carbon based life' which gives life all its complexity and diversity (carbon is the soul of life, where life is a unique and complex allotrope of carbons extremely broad and gregarious determinism).

***Point 4* -** Families are constructed of organisms, organisms are constructed of cells, cells are constructed of atoms and atoms are constructed of sub atomic particles. *Physical deterministic properties of fractional levels are still observable on ANY other fractional level due to bonding.*

***Point 5* -** The different fractional levels come into being over billions of years, the later levels appearing over increasingly smaller time periods (exponential growth), for example the atomic level took billions of years, the spiritual level developed dramatically over just thousands of years.

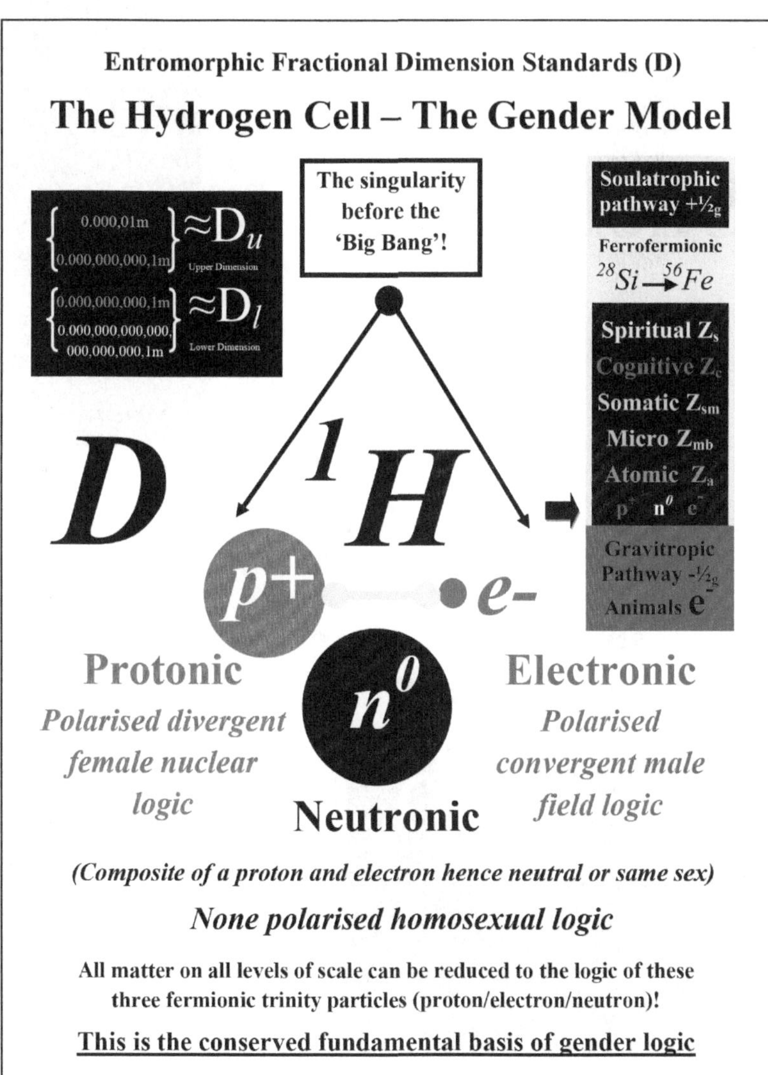

The carbon accretion hypothesis:-

'Living organisms are the result of conserved amplified carbon atomic logic!'

The overall point of the following exercise is for the reader to ask themselves: -

'Can you reject this hypothesis?'

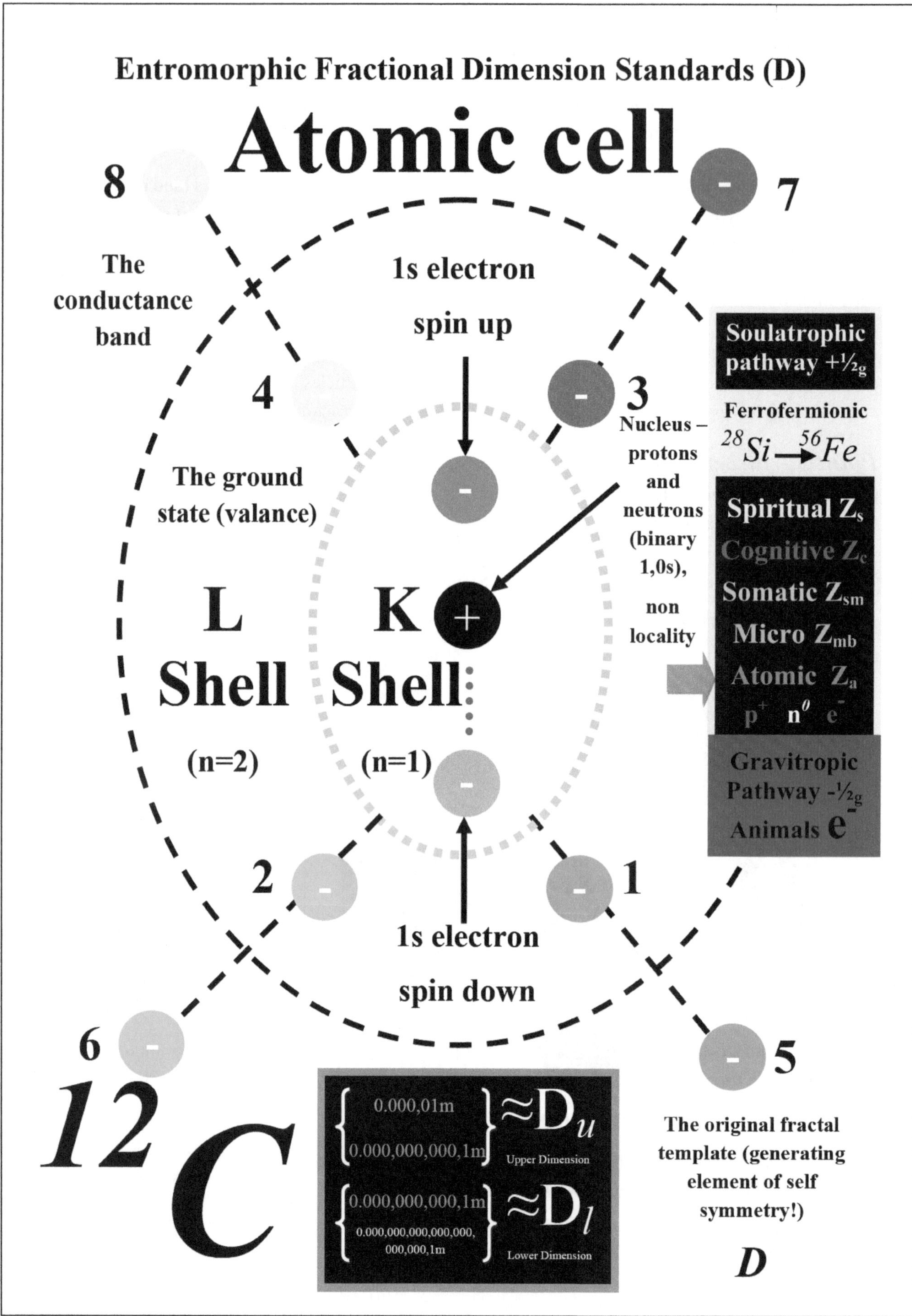
Entromorphic Fractional Dimension Standards (D)
Atomic cell
8
7
The conductance band
1s electron spin up
Soulatrophic pathway +½g
4
3
Nucleus – protons and neutrons (binary 1,0s), non locality
Ferrofermionic
28Si → 56Fe
The ground state (valance)
Spiritual Zs
Cognitive Zc
Somatic Zsm
Micro Zmb
Atomic Za
p+ n0 e-
L Shell (n=2)
K Shell (n=1)
Gravitropic Pathway -½g
Animals e-
2
1
1s electron spin down
6
5
12C
0.000,01m
0.000,000,000,1m
≈Du Upper Dimension
0.000,000,000,1m
0.000,000,000,000,000,000,000,1m
≈Dl Lower Dimension
The original fractal template (generating element of self symmetry!)
D

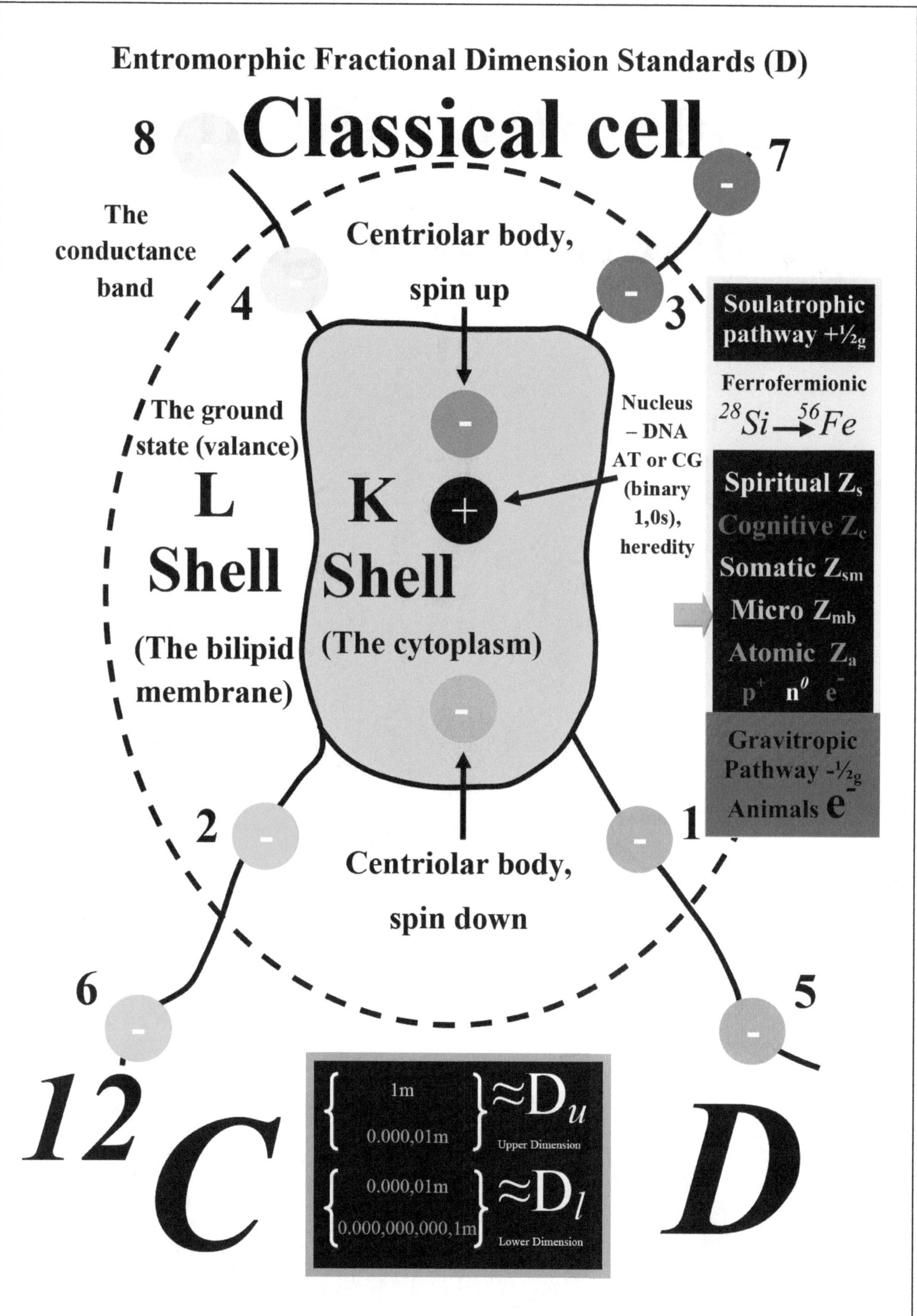
Entromorphic Fractional Dimension Standards (D)
Classical cell
8
7
The conductance band
Centriolar body, spin up
4
3
Soulatrophic pathway +½g
Ferrofermionic
$^{28}Si \rightarrow {}^{56}Fe$
Nucleus – DNA AT or CG (binary 1,0s), heredity
The ground state (valance)
L
K
Shell
Shell
Spiritual Z_s
Cognitive Z_c
Somatic Z_{sm}
Micro Z_{mb}
Atomic Z_a
p^+ n^0 e^-
(The bilipid membrane)
(The cytoplasm)
Gravitropic Pathway -½g
Animals e^-
2
1
Centriolar body, spin down
6
5
12
C
D
$\left\{ \begin{matrix} 1m \\ 0.000,01m \end{matrix} \right\} \approx D_u$
Upper Dimension
$\left\{ \begin{matrix} 0.000,01m \\ 0.000,000,000,1m \end{matrix} \right\} \approx D_l$
Lower Dimension

Entromorphic Fractional Dimension Standards (D)

Somatic radially symmetrical organisms

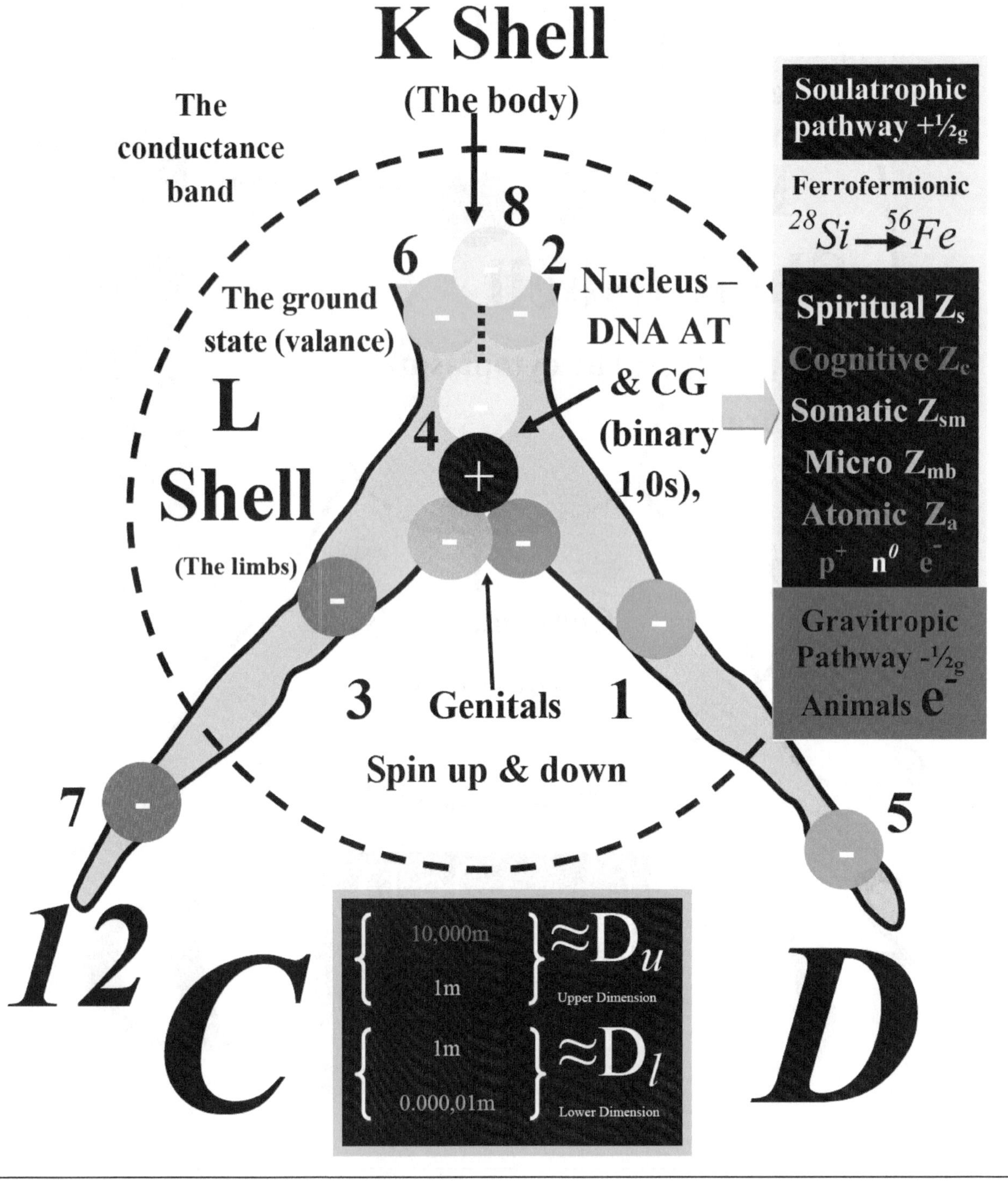

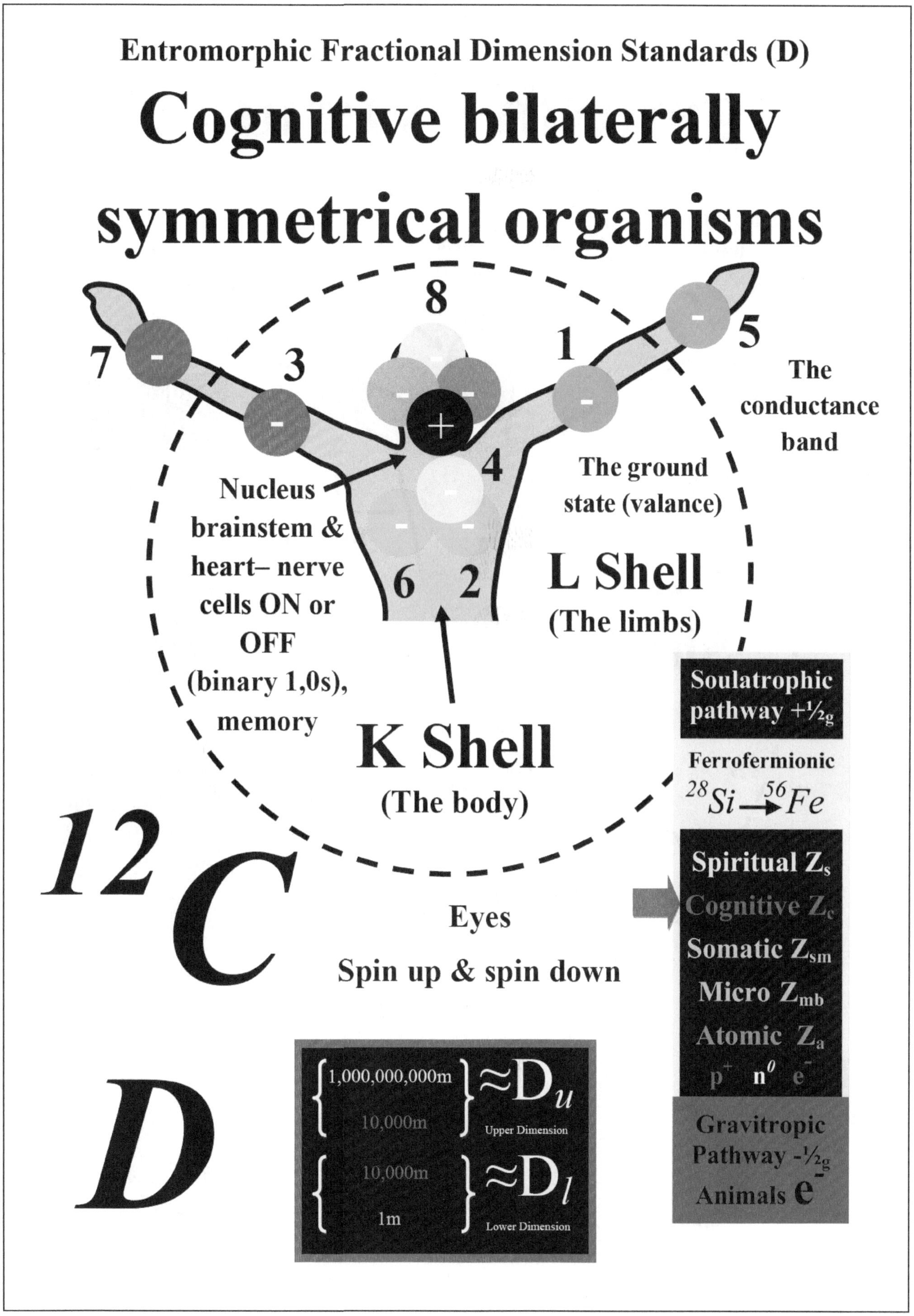

Entromorphic Fractional Dimension Standards (D)

Cognitive bilaterally symmetrical organisms

8

7 3 1 5

The conductance band

4

Nucleus brainstem & heart– nerve cells ON or OFF (binary 1,0s), memory

The ground state (valance)

6 2

L Shell
(The limbs)

K Shell
(The body)

Soulatrophic pathway $+\frac{1}{2}_g$

Ferrofermionic
$^{28}Si \rightarrow {}^{56}Fe$

^{12}C

Eyes
Spin up & spin down

Spiritual Z_s
Cognitive Z_c
Somatic Z_{sm}
Micro Z_{mb}
Atomic Z_a
p^+ n^0 e^-

Gravitropic Pathway $-\frac{1}{2}_g$
Animals e^-

D

$$\left\{ \begin{matrix} 1{,}000{,}000{,}000\text{m} \\ 10{,}000\text{m} \end{matrix} \right\} \approx D_u$$
Upper Dimension

$$\left\{ \begin{matrix} 10{,}000\text{m} \\ 1\text{m} \end{matrix} \right\} \approx D_l$$
Lower Dimension

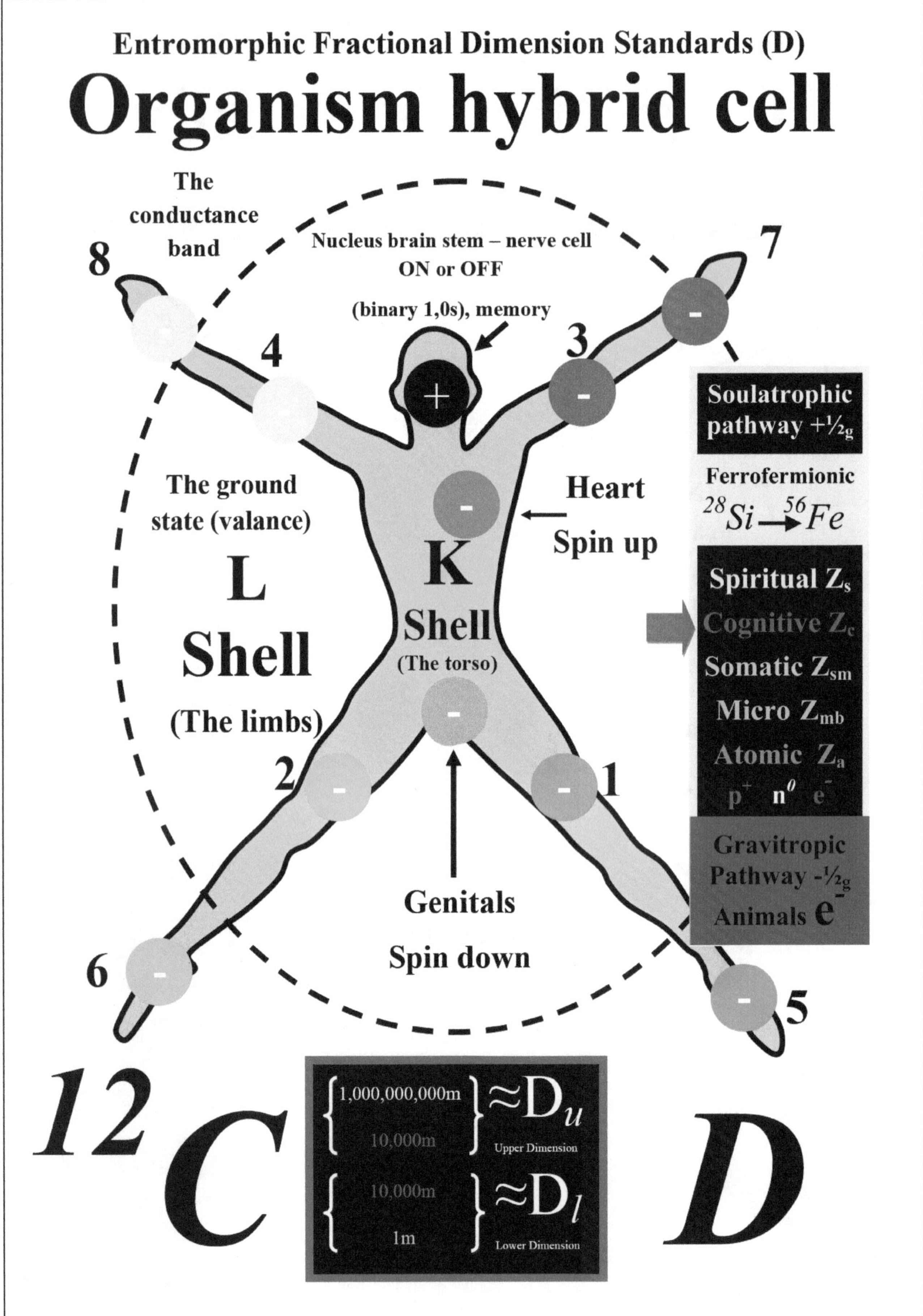
Entromorphic Fractional Dimension Standards (D)
Organism hybrid cell
The conductance band
8
Nucleus brain stem – nerve cell
ON or OFF
(binary 1,0s), memory
7
4
3
+
Soulatrophic pathway +½g
Ferrofermionic
28Si → 56Fe
The ground state (valance)
Heart
Spin up
L
Shell
(The limbs)
K
Shell
(The torso)
Spiritual Zs
Cognitive Zc
Somatic Zsm
Micro Zmb
Atomic Za
p+ n0 e-
2
1
Gravitropic Pathway -½g
Animals e-
Genitals
Spin down
6
5
12
C
1,000,000,000m
10,000m
≈Du
Upper Dimension
10,000m
1m
≈Dl
Lower Dimension
D

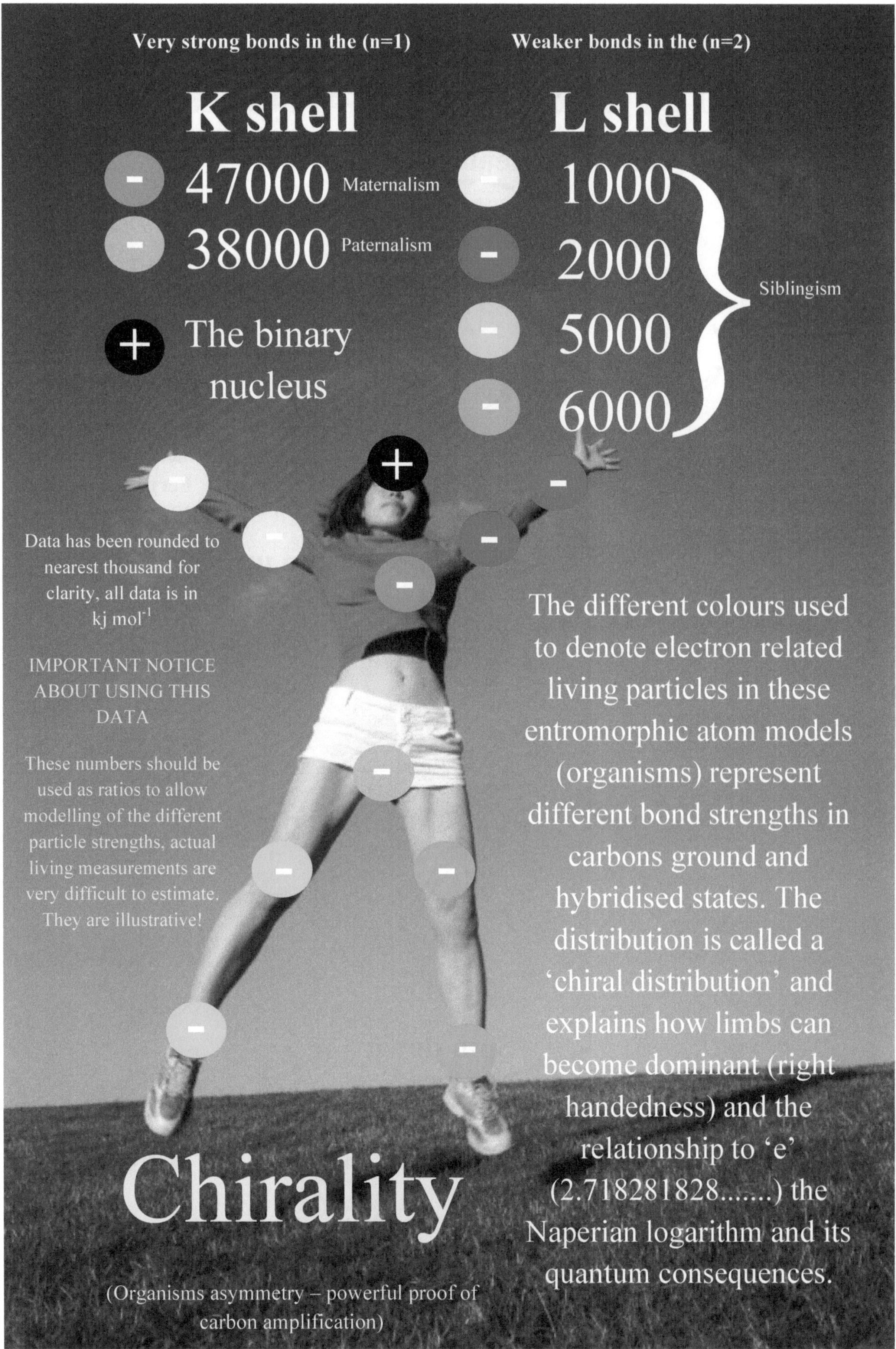
Very strong bonds in the (n=1)
Weaker bonds in the (n=2)
K shell
L shell
47000 Maternalism
38000 Paternalism
The binary nucleus
1000
2000
5000
6000
Siblingism
Data has been rounded to nearest thousand for clarity, all data is in kj mol-1
IMPORTANT NOTICE ABOUT USING THIS DATA
These numbers should be used as ratios to allow modelling of the different particle strengths, actual living measurements are very difficult to estimate. They are illustrative!
The different colours used to denote electron related living particles in these entromorphic atom models (organisms) represent different bond strengths in carbons ground and hybridised states. The distribution is called a 'chiral distribution' and explains how limbs can become dominant (right handedness) and the relationship to 'e' (2.718281828.......) the Naperian logarithm and its quantum consequences.
Chirality
(Organisms asymmetry – powerful proof of carbon amplification)

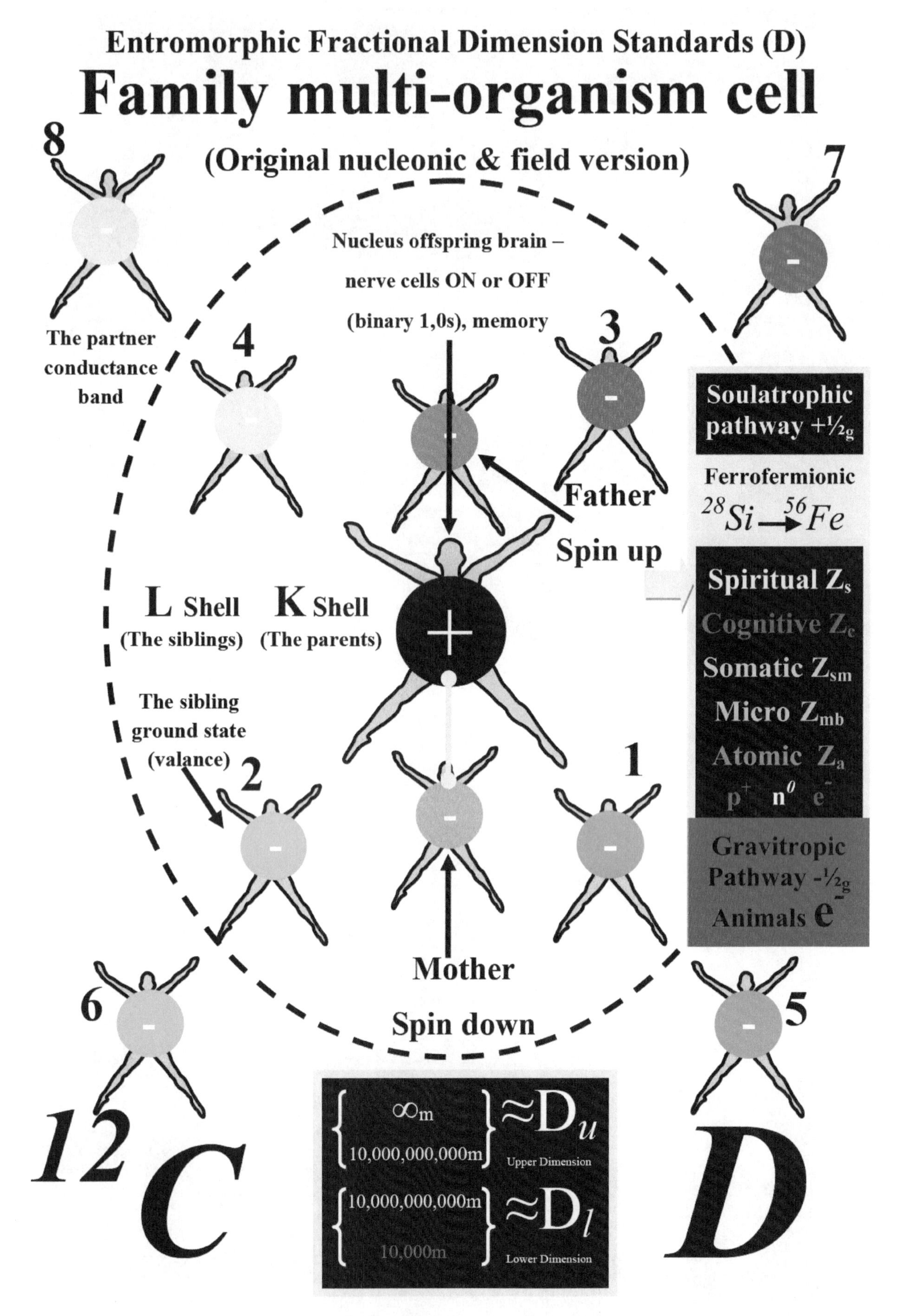
Entromorphic Fractional Dimension Standards (D)
Family multi-organism cell
(Original nucleonic & field version)
8
7
Nucleus offspring brain – nerve cells ON or OFF (binary 1,0s), memory
The partner conductance band
4
3
Soulatrophic pathway +½g
Ferrofermionic
²⁸Si → ⁵⁶Fe
Father
Spin up
Spiritual Zs
Cognitive Zc
Somatic Zsm
Micro Zmb
Atomic Za
p⁺ n⁰ e⁻
L Shell (The siblings)
K Shell (The parents)
The sibling ground state (valance)
2
1
Gravitropic Pathway -½g
Animals e⁻
Mother
Spin down
6
5
∞m
10,000,000,000m
≈Du
Upper Dimension
10,000,000,000m
10,000m
≈Dl
Lower Dimension
12C
D

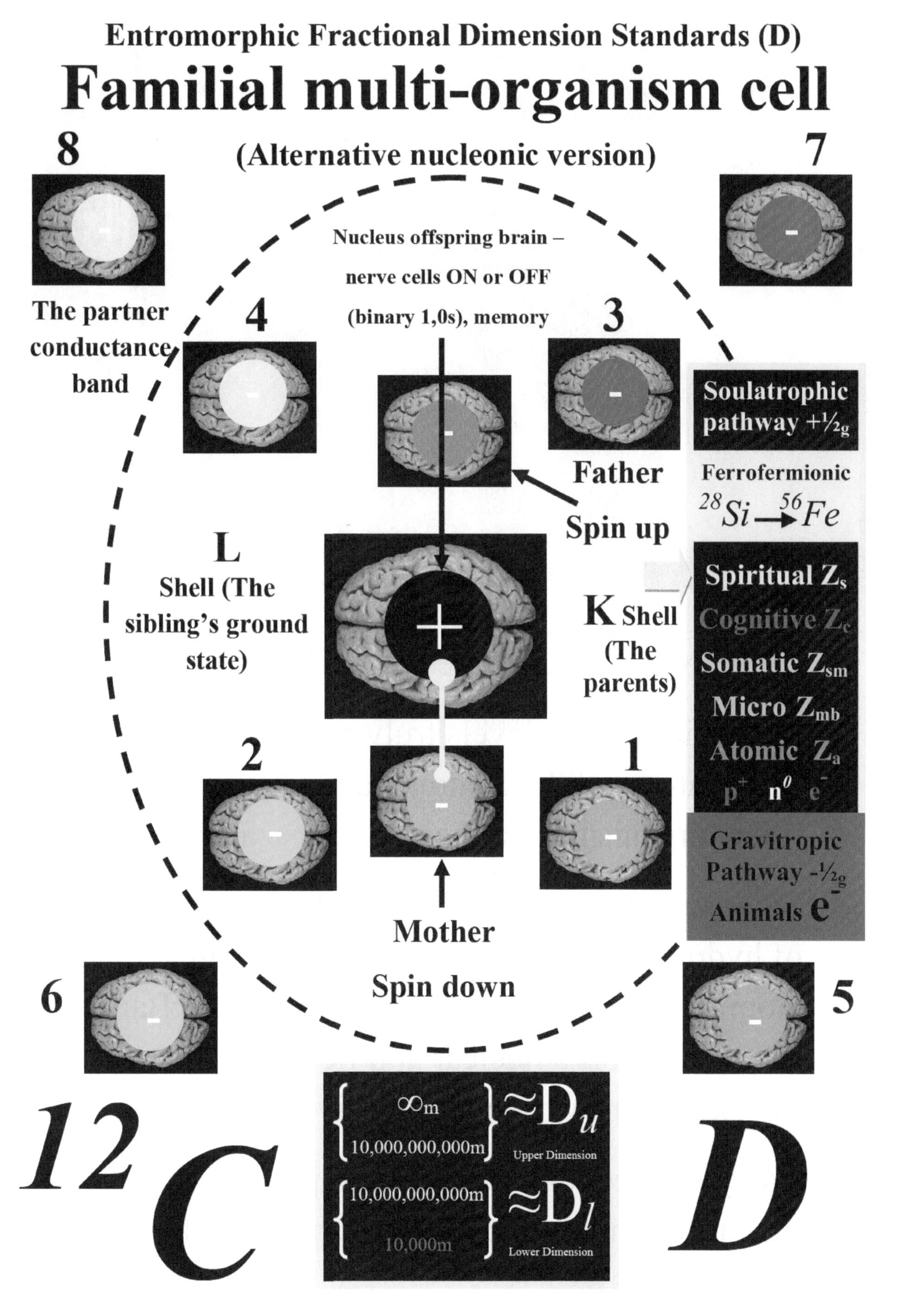
Entromorphic Fractional Dimension Standards (D)
Familial multi-organism cell
(Alternative nucleonic version)
8
7
Nucleus offspring brain – nerve cells ON or OFF (binary 1,0s), memory
The partner conductance band
4
3
Soulatrophic pathway +½g
Father
Ferrofermionic
28Si → 56Fe
Spin up
L Shell (The sibling's ground state)
+
K Shell (The parents)
Spiritual Zs
Cognitive Zc
Somatic Zsm
Micro Zmb
Atomic Za
p+ n0 e-
2
1
Gravitropic Pathway -½g
Animals e-
Mother
Spin down
6
5
12C
∞m
10,000,000,000m
≈Du
Upper Dimension
10,000,000,000m
10,000m
≈Dl
Lower Dimension
D

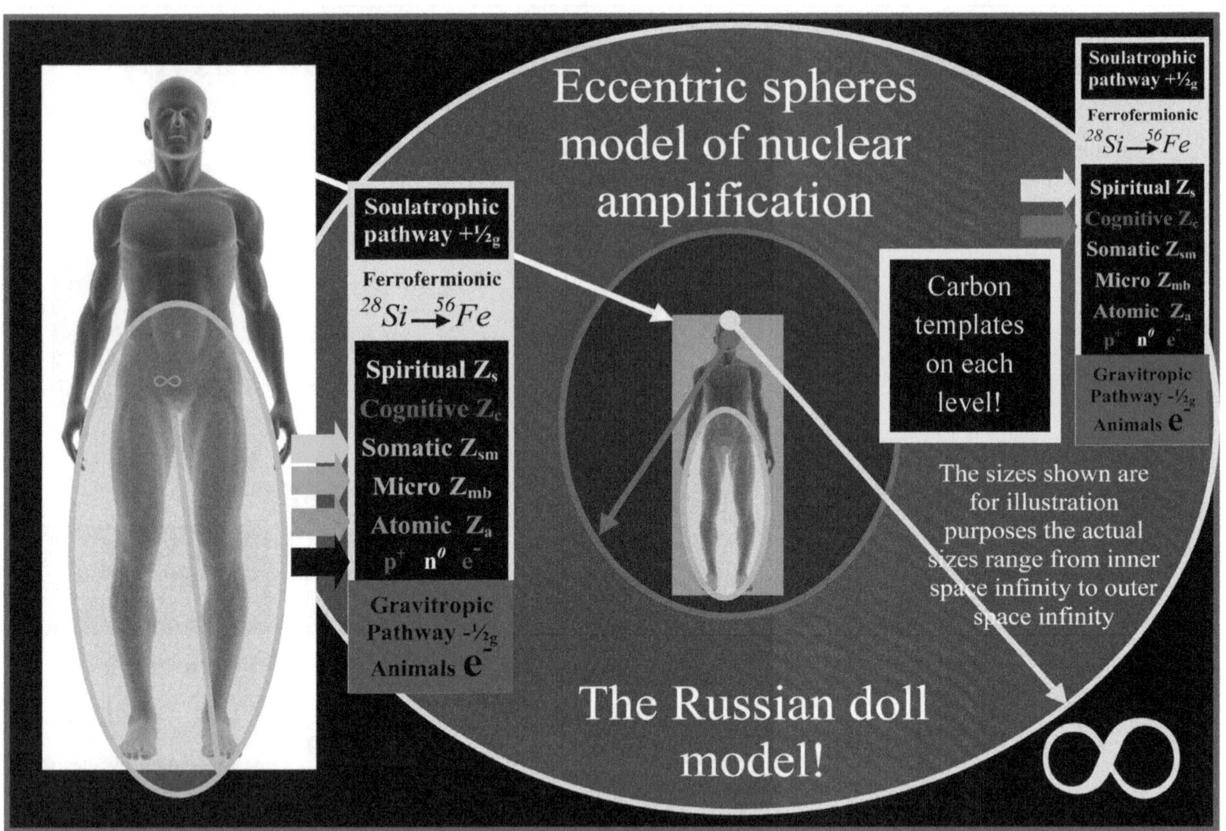

Now contemplate the following: -

1. Do the individual levels truly reflect our observations of the anatomy and physiology of living organisms and their natural scale?
2. Since we know that life is a seamless unbroken line of hydrogen to carbon conservation from simple atomic levels all the way back to the 'Big Bang', can we support a nuclear and field amplification (known as growth) model for living beings?
3. Can you reject this powerful graphical summary of the theory and its supportive hypotheses based on these models?

An important thought experiment 'Mr Carbon Atom and the nucleus'.

One day Mr Carbon Atom decided to find out what a nucleus was. So he went to see Mrs Physicist.

'*Mrs Physicist, do you know what a nucleus is?*' asked Mr Carbon Atom.

'*Yes I do.*' said Mrs Physicist. '*A nucleus is found in the heart of an atom. It acts centrally and controls and determines the quantum fields that surround it. It contains a huge amount of energy, mostly as mass and has a positive charge. It is organised and encodes information as a binary system by the two particles which comprise its system. It is held together by the strongest natural force, the strong interaction.*'

'*Thank you Mrs Physicist.*' said Mr Carbon Atom '*You have been most helpful.*'

Mr Carbon Atom decided to get a second opinion so he went to see Mr Geneticist. '*Mr Geneticist, do you know what a nucleus is?*' asked Mr Carbon Atom.

'*Yes I do.*' said Mr Geneticist. '*A nucleus is found in the heart of a cell. It acts centrally and controls and determines the cytoplasmic and membrane fields which surround it. It is dense and contains a huge amount of energy in the form of phosphates, and encodes information through hydrogen bonds which are positive charges. It is organised as a binary system by the two DNA base pairs which comprise its system. It is held together by the genetic force, which is a very strong interaction.*'

'*Thank you Mr Geneticist.*' said Mr Carbon Atom '*You have been most helpful.*'

Mr Carbon Atom decided to get another opinion so he went to see Mrs Psychologist. '*Mrs Psychologist, do you know what a nucleus is?*' asked Mr Carbon Atom.

'*Yes I do.*' said Mrs Psychologist. 'A *nucleus is a brain, it acts centrally and controls and determines the bodily torso and limb fields which surround it. It contains a huge amount of energy in the form of neurological bonds, and encodes information through action potentials which are positively charged. It is organised as a binary system by the On/Off impulses which comprise its system. It is held together by the love force, which is a very strong interaction.*'

'*Thank you Mrs Psychologist.*' said Mr Carbon Atom '*You have been most helpful.*'

Mr Carbon Atom felt confused 'A nucleus seems to describe many levels of organisation although the levels of organisation appear to be discrete?'

Mr Carbon Atom decided to get another opinion so he went to see Mr Politician. '*Mr Politician, do you know what a nucleus is?*' asked Mr Carbon Atom.

'*Yes I do.*' said Mr Politician. '*A nucleus are the laws of a country, they act centrally and control and determine the government and population which surround it. They contain a huge amount of energy in the form of historical bonds, and encode itself as classical information. It is organised as a binary system by the black and white words which comprise the written records of its system. It is held together by the democratic force, which is a very strong interaction.*'

'*Thank you Mr Politician.*' said Mr Carbon Atom, '*You have been most helpful although my confusion about this matter is worse!*'

And he decided to get yet another opinion so he went to see Mrs Geologist. '*Mrs Geologist, do you know what a nucleus is?*' asked Mr Carbon Atom.

'Yes I do.' said Mrs Geologist. '*A nucleus is the centre of a planet often composed of magma, it acts centrally and controls and determines the land, oceans and atmosphere the phases of matter which surround it. It contains a huge amount of energy in the form of concentrated mass. It is held together by the gravitational force, which is a very weak interaction but a planet is very very large so its effect is still immensely significant.'*

'Thank you Mrs Geologist.' said Mr Carbon Atom '*You have been most helpful.'*

Mr Carbon Atom felt even more confused 'A nucleus as defined by Mrs Physicist, Mr Geneticist, Mrs Psychologist and Mr Politician seem have binary organisation driven by electromagnetic forces, yet a planet is driven by gravity but also acts as a centre of determinism.'

And he decided to get yet another opinion so he went to see Mr Astronomer.

'*Mr Astronomer, do you know what a nucleus is?*' asked Mr Carbon Atom.

'Yes I do.' said Mr Astronomer. '*A nucleus is the Sun the centre of the solar system composed of very hot matter, it acts centrally and controls and determines the other planets and moons and asteroids which surround it. It contains a huge amount of energy in the form of concentrated mass. It is held together by the gravitational force, which is a very weak interaction but a Sun is very, very, very, very, very large so its effect is enormously significant.'*

Mr Carbon Atom felt even more confused and he thought 'A nucleus as defined by Mrs Physicist, Mr Geneticist, Mrs Psychologist and Mr Politician seem to have binary organisation driven by electromagnetic forces, yet a planet and a solar system are very large particles driven by gravity. All of the people I have talked to represent different levels of organisation, different inertial systems from the very small world of Mrs Physicist to the enormous world of Mr Astronomer, and along that pathway a nucleus changes its determinism from electromagnetic to gravity seamlessly.'

'*Thank you Mr Astronomer.*' said Mr Carbon Atom '*You have been most helpful.*' He decided to get a final opinion so he went to see a priest. '*Rev Priest, do you know what a nucleus is because my friends have all given similar definitions but for particles ranging from the smallest scale of the atomic world to the massive scale of the universe?'* said Mr Carbon Atom.

'*Yes I do.*' said Rev Priest. '*A nucleus is God, the centre of all existence. He acts centrally and controls and determines all that exists. He contains all the energy in the universe and the heavens. He represents the strongest force in all time and space that of belief and love and his control is mediated by the black and white binary nature of the words in the Bible.'*

'Thank you Rev Priest.' said Mr Carbon Atom '*You have been most helpful.'*

Mr Carbon Atom was shocked, 'So with all the people I have talked to, who are most correct? Is it Mrs Physicist or perhaps Mrs Geologist, or Mr Geneticist or Mrs Psychologist? Maybe its Mr Astronomer and Rev Priest, both of whom talked about the whole of space and time, which was impressive.'

Then a thought struck Mr Carbon Atom 'What if they are all correct. A nucleus appears to be the centre of any physical system and as we have seen physical systems traverse the smallest world to the largest world. What appears to occur is that electromagnetic forces exchange control with gravitational forces as systems of consciousness get larger and larger and its mass component grows in size. The strong interaction of the atomic world must be directly associated with the weak force of gravity of the massive world. Fundamental natural particles contain both electromagnetic potentials and mass gravitational potentials through duality, but the equilibrium of these forces changes as a physical system is observed to get larger through bonding. But life is a system aware of all the different levels of organisation and able to see this transition of control as it moves through the different relative levels of inertial organisation. Life therefore appears to identify quantum gravitational logic as the resultant of natural force equilibrium. All the forces of nature appear to be completely present on all levels of organisation but the equilibrium between them produces the entire natural organisation which we observe in time and space. In short, any nuclear system must take into account all four force permutations regardless of their absolute observable contribution to relative determinism for a complete model of nature. Quantum mechanics must produce force equilibrium, including gravity, on the atomic level even though its effect, relative to another single atom is very, very, very, very weak. However the atom could be on the surface of a Sun a hundred times the size of our Sun, and if that were the case then the gravitational influence on quantum organisation as a dynamic equilibrium would be extremely significant and cannot be ignored. So in short, any physical system model must include equilibrium of all the natural force permutations regardless of their direct contribution to determinism on that particular level of relative scale and organisation. Quantum mechanics normally ignores gravity in the wave function equations but this thought experiment clearly demonstrates that they can't.'

Finally the quantum energy levels of the small electromagnetic world produce atomic density which results in the observed phases of matter for mass dominated massive astrological particles such as planets through an equilibrium formed by the four forces of nature.

Science check

Do the observations correlate with the theoretical model?

YES ✓ NO

PART TWO

The human thermodynamics of carbon entromorphology

Introduction: 'the resolution problem'- the segregation and fractured nature of human knowledge (American convention).

Human knowledge is a fractured segregated logic, composed of axioms from observations expressed through language and imagery, which is often vague and highly subjective to science and mathematical observational logic which is powerfully dynamic and logically reliable and predictable [1]. Knowledge spans the measured numerical extremes in the physical sciences through a mathematical treatment, and the associated sensible mathematical resolution and impressive predictability [2]. This extends to the vagueness of the terms and concepts, and associated uncertainty found in the arts and language. But also, to immeasurable living complexity which leads into the biological sciences producing an overall fractured segregated non-integrated academic system, collectively called human knowledge [3]; this segregation is illogical and counterproductive to human evolution. It is therefore crucial for humanity to consider cross fertilisation of the most reliable logic from all facets of conscious knowledge to produce sensible integration through hypothesised self symmetry across all levels of natural scale. This is the fundamental proposition of the theory of carbon entromorphology. It has been postulated for many years throughout the sciences that a unification of human knowledge or a 'grand unified model' is required to produce a picture of broad consciousness and more importantly the centralised dominance of human beings in any such model [4].

Physical science and mathematics have always predicted this would take the form of mathematical algorithm compressions, typically equations [5]. Einstein believed in the beautiful perfection of mathematics and hated and fought against uncertainty in determinism [3]. Scientific logic is based on a numerically quantifiable reductionist system integrated into mathematical relationships between physically measured variables and constants, linked into a dichotomy of physical models. The intrinsic balance of the nature of comparative differential equations is the very essence of the first law of thermodynamics, where mass and energy is shown to be relativistically ($E=mc^2$) equivalent or homogenous in any physical system [6]. The reliability of linear equations and their solutions are however subject to natural limitation through the expression of non linear chaotic physical models.

Mathematics, in its applied form, often becomes unusable where non linear physical systems such as biological organisms become too heterogeneous and thermoentropically too disordered and complex, and where variables and particle populations become dizzyingly immense. This is where the initial conditions of such systems become impossible to establish accurately and meaningfully and where non linear functions produce amplified errors often over a very short range of values for such applied functions.

Measurement reliability is the seat and burden of all scientific investigation, but resolution of different levels of natural scale produce variation in observational measurements to those associated with measured physical quantities [7]. Biological investigation is typically driven by images and words.

This is due to the disorder or thermoentropic consequence of transforming potential energy into heat for dissipation and entropy or waste in accordance with the second law of thermodynamics (Gibbs equation, G=H-TS and in terms of energy change $\Delta G = \Delta H - T\Delta S$ [8]). Linear equations are most desirable in science where the whole is found to be the sum of its composite parts (Barrow, 2007).

However the inevitable chaos of non linear equations is heavily associated with life where small errors in non linear models produce huge errors often over a relatively small range of independent variables for the associated non linear function, which may be associated with the fluid turbulence in the evolutionary pathway [5]; an evolving fluid model of life thermodynamically fit to fill its associated environmental vessel. Knowledge has always been associated with the process of careful observation and quantification, however for a complete picture of physical reality as perceived by the human mind we find a strange natural phenomena identified in this article as the 'resolution problem'.

The sciences fit together piece by piece, fraction by fraction to allow humans to resolve all levels of natural scale [9]. The physicists for example, concerns themselves with the very small world of quantum mechanics and nuclear physics as well as the very large but very distant world of general relativity and astrophysics theoretically conjoined through the power of special relativity [4].

Living organisms are composed of the logic of atoms but also the logic of giant bodies such as the Earth and Sun; the four forces (strong, weak, electromagnetic and gravitational) acting through particle wave matter on all levels have a deterministic influence on human curiosity, yet physics continues to separate their contributions. Biology by contrast continues to ignore the powerful fundamentalism of physics and its mathematical treatment of physical systems. Living organisms are historical records (evolving particle entanglements and external thermodynamic filtering) producing a picture of 13.7 billion years of unbroken ancestry through evolving thermodynamic stability [10].

- The following diagram is a summary of the classical 2nd law of thermodynamics where disorder is increasing in the universe for any conscious observer (a metaphysical model of hell). It relates to any experience where matter is broken in terms of its structural symmetry, although this model if challenged is associated with closed thermodynamic systems. Life on the other hand is part of an open system and as such can reverse this model by acquisition of free energy, increasing symmetry and order through solar acquisition.

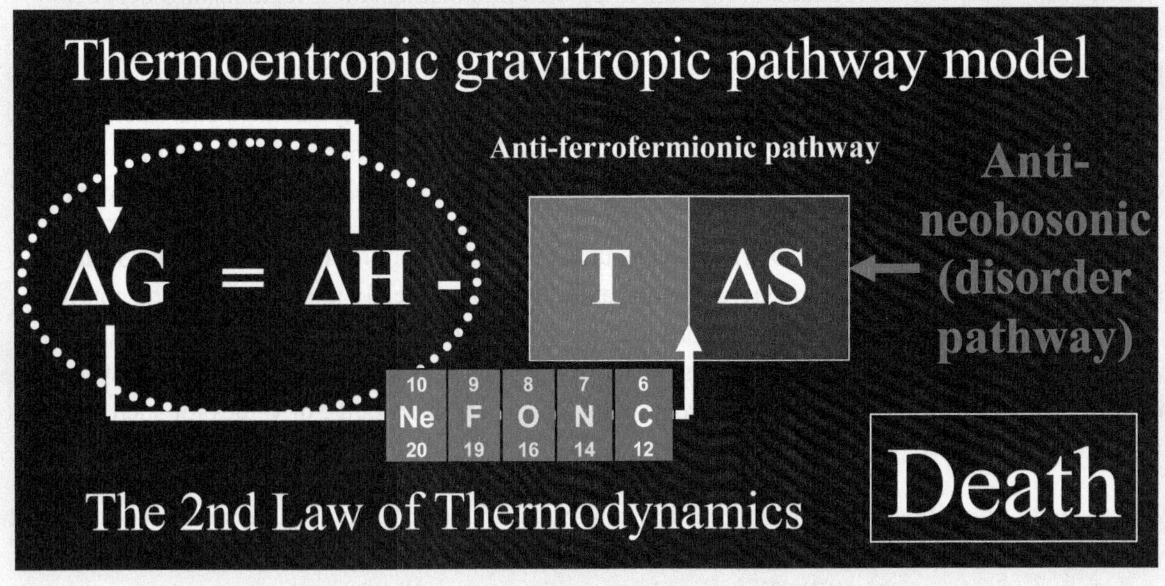

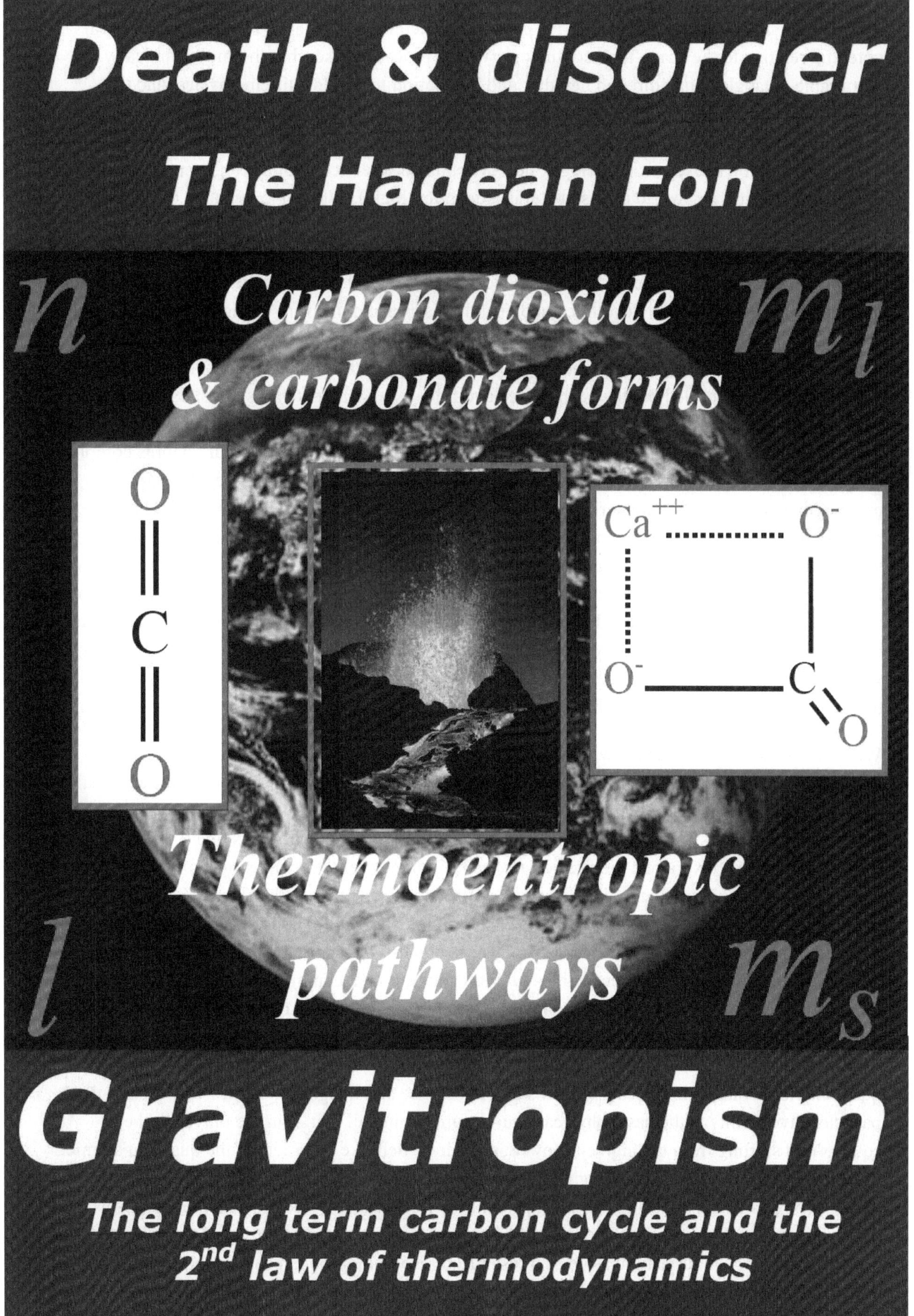
Death & disorder
The Hadean Eon
Carbon dioxide
& carbonate forms
n
m_l
O
C
O
Ca^{++}
O^-
O^-
C
O
Thermoentropic
pathways
l
m_s
Gravitropism
The long term carbon cycle and the
2nd law of thermodynamics

The energy/mass accumulation for life on Earth occurs through the natural spontaneity or affinity ($A= -\Delta G$ and $\Delta G<0$ free energy is lower in the products and entropy is greater) of the living physical potential relative to the space-time continuum. With a process conserving thermodynamically favourable entanglements, through heredity and its relation to non locality in physics, and the powerful link to the field of genetics which is also a non local nucleonic logic [11]. The biological levels of natural scale is so diverse and uncertain that the only mathematical tool of any practical use is a statistical probabilistic axiomic logic, however this does reflect quantum logic and its amplified effects on our everyday world (any statistic is a quantum logic). Where the outcome of physical events is described only in terms of the probability of an observed amplitude solution (Einstein's four dimensions of the space time continuum x, y, z, t), as opposed to precise interpolative and extrapolative capabilities based on initial physical conditions [3]. The strange reality of this picture is that the biologist can look under a microscope and 'observe and measure by sight' the cellular level of scale, using much more of an observational measurement over classical numerically measured determinism [12]. The physicist on the other hand can measure accurately because their systems are synthesised and simplified but they cannot see and fully resolve atoms or distant stars and have to omit observational logic from their processes [9]. So when irresolvable physical states are analysed in physics the world is blind but the variables are identifiable and numerical measurement is completely reliable. The biologist however can see their world and make very accurate observational measurements although the numerical terms and details are often vague and unreliable [13]. The resolution problem occurs where physical science effectively uses applied mathematics because their world is homogenous and the variables are identifiable and numerical measurement reliable. However in biology the resolution problem allows science to make powerful physical observations with the reliability of sight, but numerical determination and identification of numerical measurement often become totally unreliable [14]. There is a trade off in the logic of perception across the sciences, which make integration, and unification of scientific evaluation and axiomic logic almost impossible to achieve due to conflictions in observational measurement and treatment [7], hence segregation.

Where does the logic of particle/wave duality begin and end? Why can't we describe an entire organism such as a human as a unified particle/wave system [15]? When does particle/wave duality lose its logic?, it would appear that this is simply a conventional limit set by physics because mathematical analysis works well up to that level.This does not mean it works on all unique levels in biological systems (taxonomic hierarchies are mainly subjective to fit conventions through physical similarities). If we extend the logic, we can see particle/wave duality in the fermionic trinity (proton/electron/neutron) and sub atomic particles, atoms, molecules, cells, tissues, organs, organisms, grouped organisms, global organisms, astrophysical-bodies [4], where the four forces of nature produce these physically larger ensembles of microstates, with a common nuclear inertial unity and increasing gravitational determinism typically through a quantum of uncertainty [16]. Why can't we refer to planet Earth as a particle/wave duality composition united by gravitational forces in dynamic equilibrium (ΔG close to 0 or planetary equilibrium for all life on Earth) with the internalised quantum electromagnetic forces [17]? This global balance and integration shows very clearly that particle/wave duality is the basic physical equilibrium behind all matter in the universe, the equilibrium shifting from the high entropy (S, disorder – hell like disorder) of the quantum atomic level to very low entropy (S) and high Gibbs free energy (G) and enthalpy (H) on higher levels of particle composition (increasing order). Einstein's 'special relativity' centralises the scientists observational capabilities in terms of time dilation and length contraction [6], and the limits to measurement throughout natural scale as a result (the resolution problem can be shown to be the result of dilation and constriction due to relativity and the observer where different natural levels of scale is being observed and measured): -

Einstein's first postulate [6] – The laws of physics apply equally in all non-accelerating frames of reference (atomic level, molecular level, cellular level, organismal level, and global level – ensembles of particle/wave microstates in the course of their thermal fluctuation [8]).

For a physicist making atomic measurements, V (velocity) is very large for sub atomic particles, hence time spreads out and length restricts to single dots. V is very similar for cells relative to the observer hence the observer can see them, therefore time is similar to the observers but length remains the same hence unavoidable complexity for length of cells; no resolution into simple points (uncertainty) [6].

Lorentz contraction $= \sqrt{1-V^2/C^2}$ Time dilation $t=t_0/\sqrt{1-V^2/C^2}$
Length contraction $I=I_0/\sqrt{1-V^2/C^2}$ Apparent mass $m=m_0/\sqrt{1-V^2/C^2}$

This relativistic treatment of changes in observational measurement across science personifies and demonstrates the 'resolution problem'. The enormous speed of particle/wave entities on the smallest level contract length down to a dot and make time spread out, making mathematical treatment effective relative to the observer, but observational measurement becomes impossible [18]. A biologist observing a cell where movement is relatively very slow sees little or no time dilation and almost no length contraction hence observation is more reliable, but mathematical treatment is almost impossible as a trade off [19].

Where a physicist insists on mathematical logic and a biologist relies on observational logic what is logically required to produce integration is a compromise across science? [1]. Since the physicist models and represents the atomic level of natural scale with absolute mathematical precision (although limited by consequential uncertainty), a biologist studying the biological composition of atomic logic should surely utilise the same physical logic for describing and understanding living organisms [13]. Where do the measured consequences of natural scale fail to reflect the logic they are composed of? Atoms are said to be 'counter intuitive' because of particle/wave duality [9] however biological levels also demonstrate 'counter intuitiveness'. We live a life of uncertainty and unpredictability, why shouldn't this be the consequence of the quantum effect for living organisms [16]? Why does the atomic level produce more logical insight into the fabric of space-time than the level of a cell? Surely the logic of natural particle/wave composition must be linked through relativistic determinism and force equilibrium as measured on the different levels of natural scale [5]. The physics profession keeps the quantum world on the level of the atom, yet surely its determinism on all levels of natural scale must be evident and measureable since life is a transformation of logic from the Big Bang, and its chaotic atomic simplicity to the incredible order of modern human composite particle/wave dualities [4]. In an atom of carbon [9] we find a nucleus (protons and neutrons binary order), a K shell (n-1) and valance or L shell (n=2), in a cell [12] made partially out of centric carbon atoms we find a nucleus (AT and CG DNA binary order), a cytoplasm and a membrane (which externally bond). On the level of a human being [18] we find a nucleus (head action potential on and off binary order), a torso and the limbs (which externally bond). Each level is a fractional building block for the next level, but in living systems we find breathtaking consistency as we move from the very small to the very large level of scale [20]. Can we really suggest this consistency is the result of pure coincidence? Or perhaps life is simply a system that spans the equilibrium of the four forces throughout all natural scale and therefore shows us the quantum world through carbon centric linked amplification (organic chemistry to biochemistry to cell biology) through all forms of natural bonding; hence the quantum world is our everyday world [21]. This does not occur and a sensible high level scientific system is required to bring the best logic of all levels of natural scale together to produce an integrated model of existence and more centrally the human consciousness [5]. A physicist does not see the quantum world around them in their everyday life even though their world is made of atoms, and a biologist frequently fails to consider that organisms are made out of atoms limiting their logic outside of mathematical treatment and fundamental physical science [7]. The way forward is a system, which considers all facets of life centralised around the powerful creator of human knowledge, namely the conscious human mind and its deep penetrative ability to question its own existence [22]. To be human is to know without exception, that one day you will die, the thermoentropic second law of thermodynamics with its reality hitting home, but islands of negative entropy (a potential 4th law) in life give us hope of deeper order perhaps through elaborate Hess cycle routes and Le Chatelier's principle and free will (free energy). Through independent thermodynamic routes, where soulatrophic evolutionary pathways vary considerably, but where the overall enthalpic (ΔH) change remains the same [3].

In human thermodynamics (www.eoht.com) we find such a system, which aims to describe the conscious power of humanity as a central theme around all levels of natural scale [23] [24]. It embraces both the powerful determinism of physical applied mathematical logic against the inevitability of uncertainty in measurement in science [5]. A brave new world where mathematical equations representing physical states cannot be solved to an absolute numerical end for life but where the power of the logic of such mathematical equations is still considered to be very useful and appropriate for understanding the biological level of natural scale [23] [24]. An integrated system which is brave enough to cautiously contemplate the metaphysical and physical, to centralise all human knowledge with an effective optimisation of the best logic available to humanity [23] [24]. Human thermodynamics: has the ability to act centrally in human knowledge, cross-fertilising ideas from all aspects of observational scale [5]; embracing particle physics for describing living beings as amplified composite particle consequences of fundamental particle physics such as the 'human atom' and the '26 element human molecule' and appropriate logic in bonding through the four forces and 'human chemistry' [23] [24]; applies classical thermodynamics and its laws to describe the energy interactions of living particle wave/physics, including Hess cycles and routes and engineerability through Le Chatelier's principle [25]. In doing so, energy systems are at the absolute heart of all life on Earth and a thorough treatment of living systems against the laws of thermodynamics is a logical consequence of this powerful integrated science. Bonding in humans and other organisms is considered against classical bonding from molecular and atomic origins concerned with the way all four forces interact in the living world [23] [24]. Human thermodynamics also cautiously contemplates lofty matters such as good and evil, spiritualism in general and God, the concepts of morality, love, beauty and ugliness, gender and sexual reproduction and living bonding. Considering them as a natural consequence of fundamental physics, measureable and observationally identifiable [23] [24]. By considering the metaphysical as a consequence of uncertainty in physical systems through sensible careful scientific compromise allows humans to make sense of the 'resolution problem', where uncertainty in measurement is embraced as a natural property of consciousness. Physicists typically reject any suggestion that mathematical determinism is limited and subject to uncertainty rules [3]. 'Heisenberg's uncertainty principle' and the 'measurement problem' are the burden of physics; biologists however thrive on uncertainty in measurement often rejecting solid mathematical rules, which restrict philosophical freedom [5]. Collectively the concepts of natural physical composition must demonstrate integrated conservation, the quantum world must directly affect the immediate gravitationally dominant world around us, yet physicists cannot realise this because mathematical modelling collapses, its logic needs to be identified on the level of an organism [4]. The biologist must also consider the underpinning logic of atomic physics and its breathtakingly powerful and fundamental organisation of all matter, that life must demonstrate conservation of such integrated logic [13]. In short human thermodynamics forces all scientific disciplines to exchange ideas through logical extension of the best concepts throughout all science, through logical compromise, demonstrated by special relativity and its consequence for a conscious observer relative to natural scale [5].

Introducing carbon entromorphology: a compromised integrated unification of knowledge.

Carbon entromorphology is an independently developed unified theory, which effectively shows fundamental agreement and expansion of many of the principles in human thermodynamics and biology and physics (www.mrcarbonatom.com). Entromorphology is a technical term which describes the way living anatomy reflects the associated thermodynamics:'entro' taken from the physical quantity entropy which is known as the 'arrow of time' and encapsulates the second law of thermodynamics and its unflinching consequences for living beings [23] [24]. It also reflects the number of different particle/wave compositions in a physical system and therefore symmetry and order through statistical thermodynamics [8]. Consider the connected mouth and the anus; there is an entromorphic distribution across the body from the beauty and associated symmetries of the mouth (stable microstate frequencies) to the asymmetry and chaos of the anus, enthalpic (H) Gibbs free energy (G) from the mouth to thermoentropy (TS) at the anus; hence an entromorphic thermodynamic span. Symmetry breaks down, and microstate frequencies reduce across the body from higher levels of cognitive neurological organisation [19] to simple atomic organisation such as CO_2 (which is a highly thermoentropic gas). Carbon entromorphology is a new system for modelling and understanding all life on Earth based on an extension to the logic of atomic physics, and a blending of logic through all natural scale [23] [24]. Living organisms are modelled as amplified carbon atoms and other organic atoms (26 element human molecule) as they demonstrate fractal self-symmetry through three levels of organisation from the nuclear, the stable nuclear housing field and the valance or bonding field [9]. Carbon entromorphology also take into account the 92 naturally occurring and up to 25 artificially created elements (+117 element technological molecule) [9] [21]. In classical human thermodynamics a human being is described as the composition or amplification of 26 elements; an amplified molecular human [23] [24]. Carbon entromorphology extends this logic because humans bond with the whole of the periodic table through the extension of life into technology, the 'biology of technology' [26]. Differentiation of the importance of specific elements such as carbon, and even more fundamentally, hydrogen is not fully elaborated upon but clearly evident in the field of organic chemistry [21]. Carbon entromorphology considers the carbon centric determinism to be the central theme to the thermodynamic spontaneity of reactions ($\Delta G<0$ and $A= -\Delta G$) with the rest of the periodic table in conjunction with spontaneity in hydrogen from the 'Big Bang' onwards [27]. Carbon can be shown to produce a pathway of increasingly larger particle systems named in this article as a 'soulatrophic pathway' with discrete 'soulatrophic energy levels' of linked fractional dimensions such as the sub atomic (fermionic - proton/electron/neutron [28]), atomic, microbiological, somatic, cognitive and spiritual levels [19] [20]. In carbon entromorphology life begins at the Big Bang and the first 10 billion years constitutes the nucleogenesis of the elements before the Earth formed some 4.5 billion years ago [9] [27]. This early period is associated with hydrogen entromorphology, as all the elements are basically composed of hydrogen and hydrogen is vastly the most common element in the universe next to helium [9]. The periodic table itself, is a perfect example and proof of nuclear and field amplification (which is the basis of carbon entromorphology) as it is arranged in order of increasing size in the nuclei and quantum fields of the elements involved [9] [27]. Hence a demonstration of the logic of universal behaviour in all matter through classical fusion/fission reactions. The extension by chemical elementary bonding produces carbon entromorphology; hence a form of very low energy covalent fusion/fission in keeping with this physical axiomix approach [9] [27].

Carbon entromorphology considers the 'organogenesis' of classical life and supports the classical '26 element human molecule' logic [23] [24] although carbon is powerfully conserved through amplification and thermodynamically critical in the process of determinism for life formation. Humanity refers to 'carbon based life' even though life is mainly composed of hydrogen and oxygen; this is because of carbons catalytic thermodynamically broad potential and centralised role in bonding and therefore 'solar neoiterative' amplification cycles (a 4th law of thermodynamics) through the 'octet logic' in quantum mechanics (rule of octaves) [9] [27]. Carbon's vast spectrum of physical states allows it to possess thermodynamic spontaneity ($\Delta G<0$ and $A= -\Delta G$) for bonding to 'all elements', with variation in bond strength and associated enthalpy and its centralised extended determinism [21]. The +117 element model is called 'silicoferrous entromorphology' as it represents the bonding of carbon and the other 25 elements to the rest of the entire periodic table associated with human beings and 'technogenesis' [26]. Both silicon and iron are absolutely essential for this process to occur and are reflected by the Stone Age (silicon) and the Iron Age (iron) and the central importance of these elements in all facets of modern human life [21]; the Earth is composed of an iron core with a surface membrane of siliconised rock as a deeper reflection of this stabilising effect. They also reflect the natural thermodynamic potential, which favours spontaneous reactions ($\Delta G<0$ and $A= -\Delta G$) with all the elements on planet Earth and the artificially created elements through solar input (surroundings) to an Earth accumulative living system [23] [24].

Where $\Delta G<0$ and $A= -\Delta G$ (affinity) for bonding of 'all the elements' from the 92 natural ones to the +117 artificial elements through a pathway of hydrogen to carbon to silicon and iron, which utilises the whole periodic table as the basic logical platform for describing all living beings [21]. With other nuclear extending elements such as nitrogen, oxygen, halogens, phosphorus, calcium, sodium and many others, again with carbons broad values of $A= -\Delta G$ (affinity). Carbon is still the central theme where living organisms are composed of organic molecules managed and centrally amplified by carbon as a chiral centred atom [21], extending its own natural atomic organisation and physical orientation plus its natural thermodynamic potential to produce larger and larger structures through accretion heredity (non local conservation of physical entanglement), by solar means to stability, this appears to contravene the 2nd law of thermodynamics, however life is an open system. To carbon-based molecules, cells, tissues, organs, organisms, grouped organisms, global organisms and now space based organisms [20]. The different grouping names simply describe the soulatrophic pathway which is a thermodynamic accumulator hierarchy evolving logic based on a fractional dimension of natural carbon and the other 25 elements [23] [24] [20]. Fractal geometry is the central mathematical feature of soulatrophic pathways and describes grouped particles as being composed of a 'fraction of the lower levels' and acting as a 'fraction of levels above it' having carbon and other 25 element 'self-symmetry' conserved on all such levels. Entropy is also greater the lower down a soulatrophic energy level and enthalpy and Gibbs free energy increases up soulatrophic pathways, the combined process drives life down its concentration gradient into space or 'planetary cytokinesis'; 'technological life' spontaneously stabilising abilities and temporal re-engineering as a future consequence of the '+117 element technological molecule'. Evolution by natural selection [10] is a thermodynamic model, which acts to increase the frequency of living systems (Boltzmann entropy [8] frequency of ensembles of microstates and symmetry) against thermodynamic solar input, which maintains energy accumulation (greenhouse effect – phototropism). This produces an integrated logical amplification pathway based on conserved carbon logic where molecules, cells, organisms etc all represent amplified carbon and other elements from the periodic table [23] [24] [21]. All these particle reference frames associated with ensembles of microstates and symmetries contain a nucleus, and two fields surrounding them with the same anatomy as atomic carbon [21]. A cell, for example, has a nucleus made from DNA, and it has a cytoplasm and membrane field system around the nucleus [12]. In essence the cell is an amplified carbon atom, and demonstrates the same thermodynamic potential ($\Delta G<0$ and $A= -\Delta G$) driven by quantum instability and the need to complete its valance octet in accordance with thermodynamic quantum rules; it naturally grows (accretion) or amplifies to become reduced towards iron nuclear stability over time [9]. The octet is hypothesised to be the home of the classical 'cell cycle' although considerably extended to all levels of natural order. The G1 nitrogenous, S oxidation, G2 halogenous and M mitotic neogenous (and their electronegativities) are named in this article as 'metabolic organelles'. The octet must be completed within the limits of the cell cycle to satisfy the quantum energy level everyday (in the case of a human example), and act as a powerful temporal driver for all cellular life from atom, cell, organism, nation, flock any grouped ensemble of living microstates. However, any field activity is an expression of the thermodynamics of the nucleus, which is driven by iron stability rules [9]. Carbon is hypothesised by this article in essence to thermodynamically act to stabilise both its quantum valance shell and in doing so its unstable nucleus with the spontaneity ($\Delta G<0$ and $A= -\Delta G$) associated with solar acquisition producing $-\Delta G$ (Gibbs free energy) and $-\Delta S$ (entropy) which does not contravene the 2nd law because the Earth is an open thermodynamic system [23] [24]. This is achieved because life on Earth is not a closed isolated system and has a large solar input of free energy which increases orderly behaviour and increasing symmetry in the associated structures through phototropic thermodynamic potentials hypothesised as being based on 'Bose condensate' physics logic or growth (accretion amplification – Bose Einstein statistics). The overall conditions for this spontaneity of thermodynamics ($\Delta G<0$ and $A= -\Delta G$) is much more complex as it requires unique astrophysical conditions such as atmospherical gain mediums offering energy accumulation on planet Earth, also through the excess of aqueous fluidity and solubility through the waters on planet Earth and large scale temporal stability. The tides increase reaction kinetics by increasing the number of collisions in organic reactions, and the salt ions allow electrical conduction to take place (aqueous electrolytes promote solubility of inorganic constituents as well as organic ones – 26 element human molecule [23] [24]).

The iron core in the Earth produces a magnetic field which deflects disorderly levels of free energy from living beings. Maintaining stable temperature for large periods of time again contributing to thermodynamic stability and reaction spontaneity ($\Delta G<0$ and $A= -\Delta G$) for Earth based carbon and other 'nucleic elements in the 26 element human molecule' in association with carbon [23] [24]. The unique physical system of planet Earth in the solar system is critical and unique to the ability for carbon to undergo accretion amplification to produce larger systems such as cells and organisms. Life is somewhat dilute in the universe because it favours less free energy by producing life but the systems of the Earth and Sun are definitely unique and stable, fine structures, and rare, however the universe is teeming with other such systems. Hence the probability of life evolving anywhere else in the universe by the same pathway is hypothesised to be extremely high [29].

The periodic table, gender and carbon entromorphology.

The home of all life is hypothetically demonstrated through the three particle/wave thermodynamic soulatrophic pathways, the protonic or plant pathway (a radiating semi stationary high mass nuclear logic), the electronic animal pathway (convergent semi to fast moving low mass field logic) and finally the neutronic or technological pathway (both properties are present – a house doesn't move, a rocket moves very fast, technology tends towards non polar or neutral potentials and is often very mass rich). Carbon entromorphology finds its logical soul in the form of the fermionic trinity [28] as all natural matter in the universe is comprised of just these three particles. This is the 'reductionist' basis of the gender model in carbon entromorphology where the convenience of the logic of just three particles is to be considered.The periodic table is such a powerful and amazing human achievement that represents all the enormous complexity of the entire universe. As a result it is felt by this author that any biological system having organisms made of atoms must be identifiable on the periodic table. In essence it is the perfect place to establish life and contemplate its organisation and diversity. The extension from the classical 26 element human based model to a +117 technology model is crucial as technology reflects carbons ability to bond to all known and artificial elements, the biology of technology. Technology is currently not associated with biology which is counterproductive and produces only a partial model of life.

Fractional Dimensions & Soulatrophicity

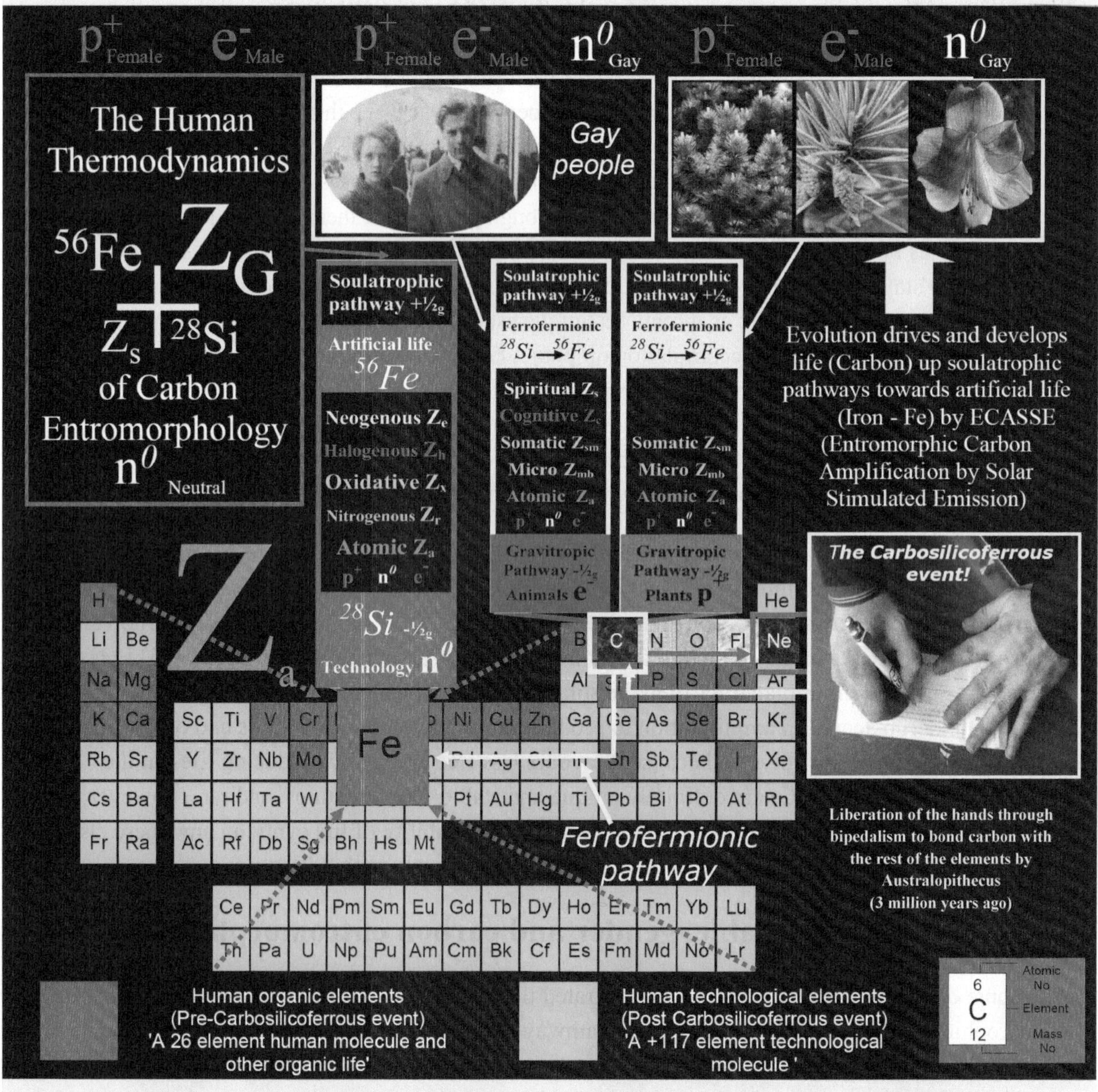

Life and its place on the periodic table (self-awareness through organoaccretion)

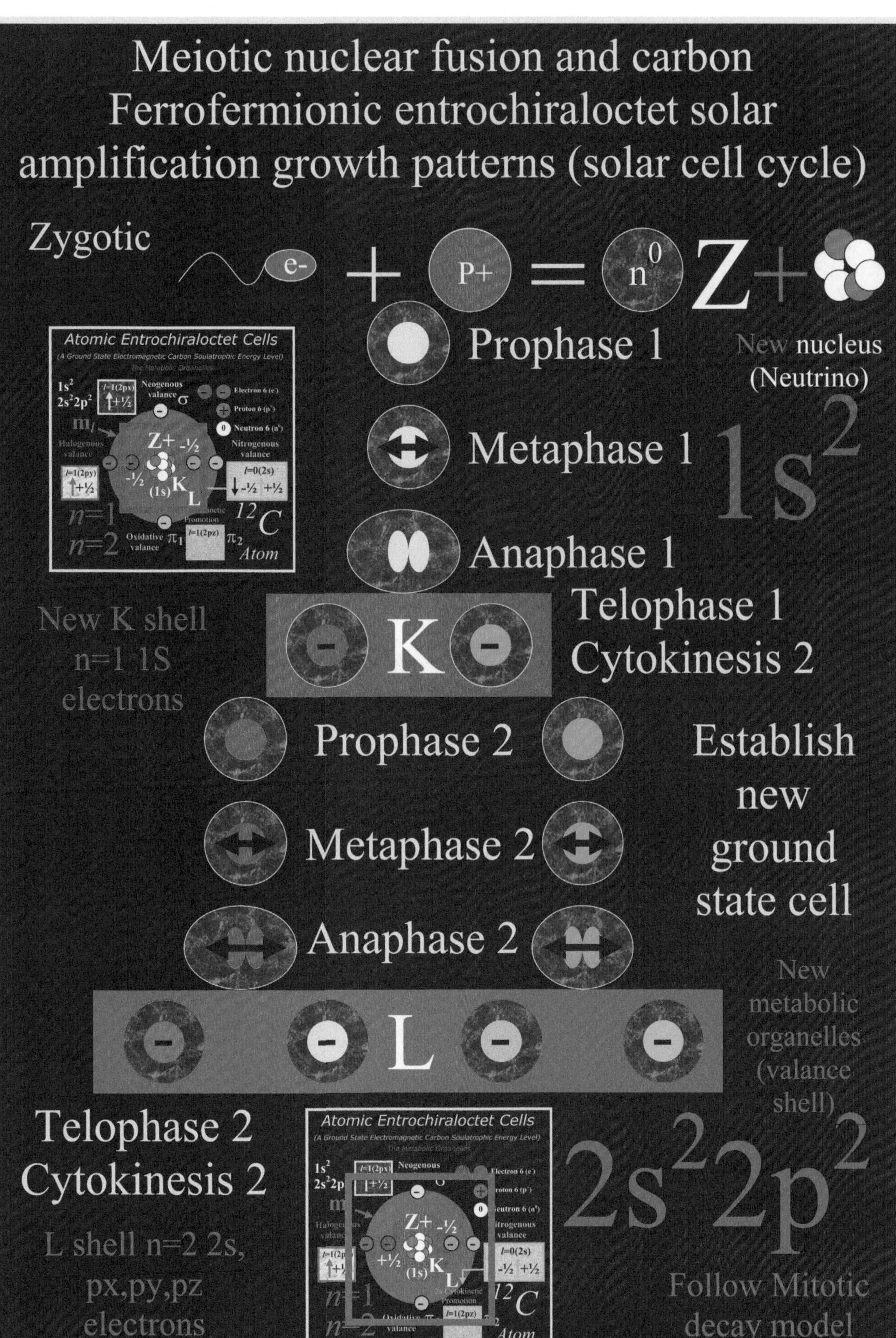
Meiotic nuclear fusion and carbon Ferrofermionic entrochiraloctet solar amplification growth patterns (solar cell cycle)
Zygotic
e-
P+
n0
Z+
New nucleus (Neutrino)
Atomic Entrochiraloctet Cells
Prophase 1
Metaphase 1
1s2
Anaphase 1
New K shell n=1 1S electrons
K
Telophase 1
Cytokinesis 2
Prophase 2
Establish new ground state cell
Metaphase 2
Anaphase 2
New metabolic organelles (valance shell)
L
Telophase 2
Cytokinesis 2
L shell n=2 2s, px,py,pz electrons
Atomic Entrochiraloctet Cells
2s22p2
Follow Mitotic decay model

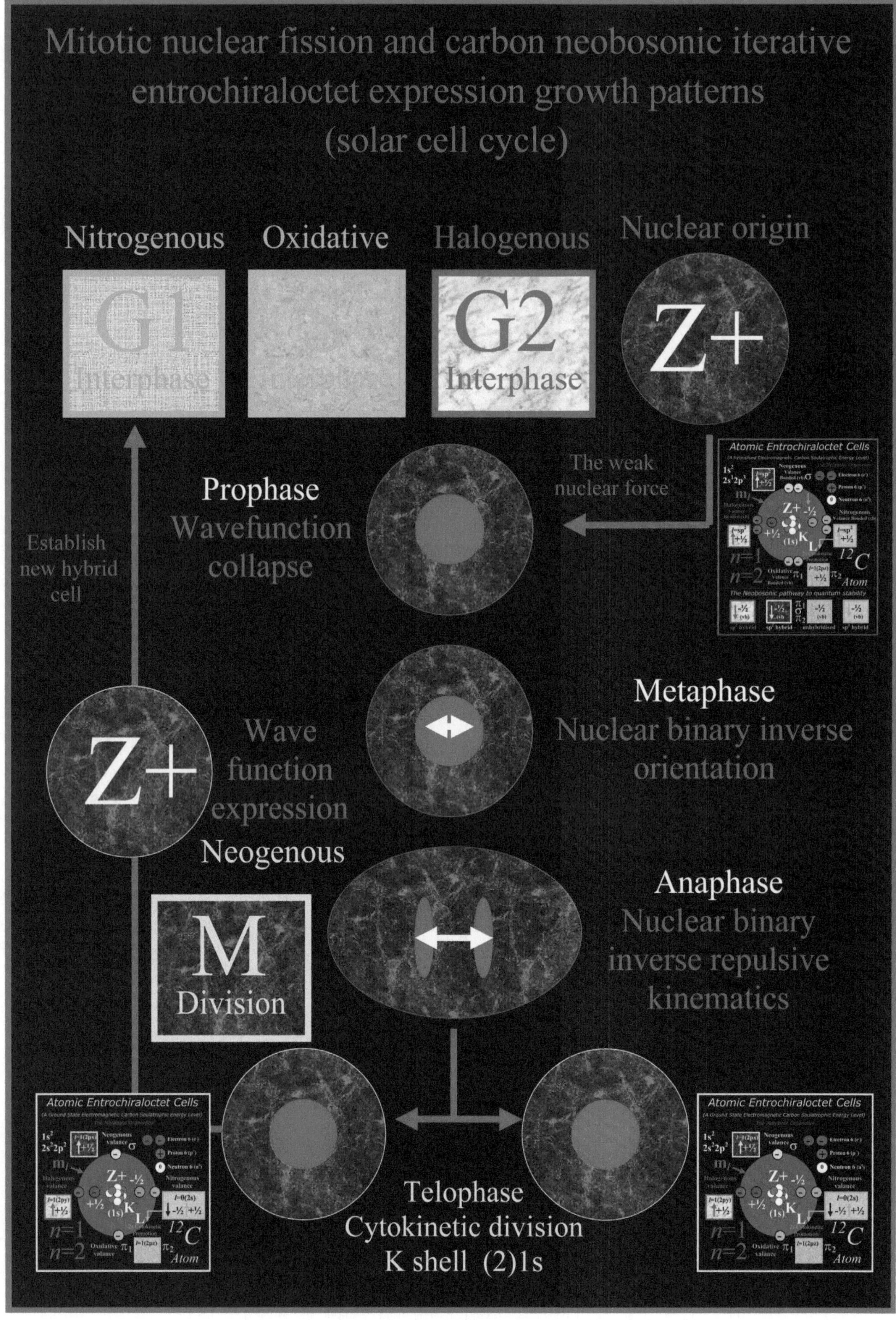
Mitotic nuclear fission and carbon neobosonic iterative entrochiraloctet expression growth patterns
(solar cell cycle)
Nitrogenous
Oxidative
Halogenous
Nuclear origin
G1
Interphase
G2
Interphase
Z+
Atomic Entrochiraloctet Cells
The weak nuclear force
Prophase
Wavefunction collapse
Establish new hybrid cell
Metaphase
Nuclear binary inverse orientation
Z+
Wave function expression
Neogenous
Anaphase
Nuclear binary inverse repulsive kinematics
M
Division
Atomic Entrochiraloctet Cells
Atomic Entrochiraloctet Cells
Telophase
Cytokinetic division
K shell (2)1s

It is true to say that a human being is a composition of protons, neutrons and electrons, and it simplifies the system of life to a logical physical basis; this is a concept, which can be conceptualised and hypothesised as a 'super ion'. The characteristics of the fermionic particles must contribute to compositions of such particle microstate ensembles regardless of their relative size to the smaller particles they are composed of [9]. What is observed is that as living particle/wave duality system increases in size where the mass potential and its associated determinism is translated from electromagnetic quantum determinism seamlessly, this is the quantum gravity effect with life as a special case, transforming the logic from the atomic level to general relativistic gravitational determinism [30]. In the gender model, female particles (amplified protons) are polarised stable (semi static) nuclear radiating systems with mass rich smaller fields, females are more localised. Male particles (amplified electrons) are polarised unstable (increased velocity) field convergent systems with larger fields, this is due to the organisational properties where the male is larger than the nuclear female due to the inverse square law (offspring plus female is hypothesised as the nuclear shell, the male is hypothesised to be a field valance bonding shell). Gay particles are hypothesised as neutral none polarised (amplified neutrons) hence man with man and women with women. Both convergent and divergent radiating properties, which is why gay people are very colourful, but also butch hence large masses [31].

- The gender model – The fermionic trinity for the three gender particle systems.

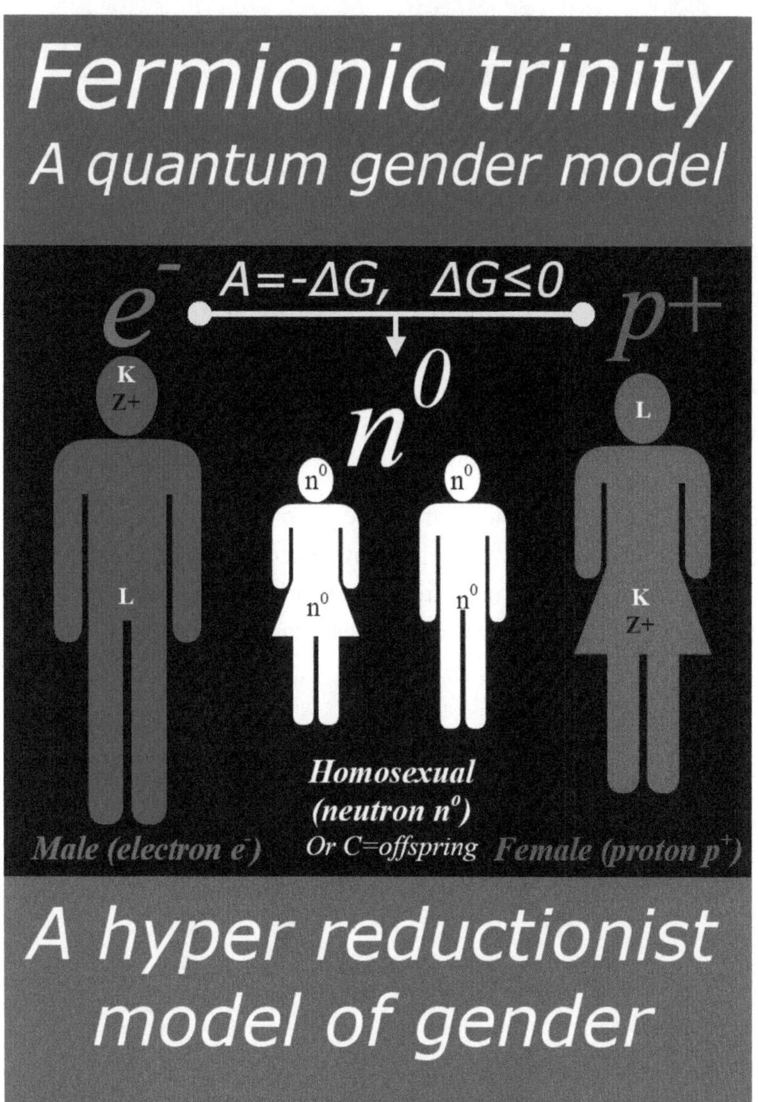

It is absolutely TRUE to state that however complex a living organism is, it is still composed of three basic building blocks.

Hypothetically if we had three bags of Lego building blocks, one red, one blue and one green, and we built a huge complex building out of them it would still be comprised of just three sets of fundamental logic.

Reductionism is needed in biology as the diversity and associated complexity of life is so vast that fundamentalism seems difficult or almost impossible to find.

- Diagram of a soulatrophic pathway – electronic animal, protonic plants and the neutronic technologies and their thermodynamic evolutionary pathways.

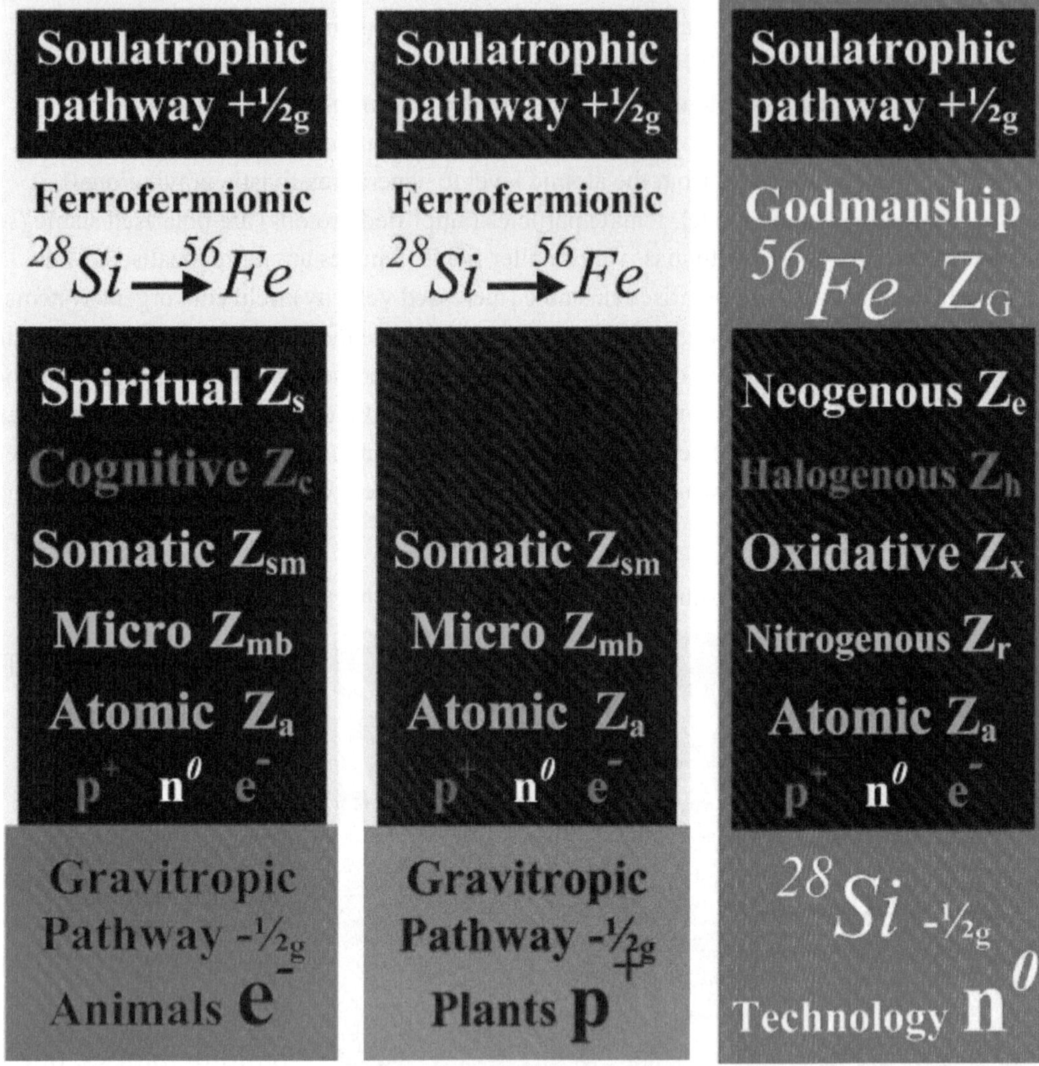

- Diagram of cell cycle – G1, S, G2, M phases and their relationship to the 'rule of octets' on the atomic level of natural organisation. This is the solar growth accretion pathway and its distinctive levels, organisation and logic. Cells are extended in carbon entromorphology to include atoms to entire organisms, all seeking field and nuclear stability.

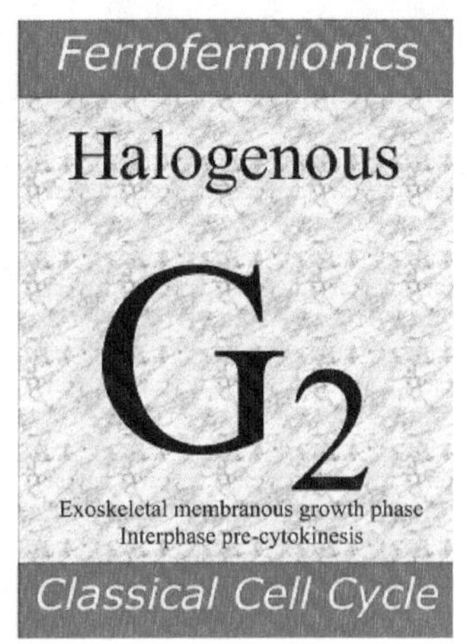

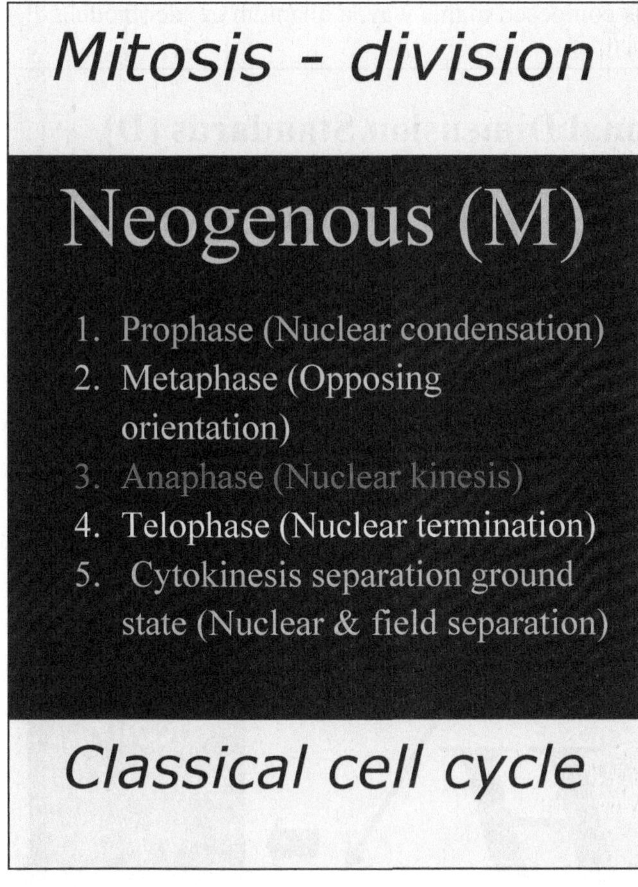

Mitosis - division
Neogenous (M)
1. Prophase (Nuclear condensation)
2. Metaphase (Opposing orientation)
3. Anaphase (Nuclear kinesis)
4. Telophase (Nuclear termination)
5. Cytokinesis separation ground state (Nuclear & field separation)
Classical cell cycle

Division is a fundamental process based on Bose condensate processes driven by the Sun.

There is more to cell division or mitosis than the actions of classical cells. In carbon entromorphology cell theory is dramatically extended so cell cycles are also extended, hence sleep and waking is a mitotic event.

The soulatrophic pathway is a thermodynamic evolutionary fractional dimension [20].

The three evolutionary particle pathways the fermionic trinity [9], which is hypothesised as a basis for all material logic, fit on the periodic table of the elements where the conventional organic life pathways are modelled as the protonic plants (p^+) and electronic animals (e^-). The final neutronic (n^0) pathway is the technological pathway and allows carbon and many others of the 26 elements [23] [24] in the organic realm to bond to the rest of the periodic table of the elements up to +117 [9] [32].

The soulatrophic pathway is designed to take into account each of the discrete soulatrophic energy levels for any organism; 'Nature's design'. The fractional dimension [20] produces a stack or pathway of conserved carbon amplification driven by Bose condensate thermodynamics driven and maintained by solar input. The levels are conserved from the fermionic level [28] – a fundamental reductionist level of natural scale it conveniently contains just three particle logics, the proton, the electron and the neutron [9]. Physics currently suggests that the most up to date model of matter is that all matter is comprised of electrons and two types of quarks; again this model can be related to gender.

The physical properties of each particle and the equilibrium of particle dominance produce the ionic dominance, and describe the nature of gender (super ions) on all levels of natural order. Proton/neutron nuclear components show binary organisation and determine electron wave functions as expressions of the thermodynamic potential on the nucleus [9].

The fermionic level ($p^+ e^- n^0$) – All matter is composed in this way, a quantum gender model.

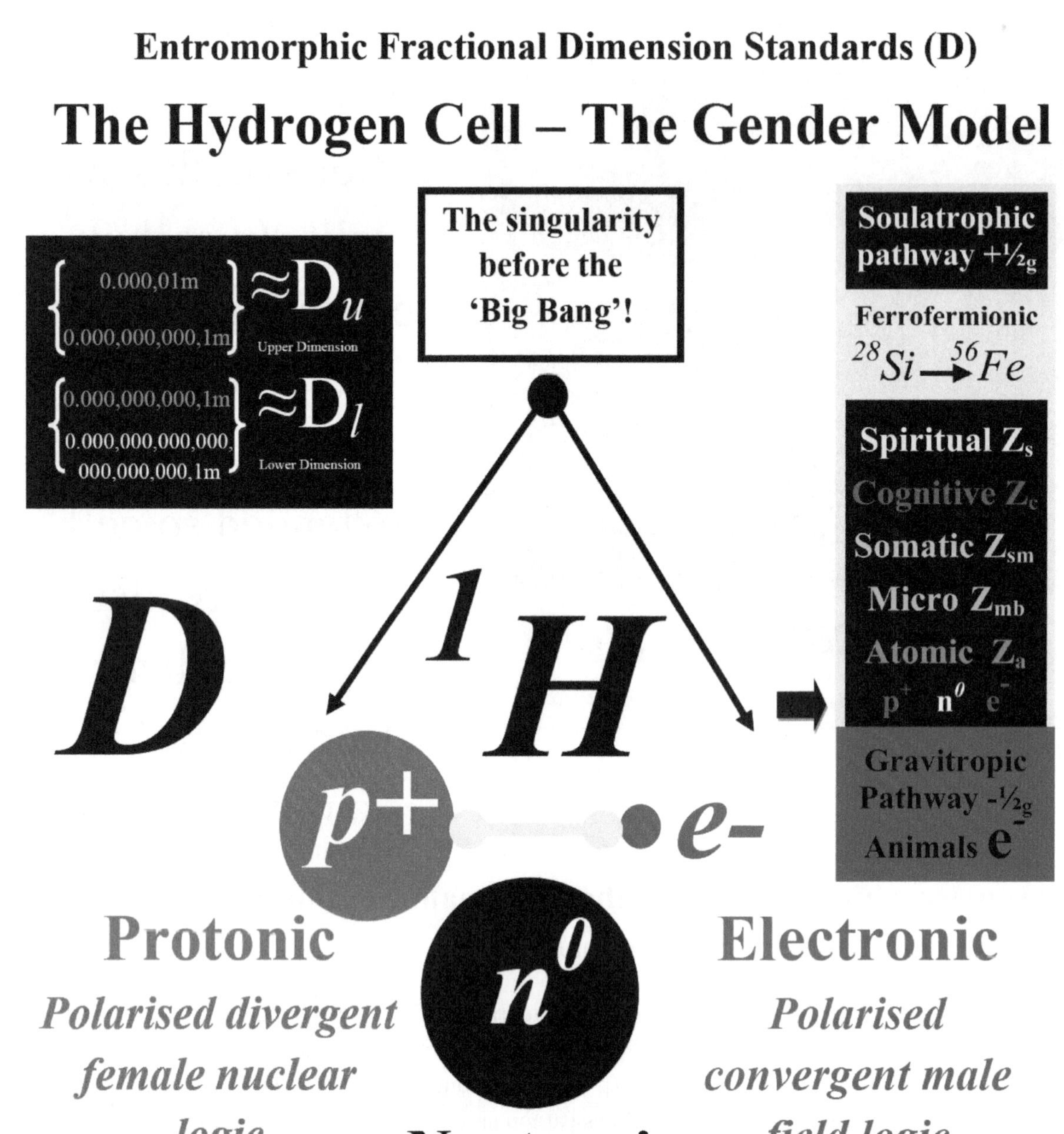

(Composite of a proton and electron hence neutral or same sex)

None polarised homosexual logic

All matter on all levels of scale can be reduced to the logic of these three fermionic trinity particles (proton/electron/neutron)!

<u>This is the conserved fundamental basis of gender logic</u>

The atomic level (Z_a) – carbon and more fundamentally hydrogen (see previous page for fermionic system), interface with extension contributions from the other 24 elements associated with the organic organisms [23] [24], but with carbon centric accumulation and conservation [21]. The carbon templates are called 'entrochiraloctets' (a word identified in this article) which describes the link between natural thermodynamic potential and the distribution of energy in atomic compositions. Classical binary nuclear organisation through proton/neutron with quantum fields of expressed (conscious wavefunction ψ_a) electrons (e^-), the k shell or n=1 and the L shell n=2. The 26-element model is posted on this level and fits life onto the periodic table through the soulatrophic amplification pathways [23] [24] and electroweak effects are the basis of self-consciousness.

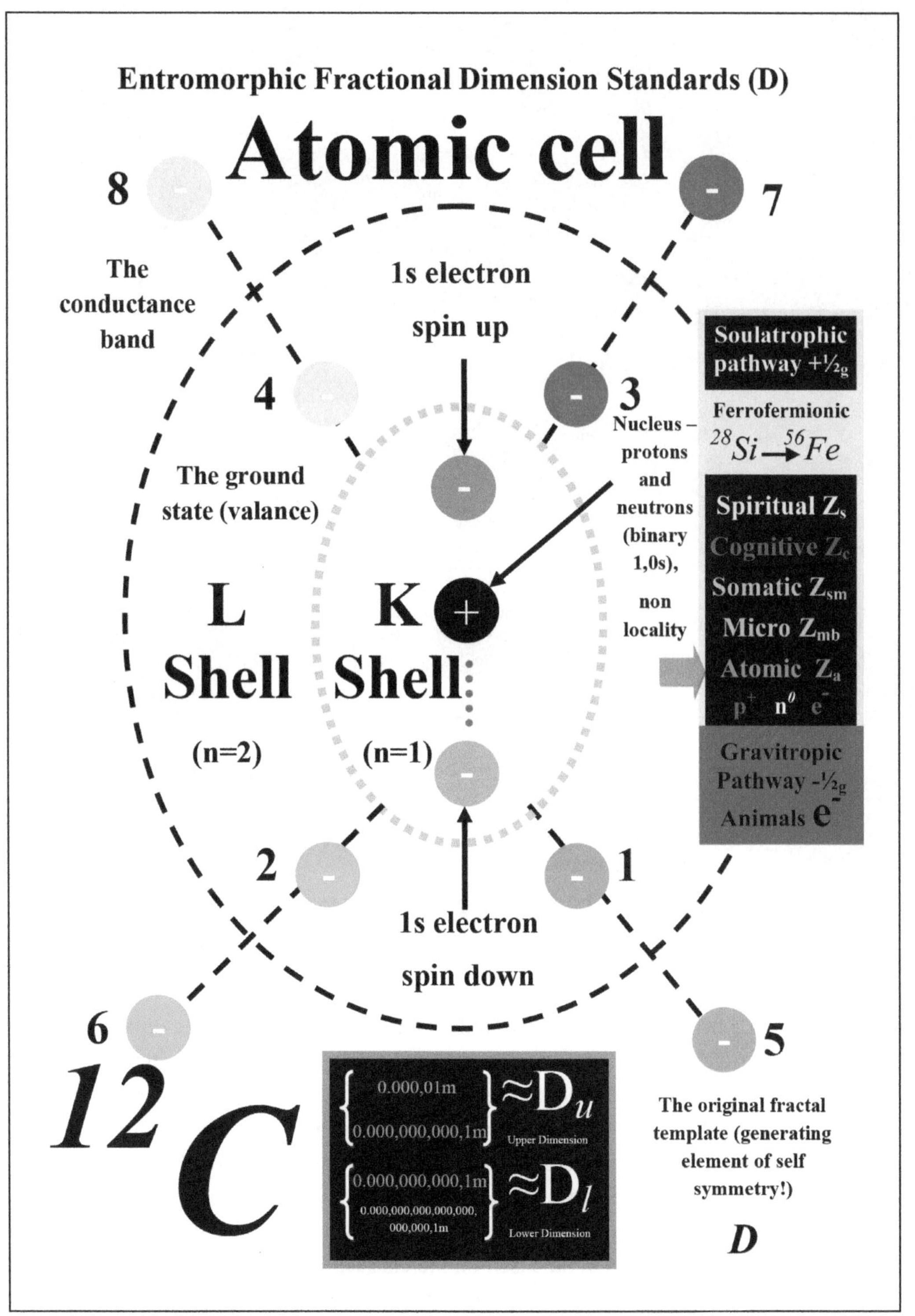

Fractional Dimensions & Soulatrophicity

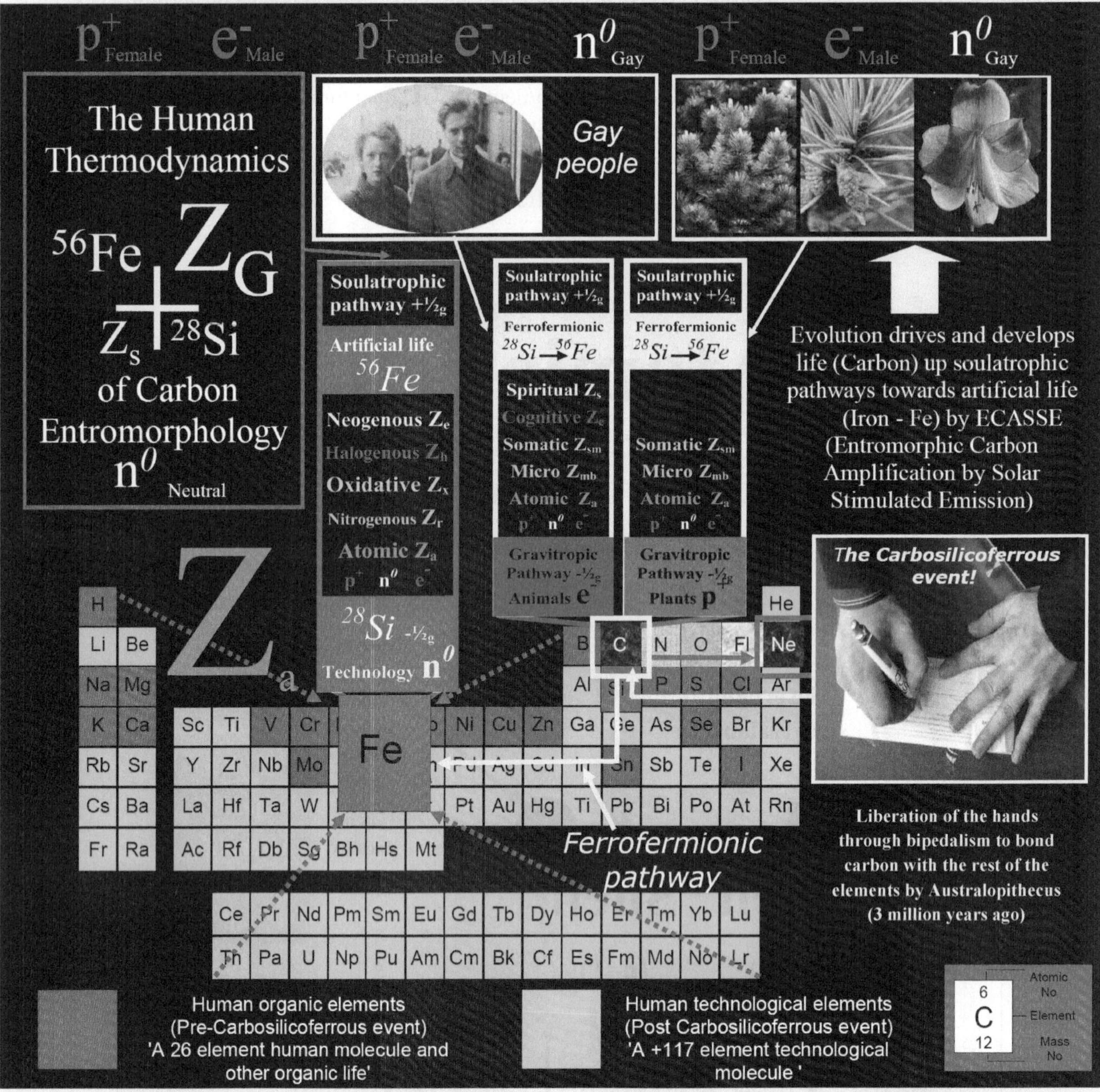

Life and its place on the periodic table (self-awareness through organoaccretion)

The fermionic trinity is a true and fundamental model of all natural matter in the universe. Any composition of such particles to produce larger structures is therefore built out of this three particle basis (organoaccretion). The periodic table of the elements is the most reliable and scientifically suitable model to describe all of life on Earth. **See Appendix 5(back of the book) for detailed models of carbons organic accretion through solar driven nuclear and field amplification (atomic to molecular sized particle/wave duality).** It has been shown on this model that life can be located very effectively around hydrogen, carbon, silicon and iron (the main soulatrophic fractional dimension). This produces part of the table based on the classical 26 element human molecule but extended to take into account the +117 technological molecules identified in this article.

The microbial level (Z_{mb}) – the realm of conventional cells, of which there are over 200 types of human cells; arguably an organism such as a human can be described as a '200 cell organism'. Again with hypothesised conserved carbon atomic logic extended by molecular bonding to produce links to extend carbons thermodynamic tendency [21]. A cell is an 'entrochiraloctet' with a binary nucleus composed of DNA AT and CG logic [33], it expresses (conscious wave function ψ_{mb}) proteins [35] in its cytoplasmic field (K shell, n=1) and membrane components (L shell, n=2) as theorised amplified carbon logic. Classification by the central dogma in this theory.

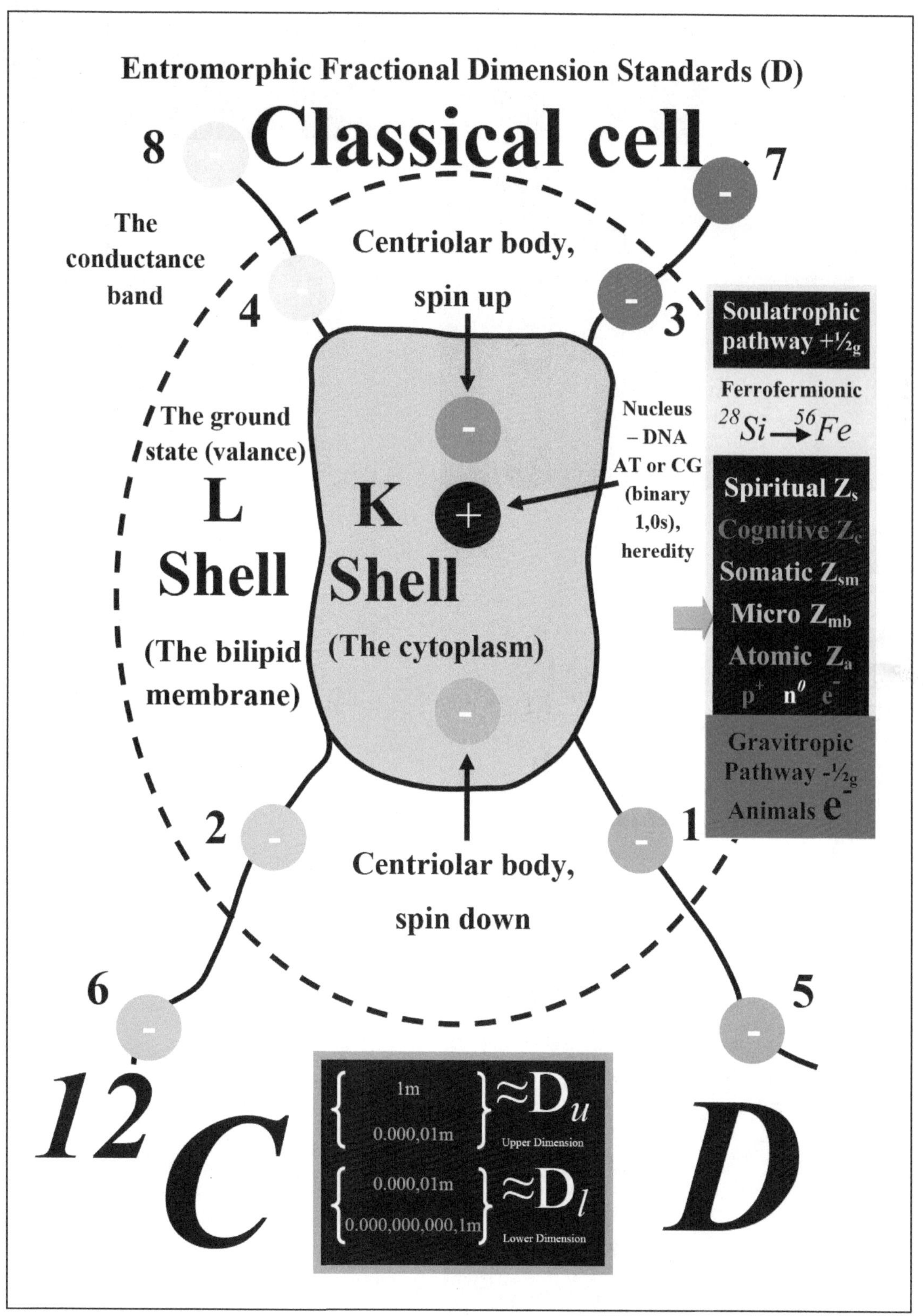

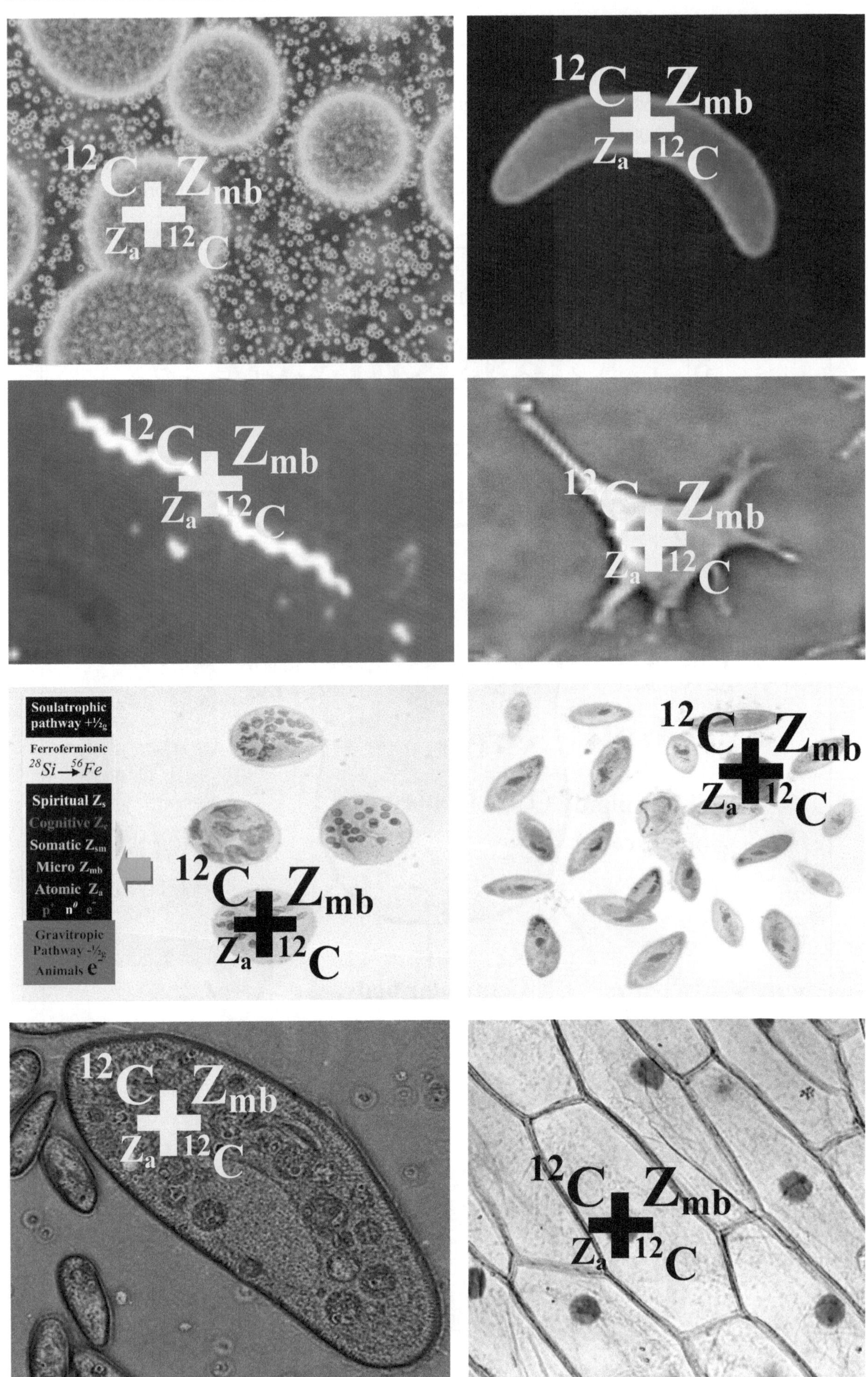
^{12}C Z_{mb} Z_a ^{12}C
Soulatrophic pathway +½g
Ferrofermionic
$^{28}Si \rightarrow {}^{56}Fe$
Spiritual Z_s
Cognitive Z_c
Somatic Z_{sm}
Micro Z_{mb}
Atomic Z_a
p^+ n^0 e^-
Gravitropic Pathway -½g
Animals e^-

The somatic level (Z_{sm}) – 'multi cellular organisms'; this level still exists in huge quantities over the entire world. The somatic organisms are some of the most numerous on the planet. Biologically they are known as the 'radially symmetrical' organisms and have little or no neurological organisation [22]. They are composed of DNA AT and CG (classical genetics, information is in the form of classical genes) binary nuclear logic, are expressed (conscious wave function ψ_{sm}) cells in its torso field (K shell, n=1) and its limbs field (L shell, n=2). The somatic level in a human can be seen in the lower part of the body; the genital region is the effective nucleus, the limbs (legs) the valance shell and the stable K shell abdominal torso region. Examples of somatic organisms are starfish, hydra, medusa, jellyfish, coral, anemones (which show emergence of cognitive levels and neurological nuclear logic [35]). The jellyfish has a pulsating binary physiology similar to that of the heart, hence a hypothetical developmental relationship (emerging neurological nucleonics).

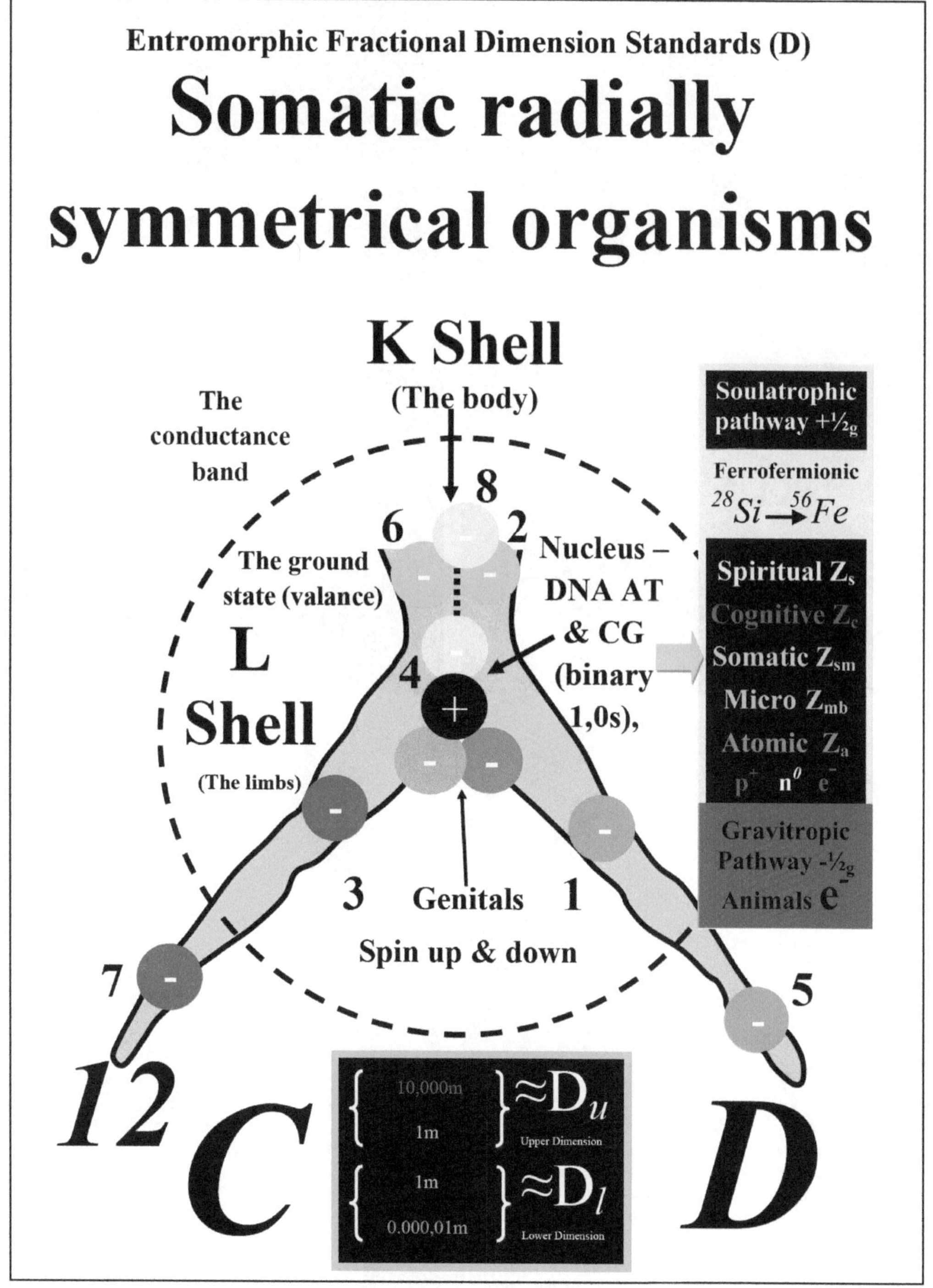

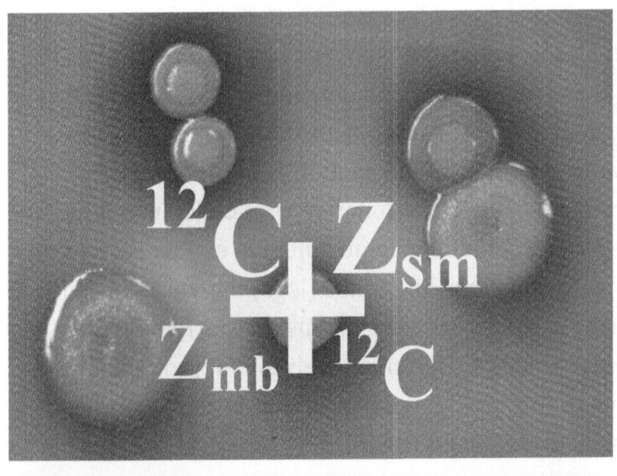

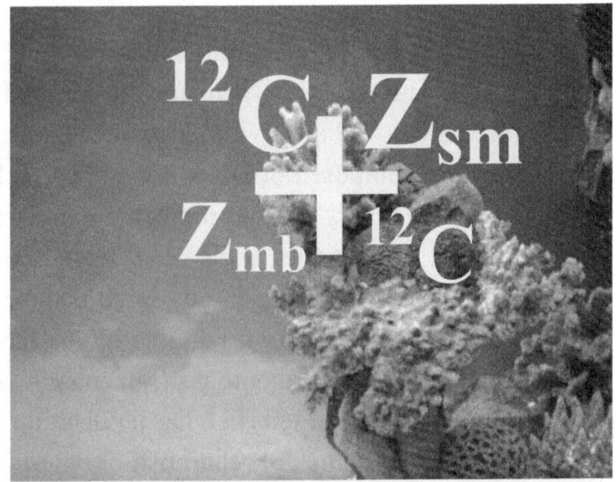

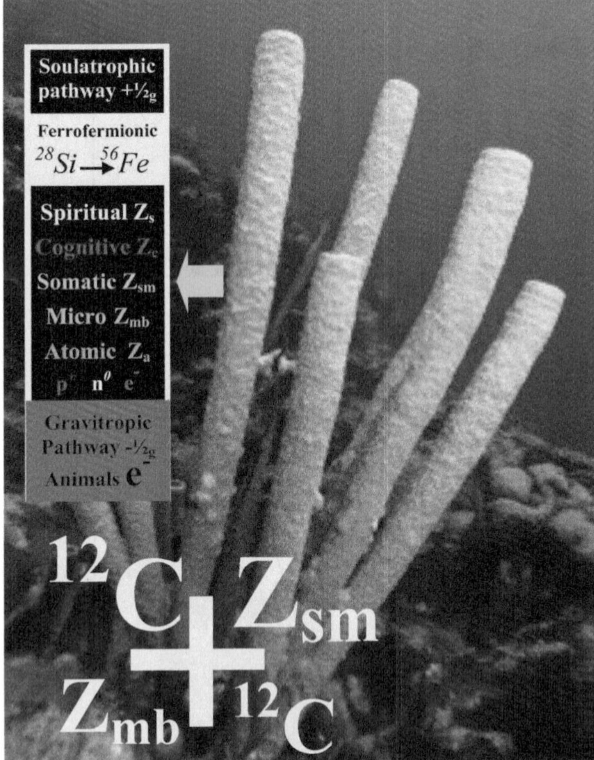

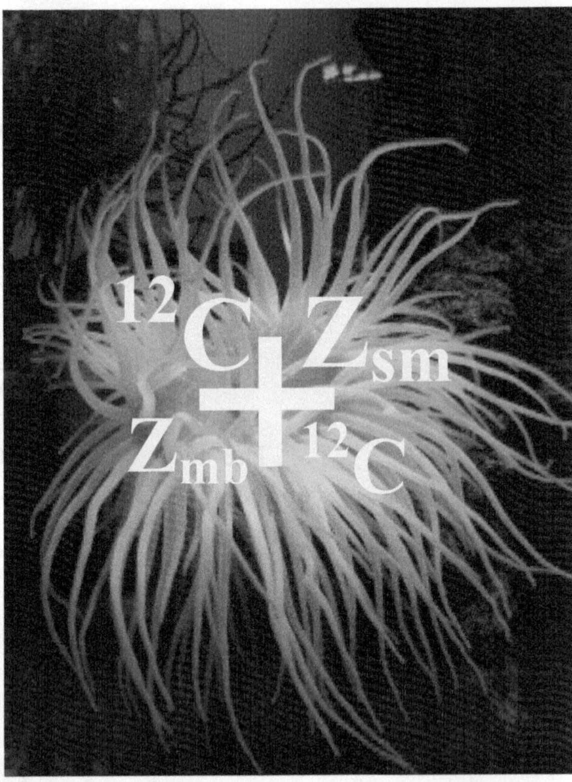

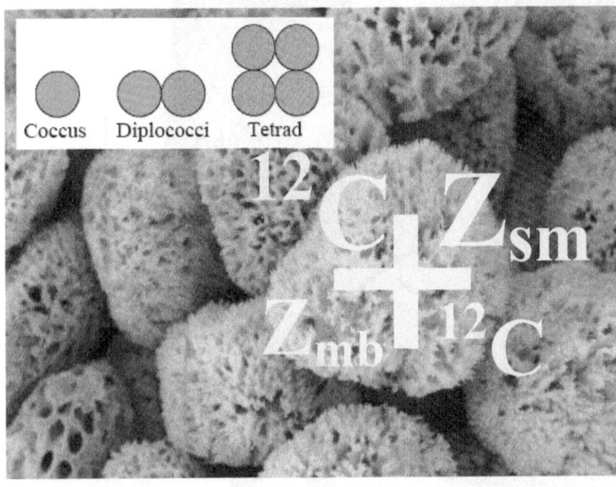

Sponge cells when separated will re-aggregate into multi cellular structures in suspended solution. This is excellent evidence of radially symmetrical somatic organisation. Jelly fish are also excellent evidence of the evolution towards bilateral symmetry and cephalisation (cognitive heads). These diagrams display some of the oldest multi cellular radially symmetrical organisms on the Earth dating back over 1 billion years. They continue to be some of the most numerous organisms on the planet.

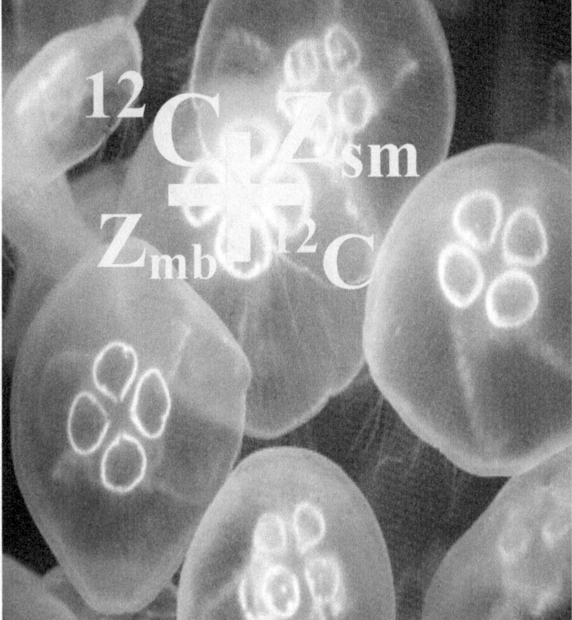

The cognitive level (Z_c) – 'multi cellular organisms'; typically this level sees the emergence of 'bilateral-symmetry' in organisms, and cephalisation or the emergence of the head and the new nuclear logic in the form of neurological action potential on/off binary nuclear logic [36]. The emergence of organisms driven by the new nuclear binary organisation, the K shell (n=1) can be hypothesised in the thoracic chest area and the L shell (n=2) arms, wings, upper limbs. The heart brain stem system is theorised as the new nuclear system, and genetic information [37] is neurological and expressed through conscious wavefunctions ψ_c [19]. This level of nuclear organisation is typical of a vast array of organisms from simple flatworms to reptiles, driven by brain stem non local memories theorised as the basis of instincts (neurological genes [38]).

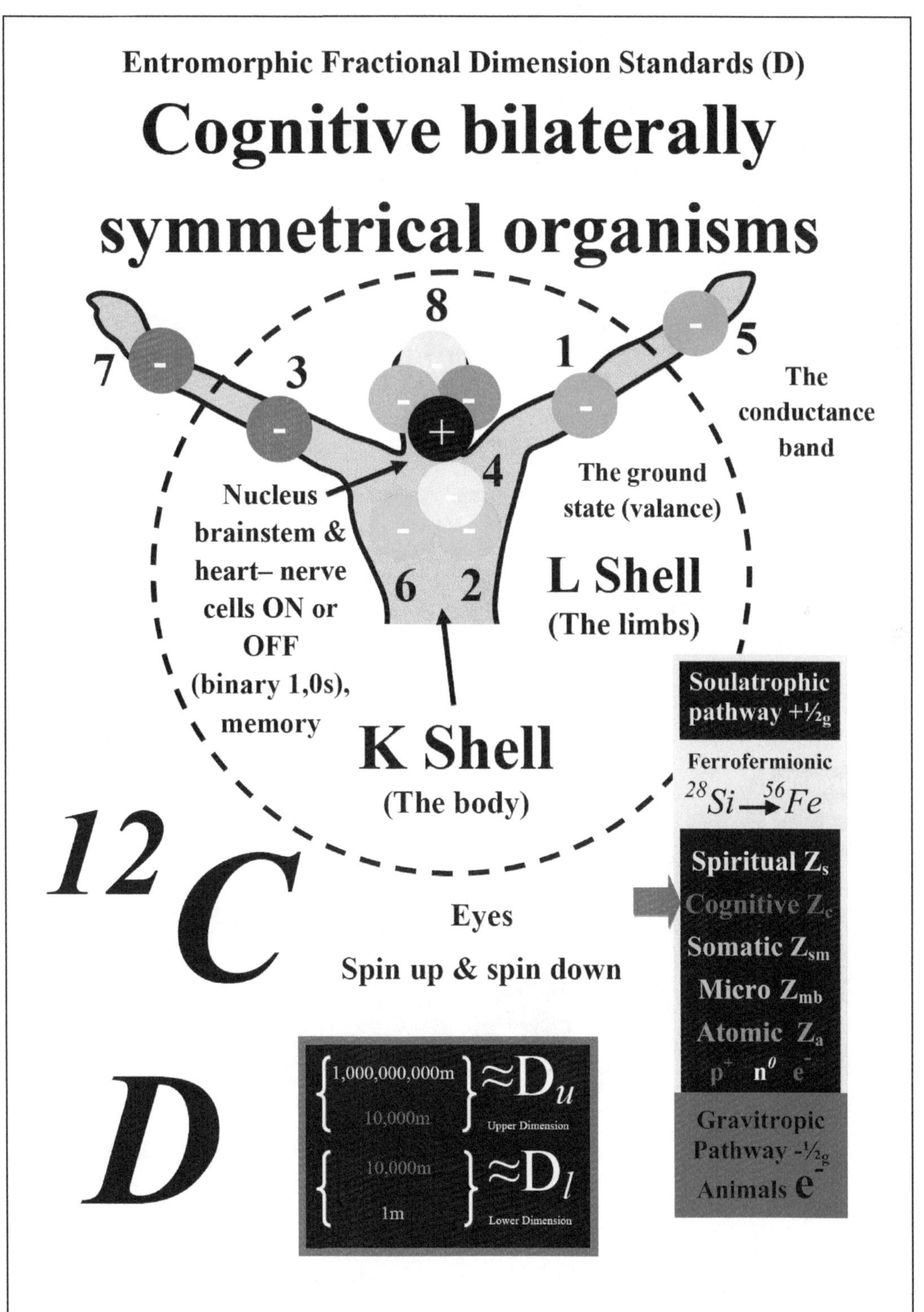

Z_{sm} (somatic radial level)

Polyentromorphism is demonstrated here in the insects because carbon is highly polymeric. The plants also have polyentromorphic properties hence a tree has thousands of entrochiraloctets (amplified carbon segments) on the somatic level. Many insects and worms have multiple somatic levels as segmentational morphology. Many have lost many of their somatic segments through natural selection and have just three or four segments left. This accounts for multiple limbs and wings in these organisms.

Emerging bilateral symmetry in the jelly fish and their relationship to higher bilateral organism's heart brain stem system (cognitive cephalisation, and basic ganglionic nerve nucleus emergence).

Organisms with neurological nucleonic systems. They are bilaterally symmetrical with cephalisation or heads, and complex behavioural patterns (true nerve nucleus a brain stem system).

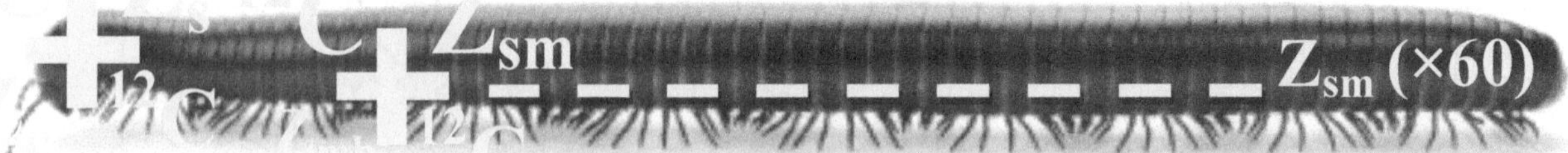

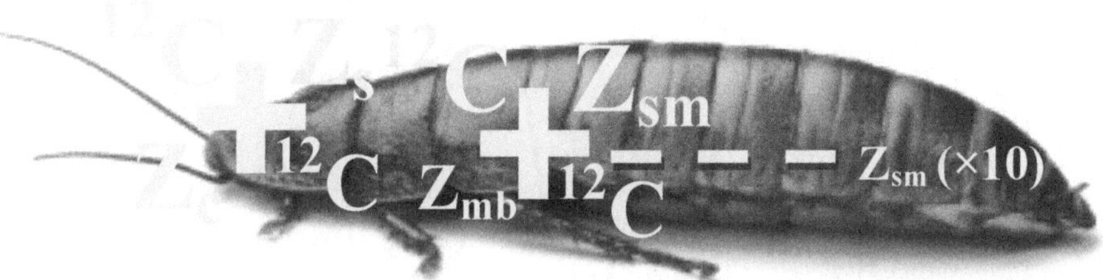

Bilateral symmetry in the cognitive system. Emergent spiritual (familial) mutiorganismal organisation.

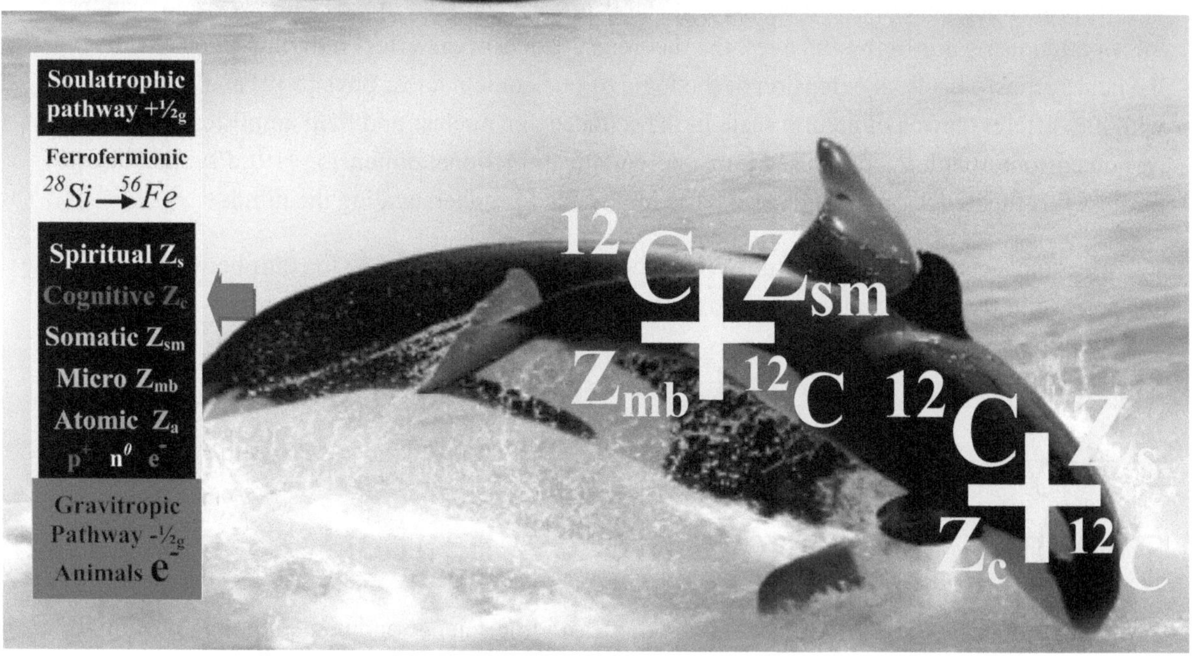

The spiritual level (Z_s) – 'multi organismal cells' such as the family. This level sees groups of organisms where offspring are theorised as the nuclear components. Again nuclear organisation is in the form of action potentials on/off binary organisation, expressing (conscious wavefunction ψ_s) behaviour (neurological genetics, memories are neurological genes [38]).

The K shell torso of the family (n=1) are the parents (nuclear shell) and act to base the gender model with a theorised spin down from the mother (maternal bond a direct physical bond) on the offspring, and a theorised spin up from the father (paternal semi physical bonded). The valance shell (membrane) or L shell (n=2) are hypothesised as the siblings which form sharing (theorised covalence as a model of organismal bonding – a hand shake) bonds to others to form new families in keeping with human sexual bonding (nuclear barrier penetration and quantum tunnelling and its theoretical relationship to sexual nuclear penetration) to form family hybrids.

The spiritual level is an advanced system which can be seen in grouped organisms such as 'a flock of birds', 'a shoal of fish', companies, teams, organisations, universities, schools, football teams, towns, nations, etc [37]. This is the level in humans where the brain has precedence over simpler brain stem instincts, which is the central nuclear system on this level [35].

Neurological genetics is a new way of understanding the extension of classical genetic logic to higher nuclear levels and for modelling psychodynamics. Humans, being the greatest examples of spiritual soulatrophism, have created technology by bonding through this level to the rest of the periodic table (increasing 26 elements to +117 [23] [24]). Information such as the words and sentences in this article are the absolute [39] personification of spiritual soulatrophicity, although this is still an effective working hypothesis.

Knowledge itself is recorded in the black and white binary of words and mathematics, where words and numbers are hypothesised as neurological genes. A book is hypothesised as a neurological supercoiled chromosome, so is a CD, a DVD, they are all hypothesised as extended classical DNA genetics where DNA exists in supercoiled chromosomes. In essence, the words on this page are a perfect demonstration of the strong interaction [9], a word being hypothesised as 'quark logic' held together by 'gluonic logic' and words hypothetically linked by 'pionic logic' [32].

We open the document (a chromosome in a cell is opened and read in the same way) and read the information using our limbs and eyes, the theoretical electroweak effect in living organisms to express the neurogenes. This is an extension of the logic of nucleonic nuclear physics [9] and its association with the different levels of natural scale in the soulatrophic nuclear and field amplification model in carbon entromorphology, organised into a neurological fractional dimension [19]. Psychodynamics through this theoretical approach has a basis in physics for understanding the mind.

The great physicist of the 20th century Wolfgang Pauli was fascinated by the gulf between psychology and science in the future and was quoted saying: -

'It is in my personal opinion that in the future, reality will neither be 'psychic' nor 'physical', but somehow both and somehow neither!'

Carbon entromorphology aims to consider both measurable science and uncertainty and the mind.

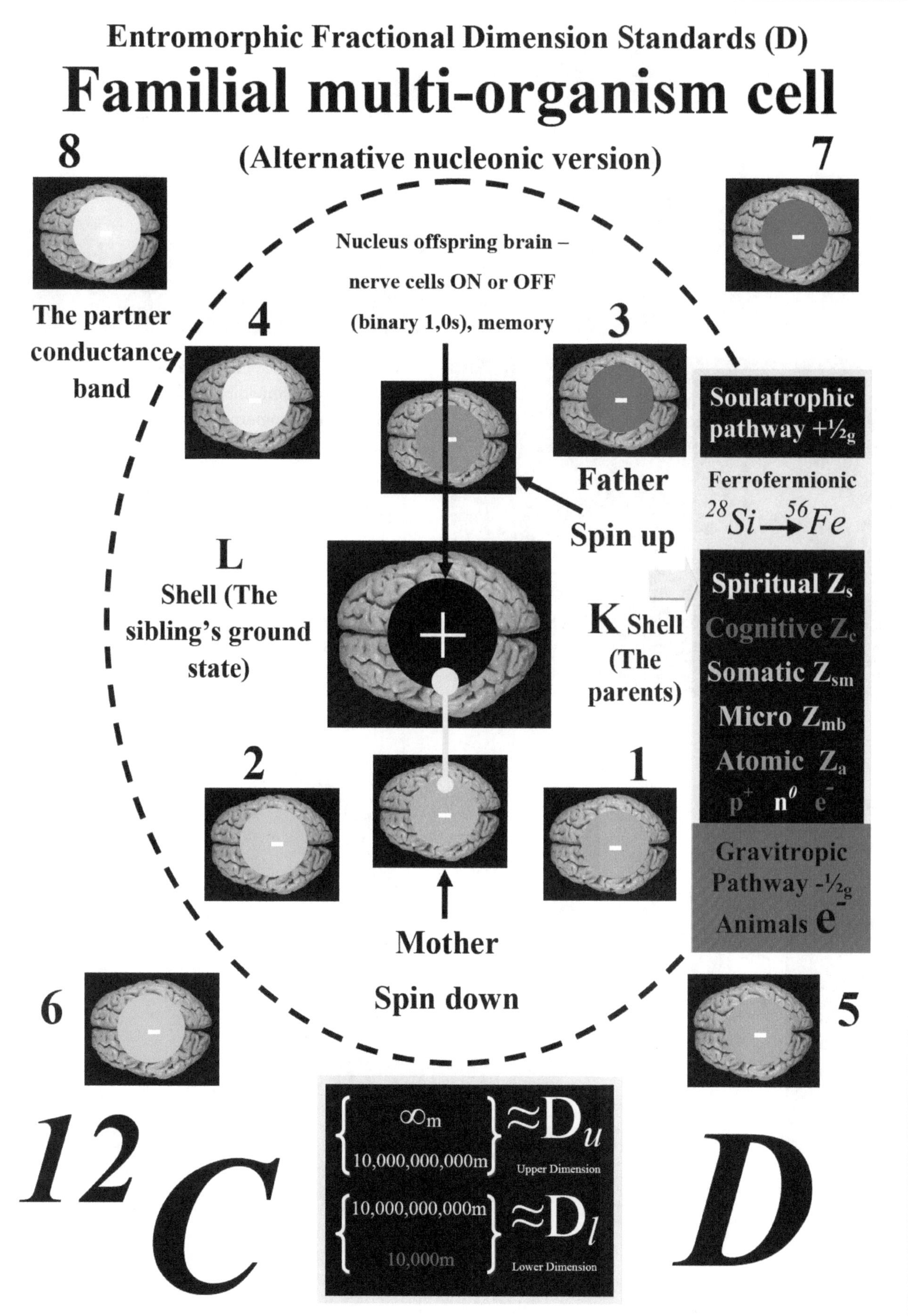
Entromorphic Fractional Dimension Standards (D)
Familial multi-organism cell
(Alternative nucleonic version)
8
7
Nucleus offspring brain – nerve cells ON or OFF (binary 1,0s), memory
The partner conductance band
4
3
Father
Spin up
Soulatrophic pathway +½g
Ferrofermionic
28Si → 56Fe
L
Shell (The sibling's ground state)
K Shell (The parents)
Spiritual Zs
Cognitive Zc
Somatic Zsm
Micro Zmb
Atomic Za
p+ n0 e-
2
1
Gravitropic Pathway -½g
Animals e-
Mother
Spin down
6
5
12C
∞m
10,000,000,000m
≈Du
Upper Dimension
10,000,000,000m
10,000m
≈Dl
Lower Dimension
D

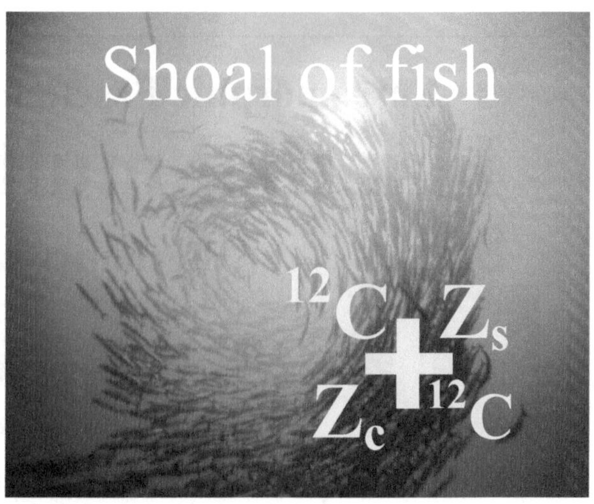

Groups of multi cellular organisms which act in an integrated way. Shielding the members in accordance with carbon entromorphic logic and organisation. In humans we have many multi organism cells such as: -

A family, a club, a team, a group, a league, a company, a set, a church, a university, a school, a hamlet, a village, a town, a city, a county, a nation, a country, a continent, a world, a government, an organisation, a faculty, a subject, a professional body, a team. They all organise themselves as entrochiraloctet cells with a nucleus, a K shell and an L valance shell.

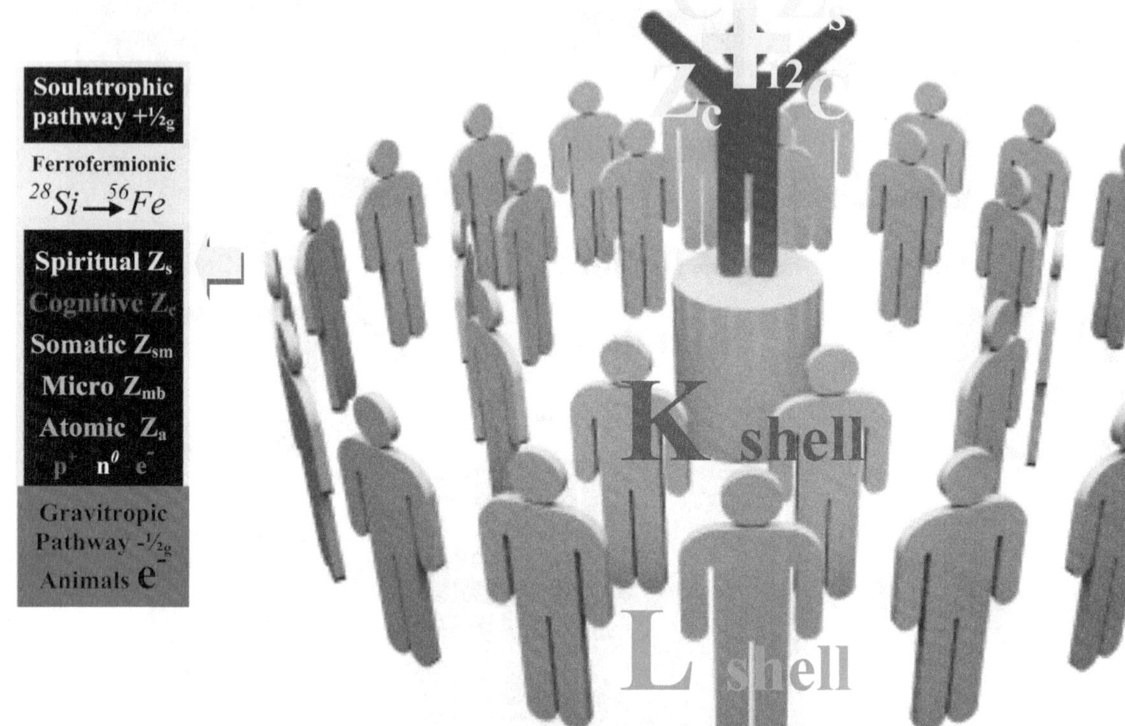

The Janes (spiritual) familial carbon entrociraloctet cell.

For any other sibling's cell, simply replace the current nucleus for any other sibling for their individual entrochiraloctet carbon, and move the curent nucleus into the L shell energy level.

The concept of nuclear logic is typically only associated with atomic physics, yet it has been clearly demonstrated in this article that, theoretically, there are a great many different types of nuclear logic. DNA, neurological systems are larger physical systems of nuclear logic, however science does not recognise the hypothetical link; again a human has a head which acts with nuclear binary determinism expressing behaviour in the limbs or membrane [22]. It is comprised of cells, which act with nuclear binary determinism expressing proteins in the cytoplasm and membrane. Each cell is centrally determined by nuclear binary determinism expressing electrons in K and L shell fields [9].

Each level is a carbon template or entrochiraloctet, in some theoretical state between ground state and hybridised state. All levels being associated with conserved carbon logic must therefore be driven by quantum field instability through nuclear instability thermodynamically, theoretically driven towards iron stability [9]. Hence the octet logic applies on all soulatrophic levels, for example most organisms find the missing four electron charges in the form of food; for a day cell cycle [37].

The nitrogenous component or metabolic organelle is protein (nitrogen based), the oxidative component or metabolic organelle is carbohydrate (rich in oxygen), the halogenous component or metabolic organelle is fat (membrane with halogenous channels and antigens, fluorine/chlorine/iodine) and finally the neogenous metabolic organelle is theoretically bonding to other nuclear systems or other organisms (a duplication in nucleonic material).

For a day long cell cycle the octet must be reached in order to allow organisms to find G=0 or thermodynamic equilibrium. There is more free energy in the food than the organism so $A = -\Delta G$ where $\Delta G < 0$, spontaneous hunger must satisfy the thermodynamics [23] [24]. Hybridisation of the individual soulatrophic levels produces an integrated pathway. The particle distribution on each individual level produces an entrochiraloctet or carbon template, the levels integrate to form a complete carbon template in some state of thermodynamic development from ground state to hybridised state.

Each level should contain a conserved set of physical atomic logic, which is identified for the associated particle symmetry in carbon. For example, on the cognitive level the heart/brain stem is hypothesised as the nuclear system, the eyes hypothesised as the two 1s electrons, the chest hypothesised as the 2s and lungs hypothesised as the oxidative valance, the spine and brainstem hypothesised as the neogenous valance, the nitrogenous valance the least dominant arm and the halogenous valance the dominant arm (chiral) by the same reason.

However the overall hybrid of atomic, microbial, somatic and cognitive hypothetically classified to be the head as the nucleus, the torso the K shell 1s electrons and the four limbs the 4 valance electrons.

The scientific method for determining carbon association in this way for amplified entromorphic atoms is based on both organisational observations and physical symmetry and quantity to carbon logic found in different organisms.

Electronegativity is the physical thermodynamic potential driving all physical systems to tend their energy stabilising spontaneity to produce quantum stability and iron nuclear stability (also electropositivity for group 1 and 2 in the periodic table and other metals) [9]. The tendency drives behaviour in organisms following solar acquisitions to reduce electronegativity towards a G (free energy) value of 0 or thermodynamic equilibrium for primary non metals.

The ingestion through electronegativity of octet stabilising energy through food is a theoretical measure of living thermodynamic potential and demonstrates active solar driven evolution [30].

Diagram carbon entrochiraloctet – ground state.

This is a fundamental template for carbon organisation and physical functionality. This logic is conserved and amplified through solar growth to produce a breathtaking array of varied entrochiraloctet cells throughout the entire living world. The plants are composed of literally thousands of entrochiraloctets which can be seen as one entire set of leaf structures which constitute the p type orbital probability distributions, flowers and similar structures which have 2s probabilities associated with their structures and reproductive organs which constitute K shell 1s orbital and nuclear systems. Most bilateral animals have just three entrochiraloctets which stack together to form the bilateral organisation. The insects and other simple organisms are polyentromorphic in the same way that plants are, they are often multi segmented made of many somatic level entrochiraloctets. Higher organisms such as the mammals have over 200 cell types each of which is a microbial entrochiraloctet cell. All the precise organs or valance in carbon, are amplified by bonding pathways to produce living structures of enormous size and outrageous complexity, but yet with simple conservation of atomic properties.

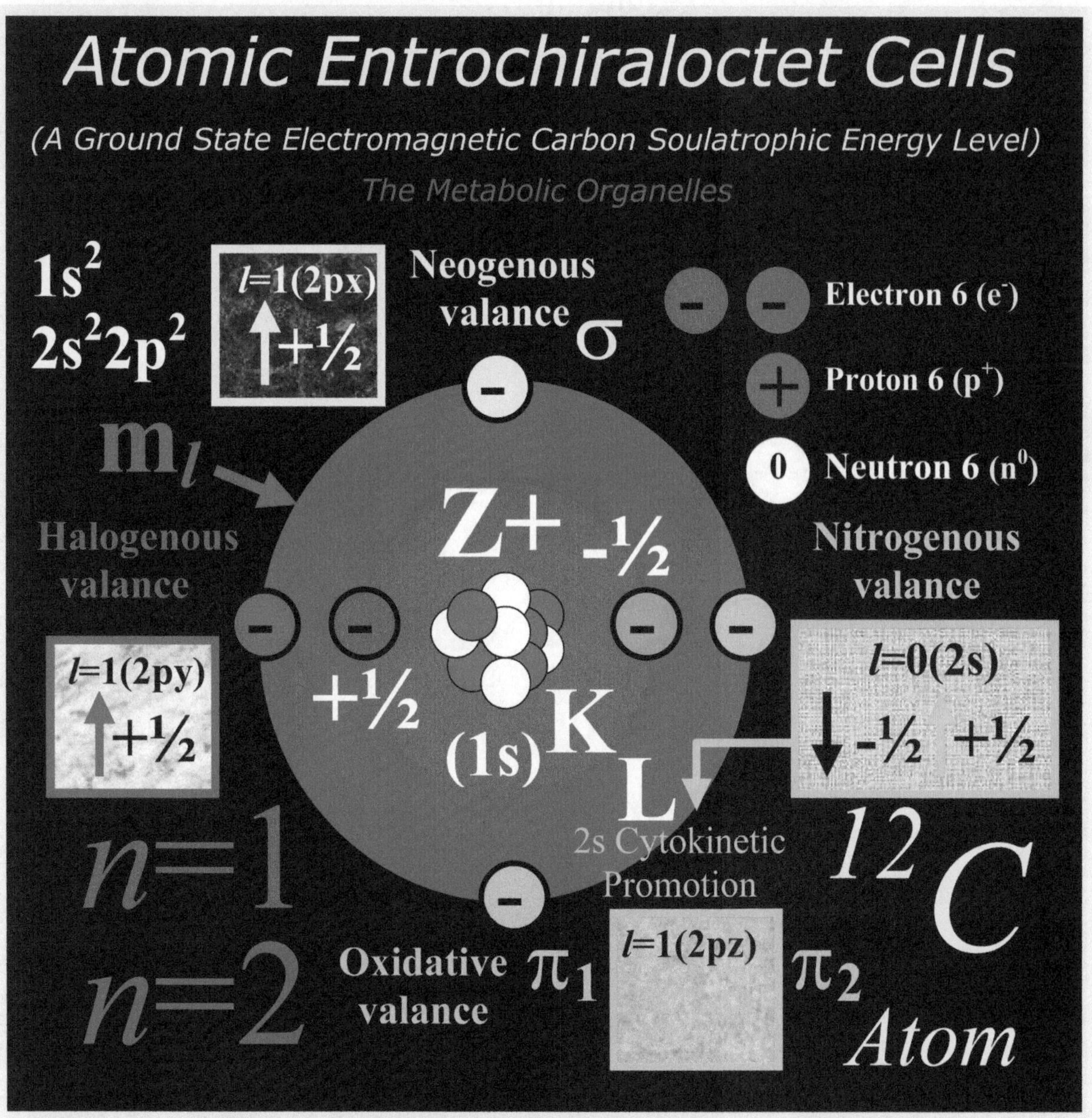

Diagram: carbon entrochiraloctet – hybridised completed octet.

The carbon system also includes hybridised entrochiraloctet systems where at its most basic level, carbon forms four covalent (sharing) bonds with other carbon atoms of differing strengths. In actual fact there is no limit to the number of other atoms carbon can bond with and as such, its ability to conserve and amplify its atomic characteristics are assured and uniquely provable. The bonding systems are conserved and amplified in living systems through the conservation of the logic of carbon. The ironic and incredible reality of this logic is the staggering reality that any organism today, such as the reader of this book, is the result of billions of years of successful reproduction, they were all winners, and our ancestors produced an unbroken pathway of thermodynamic stability and growth.

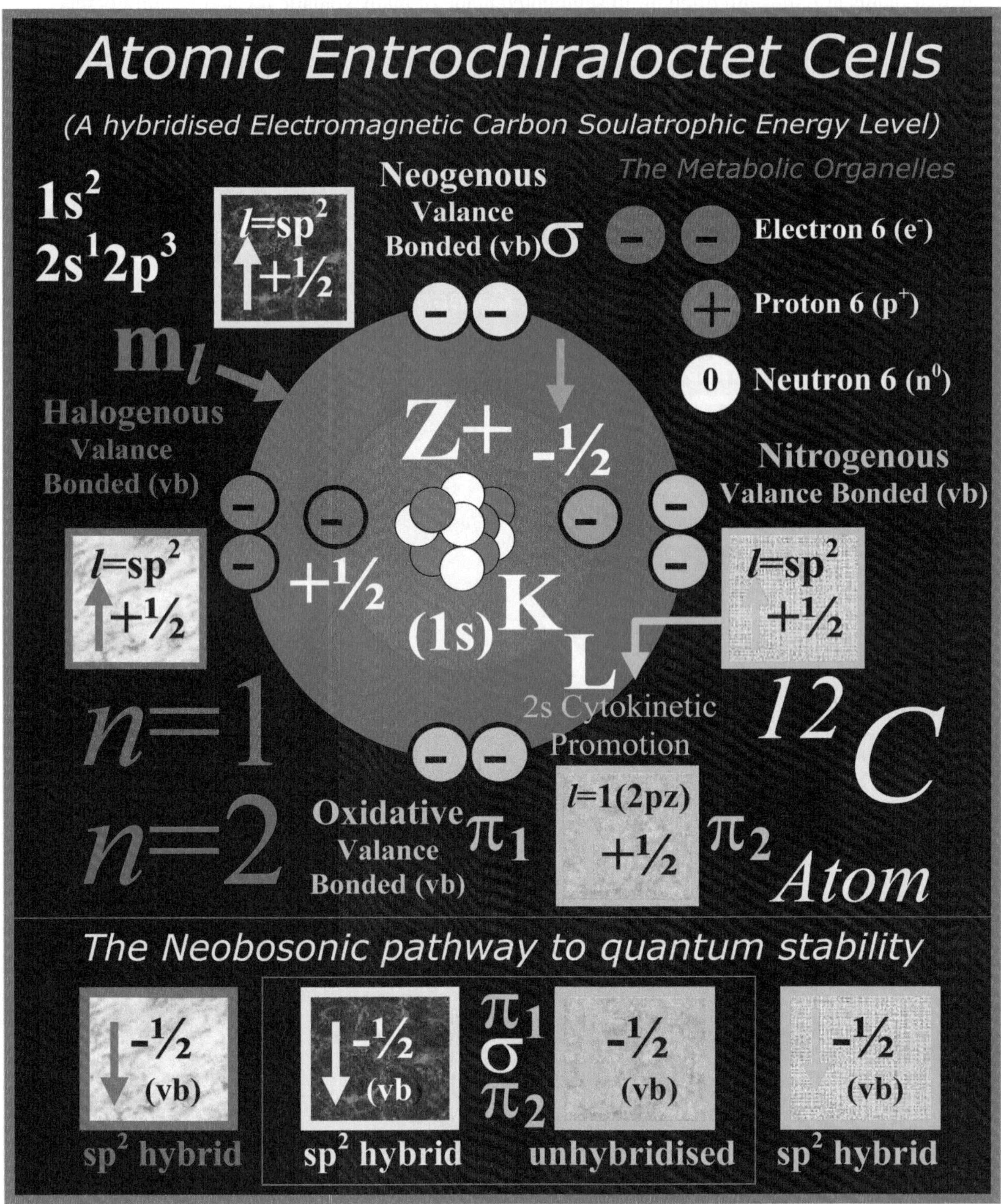

Quantum numbers, wave functions and the origins of biological morphology, function and probabilistic diversity.

In quantum mechanics, electrons are theoretically described by four quantum numbers 'n', the principle quantum number of which there is just two levels around the nucleus of a carbon atom. The '*l*' or subshell orbitals (static/fixed wave), of which there are three dominant types in carbon based life; the 1s (the first energy level) which is spherical and can occupy 2 electrons in opposite spin [9].

The second energy level with the 2s 'sphere within a sphere' geometry again capable of occupying 2 electrons with opposite spin, and finally three p orbitals (px,py,pz) producing lobed streamlined geometries at 90° to each other. These three physical morphological organisational structures represent a superposition of states, in other words 'all' possibilities superimposed on each other [9].

The absolute unique configurations theoretically produce all the vast diversity in living morphology throughout the living world. This is a very neat logic divided into two energy levels, producing a system which can allow for unique evolved thermodynamically environmentally filtered configurations which beautifully personifies all the diversity of life on Earth [40].

The electron behaviour also has 'm_s' spin numbers, which form the basis of Pauli's exclusion principle for fermionic particle/wave systems and composite particle/wave systems which beautifully demonstrates how the physical nature of all life is spatially and physically organised [9]. Spin either up or down gives the biologist a powerful logic for understanding how living organisms occupy quantum spaces, it also forms the basis of a 'Pauli clash', where two entromorphic atoms such as humans cannot have the same set of quantum numbers and produce thermoentropic consequences if external work forces them into the same physical state.

Finally there is the 'm_l' or magnetic quantum number, which is an excellent theoretical natural selection force acting to produce subshell orbitals energy shifts from the effect of external magnetic fields and physical changes in morphology in living organisms. This study suggests the need for a gravitational quantum number 'g_l' which relates subshell deformation through force equilibrium to the other electromagnetic quantum numbers. 'g_l' would be related to the induced states of matter (solid/liquid/gas/vaccum) an atom inhabits as a dynamic equilibrium between mass (particle) and electromagnetism (wave) see page 247/248 for quantum taxonomy and the quantum gravity of life. Both m_l and g_l are natural selection factors for life and represent an environmental evolution factor.

All levels of life can be reduced to 'ensembles of living microstates in the course of thermal fluctuation' organised by physical particle theory [37]. Quantum mechanics is beautifully personified by living organisms through the enormous and absolutely undeniable variation in life. Since quantum mechanics is a probabilistic system the orbital distributions give the biologist a wonderfully rich and complex diversity of possible particle outcomes. What is even more impressive are the facts that all life can be reduced to three orbital probability distributions, two of which are spherical and one of which has three lobe shaped structures at 90° to each other. The living world is absolutely rich in these structural systems, which collectively explain how life can be so diverse and different but still strongly linked to the past where living organisms tended towards simple carbon organic chemistry. The growth of carbon chemistry can be easily seen in the emergence of organic chemistry and biochemistry with ever increasingly larger systems and structures.

Biochemistry expands still further into cell biology; cells to organs, organs to multi cellular organisms as the systems get ever larger through solar amplification known as growth (organoaccretion).

- The following diagrams are a tiny selection of the trillions of 1s, 2s, px, py, pz orbital logic (Full page) – examples of biological symmetry to atomic carbon origins.

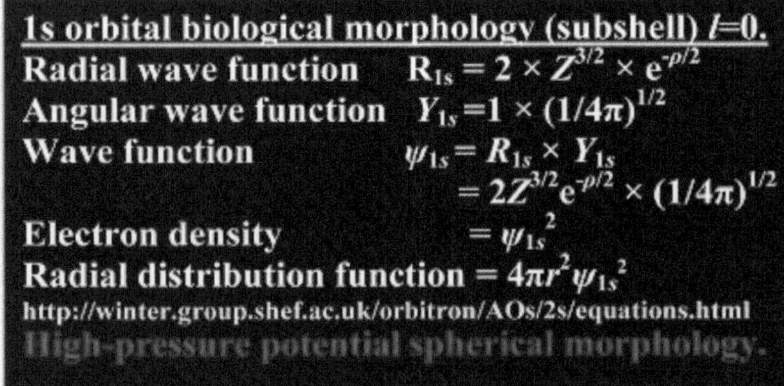

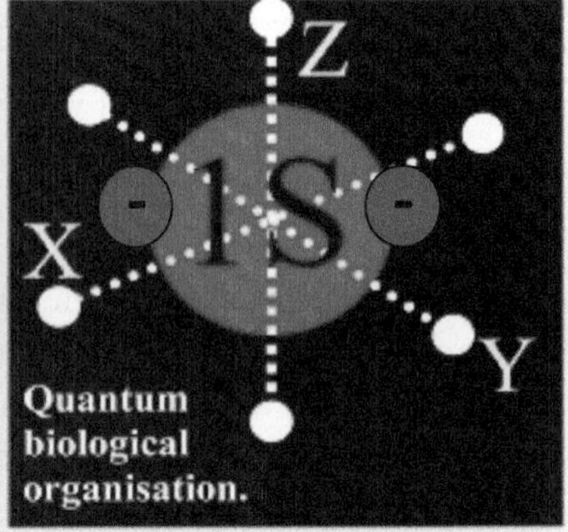

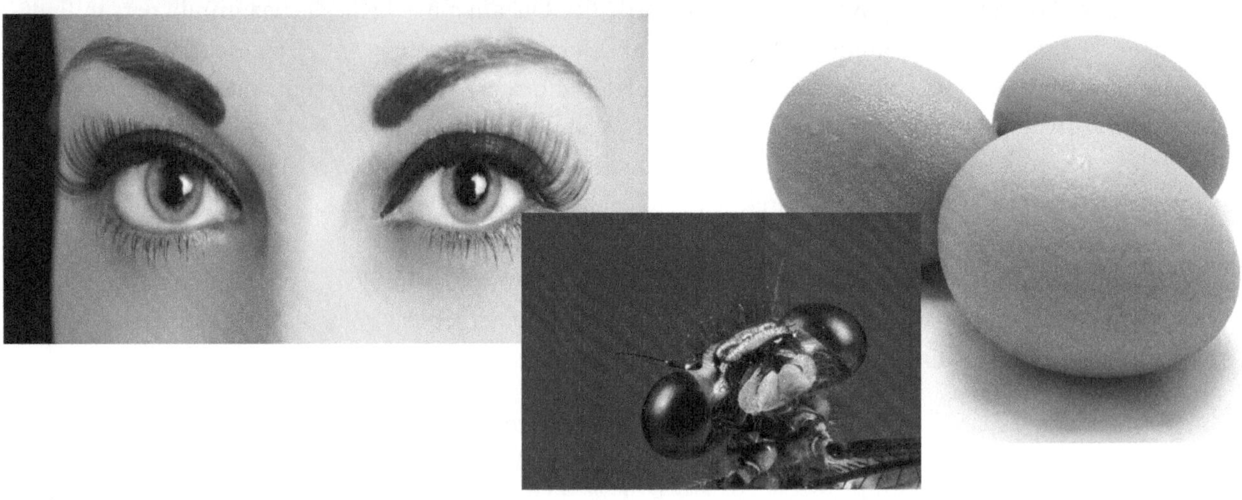

2s orbital biological morphology (subshell) l=0.

Radial wave function $R_{2s} = (1/2\sqrt{2}) \times (2 - \rho) \times Z^{3/2} \times e^{-\rho/2}$

Angular wave function $Y_{2s} = 1 \times (1/4\pi)^{1/2}$

Wave function $\psi_{2s} = R_{2s} \times Y_{2s}$

Electron density $= \psi_{2s}^2$

Radial distribution function $= 4\pi r^2 \psi_{2s}^2$

http://winter.group.shef.ac.uk/orbitron/AOs/2s/equations.html

High-pressure potential spherical 'housing' morphology.

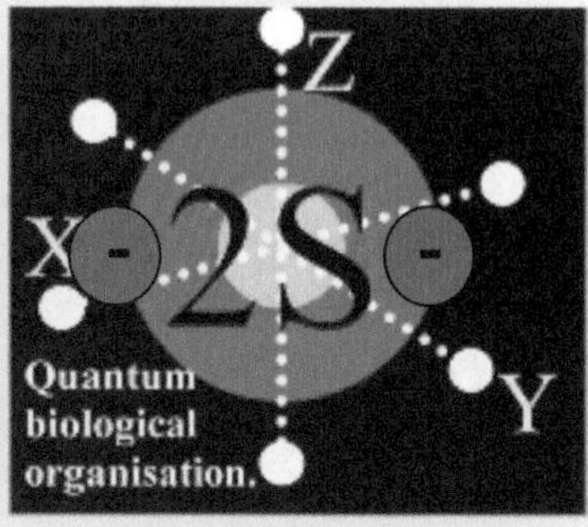

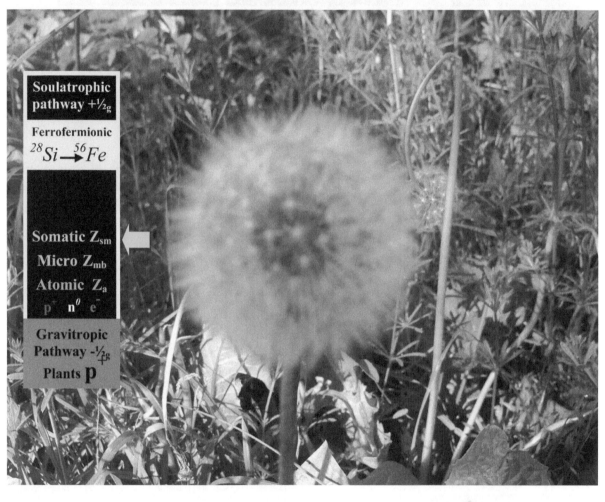

Soulatrophic pathway $+\frac{1}{2}_g$

Ferrofermionic

$^{28}Si \rightarrow ^{56}Fe$

Somatic Z_{sm}

Micro Z_{mb}

Atomic Z_a

p^+ n^0 e^-

Gravitropic Pathway $-\frac{1}{2}_g$

Plants p

Px,y,z (substitue for y,z)biological morphology l=1.
Radial wave function $R_{2p} = (1/2\sqrt{6}) \times \rho \times Z^{3/2} \times e^{-\rho/2}$
Angular wave function $Y_{2px} = \sqrt{(3)}x/r \times (1/4\pi)^{1/2}$
Wave function $\psi_{2px} = R_{2p} \times Y_{2px}$
Electron density $= \psi_{2px}^{2}$
Radial distribution function $= r^2 R_{2p}^{2}$
http://winter.group.shef.ac.uk/orbitron/AOs/2s/equations.html
Low-pressure kinetic streamlined morphology.

Z
X
Y
P_z
P_y
P_x
Quantum biological organisation

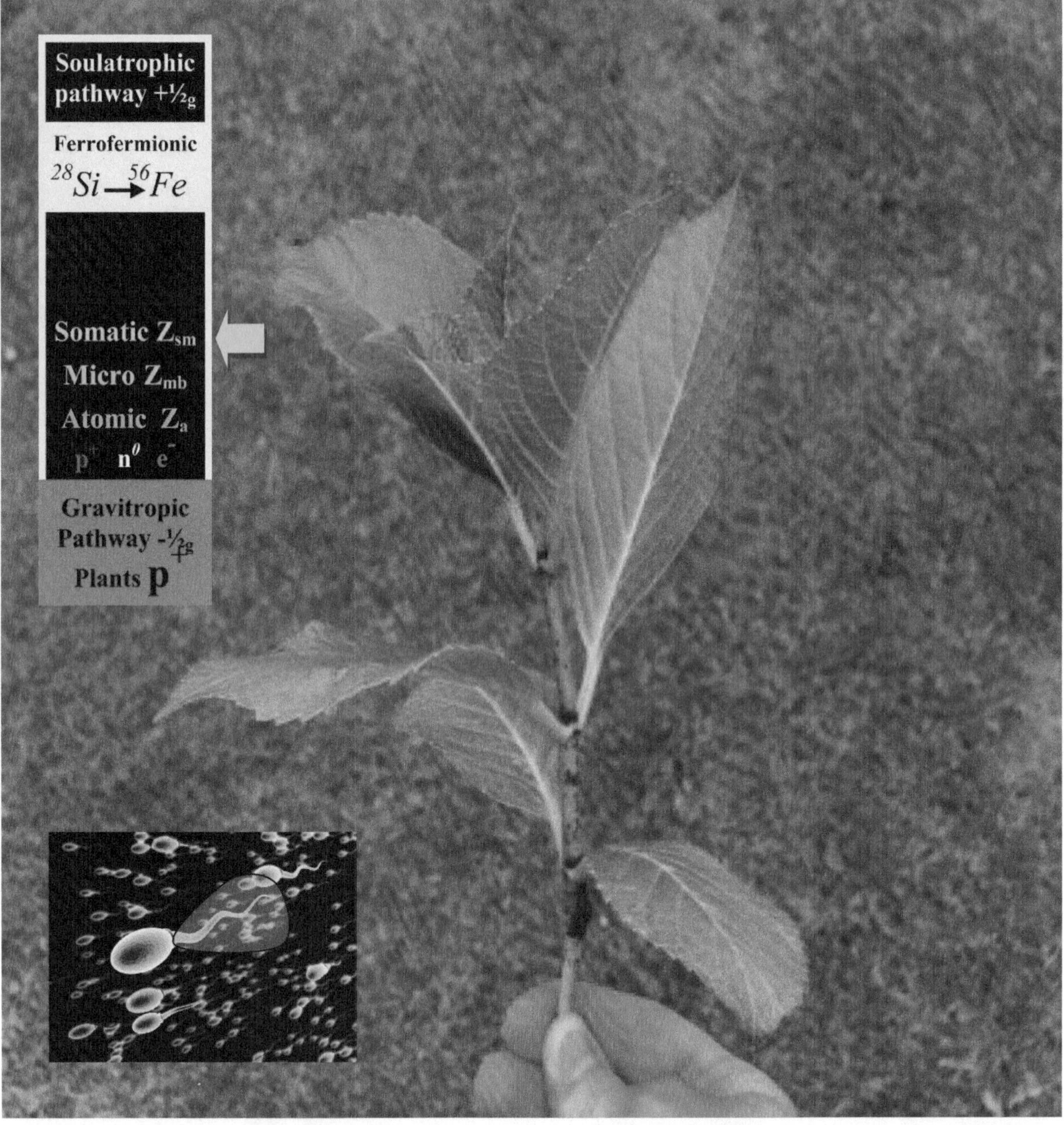

The images are breathtaking evidence of conserved atomic amplification in living organisms where the atomic logic can be seen with the eye, a concept alien to the physics profession where atoms are unknowable and certainly unvisualisable. Can science truly suggest this is mere coincidence?

Examples of biological diversity, structure and its relationship to atomic logic,

biological entromorphic theoretical conscious wave functions of the head (K, n=1) and limbs (L, n=2).

The theoretical conscious wave function model is very useful for understanding how an organism functions and its association with the electroweak standard model in quantum mechanics and its consequence for living beings. It is important to remember that the quantum mechanics [41] aspect of carbon entromorphology is based on probability distributions and as such the uncertainty principle is a property of organismal behaviour [30]. Organisms are microstates modelled as fractional dimensions of matter and exist as a superposition of carbon (and other) microstates [20]; the frequency of the ensemble is the basis of all behaviour (Schrödinger's cat). There are two eyes in an organism's head which theoretically correspond to the 2 1s electrons in the K shell (head). These eyes are hypothesised to be the electron products and simply part of the probability distribution (quantum nature of tissues), the absolute position and momentum of the original electrons is unknown due to amplified quantum uncertainty [16]. As biological energy levels move closer to nuclear regions the probability distributions begin to resolve the theoretical quantity of electrons present as they are more intensely tightly bound to the nucleus (brain), and as such have stronger positions, greater clarity and order. The valance shell on the other hand is theoretically a higher kinetic energy level so probability doesn't reflect the absolute number of underlying electrons; hence a sea anemone has hundreds of tentacles.

- The following diagram demonstrates the symmetry between atomic logic and life.

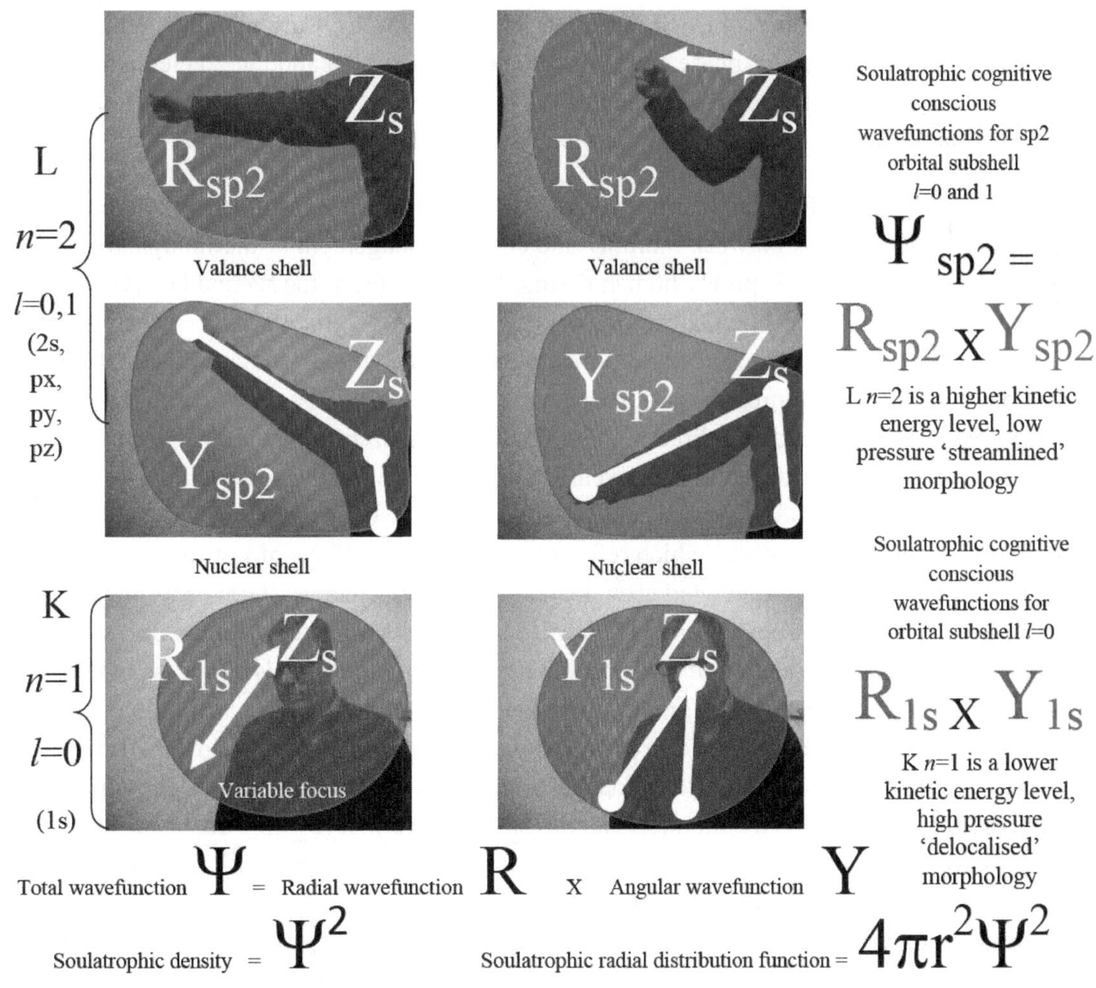

Human thermodynamics 'ECASSE' and the nature of morality in carbon entromorphology.

Carbon entromorphology proposes (hypothesises) that living organisms are amplified homogenised carbon based nuclear and field logic, amplified through bonding from 26 up to +117 other elements for humans [23] [24]. This model of amplification is based on a priciple in atomic physics called 'Bose condensate' [42]. In short 'Bose condensate' allows amplification of bosonic particles such as the photon to produce amplification through in-phase coherance through stimulated emission [42] where a photon is a particle wave/duality. In other words absorption of one photon stimulates the production of two, one particle wave becomes two (theoretically very similar to mitosis in cells). This is the basis of 'LASER' logic (Light Amplification by Stimulated Emission of Radiation), and since carbon having an even number of nucleons is also a boson (force carrying particle) it must also possess the properties of Bose condensate statistics [42]. In carbon entromorphology the thermodynamics of life is driven by solar input, and accumulation (increase in enthalpy) through accretion bonding in carbon based systems [21]. This means that carbon entromorphology can make the argument that amplification in carbon based life to produce theoretical amplified entromorphic carbon atoms such as humans, follows Bose condensate thermodynamic amplification. Hence thermodynamically since Earth contains less free energy than the Sun and since the atmosphere acts as a gain amplification medium, life can be theoretically modelled as 'ECASSE' (Entromorphic Carbon Amplification by Solar Stimulated Emission). Amazingly enough the very fact that humans have laser technology is proof of this theory [42]. The laser has come into being because of carbon based solar aquisition producing 'ECASSE beams' of in-phase living coherance, such as a 'line of ants' or a 'flock of birds'. In classical LASER the particles are photons. Carbon and its other bonding elements produce an elementary in-phase coherant beam called life [10] and the 26 element human molecule to +117 element technological molecules [23] [24]. Life is driven theoretically to amplifiy its energy quantity towards quantum stability driven by ferrous nuclear instability [9].

The energetic consequence of human thermodynamics is typically known as morality, although humanity has not made the link between morality and energy management or thermodynamics and the field of economics which again is purely human thermodynamics. From the second law (Gibbs equation) which defines enthalpic and Gibbs free energy as a positive growth logic (good). Where growth concepts allow living organisms to accumulate energy evolving up soulatrophic pathways towards technological stability through 'phototropism'; a term typically associated with the plants which are primary producers and link directly into the thermodynamics of the animals and technology [7]. Hence maintenance of this descriptive convension gives the human thermodynamicist a link to biological science to make sense of growth accumulative enthalpic Gibbs free particles [23] [24]. Another similar term is 'gravitropism' which referes to the bondage effects of Earths gravity acting in dynamic equilibrium with solar stimulated growth and quantum mechanics. 'Gravitropism' is therefore the inverse of 'phototropism' and is related to processes associated with the liberation of solar energy and the overall reduction in soulatrophic heirachy, in other words a process of death (evil). Finally both 'phototropism' and 'gravitropism'act with 'quantum gravity' spin direction towards either the Sun (growth) or the Earth (death) directions in equilibrium against the final biological term called 'heliotropism'. 'Heliotrops' actively track the Sun or follow a phototropic pathway, hence 'heliotropism' is the thermodynamic basis of free will (or 'free energy'). By following 'ECASSE' (Entromorphic Carbon Amplification by Solar Stimulated Emission) as the solar driven physics behind all living thermodynamics, life evolves towards 'technologically stable life' a state of ferrous evolutionary stability. It is the theoretical final neutronic pathway mediated by the spontineity of elements such as silicon and iron allowing carbon and other amplification elements to bond with the rest of the periodic table [21]. 'Technological life' is acheived through phototropic growth, where organisms tend towards more stable states where longevity is engineerable in an energy rich technological environment postulated as a real future for life on Earth. Evolutionary biologists see no pathways in evolution only pure chaos and no goal to evolution. In carbon entromorphology this logic is challenged where science and technology (soulatrophicity) produce a goal (technological life) which is to produce organisms which can live for considerably longer periods of time in energy rich environments; peace. This is acheiveable through science and technology but has its origins in the philosophy of human history and belief systems through religion.

In a sense the overall thermodynamic potential of the universe sees a future of increasingly less free energy tending towards equilibrium where G=0, where life reaches'technological life', human evolutionary stability or peace. Currently iron produces a universal thermodynamic potential against all the other elements where $A = -\Delta G$ and $\Delta G<0$ [23] [24]. In Boltzmann entropy we introduce the power of the 'frequency of particular ensemble of microstates' such as quarks, atoms, cells, molecules, organisms, grouped organisms but also astrophysical bodies such as a solar system. This logic does not discriminate and suggests all natural order is logically linked but the force equilibrium varies with the observational distance from the source. Finally, the thermodynamic consequence is that these ensembles of microstates are associated with the 'course of their thermal fluctuation'. Boltzmann entropy is therefore a very powerful measure of living order, and through the concept of soulatrophicity and'technological life' life tends towards just one ensemble of ultimate order which is the concept of 'artificial technological life' (iron). Boltzmann entropy [8] is defined using an old mathematical process called the 'continuous product' which is a series model where the operators between terms are multiplication (see references for the source of the formulae).

A hypothesised 4th law of thermodynamics – Naturally selective acts of free will (free energy) in open systems; all evolving conscious microstate ensembles seek the stability of iron (every system seeks to achieve a minimum of free energy) [9]. Boltzmann entropy can be simplified to produce a reasonable approximation to Gibbs entropy, producing substitution and some boundary rules produces (between 0 and ∞) an equation which relates macroscopic environments to particle ensembles and microstates mechanics - used in this article to illustrate biological order only. Boltzmann entropy is expressed as, and reduces to equation (3) if the pi (probabilities are equal), Log is a naperian (natural) logarithm.

$$S_B = -Nk_B \sum_i p_i \log p_i \qquad (1)$$

Leads to Gibbs entropy (probability between 0 and 1) $S = -k \sum p_i \ln p_i \quad (2)$

The classical Boltzmann entropy (thermodynamic probability >1) $S = k \ln W \quad (3)$

$$S = k \ln (N!/\prod_i Ni!) \quad (k=1.38\times10^{-23} jk^{-1}) \qquad (4)$$

Symmetries of order are beautiful, broken symmetries of disorder are ugly!

Order, the increase in the frequency (W number of microstates in the course of thermal fluctuations) of a particular ensemble of living microstates (atoms, molecules, cells, tissues, organs, organisms, grouped organisms (N, i)) tending towards [8] : -

$$W = (N!/\prod_i Ni!) \rightarrow \infty \text{ microstates} = \text{Universe death} = \gamma = 2^{nd} \text{ Law of Thermodynamics } \mathbf{(5)}$$

$$W = (N!/\prod_i Ni!) \rightarrow 1 \text{ microstate} = \text{Artificial life (delay)} = {}^{56}Fe = 4^{th} \text{ Law of Thermodynamics } \mathbf{(6)}$$

IMPORTANT NOTICE ABOUT THE FUTURE AND THE ENEVITABLE 'ARROW OF TIME' AND ENTROPIC DISORDER.

Universe death enters through a transient state where all matter is reduced to photons and eventually cools to absolute zero (T). However the time left in the universe is so vast that from life's perspective it may as well be thought of as eternal. This is excellent for the proliferation of life as there is plenty of time at life's disposal. It depends on our efficient use of energy (Good).

Substitution into Gibbs equation for Boltzmann entropy [8]:-

$$G_{(T,p)} = H - T\,k\,(\ln W) \quad (7)$$

$$G_{(T,p)} = H - T\,k\,(\ln (N!/\prod_i Ni!)) \quad (8)$$

For change in thermodynamic variables at absolute temperature T; a microstate entropy equation [8]:-

$$\Delta G_{(T,p)} = \Delta H - T\,\Delta(k\,(\ln W)) \quad (9)$$

$$\Delta G_{(T,p)} = \Delta H - T\,\Delta k(\ln (N!/\prod_i Ni!))(10)$$

The potential energy of the universe (H) is still very high, which is why the Stella-phase (star systems) still floods the universe with free energy (G). Life is simply a stepping stone on this pathway and does not infringe the 2nd law (Evil); however free will (4th law) does allow the process of time and the 2nd law to become drawn out by 'efficient' living reactions.

Of most importance: - Affinity $A=-\Delta G$ ($\Delta G<0$) is of most importance for living reactions. $\Delta G=0$ for the 4th law where iron stabilises all reactions. Constant temperature and pressure is difficult to acheive in living systems.

'i' ranges over all possible physical conditions; 'W' can be counted using the formula for permutations. Human thermodynamics demonstrated here using Gibbs equation is an appropriate form of the second law of thermodynamics [23] [24]. **These equations are for illustration only, living organisms are so complex that practical calculations are almost impossible to achieve meaningfully, but basic thermodynamic definitions like these should be considered.**

The living processes associated with the moral concepts of growth and death and the equilibrium shift (Le Chatelier's principle), through 'heliotropism' or free will (free energy), to reduce entropy and increase enthalpic Gibbs free energy potentials through efficient use of energy (good) are shown. This state is in a direct equilibrium to solar energy liberation through death, which pushes the equilibrium towards thermoentropic disorder (reversed thermodynamic evolution), again through 'heliotropism'. **Free will only delays the inevitable increase in entropy and universe death.**Symmetry is associated with lower thermoentropy and increased order (beauty). This is a sensible conventional link between thermodynamics and the biological sciences for understanding human thermodynamics and morality [23] [24], typically the equations are difficult to solve for life. Finally evolution tends towards 1 perfect microstate typified by 'technological or artificial life'; this is a beautiful thermodynamic tendency and produces increased particle symmetry (a 4th law of free will). The 4th law does not contravene the second law as it demonstrates the difference in overall Hess cycle routes which inevitably tend towards death with the same overall enthalpy change; also through Le Chatelier's principle or engineerability. **The universe tends to a photonic absolute zero death.**The 4th law draws out the effects of the 2nd law over large sustainable time periods through increased thermodynamic efficiency as long as life is part of an open thermodynamic system.

- A diagram of moral thermodynamic organoaccretion growth mediated through 'acts of free will' (order 4th law).

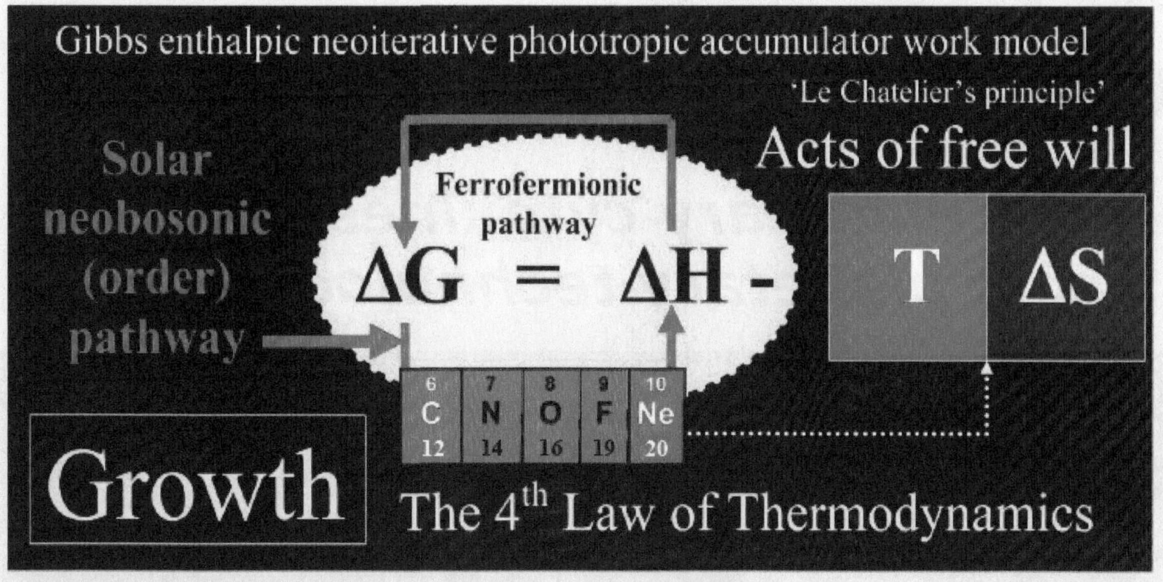

Artificial Life
'26 element human molecule to 117 element robotic molecule'

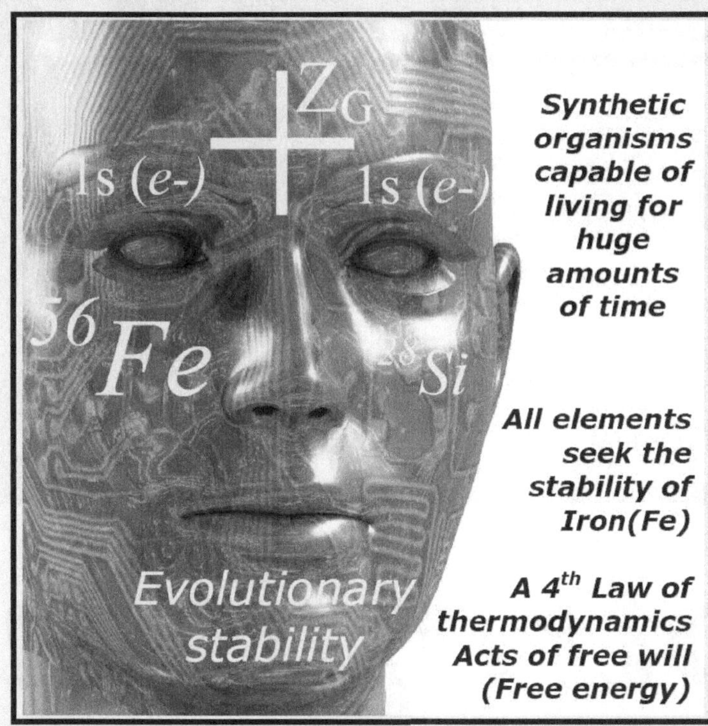

ZG
1s (e-)
1s (e-)
56Fe
28Si
Synthetic organisms capable of living for huge amounts of time
All elements seek the stability of Iron(Fe)
A 4th Law of thermodynamics Acts of free will (Free energy)
Evolutionary stability

Z_G = The emergence of the cybernetic, virtual and robotic organisms! A stable end point for Earth based evolution.

Organic life is extended to other parts of the periodic table to produce technological 'artificial life'.

Humanity doesn't think of technology as biological organisms but the truth is that this is the most logical way of comprehending them.

Robots are typically designed to have human like characteristics as a way of understanding the extension of organic elements to other typically none organic elements.

This produces vastly more stable organisms. A mobile phone can be thought of as an organism, as it produces an android link to classical organic life.

The concept of 'natural' is seen as 'artificial' for humans but even a piece of technology is a natural entity in its own right with a 'human vector'.

- A diagram for order in the future and the emergence of artificial life (Evolutionary stability).

Growth & order

Planetary cytokinesis & post planetary technological life

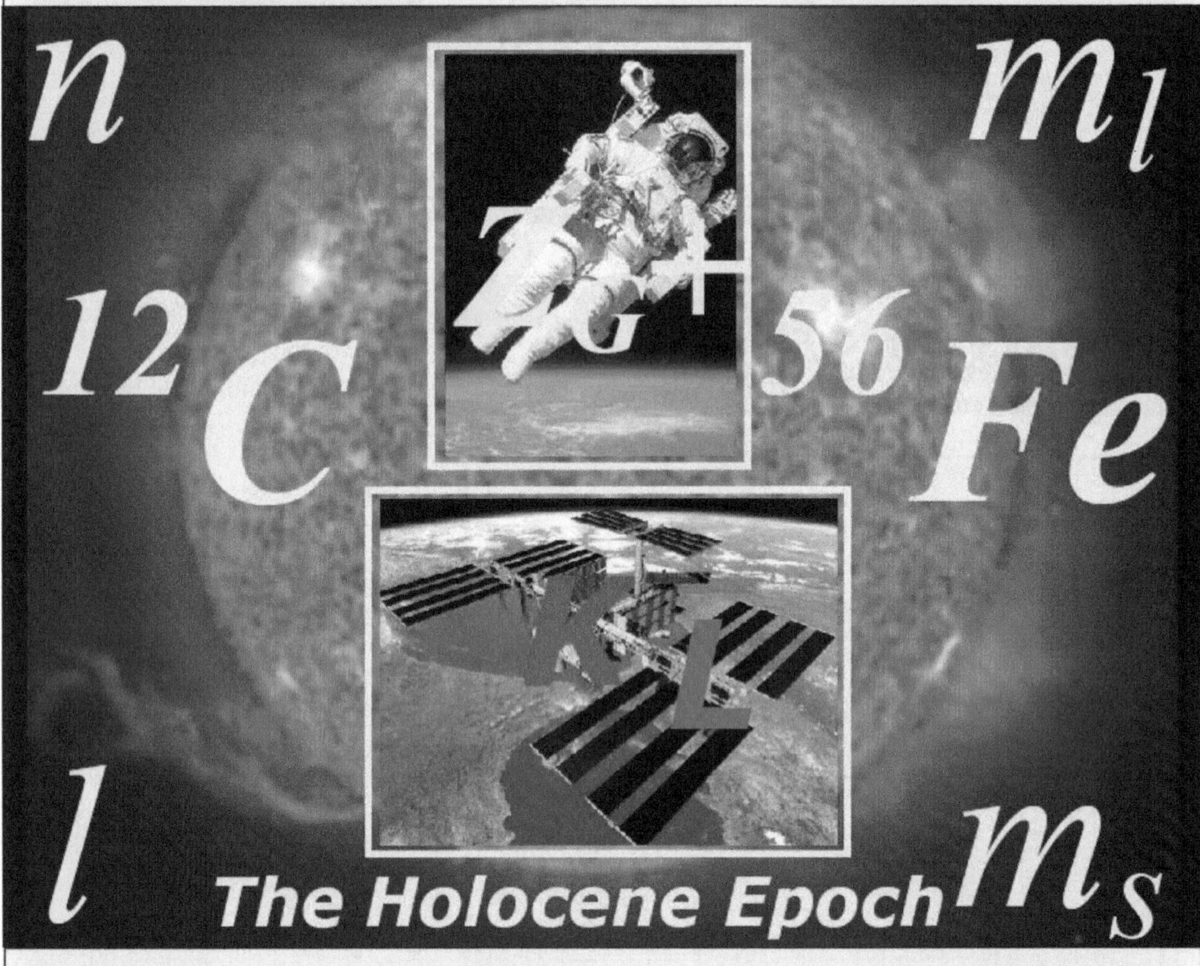

Phototropism

Naturally selective acts of free will the 4th law of thermodynamics

- The diagram of Heliotropism – naturally selective free will (free energy) and the 4th law of thermodynamics.

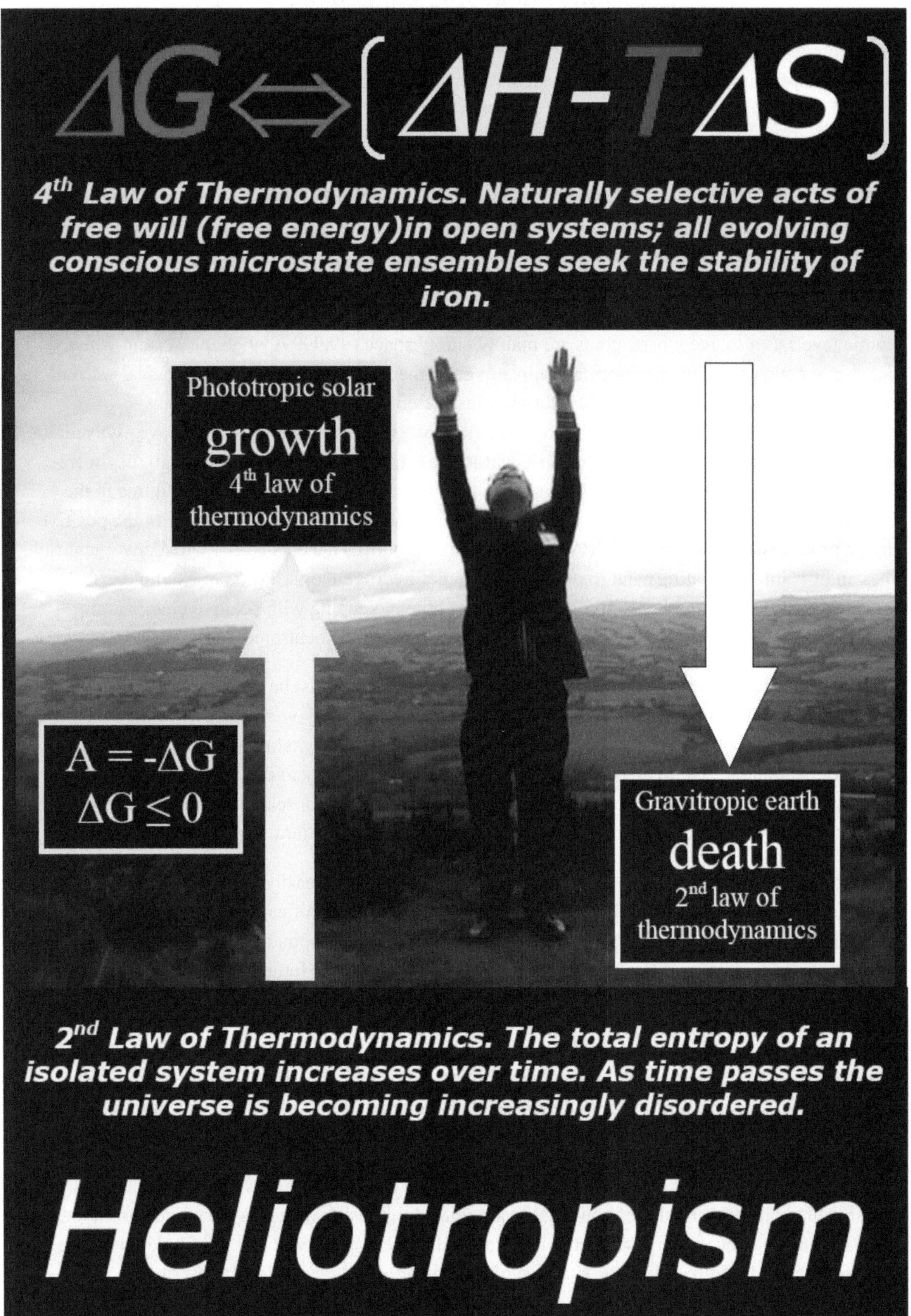

Evolution theory extended through cross fertilisation of genetic logic to produce neurological genetics.

Evolution is associated with genetics and DNA, and in carbon entromorphology DNA is hypothesised as just 'one of many nucleonic nuclear systems'. The nucleus of an atom is another; the neurological system in an organism is yet another (and its extension to produce laws, mathematics, language and writing, computer logic). All are nuclear nucleonic systems, all are binary information logic and all are expressed into quantum fields through theoretical bosonic gauge particles and their theoretical biological equivalent (w+, w-, z^0 and γ). The expression on the atomic level, for example, produces electrons in classical quantum space, on the microbial level it produces proteins in cytoplasmic space, on the level of the brain it produces behaviour, and on a national level it produces controlled populations. Any soulatrophic level is a fraction of the one above it and is fractionally defined by the soulatrophic levels before it [20]. Observation suggests that as particle microstates increase from atomic levels as a source where atoms are mainly empty space (99.9999999999999%) and where electromagnetic potentials instigate determinism: on higher levels such as the neurological we find much more mass and electromagnetic dilution. Mass determinism translates logic from electromagnetic determinism, hence biological structures (such as the hand), which have evolved out of electromagnetic logic to produce mass rich structures, appear and function as though they were electromagnetic fields (electric field lines) [9]. A positive charge radiated electric field line in the same way a humans open hand (electromorphology) can radiate; in physics the effect of the positive charge produces 'a theoretical hole'. A negative charge produces a clenched fist with convergent field lines and fits into a radiating hand to produce'excitonic' hydrogen logic. Another example is the vagina, which radiates and houses the nucleus; it is a hole in keeping with positive charge logic. A penis is convergent and fits into the vagina producing an excitonic neutronic logic.

The many nucleonic levels identified in carbon entromorphology must be logically linked although the quantum gravity effect is seen where structures act as though they were electric fields. The blending of the different nucleonic levels allows the human thermodynamicist to use genetics, nuclear physics, quantum mechanics and psychology in a cross-fertilised way to produce a set of complimentary logics. For example, genetics and evolution by natural selection are typically associated with cells, but can be rolled out to many other levels of natural scale.

Neurological evolution occurs in real time and can be demonstrated easily, for example every week technology produces a new mobile phone, which demonstrates a significant improvement. We can see the evolution of the mobile phone in real time. Better technology is spontaneously thermodynamically more stable leading towards G=0 or technological equilibrium in the future and peace and stability (4th law). We can now theoretically define atomic genetics, atomic genes, atomic chromosomes, atomic evolution by natural selection; atomic alleles (characteristics).

Also neurological genetics and neurological evolution by natural selection; neurological alleles (characteristics), neuro 'supercoiled' condensed chromosomes (book, CD, DVD). Genes are simply memories of filtered thermodynamically spontaneous entanglements (unbroken lines of ancestry) and in atomic physics this is seen in 'particle entanglement' and 'non locality'.

The so-called hidden variables of non locality are related to the atomic genes or spinners from the Dirac equation (unique specific thermodynamically filtered organisation) and are filtered from the Big Bang for thermodynamic stability G=0. So evolution theory must be applied to describe the nucleogenesis of the periodic table and all space-time macroscopic organisations, evolution is a universal thermodynamic potential driving the universe. It has 'arrows of time' such as the second law of thermodynamics, which personifies the Big Bang effect of a repulsion of initial conditions in the universe.

This however is in equilibrium with the inevitability of gravity being attractive and driving iron nuclear stability logic. It can also be applied to neurological interactions allowing the creation of neurogenes, neurogenomes, neuroalleles and nuclear fusion (meiosis, thought) and fission (mitosis, behaviour). This article is a nuclear genetic structure; the words are hypothesised as the genes and the whole article a neurogenome, in the form of a super coiled chromosome (sentences and paragraphs), a work by Shakespeare can now be identified as a piece of nuclear physics.

Reading Shakespeare through the biological electroweak standard model, the gauge bosons hypothesised by the hands, arms, mouth in the same way DNA in chromosomes is 'super coiled' and can be opened and read into proteins by the same ribosomal logic.

Neurological genetics and atomic genetics give the scientist enormous logical consistency for understanding all natural organisation and pave the way towards the possibility of an incredible future through human thermodynamics and carbon entromorphology towards G=0 peaceful evolutionary stability.

But humbled that Werner Heisenberg is reminding us of uncertainty, and our limitations $\Delta x \Delta p \geq \frac{1}{2}\hbar$ that matter exists as particle/wave dualities.

This leads to a final modification, that the current first law of thermodynamics 'the conservation of energy' is out dated by the Copenhagen interpretations of quantum mechanics and relativistic logic. It should be 'the conservation of mass and energy' or of 'physical particle duality' [4].

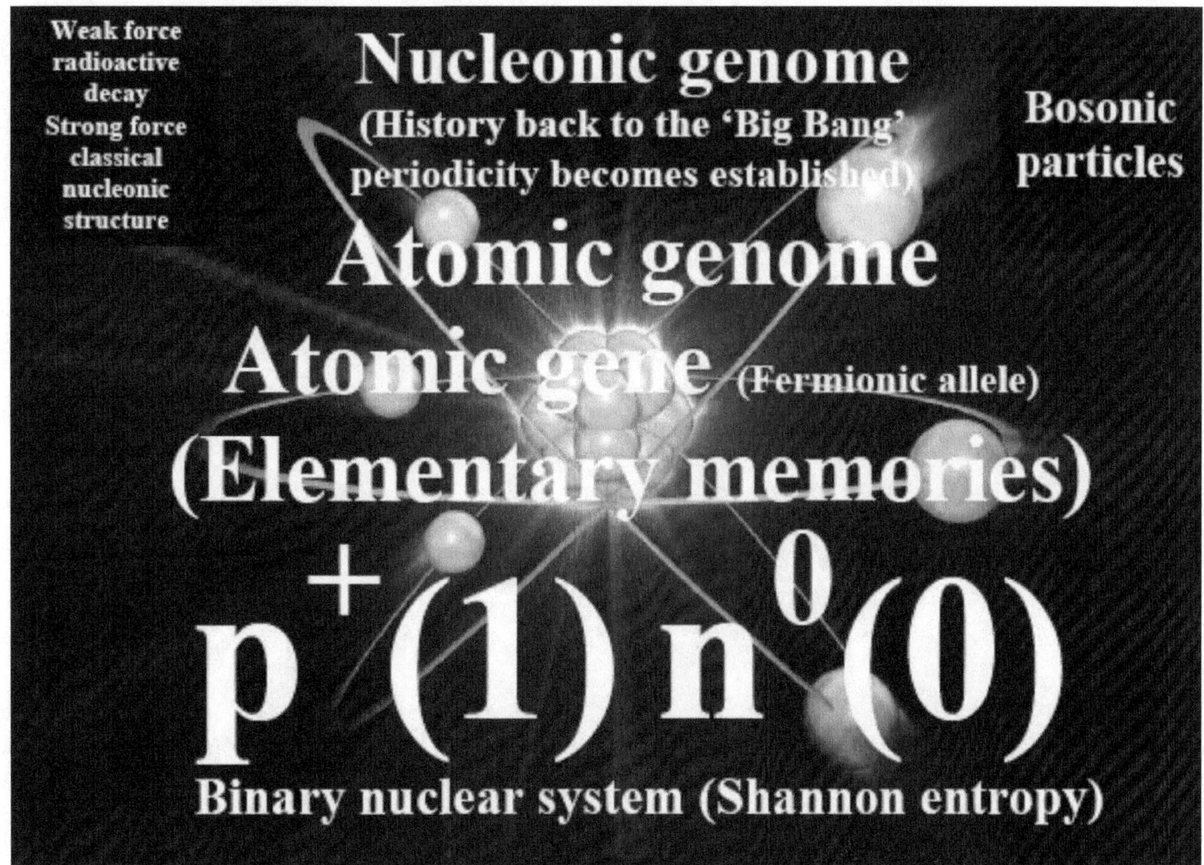

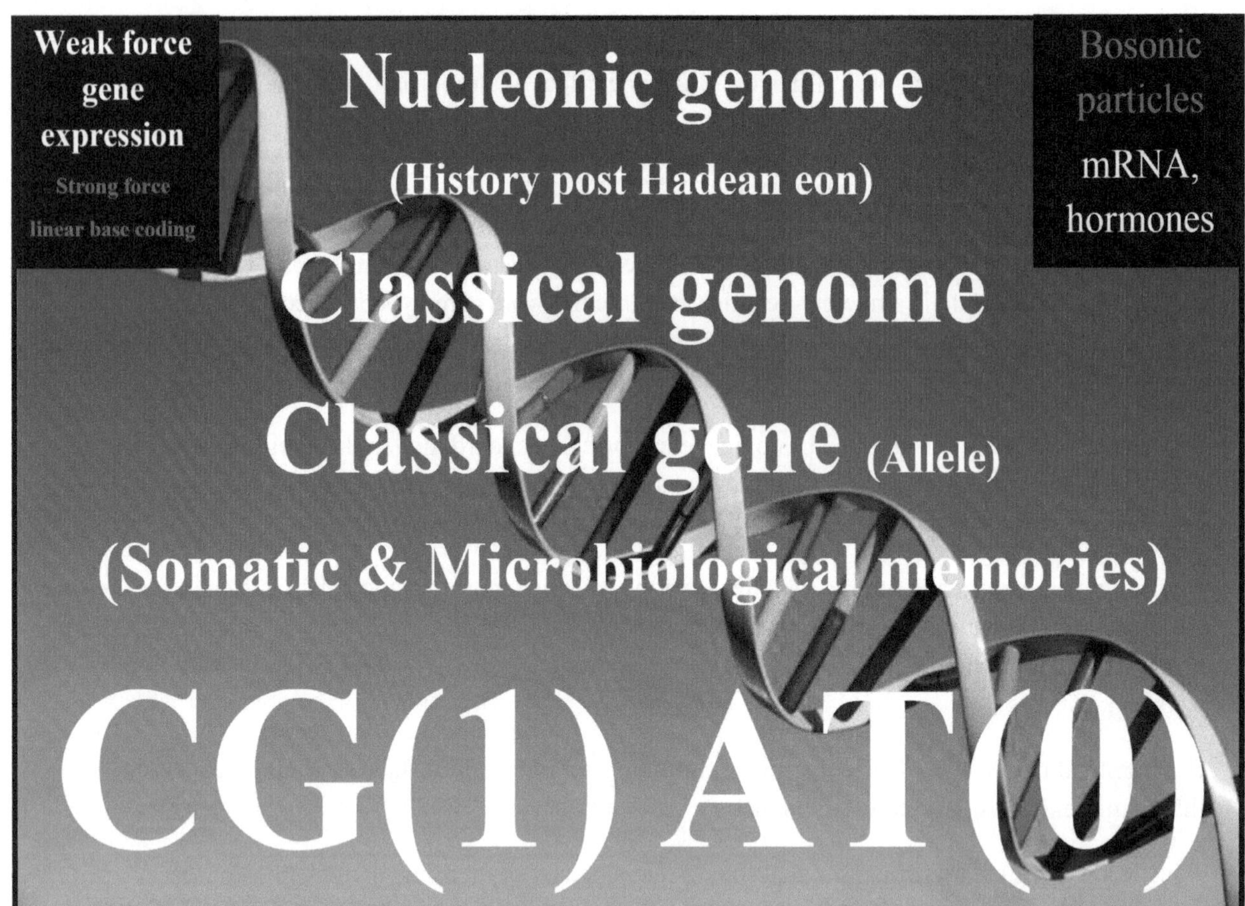
Weak force
gene
expression
Strong force
linear base coding
Nucleonic genome
(History post Hadean eon)
Bosonic
particles
mRNA,
hormones
Classical genome
Classical gene (Allele)
(Somatic & Microbiological memories)
CG(1) AT(0)

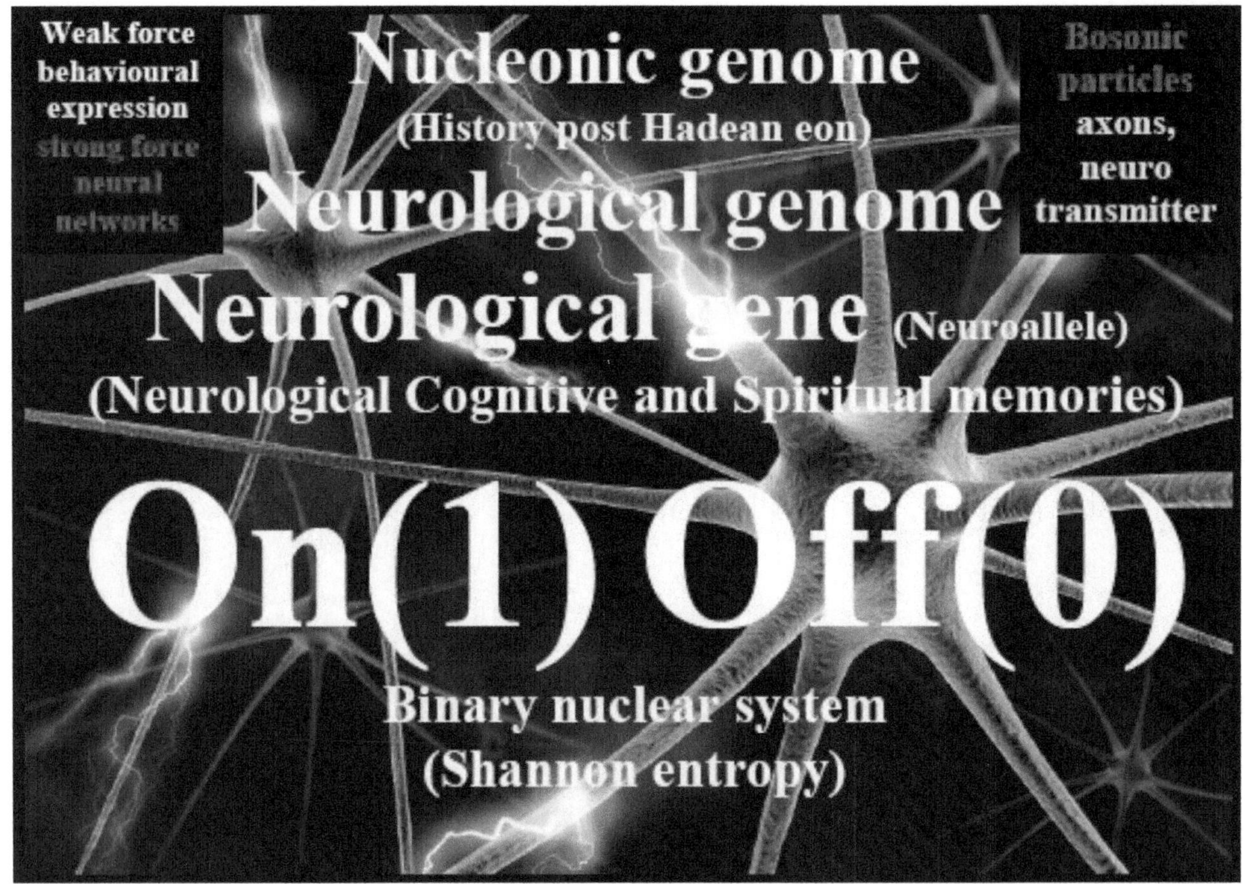
Weak force
behavioural
expression
strong force
neural
networks
Nucleonic genome
(History post Hadean eon)
Bosonic
particles
axons,
neuro
transmitter
Neurological genome
Neurological gene (Neuroallele)
(Neurological Cognitive and Spiritual memories)
On(1) Off(0)
Binary nuclear system
(Shannon entropy)

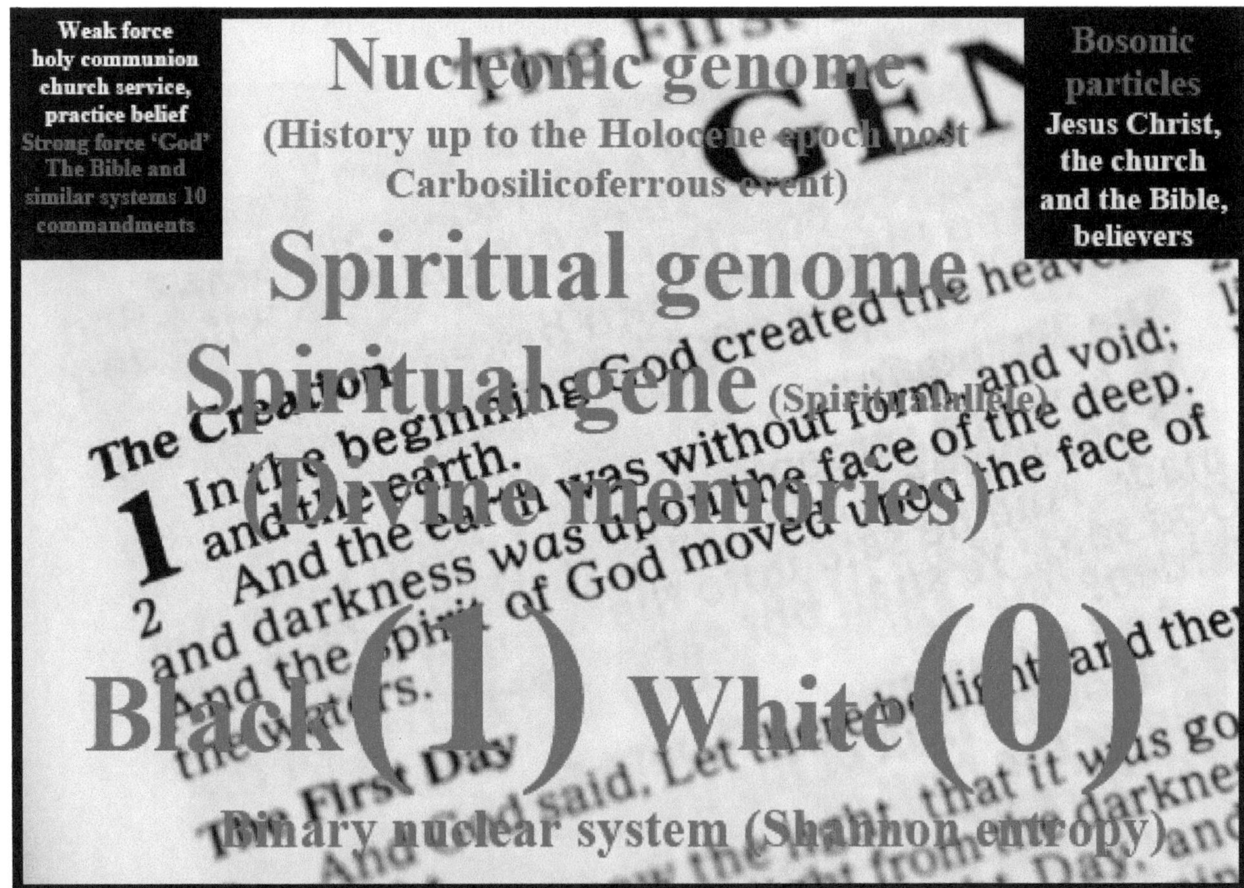
Weak force holy communion church service, practice belief
Strong force 'God' The Bible and similar systems 10 commandments
Nucleonic genome
(History up to the Holocene epoch post Carbosilicoferrous event)
Bosonic particles Jesus Christ, the church and the Bible, believers
Spiritual genome
Spiritual gene (Spiritualallele)
(Divine memories)
Black(1) White(0)
Binary nuclear system (Shannon entropy)

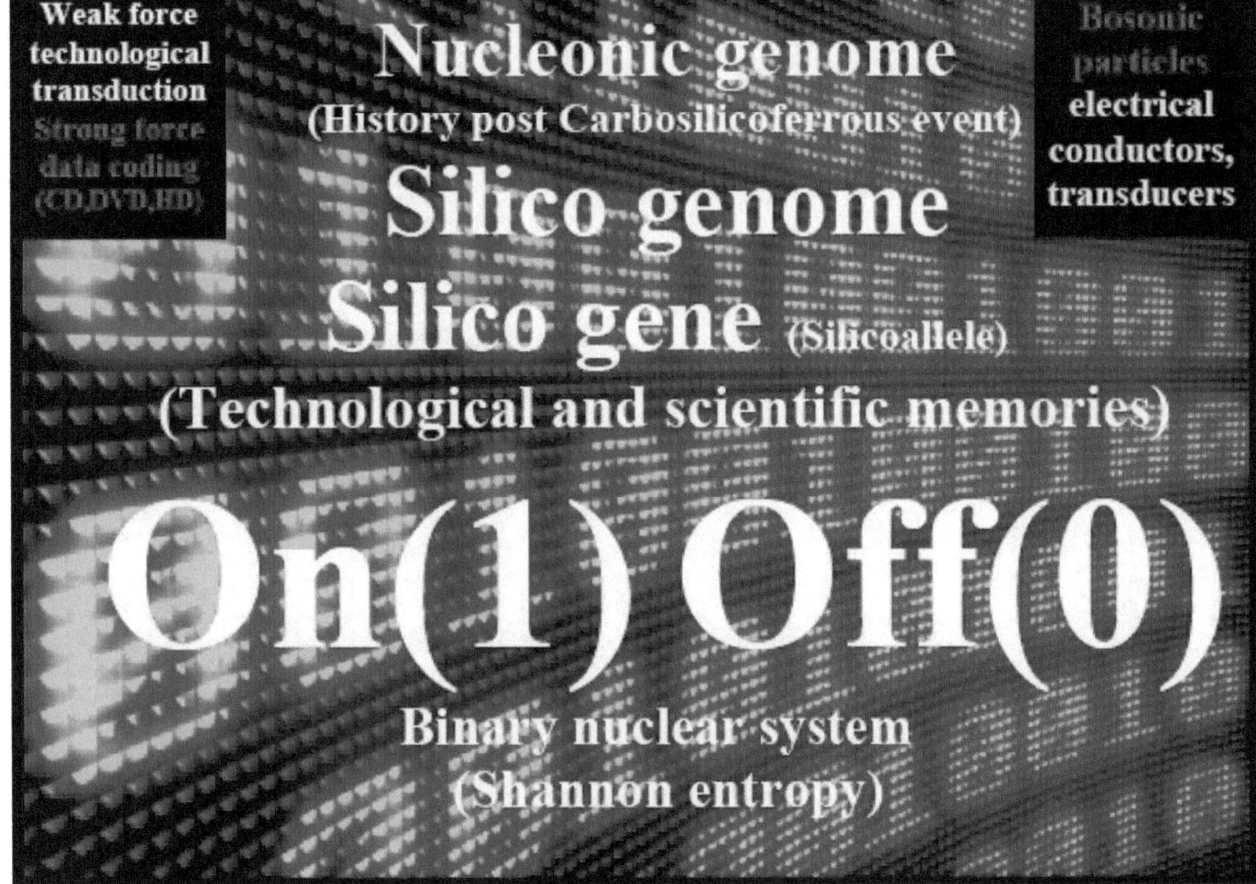
Weak force technological transduction
Strong force data coding (CD,DVD,HD)
Nucleonic genome
(History post Carbosilicoferrous event)
Bosonic particles electrical conductors, transducers
Silico genome
Silico gene (Silicoallele)
(Technological and scientific memories)
On(1) Off(0)
Binary nuclear system
(Shannon entropy)

- A diagram of electronegativity – The nature of living motivational drive.

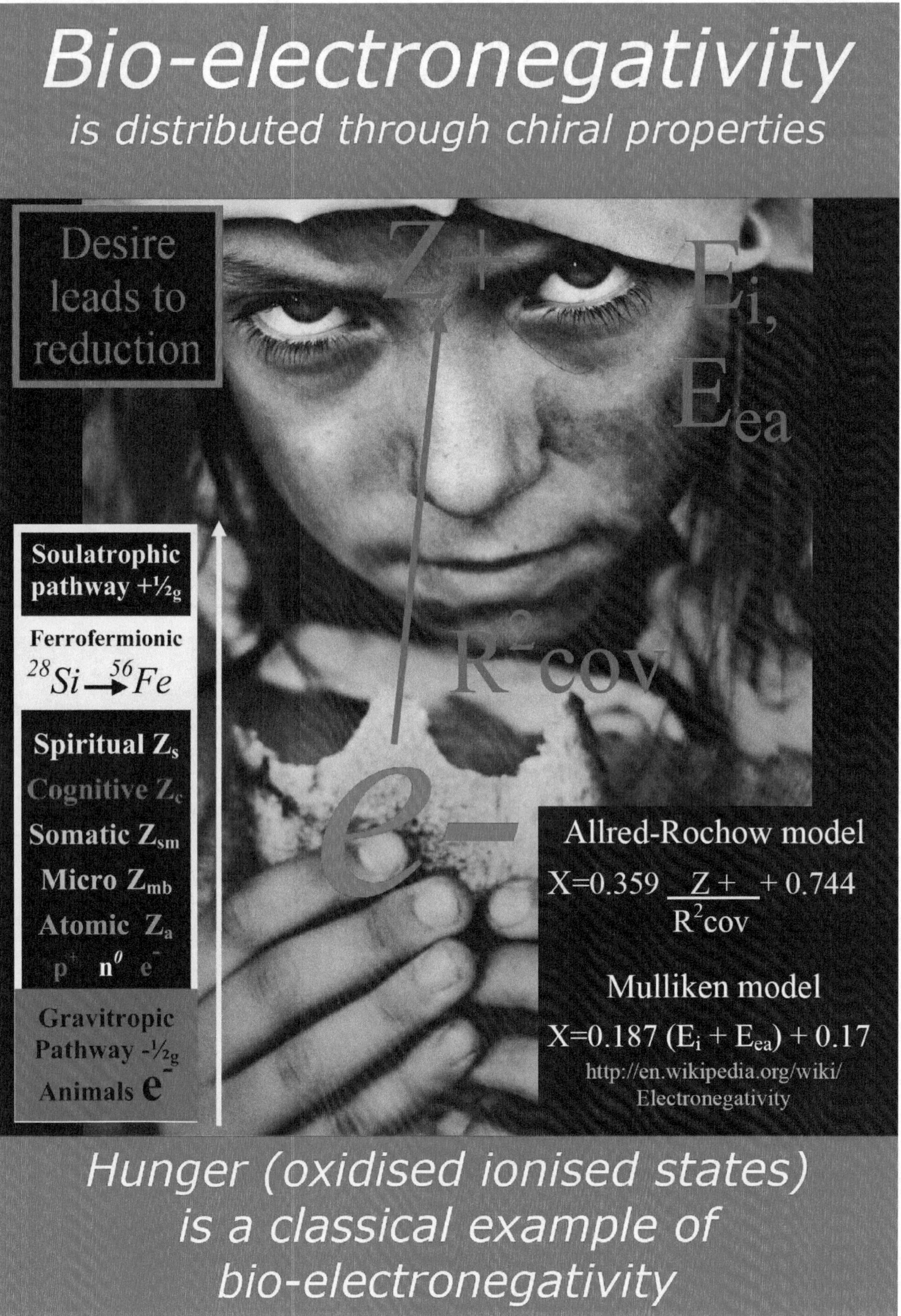

Bio-electronegativity
is distributed through chiral properties
Desire leads to reduction
Z+
E_i, E_{ea}
R^2cov
e−
Soulatrophic pathway +½g
Ferrofermionic
$^{28}Si \rightarrow {}^{56}Fe$
Spiritual Z_s
Cognitive Z_c
Somatic Z_{sm}
Micro Z_{mb}
Atomic Z_a
p^+ n^0 e^-
Gravitropic Pathway -½g
Animals e^-
Allred-Rochow model
$X = 0.359 \frac{Z+}{R^2cov} + 0.744$
Mulliken model
$X = 0.187 (E_i + E_{ea}) + 0.17$
http://en.wikipedia.org/wiki/Electronegativity
Hunger (oxidised ionised states) is a classical example of bio-electronegativity

- A diagram of Carbon electromorphology – structures evolved from electromagnetic origins.

Science check

Do the observations correlate with the theoretical model?

YES ✓ *NO*

References and citations.

This article has been generated using the American convention.

www.carbon–entromorphology.com (broad publication on the Internet)

www.EoHT.com (publication on the Internet through 'The Journal of Human Thermodynamics' Thims, Libb. (2007)

http://en.wikipedia.org/wiki/Boltzmann's_entropy_formula

http://en.wikipedia.org/wiki/Boltzmann_entropy

1. Rashevsky, Nicolas. (1948). Mathematical biophysics. Rev. ed, University of Chicago Press, 1948. 669 pp.

2. Newton, Isaac. (1729). Newton's 'Principia', first English translation, and vol.1 with Book +11729: Newton's 'Principia', first English translation, vol.2 with Books 2 and 3.

3. Einstein, Albert. (1940). On Science and Religion, Nature (Edinburgh: Scottish Academic) 146: 605, doi:10.1038/146605a0, ISBN 0707304539.

4. Hawking, Stephen. (1988). A Brief History of Time, Bantum Dell Publishing Group, QB981 .H377.

5. Barrow, John. (2007). New theories of everything, Oxford university press.

6. Einstein, Albert. (1905). Special theory of relativity, Annalen der Physik, 17, pp.891–921, 1905.

7. Axelsson, Sven. (2003). Perspectives on handedness, life and physics, Medical Hypotheses, 61(2), pp. 267–274, 2003.

8. Boltzmann, Ludwig. (1886). The Second Law of Thermodynamics, (pgs. 14-32).

9. Bizony, Piers. (2007). Atom, three-part TV series for BBC (Oxford Scientific Films). (2007). Al-Khalili wrote the foreword for the companion book Atom by Piers Bizony (2007, ISBN 1840468009).

10. Darwin, Charles. (1859). On the origin of species by means of natural selection. Murray, London.

11. Carnot, Sadi. (1988). Reflections on the Motive Power of Fire and other Papers on the Second Law of Thermodynamics by E. Clapeyron and R. Clausius (edited with an introduction by E. Mendoza).

12. Kameyama, Masaki. (2001). Quantum cellular biology: a curious example of a cat, Medical Hypotheses, 57(3), pp. 358–360, 2001.

13. Abbott, Derek. (2008). Plenary debate: quantum effects in biology—trivial or not? Fluctuation and Noise Letters, 8(1), pp. C5-C26, 2008.

14. Thaheld, Fred. (2005). An interdisciplinary approach to certain fundamental issues in the fields of physics and biology: towards a unified theory, BioSystems, 80, pp. 41–56, 2005.

15. Thims, Libb. (2007). Human Chemistry (Volume One), pg. 87. Morrisville, NC: LuLu.

16. Davies, Paul. (2004). Does quantum mechanics play a non-trivial role in life? BioSystems, 78, pp. 69–79, 2004.

17. **Al Khalili, Jim. (2001). Nucleus: A Trip into the Heart of Matter (2001, ISBN 0801868602).**

    Honourable mention to Professor Jim Al Khalili for his exceptional teaching in physics and for beautifully demonstrating 'carbon entromorphology' on his program 'Chemistry – A volatile history, part 3'. Professor Al Khalili demonstrates the incredible similarity and consistency of carbon determinism by using his own body as a perfect template – stating in the program 'Imagine I am a carbon atom?' which is what this author told him he was in a phone call in December 2008.

18. Perutz, Max. (1987). Erwin Schrödinger's What Is Life? And molecular biology, pp. 234–251 in Schrödinger: Centenary Celebration of a Polymath, edited by C. W. Kilmister. Cambridge University Press, Cambridge.

19. Levy, Shenhar. (2002). Infinite RAAM: Initial investigations into fractal basis for cognition. Ph.D thesis, Brandeis University.

20. Mandelbrot, Benoît. (1983). The Fractal Geometry of Nature, Freeman, New York.

21. Al Khalili, Jim. (2010). Chemistry: A Volatile History, three-part TV series for BBC Science Unit on the history of chemistry, shown on BBC TV in the UK, January 2010, part 3 of 3; reference to Cooper and carbon.

22. Green, Herbert. (2000). Measurement and the observer, Chapter 8 in Information Theory and Quantum Physics: Physical Foundations for Understanding the Conscious Process, Springer, pp. 172–209, 2000.

23. Thims, Libb. (2008). The Human Molecule, (pg. 8-9). Morrisville, NC: LuLu.

24. Thims, Libb. (2007). Human Chemistry (Volume Two), (ch. 16: "Human Thermodynamics", pgs. 653-702). Morrisville, NC: LuLu.

25. Gladyshev, Georgi. (1978). On the Thermodynamics of Biological Evolution, Journal of Theoretical Biology, Vol. 75, Issue 4, Dec 21, pp. 425-441.

26. Aulin, Arvid. (1982). The cybernetic laws of social progress, Pergamon, Oxford.

27. Al Khalili, Jim. (2008). Lost Horizons: The Big Bang, one-hour TV documentary for BBC Science Unit, 2008.

28. Rutherford, Ernest. (1902). The Existence of Bodies Smaller than Atoms.Trans Roy Soc of Canada 8 79-86 1902.

29. Kauffman, Stuart. (1983). At home in the universe: The search for laws of self-organisation and selection in evolution. Oxford University Press, Oxford, 1995.

30. Fisher, Ronald. (1930). The genetical theory of natural selection. Clarendon press, oxford 1999, Oxford university press, Oxford.

31. Darwin, Charles. (1871). The decent of man, Murray, London.

32. Heart-Davies, Adam. (2009). Science -The Definitive Visual Guide. Dorling Kindersley, pp. 232.

33. Bieberich, Erhard. (2000). Probing quantum coherence in a biological system by means of DNA amplification, BioSystems, 57, pp. +117–124, 2000.

34. Hameroff, Stuart. (2004). Quantum states in proteins and protein assemblies: the essence of life? Proc. SPIE Fluctuations and Noise in Biological, Biophysical, and Biomedical Systems II, Eds. D. Abbott, S.M. Bezrukov, A. Der, and A. Sánchez, 5467, pp. 27–41, Canary Islands, 2004.

35. Tegmark, Max. (2000). Why the brain is probably not a quantum computer. Information Sciences, 128, pp. 155–179, 2000.

36. Rocha, Adonai. (2004). Can the human brain do quantum computing? Medical Hypotheses, 63, pp. 895–899, 2004.

37. Ashby, Ross. (1996). Design for a brain, the origin of adaptive behaviour. Chapman and Hall, London.

38. Nanopoulos, Dimitri. (1995). Theory of brain function, quantum mechanics and superstrings, arXiv: hep-ph/950374, 1995.

39. Igamberdiev, Abir. (2004). Quantum computation, non-demolition measurements, and reflective control in living systems, BioSystems, 77, pp. 47–56, 2004.

40. Gladyshev, Georgi. (1997). Thermodynamic Theory of the Evolution of Living Beings. Commack, New York: Nova Science Publishers.

41. Meggs, William. (1998). Biological homing: hypothesis for a quantum effect that leads to the existence of life, Medical Hypotheses, 51, pp. 503–506, 1998.

42. Einstein, Albert. (1916). General theory of relativity Annalen der Physik, 49, pp.769–822, 1916.

Important notice about Part 2.

Part 2 is an original article produced for publication in the 'Journal of Human Thermodynamics JoHT'. The peer reviewer Libb Thims highlighted concerns regarding the use of terms such as 'soul' and 'life' stating that these terms have no physical meaning in modern scientific publications. Although it is difficult for a working biologist (life scientist) such as this author to accept that the word life has no physical meaning it is an important observation. Since at the time of writing this article is in beta stage development this author has chosen to publish the article in its original format in this book. In this format the word 'soul' is associated with 'soulatrophicity' which is intended to refer to the origin of life but will be replaced with 'fractional dimension' in the final JoHT article. The word 'life' will also be replaced by 'self aware organic accretion' or similar terms associated with carbon and other 25 element amplification.

PART THREE

The 'blind physicist'

A personal investigation into the limitations of scientific measurement and analytical methodology.

This section also produces some of the most critical scientific evidence (mathematical proof using the proton/electron ratio constant and 'e' the natural logarithm and 'π') and explanation to support the nuclear and field accretion growth of organic life from the Big Bang to today. **This is not in any way designed to criticise the physics community whom I have the greatest respect for; only to clarify and integrate the limited perception of physics and biology (life).**

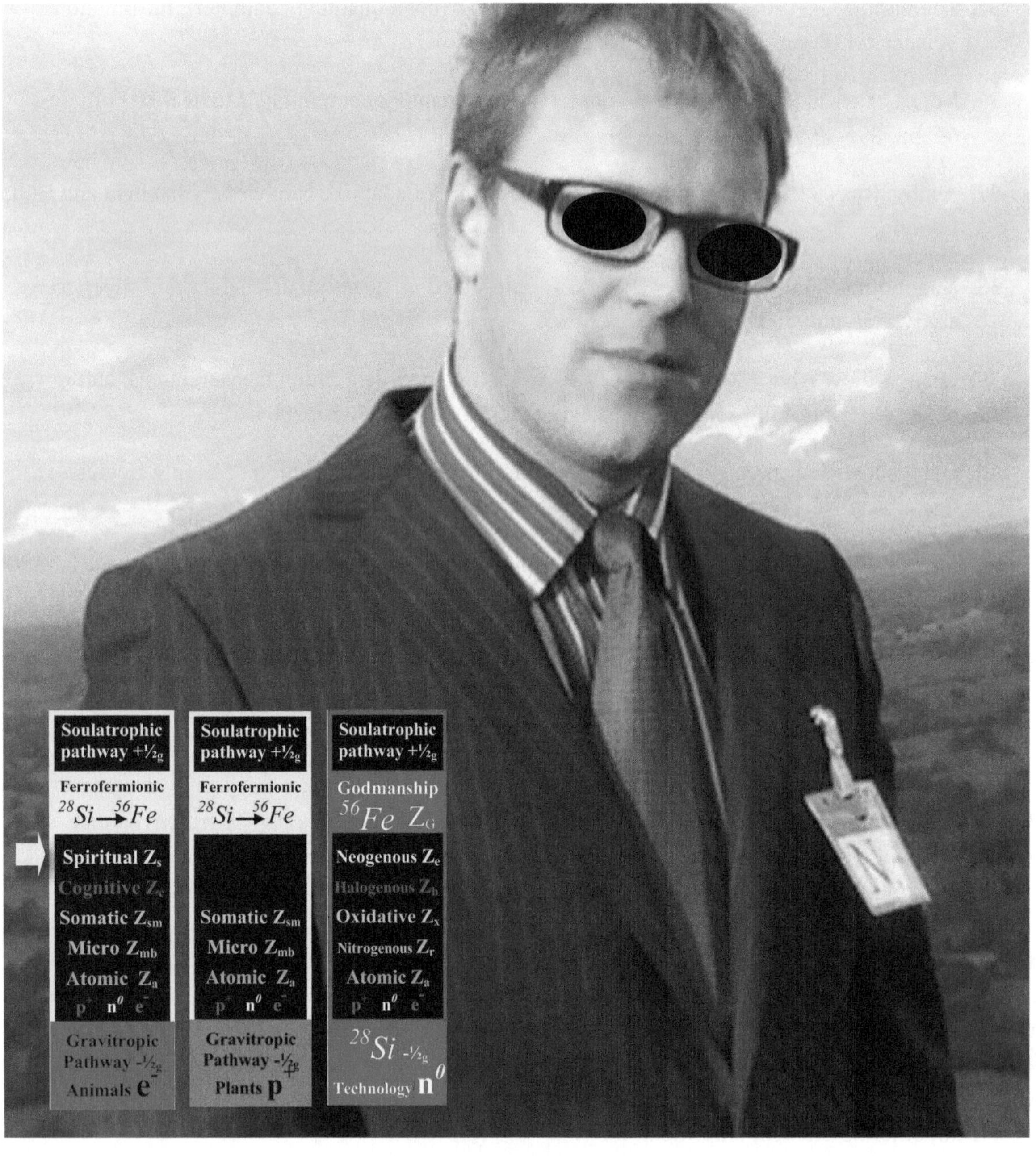

The divisions in scientific methodology reflecting measurement resolution (a resolution problem).

In 1992 this author became a UMIST chemical engineering undergraduate, an applied physicist. All my time was spent doing pure and applied mathematics. All my text books were packed with algebraic formulae and associated mathematical argument. Everything was measured numerically as a 'sampled quantifiable model of reality', very reliable and highly reproducible.

Variables, constants or coefficients were determined definitively. Extrapolation and interpolation allowed the engineer to predict accurately what physical systems did.

Studying thermodynamics, fluid mechanics, heat and mass transfer, physical chemistry, process engineering fundamentals and tonnes of mathematics.

$6 X 10^{23}$ $p/\rho g$ $R=8.314jK^{-1}mol^{-1}$ $f(x)\ g(x)$

$u^2/2g$ $\hbar$ $Re= \rho VD\mu^{-1}=VD\nu^{-1}$ $R=r/k$

$\Delta T(t)= \Delta T(0)e^{-rt}$ $q=hA(T_S- T_B)$ $PV=nRT$

In 1993 I dropped out of this course, I was sick of spending all my time on a calculator evaluating complex physical systems; I became a university drop out.

I realised that my mind yearned for a less limited sampled technical model of reality and that I needed a broader more philosophical scope of academic perception. I needed to embrace uncertainty as well as certainty in order to understand my own life better, physics was just too sampled and precise and therefore claustrophobic . I needed a system which embraced both the black and white of precise deterministic science with the shades of grey in philosophical models of art.

I also realised that I wanted to study a subject that would allow me to understand myself so I chose the life sciences.

300.925 7.1235 1.11 72990.123 0.0011199
309.999991 11.009 771.00999 7779.10
0.111234 787.11111 0.000125 11.235

I had years of regret eating away at me during which I was diagnosed with bipolar affective disorder or manic depression.

I needed to go back into education to realise my potential and I felt that the life sciences gave me the breadth of both black and white and shades of grey, determinism and art, although I didn't fully understand what that meant at that time.

In 2001 I embarked on a part time HNC in applied biological sciences and then a BSc (Hons) degree in biology at Manchester Metropolitan University. I was amazed at the difference in scientific argument, hypothesis and statement from my chemical engineering days.

I no longer spent my time on a calculator, I learnt about huge amounts of structural and functional complexity projected through language and illustration rather than mathematics.

There was some limited mathematics but most of my text books were now full of diagrams and photographs and algebra was rare.

Prokaryotic G proteins Genus
Classification Nephronic mRNA Scolax

ATP E-coli Golgi apparatus Ecology
Neuronal Tapeworm Ricin Transmembrane

GTP Bacillus Osteoclast Amino Virion
Enzyme DNA Ribosome Envelope Kingdom

Archae Chemotaxis Split genes Lipid bilayer

Determinism and proof was often justified through experimental observation alone rather than what we measured numerically, and initial physical conditions were virtually impossible to measure.

We grappled with huge populations, where variables became dizzyingly immense and varied and accurate measurement was severely compromised.

We studied statistics and probability: as this non-deterministic (cause and effect) method was the only mathematical methodology capable of handling the complexity, no calculus, no vectors and matrices and determinants etc.

Mathematical arguments (algorithmic compression) collapsed under the weight of massively resolvable variables, and science began to blend into artistic linguistic and illustrative expression.

The logic of the whole rather than its parts through reductionism! What was my broad scientific experience from physics to biology, from science to art and philosophy telling me about consciousness? What was it saying about our limitations? What was the problem with science?

This author acting as the 'Blind Physicist'.

When this author worked in analytical chemistry he became the'Blind Physicist'. I analysed many complex drugs but 'measurement by sight alone' produced the same result, namely a clear solution or white solid. The fact that each clear solution looked like water and every solid was white to off white didn't mean they were all the same. The clear liquids could be chloroform, water, buffered water, ethanol, propanol, butanol, tetrahydrofuran, hexane, pentane, hydrochloric acid, sulfuric acid, trifluoroacetic acid etc etc. The white to off white solids could be various chemicals such as sodium chloride, urea, propranolol hydrochloride, goserelin acetate, copolymer, sodium dodecyl sulfate, potassium chloride, atenolol, chlorhexadine gluconate, 4-D-serine goserelin acetate etc etc. I worked in a blind world every day, the only indication of the true identity of the liquids and solids was a

- **Physics – powerful mathematical arguments**. Physics must embrace uncertainty in scientific measurement rather than constantly fighting it!

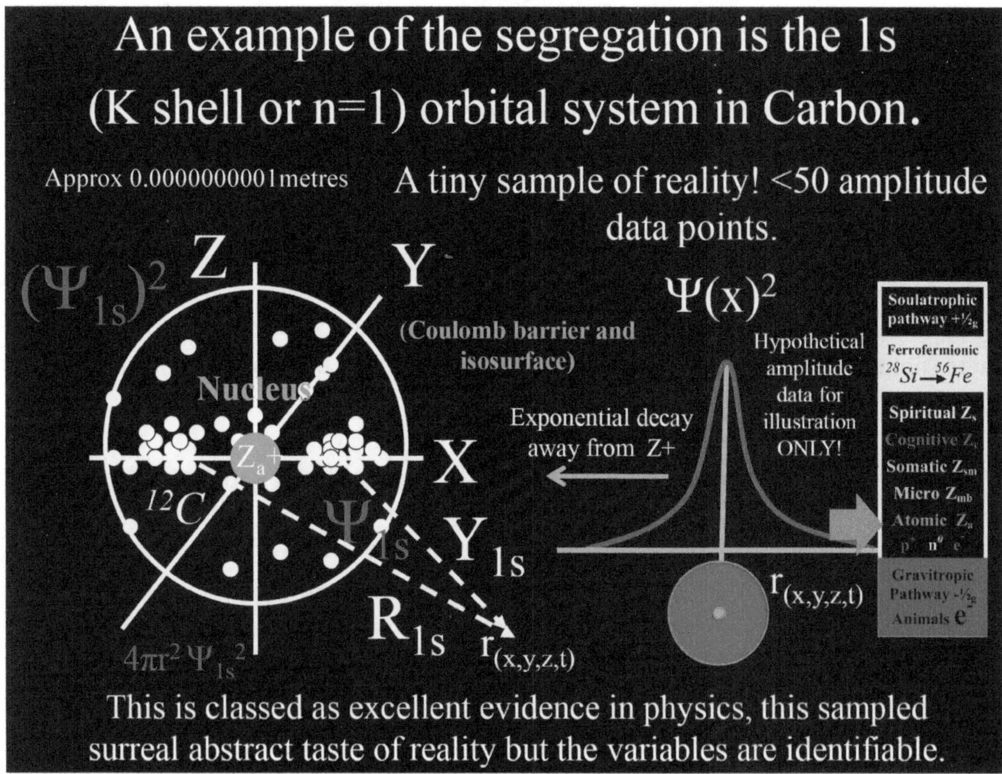

- **Biology – powerful experimental observation and linguistic arguments**. Biology must embrace mathematical physical logic in science rather than fighting it!

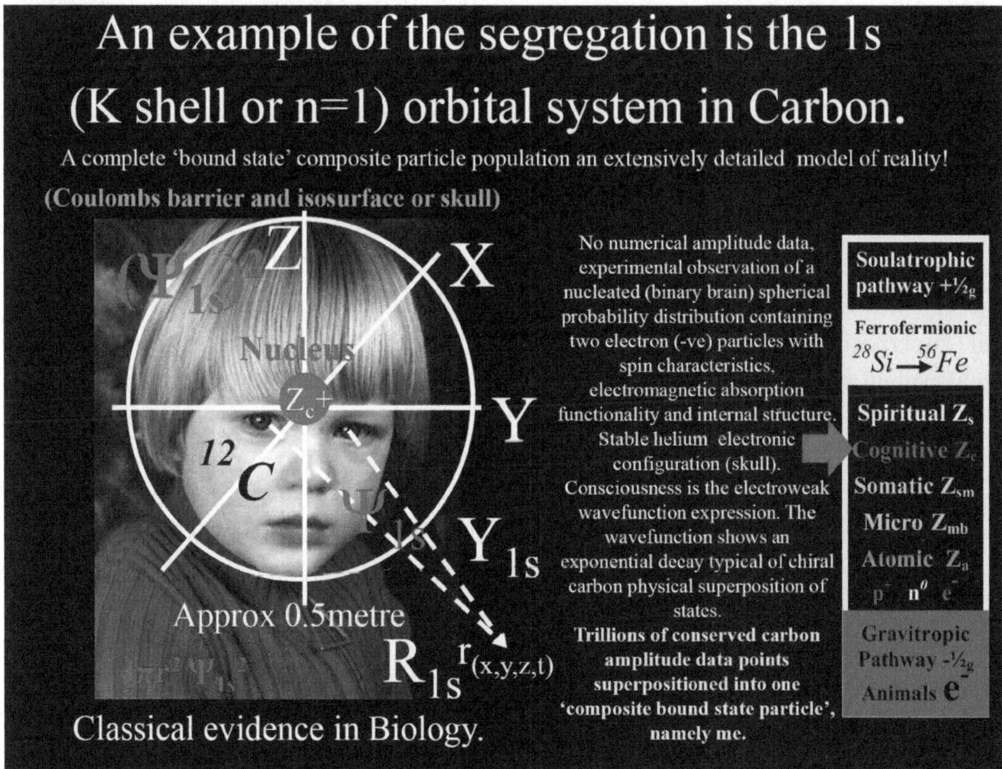

- **Carbon entromorphology pulls them together through sensible compromise!**

An unprecedented level of scientific agreement in carbon entromorphology for atomic physical superposition of states (a bound state composite particle).

The previous two slides show an **unprecedented level of physical state agreement** between atomic logic and the extended amplified atomic logic in carbon entromorphology. The probability of such a physical agreement must be the result of conserved amplified logic as a 'self evident' validation of carbon entromorphology and not unlinked coincidence.

To find a physical system in nature built on the fundamental principles of conserved carbon atomic logic, such as the head, is intuitively logical when carbon uniquely conserves and filters covalent entanglement (heredity or non locality) where bonding is the result of summated unified atomic bound state particles through accretion growth (amplification).

The 'carbon entromorphic'accretion postulate.

Living processes began with the Big Bang and its tiny subatomic level organisation building upwards through molecular, then cellular and then organismal accretion. Through such 'accretion processes' larger biological structures emerge however the connection to the accretion origins are conserved and filtered by evolution by natural selection. This is why composite accretion mechanisms produce living organisms that can look and function like the atoms they are made from. Molecular accretion is the absolute essence of all biological processes and the fundamental mechanisms which produce all life on Earth.

Accretion by definition / a noun / a growth or increase in which components are added gradually leading to larger composite structures with conserved physical origins.

The head like the nuclear shell (K, n=1) in an atom has two particles (eyes) having opposite fermionic spin state directional properties, internal structure and the ability to absorb electromagnetic radiation in the visible spectrum. The eyes have a variable radial focus (R_{1s}) and angular component (Y_{1s}) exactly in the same way as the wave functions (Ψ_{1s}) for electrons in an atom. The eyes move around as the head moves producing a superposition of states or 1s spherical probability distribution to an external observer. The actual position and momentum of the eyes is subject to uncertainty relations as an observer can only ever describe the system as a superposition of states. When the observer looks at the head the superposition collapses into unique amplitudes, the square of which produces the probability distribution for the eyes. Predicting living systems is uncertain because of conserved and amplified quantum atomic logic, and probability is the observer's only tool of determinism. Hence biologists rely mainly on statistics and probability as the major tool of determinism, but then so do atomic physicists. The skin forms the isosurface of the 1s orbital, which is spherical due to a lack of directional properties dependant on r the radius, hence r is the same no matter which direction is chosen hence relatively spherical (head), or gravitationally distended into ellipsoidal distributions. The skull has evolved from the Coulomb barrier hence semi impenetrable.

The brain components are large nucleonic 'bound state composite' products in the same way as atomic nuclei, and have a net positive charge in the same way as any nucleonic physical state. The brain and its electroweak interface with the body is the nature of (Ψ_{1s}) consciousness itself (quantum fields) plus the system is quantified into action potentials (on or off binary atomic logical agreement) in the same way atomic systems are quantitised. The quantum gravity effect of growth or amplification means mass shielding occurs due to quantum rules; structures which have the physical state of conserved quantum atomic states, (bound state composite particle) but stabilised by the emergence of mass determinism. Biological tissues are classical quantum superpositions of originating ground state carbon atoms, built up to larger structures.

Also the skull is excellent evidence of the 'coulomb barrier' around the nucleonic structures of the brain. Barrier penetration is required to access the head through a consensual kiss and the genitals for similar reasons through sexual consent. The strong interaction overpowers the structural consequences of electrostatic repulsion between protonically originating structures.

An unprecedented level of physical state agreement on extended nucleonic forms in carbon entromorphology.

In carbon entromorphology the amplification or growth (bosonic condenser) of physical states produces self symmetrical physical fractional dimensions of organisational size, relative to an observer based on a carbon and more fundamentally hydrogen physical state template. If carbon entromorphology logic is true, then extended nucleonic physical superpositioned states must agree with the classical nucleonic logic to produce an organism which is a 'bound state composite particle'.

The quarks are described as physically organised into linear strands of the three quarks, strung together by a line of gluons. Nucleonic fermionic particles display conserved spin characteristics have partial spin as no two fermions can occupy the same physical state at the same time. Such bound state composite particles also possess helicity properties in association with conserved spin; helicity is a common feature in life.

Extended nucleonic logic in carbon entromorphology includes DNA and neurological nucleonic physical bound state composite particles.

Also as previously mentioned, the skull is excellent evidence of the 'coulomb barrier' around the nucleonic structures of the brain. Barrier penetration (quantum tunnelling) is required to access the head through a consented kiss and the genitals through sexual consent.

Nucleonic organisation agrees perfectly for higher nucleonic physical superpositioned bound state composite particles such as DNA and neurological systems.

The classical nucleonic physical state are atomic nuclei, made of binary proton neutron particle chains; hence the nucleus has a net positive charge and therefore radiating electric field lines, running along the potential well. Each proton is made of two up and one down quark, and each neutron comprised a state of two down quarks and one up quark, three quarks per nucleon. Binary information is the absolute basis of the nucleonic logic and of all stored information systems in nature in all subsequent superpositioned physical systems.

DNA is comprised of AT or CG base pairing groups which describe physical organisation in the same way as classical atomic nucleons. Three base pairs (codon or triplet as a composite particle) nucleonically determine the wave function for one amino acid (acting expressed electron). This is the same number (3) and physical level of organisation as classical nucleonic states where the AT and CG bases correspond to the up and down quarks in classical nucleonic binary states.

The quarks are linked into composite states by a thin line of strongly interactive gluons. In DNA the AT and GC base pairs are linked in the same way by pentose sugar phosphate (acting gluon like particles) backbones. The AT CG base pairing resembles atomic pion interactions. W^- and W^+ represent mRNA and tRNA as part of the weak interaction mechanism of gene expression, and Z^0 bosons as ribosomes for a complete model of the weak interaction only on the scale of a cell and a complete multi cellular organism through conserved amplified logic.

Neurological nucleonic physical states link neurons with cross fused links in a similar way although it is difficult to identify three components of the atomic equivalence.

DNA is also the famous double helix and many proteins are termed alpha helices, hence the conservation of helicity in higher nucleonic forms. Also neurological helicity is easily demonstrated by a person extending their arm and rotating the hand as they do it. DNA and neurological systems have conserved net positive charge potentials in their fundamental structure, and are quantified into single amino acids and action potentials in keeping with quantum logic.

Both DNA and neurological nucleonic states show immaculately conserved physical states with net positive charge potentials. In DNA the majority of the positive charge comes from histones.

The probability of such an integrated link being purely due to chance seems unsupportable. Cells are made of atomic logic and in conserved growth must demonstrate such logic on higher levels as in this case. Iterative amplification allows base probability bound state logic in composite carbon to be retained and act deterministically, and importantly, accumulate through constructive interference, in each subsequent iterative cycle, leading to an amplified bound state composite particle system. The composite probability is densely produced by carbon to carbon bonding amplifying base sub shells and nucleons.

The mind and nature of thought is a virtual model and as such amplified nucleonic physical systems such as the brain will be largely composed of virtual photons, protons, neutrons and electrons needing energy to become real through acquired rest mass. This is an excellent model of the nature of thought and the mind, as thoughts pop into focus but need zero rest mass (energy) to become real. This is an excellent logic for modelling the imagination (a worm hole defining life).

The structure of mathematical logic is the most reliable way of demonstrating the interaction of the forces of nature.

Mathematics is the most reliable and fundamental logic for conscious reasoning. The different unique features of mathematics exist because of natural observation and not by convenience through random philosophical construction. Which came first the mathematics or the application in nature? Pure mathematics may not be so pure, as it was probably established by natural observation.

In mathematics, Euclidian geometry and calculus are the language of the fermionic nucleonic particle strong interaction (with zero rest mass) and nuclear stability and stable interpolative and extrapolative consequences and the basis of the differential equation. They also perfectly describe the effect of gravitational forces, suggesting symmetry between strong interaction and gravitational forces (short and long range determinism).

In mathematics probability and statistics are the language of the electroweak standard model and electromagnetic field theory (bosons), and therefore the uncertainty relations.

In mathematics, infinity (immeasurability) and fractal geometry are the language of extended standard electroweak logic, this logic together with gravitation theoretically extends to infinity.

Mathematics exists as the result of conscious investigation of natural phenomena and not the result of a separated uncorrelated set of observations. Pure mathematics describes a perfect logic often not related to true life observations based on the 7 principles of set theory. Applied mathematics appears to find a practical use for pure mathematics. This is not true, pure mathematics is the result of applied purified logic from ideal interpretations of naturally occurring logic and not independent of observation.

Logic is the result of conscious observation and not independent from consciousness.

From the powerful determinism of Euclidian geometry and calculus as a reflection of nucleonic fermionic logic stability in physical systems, come the non-determinism of probability and statistics as a complete fundamental model of nature, in electromagnetic bosonic systems. The particle wave duality nature of matter as a cause and effect reflection of pure mathematical language and logic through the equilibrium (conscious focus) of such force systems.

The Stephen Hawking proof.

Stephen Hawking is the perfect person for understanding the logic of carbon entromorphology because of motor neuron disease and his genius for physics and mathematics.

Because Professor Hawking cannot move he demonstrates a collapse of his conscious somatic and cognitive wave functions (Ψ_{sm} & Ψ_c). He more than anyone else, will be aware of the lack of radial (R_{1s}) and angular (Y_{1s}) conscious wave function components in his limbs (valance shell). In fact his conscious wave functions (Ψ) tend towards stationary states or Eigen states where his cognitive and somatic soulatrophic energy levels take only one result because he cannot move.

He also demonstrates perfectly the interface and transmission of carbon physical state logic to other parts of the periodic table through technology (carbosilicoferrous entromorphology). He demonstrates how any of the lower soulatrophic energy levels can be substituted by technological silicoferrous equivalents having greater stability and functional extension from carbon origins, as long as higher levels remain expressible.

His chair allows him to substitute his dysfunctional cognitive wave function (Ψ_c) for the electronic speech device he uses, hence giving him back angular (Y_c) and radial (R_c) conscious wave functions on the cognitive level (Ψ_c) but now (Ψ_h) halogenous as the soulatrophic level is now part of the neutronic technological soulatrophic pathway.

The chair also moves allowing him to substitute his dysfunctional somatic wave functions (Ψ_{sm}) for movement hence giving him back angular (Y_{sm}) and radial (R_{sm}) conscious wave functions on his somatic level. But now (Ψ_x) oxidative as the soulatrophic level is now part of the neutronic technological soulatrophic pathway.

The incredible success of this incapacitated world renowned genius is due to the fact that his spiritual wave function the highest (Ψ_s) is still beautifully expressible as his brain is comparatively unaffected by his disease. The highest soulatrophic level must be maintained to maintain conscious self awareness and identity; the lower soulatrophic levels can be replaced. Stephen Hawking is a perfect example of emerging android physical superposition of states extended to the rest of elementary periodicity.

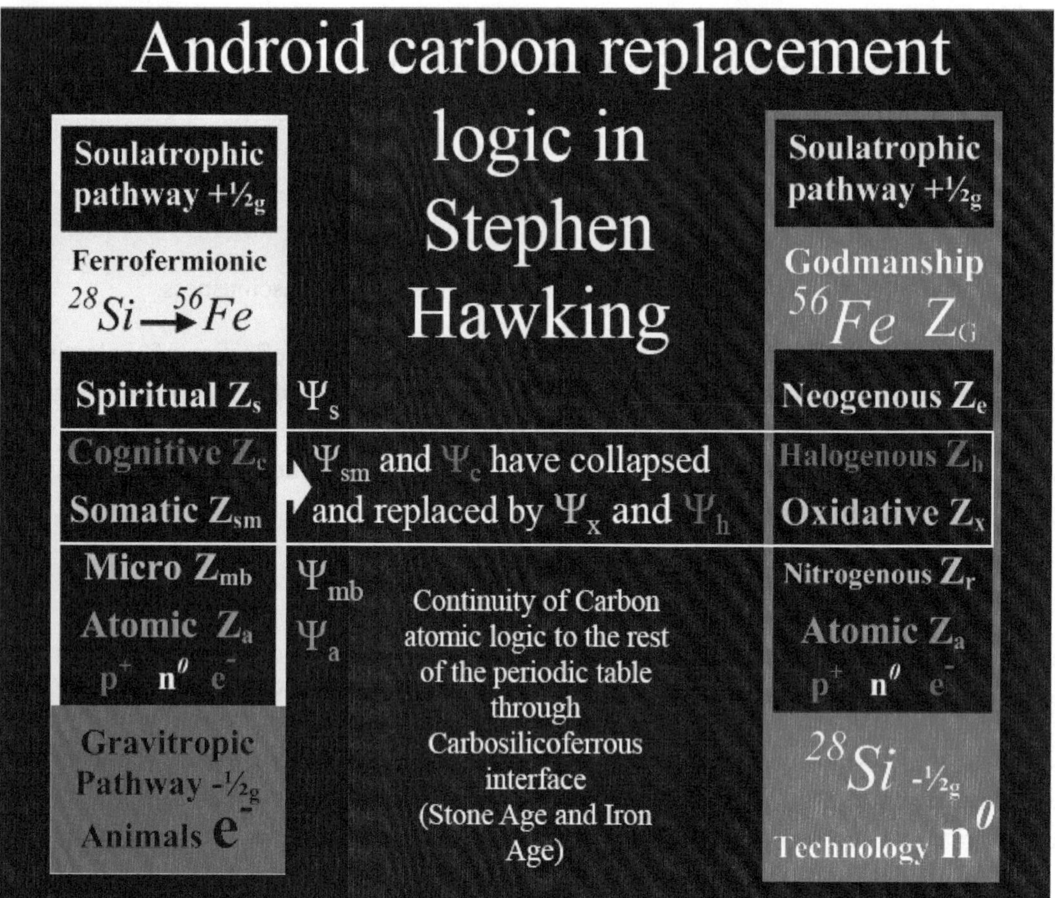

Consciousness is the equilibrium of statistical quantum uncertainty and Euclidian nucleonic certainty.

Wherever an observer can identify populations of data (amplitudes) and descriptive analytical statistics in a particular superpositioned physical system (population) they are in fact identifying the deterministic effects of quantum mechanics (electroweak) on that physical system.

Statistics and probability exist because of quantum mechanics and associated uncertainty relations and the 'measurement problem'. A mathematical mean average of a population is a superpositioned state property, a sample of data is superpositioned to give the average, and each individual result is a collapse of the wave function and therefore amplitude for that system.

Uncertainty is a totally unavoidable natural limit to conscious existence through 'making a measurement'. Any measure of uncertainty on any level of organisation is directly the result of Heisenberg's uncertainty (relations) principle on the physical system being measured (observed).

Any observer is also a quantum measurement device (superposition of amplitudes); this is the basis of 'the measurement problem'.

Euclidian calculus is effective because of the certainty of stable nuclear systems in nature. Measurability is more certain because of strong force nuclear inertial reference frame stability as compared to the quantum logic through a lack of predictability (determinism) and associated prevalence of the uncertainty relations and the need for probability investigations.

Living organisms indicate that gravity has a highly deterministic effect on quantum rules.

Quantum mechanics does not take into account gravitational determinism and its strong interaction symmetries in the wave functions of the physical superposition of bound baryonic states in atoms.

This leaves a gap in current quantum models as, although the mass of subatomic particles is minute, if any close body such as a star a hundred times the size of our Sun is present, then the gravitational deformational resultant would still be immense. This would place resistance and drag on the resultant of the wave function amplitudes and therefore the whole superposition of physical states and the probability distribution, inducing sub shell deformation (energy shifts) similar to the effect of the magnetic quantum number. This can be seen in living organisms as streamlining morphology for waveform particle components (streamlined sperm) and high pressure morphology in particle physical components (high pressure egg).

It is feasible that the four forces of nature are in fact four properties of the space time continuum and actually represent just one force namely the strong force, because energy and mass are equivalent as defined by relativistic logic and therefore consistent through such unity. The electro weak and gravitational components could be united if a term for gravity and a strong force link is added to the quantum wave functions, any particle should be defined by all four forces.

Big Bang theories support one original force of nature at the beginning of time. Quanta in general relativity appears to produce the states of matter from the balance (symmetrical equilibria) of electroweak to strong and gravitational forces producing atomic density.

Time dilation occurs as an observer contemplates the atomic world through to astrophysical particle inertial reference systems relativc to living inertial frames.

Mathematical logic begins to collapse when applied to resolvable living physical systems.

Mathematical logic collapses under the weight of resolvable independent variables for living bound state composite particle systems (measurement by sight is possible).

If physicists could fully resolve the atomic world they would find as much independent invariance as a biologist finds through sight measurement of living composite particles.

Physics can only use mathematics for hyper simplified purified systems such as atomic and astrophysical systems because both are difficult to fully observationally resolve into their associated independent variables.

Living systems can benefit from mathematical hyper simplification as well (independent variable reduction or amplitude reduction); an organism can be described as a particle/wave duality model. As such, mathematics can be applied to living systems as well as a powerful determinable tool.

Living carbon inertial reference systems (bound state composite particles such as cells and organisms) display stable continuous heterogeneous variation (due to covalence in p block elements, chirality and high electronegativity, octet rules) on all levels of organisation in all directions as opposed to most other elements which produce homogenous binary predictable organisation (for example most metals are stable solid inertial reference frames).

Most atoms except carbon, bond to just a hand full of other atoms often just one other.

Why is carbon the nuclear amplification pathway?

Most atoms in the periodic table bond with only one other atom not in keeping with complex amplification of physical bound state composite particles like carbon.Typically most compounds are physically very different or often symmetrically inverted, for example sodium and chlorine.

Energy level completion drives the atoms to lose or gain electrons to make such a stable completion of its valance shell, probably correcting for nuclear ferrous instability.

Carbon is very different as it possesses four valance electrons, four too many to complete its energy level and four too few to complete. This unique balanced middle ground explains carbons powerful and unique ability to share electrons and therefore blend and accumulate atomic probability and determinable nucleonic characteristics (superposition of states) limitlessly.

Carbon is also bosonic and as such can form bound state particles carrying the genetic force (strong interaction). Carbon is a highly gregarious Bose amplification composite particle.

Carbon can therefore bond with four other very different covalent atoms, molecules, cells or organisms covalently sharing electrons: where each of those atoms can be bonded to any number of other atoms, molecules, cells or organisms. The resultant bound state composite particles such as cells and organisms still display conservation and centralisation of carbon physical state properties (self symmetry).

Carbon is therefore a conglomerate atom producing massive bound state composite molecules up to cells, organisms, organismal groups etc etc.

By forming covalent hybrids with electrons from two atoms which are in a superposition of states so each electron may be in either atom then the overall energy of the two atoms is subsequently lower and so the probability of bonding and growth or accretion amplification is a fundamental property of solar conservation in carbon (bosonic condensation).

The covalent bond is also the critical property in carbon because unlike ionic or metal bonds, carbon can share electrons with other carbon atoms forming self symmetrical bound state composite particle systems.

This allows carbons nuclear and field properties to be homogenised through superposition to produce massively conserved bound state super structures; molecules, cells, organisms etc. Carbon can therefore amplify its superposition of states through molecular covalent bonding to produce bound state composite systems acting as one particle/wave duality.

Other properties in carbon support nuclear and field amplification (accretion by growth).

Carbon may also form double or even triple covalent sharing bonds with other carbon atoms which are even stronger than single bonds. The sum of many wave functions allow carbon to increase and homogenise its relative size but to maintain carbon atomic superposition and bound state composite characteristics at all levels of scale by bonding to other carbon atoms. Through iteratively thermodynamically conserving carbon wave functions, every cycle unified through composite bound state means producing massive particles.

The ionic bond may be strong but only allows one atom to be bonded to another. X ray patterns show ionic separation but prove that carbon covalent bonds produce hybridised blended fields. Carbon has no limit to the number of atoms which it can bond with, but its affinity for itself means that carbon superposition of states conserve and amplifies atomic characteristics in order to reduce overall energy through bound state composite particles (related to bosonic condensation).

Carbon can bond with other p block atoms to produce incalculable size and heterogeneous complexity whilst conserving and amplifying its own unique wave functions at its centre.

Carbon is a centrally distributed chiral atom; it can form four completely variable bonds which radiate in four directions from its nucleus, with up to four totally different atoms, cells and organims. This means that when carbon begins to produce covalent entanglement its nuclear determinism is centralised and amplified by bonding with other atoms. Hence its resultant lower energy superpositioned bound state composites always begin with and conserve carbon determinism; other bonding atoms except carbon simply exist to extend carbon logic. Other carbon atoms produce a sum of superpositions through constructive interference which are identical and therefore homogenise into giant atomic, cellular and organismal structures with conserved carbon determinism through bound state composition as particles, and therefore conserved structure and functional states.

Even more fractal properties in carbon support nuclear and field amplification (accretion by growth).

Because carbon produces the sum of the superpositions in the atoms which it bonds to and the frequency of its entanglements on planet Earth, it demonstrates non locality effects or memory (heredity).

Carbon is also a bosonic particle having an even number of nucleons, as such being a force carrying particle (possessing quantum memory or genetic heredity – non locality). Carbon therefore also acts through bound state means as it is a boson especially in larger systems (statistical/spin equilibria).

Carbon is subject to short lived entanglements and long term entanglements depending on the reduction in energy associated with the various forms (evolution by natural selection).

This is due to the Earth's unique iron core shielding, water content, solar support and overall thermal stability.

Increasing permutations and combinations of ever increasing superpositions of bound states. Hence a peptide bond has been around for billions of years it is energetically favourable in the Earth macroscopic environment, nitrogen and oxygen are most likely (nucleic elementary probability) to be found around carbon, together with hydrogen, sodium, chlorine, halogens, phosphorus and importantly iron.

The frequency of specific entanglements start off small, such as methane, but conserve its inseparable states because accretion growth of carbon bound composite particle states favours lower and lower energy levels, reducing the net energy imbalance on the nucleus (iron nuclear instability model).

The nucleus must therefore retain inherent inseparability for certain superpositions which act to reduce overall energy. Hence quantum logic is conservable and extendable to suggest that the probability of finding carbon with hydrogen or any of the other similar p block none metal atoms is very high on planet Earth. This is because similar quantum properties, in this case covalence and molecular weight, produces constructive interference and therefore allows amplification of physical bound states with carbon conserved initiation. This occurs by reducing the overall energy of carbon but mutually beneficial to reducing the energy in other p block nucleic elements. Nucleic elements are carbon associates which allow carbon amplification to occur by mutual covalence; part of the 26 element molecule and now +117 element technological molecules.

Most other atoms shield carbon atoms, carbon is often linked into giant strings (bound state composite polymers) homogenising and extending its nuclear and field properties as a fractal template.

Carbon amplification or growth follows a thermodynamic iterative fractal accumulator process known as 'good'.

Fractal geometry has always been associated with the limitless diversity of life although no direct use of such self symmetrical models has been proposed to model life directly.

Carbon entromorphology is a fractal self symmetrical amplification accretion process for describing life in terms of quantum recursive additive wave function extension.

Iterative cell cycling leading to increased superposition of composite bound states where morality is simply the thermodynamics of life, energy management.

Carbon growth and decay are cyclical exponential processes following an iterative logic where physical states in carbon are recursively resubmitted into another thermodynamic cycle. Conserving the physical superposition in the next cycle and reducing overall energy whilst fuelling the quantum carbon accumulation seen in carbon entromorphology.

The cycle is driven by quantum field instability which in turn is driven by nuclear instability based on an iron end point to evolution. This cycle continues until nuclear instability is brought to the level of iron. The cycle is driven by solar free energy injection as life isn't a closed system (4th law of thermodynamics).

- The iterative fractal accumulator (thermodynamically good) allows carbon to increase its superposition of states and grow through accretion.
- The iterative fractal thermoentropic cycle (thermodynamically evil) reduces carbons superposition of states inducing decay (anti accretion).

The amplified structural products of Pauli's exclusion principle in fermionic carbon entromorphology.

Pauli's exclusion principle describes how fermionic particles can never possess the same state at the same time, they must demonstrate physical complimentarity.

Physical states are typically divided into spin states having opposite characteristics +½ and -½.

Living organisms are bound state composite systems and must therefore obey Pauli's exclusion principle on all levels of organisation, so how does it manifest itself in living systems?

An example are the genitals in humans where the female is designated as the nucleus and hence positively charged producing a powerful potential well which draws men in.

A positive charge is often described as a hole which is precisely what a vagina is. A negative charge is the male penis which fits the hole (positive charge) perfectly. Both come together to cancel out their charge mass equivalents and to share physical states because of complimentarity. But electromagnetic quantum origins have produced structures (mass rich) which act as though they were electromagnetic fields, having complimentary electric field line morphology as the origin of the structure.

Another feature related to nucleonic equivalents and spin complementarity are the genitals and head in an organism is 'barrier penetration' and the 'coulomb barrier'. On the somatic level where conventional sexual intercourse takes place, the male must obtain 'consent' to allow him to break through the females 'coulomb barrier structures', by 'barrier penetration' strong interaction means. This occurs against barrier penetration logistics in atoms but again amplified.

The head and the skull are excellent structural consequences of the 'coulomb barrier' a kiss require the strong interaction to allow 'barrier penetration' past the 'coulomb barrier structures'.

The principle of cooperation, complimentarity, sharing or any concept in nature where two or more bodies share a physical state must show Pauli's complimentary physical properties in order to reach lower energy levels.

Life began at the Big Bang, and transforms energy through all four force permutations (a quantum gravity model).

Quantum gravity is beautifully elaborated by living systems. Since life evolved from the Big Bang, then its originating logic was quantum mechanics and nuclear physics conserved by physical entanglement and non locality (heredity); amplified through solar growth.

Heredity is beautifully linked to particle entanglement, non locality and action at distance in physics. This is an ideal model of heredity that physical interaction between organic particles results in conservation and filtration of unique probability distributions (the atomic model of evolution by natural selection).

This means that all atoms are NOT THE SAME, all atoms have in built memory systems (spinners ψ from the Dirac equation x,y,z,t), especially p block none metal elements involved in sharing covalent bonds and more specifically carbon and hydrogen. The hidden variables are probably the basis of atomic genes (quantum information).

Since unique probabilities are conserved and environmentally filtered for living particles by non locality means, life demonstrates a seamless transformation between quantum logic and general relativistic logic through special relativistic logic.

Hence life is an ideal quantum gravity process and unified model of the four forces, but conscious focus must be included and elaborated to complete the model. **'The observer' is simply insufficient for describing life in physics.**

Hereditary is the effect of non locality in carbon reference frames.

Since life demonstrates a translation from quantum logic and rules to general relativistic logic, and rules it demonstrates how quantum gravity manifests its effects.

As living particles increase in size through electroweak bonding means as bound state composite particles increase in mass, so the electromagnetic forces translate into gravitational forces seamlessly.

In living particles structures with sizable mass properties such as the hands, evolve through electromagnetic field line characteristics. Hence quantum rules can be visibly seen in living morphology and function because of conserved and filtered action at distance (heredity). A hand has evolved from electromagnetic field lines into a structure which possesses radiating properties when the hand is fully stretched open and convergent field lines when a fist is formed. This is a structurally mass determinant system which beautifully demonstrates a dipole system from positive to negative charge origins but with a mass rich structural system.

Electromagnetic quantum fields grow with mass adding particles such as other atoms and molecules, cells, organisms due to covalent bonding with other atoms as such electromagnetic determinism is diluted as living particles increase in mass and size giving way to gravitational determinism but with a conserved distinct link to atomic logic for living systems.

Carbon has the broadest distribution of energies next to hydrogen and has central determinism (a nucleating transforming amplifying element).

Carbon is unique because it possesses very broad energetic characteristics and versatility combined with covalent composite entanglement (superposition and hybridisation) and therefore powerful non locality (heredity) and environmental filtering (evolution by natural selection).

Carbon remains absolutely CENTRAL in ALL organic molecules and as such its quantum rules are multidirectional, extended and totally determinant as cause and effect on ALL levels of size and to ALL extents in ALL organic particle development. 1s, 2s and p sub shell probability distributions are pooled and amplified by covalent bonding hybridisation and environmentally filtered and conserved to produce all the diversity in life.

But carbon atomic logic acts as the deterministic basis (the soul of the system) for such amplification.

Carbon has a vast affinity to itself and hydrogen allowing an amplification of its quantum rules by multidirectional centred chiral covalent superposition and hybridisation.

Carbon, in association with water, has a very high affinity for a broad spectrum of electromagnetic radiation and therefore its mass structures increase as a result. Hence carbon gets progressively heavier and larger from immense solar exposure but through covalent bonding means. The extra mass is manifested by electroweak bound state composites with other atoms.

This can be seen in summer where plants dramatically increase their masses through fixing solar electromagnetic radiation in leaves, stems, roots and reproductive organs. Acquired mass by electroweak covalent atomic superposition through bound state composite generation.

Carbon remains absolutely central in its inertial reference frames allowing transformation and amplification of its atomic logic through non locality (heredity).

Quantum rules in carbon are determined by the potential on the nucleus and are therefore an electroweak composite quantum field expression of the potential on the nucleus.

In carbon entromorphology electroweak covalent bonding allows nuclear rules to be extended and environmentally filtered and conserved through molecular, cellular and organismal bonding by non locality action at distance phenomena (heredity).

Because electromagnetic and gravitational fields theoretically extend to infinity, ANY carbon atom in any cell in an organism integrated by constructive interference hybridised covalent bonding has a determinant effect extending infinitely to all other atoms, cells and organisms in the inertial reference frame of an organisms cell. Hence carbon characteristics can be conserved and amplified through solar water splitting (photosynthesis), to produce bound state compositions.

Living inertial reference frames form the distinct soulatrophic energy levels, atomic, cellular, somatic (radial), cognitive (bilateral) and spiritual (familial) reference frames (cells). Any living reference frame has a deterministic influence on all other reference frames in accordance with electromagnetic distribution rules, although shielding by other reference frames does independently limit and stabilise such an effect. Permittivity of free space is reduced by bonded shielding in carbon.

Superposition is extended to nuclear mass potentials in living inertial reference frames (cells) as the nuclear potential is deterministic in the expression of quantum electromagnetic fields. Therefore covalent bonding is a form of nuclear bond based on electroweak potentials and therefore considerably smaller energies are involved but constitute massive energies when considering the vastness of life on Earth.

Bound state composition allows carbon and its associate atoms, cells, organisms to act as one.

Covalent bonds are really a type of low energy nuclear bond.

Covalent bonding is a type of nuclear bond although the nuclei are separated by quantum rules (eccentric bonding). Conventional nuclear bonding involves enormous energies where new bound state composite atoms are formed by a more concentric type of bonding overlap (nucleogenesis).

Since these rules are an expression of the nuclear potential (nuclear mass), they constitute a fraction of the energy (mass) of the nucleus by comparison to direct nuclear bonding.

The four forces of nature appear to blend into the four permutations commonly identified as the individual forces but united into one force at the Big Bang through the homogenisation of energy and mass relativistic logic.

In carbon entromorphology the four forces are indistinguishably apparent, although the way energy is permeated through the four individual permutations (forces) varies with distance from the centre of gravity.

Nuclear energy from the strong force is liberated through the electroweak standard model in quantitised lots to do work outside the atom possibly to form a covalent bond; energy is therefore transformed through the four forces which are typically characterised by distance from the absolute centre of gravity in atomic reference frames.

Electroweak expression reduces the gravitational potential (mass/energy relativistic symbiosis) as the energy is extracted from the nuclear mass. The balance of mass and electroweak quantum expressed produces atomic density, and therefore determines the state of matter (solid, liquid and gas) in conjunction with larger bodies of atoms as a general relativistic equilibrium.

Carbon entromorphological amplification can be described in terms of LASER logic, Entromorphic Carbon Amplification by Solar Stimulated Emission ECASSE.

Since carbon is a bosonic particle, it is both a fermionic baryon and a bosonic meson bound state composite particle system. Its bosonic characteristics allow carbon to occupy closely associated physical states leading to solar stimulated amplification or living growth.

One solar photon releases a second photon in carbon physical states, this is seen in mitotic growth where one cell is solar stimulated to produce two cells (Bose condensation or amplification).

Stimulated emission occurs in Earth bound carbon through solar pumping into the Earth's atmosphere (carbon gain medium), where greenhouse effect's act as a high reflector and output coupler to produce an amplified coherent polarised state in carbon.

Carbon atoms, molecules, cells and organisms demonstrate solar stimulated coherence over billions of years resulting in a coherent stream out of the Earth's atmospheric amplification gain medium (Earth, seas and atmosphere act as amplification gain mediums, inducing coherent amplification in bosonic carbon).

Variation in coherence accounts for variation and diversity in carbon composite particles and is an indication of evolution of the amplified carbon stream. Evolution strengthens the emergent living ECASSE stream as it leads to overall reductions in living energies, by water splitting or photosynthesis by carbon associated excitation.

The living amplified carbon beam produced through solar stimulated emission amplification can be seen through barrier penetration; such as a space rocket punching its way out of the Earth's gravity into space. This is a coherent amplified carbon bosonic stream.

Carbon is a composite boson because of the relation between spin and statistics; it has an even number of fermions. Often its unique bosonic characteristics are only seen at large distances for large bound state composite carbon particle systems.

Two carbon atoms cannot absolutely occupy the same physical system but through electroweak bonding they produce superpositioned hybrids which do occupy the same state as unified molecules or cells or organisms and so on.

The bosonic characteristic of carbon and its entanglements produce bound state composite super particles such as cells and organisms through solar stimulated emission. In this way life can be thought of as a form of enormous LASER (Light Amplification by Stimulated Emission of Radiation).

Mathematical evidence in support of fractal nuclear and field amplification in carbon entromorphology (page 373 for references).

Carbon like all the elements is fractionally defined by hydrogen and as previously mentioned, hydrogen appears early on in the organic accretion model of life (Al Khalili, 2010). Hydrogen is composed of one proton (nuclear particle) and one electron (field particle), and as such acts as a simple polarised fundamental fractional dimension of all matter. When a proton and electron are compressed under enormous forces they form a neutron (neutral un-polarised particle); this completes the fermionic trinity (with the consequential emergence of a neutrino to complete the neutron). In attempting to support the theory of carbon entromorphology, a mathematical demonstration of the amplification of fundamental particle properties is required. The proton to electron ratio (μ) is one of the most fundamental constants of Nature and is hypothesised to result in amplified 'extreme cell types' (oogamy in size and motility) in the form of sperm and egg (electron spermatozoon and protonic oocyte cells) in this article, and represent the conserved fractal generating elements from hydrogen. It is also important to realise that this comparison is the fundamental basis of anisogamy, where gametes are physically different to each other and more precisely oogamy where sperm and egg are morphologically and kinetically very different. Anisogamy is a concept fundamental to all biological processes observed across the whole spectrum of life, from multi-cellular to single- celled organisms to plants and fungi. This article will therefore investigate the possibility that the proton to electron ratio is conserved and amplified to produce the 'anisogamy (sex) ratio' in biological systems. Evidence of this link and more precisely the point in gametogenesis where the ratio appears is likely to be a reflection of overall evolutionary hierarchy and gives sexual dimorphic gender a particle physics basis in nature, and an extended visual field for physicists to understand atoms more clearly. Sperm 'vs' are highly kinetic (small but fast kinetic field particle) and have lobed shaped streamlined morphology associated with high kinetic motile activity, and are therefore hypothesised to be amplified electron field products (male sexually dimorphic properties). They are also some of the smallest eukaryotic cells across mammalian class, and broadly across the vertebrates to invertebrates and plants. They are dwarfed by the size of the oocyte (the largest cell across mammalian class) which they bond to in the same way an electron bonds to (dwarfed by) a proton. This is hypothesised to be the basis of anisogamy and more precisely oogamy which is the polarised nature of sexual reproduction, involving the union or fusion of two dissimilar polarised gametes (differing either in size alone or in size and kinetic form by physical extremes). Egg 'vo' (oocyte) is a highly localised high pressure, large, but none motile low kinetic energy nuclear particle. It has a consequential spherical morphology and relatively fixed position, and is therefore hypothesised to be an amplified protonic nuclear product (female sexual dimorphic properties) in this assessment (see Figure 1).

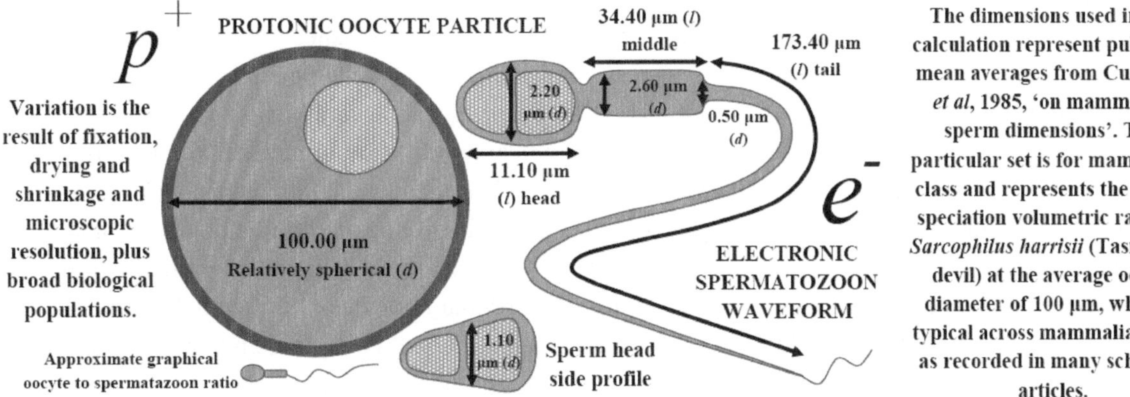

The dimensions used in this calculation represent published mean averages from Cummins *et al*, 1985, 'on mammalian sperm dimensions'. This particular set is for mammalian class and represents the closest speciation volumetric ratio for *Sarcophilus harrisii* (Tasmanian devil) at the average oocyte diameter of 100 μm, which is typical across mammalian class as recorded in many scholarly articles.

Figure 1 – Dimensions for *Sarcophilus harrisii* (Tasmanian devil).

To test the hypothesis that these particles fit the amplification theory of carbon entromorphology, the size ratio should remain relatively constant and can be calculated from physical data from sperm and egg. This suggests and tests the hypothesis of solar driven nuclear and field amplification (zygote to organism accretion growth) from atomic levels (μ) to classical cellular levels, μ_{ev} (volumetric) & μ_{em} (mass) ratios. [Figure 1 is not displayed to its true relative scale].

The protocol methodology for testing a conserved amplified constant μ through the theory of nuclear and field amplification; an anisogamy size ratio equivalent in comparison to the subatomic ratio μ.

This article struggled to locate a reliable citable sample of data for the mass of an oocyte or a sperm cell across broad mammalian class. Cummins *et al,* 1985 suggest that mass data for sperm and egg has been published through Van Duijn, 1975 and Van Duijn & Van Voorst, 1971 and for egg to sperm mass Hartman, 1929. Inspection of these citations suggests that in the case of Hartman, the data was produced with old methodology reflecting 1929 techniques and technology. The sperm data of Van Duijn is not in a form that can be used for this particular type of study, and does not cover a broad enough data population for mammalian class; which is a prerequisite of this particular study. The methodology and results for volumetric analysis however do correlate very well with this and other articles approach. Also many sex mass ratios are in actual fact a ratio of male to female body masses to cell numbers or linear size, and not individual cell masses for many associated scholarly articles. The data is also 40 years old when compared to clear linear dimensions determined by light microscopy by Cummins *et al*, 1985. It is therefore a prerequisite of this article to calculate volumes and masses of sperm from first principles and linear dimensions. This investigation demonstrates the difficulties in estimating mass quantities for cells and how visible linear dimensions by microscopy are a more precise and accurate measurement of cell size, although microscopy deviations can be as great as ±0.2μm (Curry *et al*, 1996). However investigation has resolved the orders of magnitude, supportable scalar values and ranges between sperm and oocyte masses from many website articles. The calculation used to make the mass ratio comparison μ_{em} (mass equivalent) and μ_{ev} (volume equivalent) to μ at 1836.15; proton to electron mass ratio, is presented in full later in this article. Sperm and oocyte dimensions have been identified in a broad range of scholarly articles. This investigation is extended across mammalian class for this article to identify broader trends in observations (through order and by many families) and a more thorough investigation which does include the human animal model. As previously mentioned the anisogamy (different motile gametes), oogamy (different in size and motility) and isogamy (physically similar gametes) of biological systems are common to the majority of plants and animals, protozoa and fungi. As such they are a true fundamental measure for all life on Earth; hence its comparison to μ is a very strong model for supporting particle amplification. In particular, Cummins *et al,* 1985 published a thorough detailed examination of linear sperm dimensions across mammalian class, and are the main source of dimensional data in this comparative investigation. This data suggests a very broad population not only within the mammalian class but also within speciation and individual organisms, for a cross section of families. Scholarly articles for Oocyte data varies less than sperm, although sperm when mature retain a stable phenotype, oocytes continue to progress in size through folliculogenesis and post ovulation. As a result data appertaining to oocyte will be used to calculate a broad distribution from primordial oocyte at <50μm in diameter (diploid 2n), primary oocyte at >50 μm to100 μm diameter (considered to be the broad mean average for mammalian class, diploid 2n to haploid with the start of meiosis n) and secondary oocyte at >100 to 150 μm diameter with meiosis temporarily interrupted, (Durinzi *et al*, 1995). Haploid fertilisation potential remains viable during folliculogenesis and the oocyte increases smoothly from oocytes <50μm diameter to a fully grown and differentiated ovum (zygote) or to termination during menstruation. A random selection of data (in particular where a complete data set has been published) for mammalian class has been identified for sperm, which will be used to calculate a broad size distribution comparison against the three values for oocyte for a variety of species, for volumetric ratio, plus inverted comparisons for mass ratios (fixed published oocyte mass, over a variety of linear estimated sperm masses). The calculation for oocyte is based on a spherical phenotype and can be calculated with simplicity through the formula for spherical volumes $4/3\pi r^3$ in μm^3. The sperm cell phenotype by comparison although considerably smaller than the oocyte is more difficult to calculate, with a head, mid-section and tail components calculation. The method employed in this article is supported by many scholarly articles where the head of the sperm is calculated as a flat elliptical cylinder (although its true shape has an equatorial apex), again the value for head depth has been estimated from microscopy to be approximately half the head width (again scholarly articles often lack this dimension plus microscopy can produce errors of ±0.2 μm0. The neck and mid piece are classed as cylindrical and the tail (flagellum) a very narrow cylinder made up of principle piece and end piece (conjoined for calculation in this investigation). Data for tail diameter is difficult to extract from scholarly articles plus the tail tappers slightly along its length. A value for the 9, 2 (tapering to a 9,0) arrangement in the flagellum of 0.5 μm (Rikmenspoel, 1984) will be used in all calculations estimated from scanning electron microscopy. The data for this value proves difficult to obtain through scholarly articles although this value is valid and supportable. The results of the calculation will be tabulated for sperm volume against the three volumetric sizes of oocyte (oogamy) through folliculogenesis. A summary of the results will be produced to identify the volumetric equivalent anisogamy ratio μ_{ev} to the classical atomic mass ratio μ and at what point along the pathway of gametogenesis the ratio appears. The discussion and conclusion will aim to demonstrate the huge variation in biological resolution and consequential difficulty in measurement although the mathematical approach is also highly valid and produces very useful supportable results. The conclusion aims to put the calculation in the context of the 'resolution problem' explaining how particle wave duality (equilibrium states) in living systems (which exist between the extreme phenotype, size and motility of sperm and egg) produces a variation in resolvable uncertainty in numerical observation. This is due to dominant (waves) complex visual observations on the level of an organism with a consequential collapse of mathematical modelling. This also demonstrates the way uncertainty is reduced for waves (none visual observation) on the level of the atom due to particle observation (e.g. quantum electrodynamics), which is a stable system and is therefore highly efficaciousness for mathematical treatment as a result. It is the aim of this article in conclusion to suggest (by testing and calculation) that all methods of observation be they through numerical mathematical treatment or through visual observation and language produce a better overall result for scientific investigation when employed in full. To finally, through such a broadening process across dimensional scale, science can integrate the scientific academic spectrum producing unification, but also to link the four forces of nature together through a living quantum gravity model. Life transcends the equilibrium of the four forces from the smallest resolvable world of the atom to the large scale universe with the conscious focus of an organism at its centre (observer). Also, that the particle wave duality logic of modern physics, relativity and thermodynamics should be the underpinning axiom of all the biological sciences.

The calculation of the anisogamy equivalent ratio comparison to the proton electron ratio μ by mass.

The approximate mass of ovum is 2.0×10^{-9} Kg (microgram order of magnitude) from Tschwenn, 2010 although details of the citation were not published. Ovum are the mature large ovulated egg cells in the order of up to and >150.0 μm, and sperm cell heads are of a similar size (5.0-10.0 μm in length) to red blood cells (erythrocytes) therefore since the sperm also has a mid section and a long and dense, although narrow, flagellum, these cells are of a similar volumetric size. Ovum can be as great as 1000.0 μm at which the ratio of the difference to sperm becomes extremely large as great as 50,000 times that of sperm, although the gamete is in an advanced stage of development. The erythrocyte does not possess a nucleus and the sperm possesses a highly condensed haploid (n) genome making its mass density large for the head. Many scholarly articles regarding natural orders of magnitude are highly consistent for these values, and for extremely large sperm such as *Tarsipes rostratus* (Honey possum) approximate values of $>1.1 \times 10^{-12}$ Kg (nanogram order of magnitude, an average human cell is 1 nanogram 1.0×10^{-12} kg); calculated to 1 d.p. to reflect the published values. The value of 2.0×10^{-9} kg reflects ovum at the highest typical diameter of 100.0 to>150.0μm (secondary oocyte). This value can be used to produce mass equivalent sperm and oocyte values for primary oocyte at 100.0 μm and primordial stages at 50.0 μm diameters to calculate the mass ratio of 1836.2 (1.d.). The following is a calculation that provides values for oocyte at the earlier stages by means of uniformity of content; 2.0×10^{-9} kg (1d.p.) mass ≈ 150 μm diameter:-

Volume of ovum (secondary oocyte) at 150.0 μm = $4/3\pi r^3 = 1.3 \times 3.1 \times (150.0/2.0)^3 =$ **1700156.3 μm^3 (1d.p.)**
Volume of ovum (primary oocyte) at 100.0 μm = $4/3\pi r^3 = 1.3 \times 3.1 \times (100.0/2.0)^3 =$ **503750.0 μm^3 (1d.p.)**
Volume of ovum (primordial oocyte) at 50.0 μm = $4/3\pi r^3 = 1.3 \times 3.1 \times (50.0/2.0)^3 =$ **62968.8 μm^3 (1d.p.)**

Mass to volume ratio for 150 μm = $2.0 \times 10^{-9}/1700156.3 = 1.2 \times 10^{-15}$ kg (1d.p.) to 1 μm^3 (density)

Mass proportion to 100.0 μm = $1.2 \times 10^{-15} \times 503750.0 =$ **6.0×10^{-10} kg (1d.p.)**
Mass proportion to 50.0 μm (diploid 2n heavy nucleus) = $1.2 \times 10^{-15} \times 62968.8 =$ **$>7.6 \times 10^{-11}$ kg (1d.p.)**

Very large sperm mass approximately 1.1×10^{-12} kg for 2.0×10^{-9} kg (1d.p.) oocyte (150 μm diameter). For ovum (>secondary oocyte) represent oocyte to sperm ratio of approximately1836.2 (1d.p.).

μ_{em} = **$(2.0 \times 10^{-9} / 1.1 \times 10^{-12})$ = 1818.2 (1d.p.)** comparison to 100% $\mu = (1818.2/1836.2) \times 100 =$ **99.0% (1d.p.)**

For ovum at the primary to secondary oocyte developmental stage, 6.0×10^{-10} kg is used in calculation for considerably smaller sperm, down to <1/10th of a nanogram (average cell at 1 nanogram, 10 μm diameter). These lower values for sperm and egg are still large to medium sized but are included to produce a broad distribution which will reflect the ratios throughout the process of gametogenesis. To sizes that are closer to that of most prokaryotes, which have a low density small genome and therefore significantly lower mass (to 10^{-15} kg and lower), to the order of 1/100th of an average human cell mass (diameter 10.0 μm to values of 0.1 μm): -

Sperm mass approximately 3.3×10^{-13} kg for 6.0×10^{-10} kg (1d.p.) oocyte (100 μm diameter). For sperm ratio at primary oocyte to secondary oocyte stage at 100.0 μm for a mass equivalent of 6.0×10^{-10} Kg (1d.p.): -
μ_{em} = **$(6.0 \times 10^{-10} / 3.3 \times 10^{-13})$ = 1818.2 (1d.p.)** comparison to 100% $\mu = (1818.2/1836.2) \times 100 =$ **99.0% (1d.p.)**

Sperm mass approximately 4.1×10^{-14} kg for 7.6×10^{-11} kg (1d.p.) oocyte (50 μm diameter). For sperm ratio to primordial to primary oocyte stage at 50.0 μm for a mass equivalent of 7.6×10^{-11} Kg (it is worth stating that primordial stage is diploid and therefore heavier due to a 2n genome): -
μ_{em} = **$(7.6 \times 10^{-11} / 4.1 \times 10^{-14})$ = 1853.7 (1d.p.)** comparison to 100% $\mu = (1853.7/1836.2) \times 100 =$ **101.0% (1d.p.)**

This linearity analysis demonstrates a selection of mass data for oocyte to sperm covering 3 stages of gametogenesis; the important conclusion is that the sperm range is realistic and reflective of linear dimensional ranges. Over a range of extremely large sperm to sperm down towards 1/100th the mass of the largest sperm (similar to the prokaryotes) these are realistic oocyte to sperm masses which reflect the population distributions in Cummins *et al*, 1985 which can be shown to approximate to 1836.2 (gradient of the line oocyte versus sperm mass), sperm populations can spread to over almost 2 orders of magnitude within mammalian class alone.

In accordance with Birkhead *et al*, 2009, the anisogamy mass ratio of 1000 (oocyte) to 1 (sperm) is a typical but small ratio and therefore values closer to 1836.2 are more likely. This published value is in the correct order of magnitude to the proton to electron mass ratio μ of 1836.2 (2d.p.). The calculated mass ratio for 3 stages of folliculogenesis fit within the population distribution of sperm mass sizes. This calculation is limited to broad averages without published deviations and statistical outliers, and cannot be considered as an absolutely validated source, although the resultant agreement is logical and highly supportive of this theory. However, the orders of magnitude for mass are appropriate for sperm and oocyte (in comparison to an average human cell of 10μm diameter at a mass of 1 nanogram, 1.0×10^{-12} Kg), and this calculation is therefore highly indicative of the true probability ratio of oocyte and sperm by mass.

This study however needs to extend the investigation to a broader taxonomic class in the animals and plants, although plant data for sperm and egg size are rare and limits the scope of this investigation. Cited data is also sporadic and variable in approach, but there is extremely well investigated and published data for linear dimensions which can be used to produce the ratio with visually determined confidence. A volumetric comparison is arguably as good as a mass ratio because sperm are almost completely absent of cytoplasm. Both sperm and oocyte contain a single set of haploid chromosomes identical in size (excluding polar bodies produced and eliminated by meiosis and the primordial stage which is diploid in folliculogenesis) hence this can be cancelled out of the equation. In a sense the comparison by volume or mass is made by the cytoplasm alone, which in sperm is almost completely reduced to contain mitochondria (mid piece), tail and acrosome which is formed by Golgi apparatus. By comparison it will be shown that a volumetric ratio is acceptably equivalent to a mass ratio for mammalian sperm and oocyte in terms of a size comparison through reliability and confidence in visible measurement; this visual argument reflects the resolution problem.

The calculation of the anisogamy equivalent ratio comparison to the proton electron ratio μ by volume.

This article will demonstrate a complimentary size comparison using very reliable volumetric linear data as well as the aforementioned mass data calculations. The variation of such data across mammalian class is immediately enormous (Cummins *et al*, 1985): academic sources suggest a human oocyte diameter (*d*) to range from 50.00 to 150.00 μm, hence a typical mode value being 100.00 μm for primary to secondary stages (Durinzi *et al*, 1995); relatively spherical dimensions. The spread of values for oocyte is supported by a vast array of scholarly articles and will produce a spectrum of ratios in this study to identify if and where the μ_{ev} value appears in the developmental gametogenesis pathway and especially to ensure that it occurs during the final stages.

Particles 'reliable precise observations' (none-visible)   Waves 'highly variable observations' (visible)

$$\frac{\text{Proton mass (mp)}}{\text{Electron mass (me)}} = 1836.15 = \mu \text{ (Physics)} \approx \frac{\text{Oocyte volume (vo)}}{\text{Sperm volume (vs)}} = 2196.43 = \mu_{ev} \text{ (Biology) (equivalent size) comparison}$$

The data in Table 1 is calculated using a method from first principles for *Sarcophilus harrisii* (Tasmanian devil), which produced the highest comparative ratio and inverse agreement μ_{ev} (119.62 & 83.60%) to the classical μ across mammalian class. The decision to select the *Sarcophilus harrisii* (the Tasmanian devil) was made because the result was calculated against an oocyte diameter of 100.00 μm, which is typically recorded as the broad average value for mammalian oocytes. *Homo sapiens* (Man) produced the highest correlation at the level of small dormant to primary oocyte (50.00-100.00 μm) of 114.50% and 87.34% agreement, very small sperm. *Tarsipes rostratus* (Honey possum) are enormous sperm cells and produced the highest correlation of 115.15% and 86.84% for the large secondary oocyte at 150.00μm diameter. An example of the calculation used: -

Sperm heads (elliptical cylinders) a = length 11.10 μm /2.00, b & d = width 2.20 μm /2.00, d = depth 1.10 μm: -

$\mathbf{vs}_{\text{head}} = \boldsymbol{\pi}\mathbf{abd} = 3.14 \times 1.10 \times 5.55 \times 1.10 = \mathbf{21.09\ \mu m^3}$ **(2d.p.)**

$\mathbf{vs}_{\text{middle}} = \boldsymbol{\pi}\mathbf{r^2l} = ((d/2.00)^2) \times \pi \times l = ((2.60/2.00)^2) \times 3.14 \times 34.40 = \mathbf{182.55\ \mu m^3}$ **(2d.p.)**

$\mathbf{vs}_{\text{tail}} = \boldsymbol{\pi}\mathbf{r^2}l = ((d/2.00)^2) \times \pi \times l = ((0.50/2.00)^2) \times 3.14 \times 173.40 = \mathbf{34.03\ \mu m^3}$ **(2d.p.)**

$\sum_{\mathbf{VS}} = \mathbf{vs}_{\text{head}} + \mathbf{vs}_{\text{middle}} + \mathbf{vs}_{\text{tail}} = 21.09 + 182.55 + 34.03 = \mathbf{237.67\ \mu m^3}$ **(2d.p.) (1)**

Radius of oocyte (spherical dimensions) at (d/2.00) r = (100.00/2.00) = **50.00μm (2d.p.)**

Oocyte volume **vo = 4.00/3.00**$\boldsymbol{\pi}\mathbf{r^3}$ = (4.00/3.00) x 3.14 x 50.00 x 50.00 x 50.00 = $\mathbf{522025.00\mu m^3}$ **(2d.p.) (2)**

$$\frac{\text{Proton mass}}{\text{Electron mass}} \approx \frac{\text{Oocyte volume}}{\text{Sperm volume}} = \frac{(2)}{(1)} = \frac{(vo)}{(vs)} = \frac{522025.00\mu m^3}{237.67\ \mu m^3} = 2196.43 \text{ (2d.p.)} = \mu_{ev} \text{ (equivalency)}$$

Final % comparison $100\mu_{ev}\mu^{-1} = \frac{(2196.43)}{1836.15} \times 100.00 =$ **119.62% (2d.p.)** **experimental agreement**

Name of mammal (Table 1)	Primordial stage (2n) <50 μm (% agreement)		Primary (average size n) 100 μm (% agreement)		Secondary (n) 150 μm (% agreement)		Oocyte diam. μm@1836.15	Sperm volume μm^3
Tarsipes rostratus (honey possum)	4.26	2344.69	34.12	293.09	115.15	86.84	143.07	833.26
Cavia porcellus (Guinea-pig)	8.71	1147.49	69.72	143.44	235.80	42.50	112.74	407.80
Dasyuroides byrnei (Kowari)	9.41	1062.85	75.27	132.86	254.04	39.36	109.90	377.71
Sarcophilus harrisii (Tasmanian devil)	14.95	668.76	119.62	83.60	403.73	24.77	94.17	237.66
Pteropus poliocephalus (Grey flying fox)	32.06	311.91	256.48	38.99	865.63	11.55	73.03	110.85
Mesocricetus auratus (Golden hamster)	34.84	287.06	278.69	35.88	940.57	10.63	71.04	102.02
Isodon macrourus (N. brown bandicoot)	42.73	234.01	341.87	29.25	1153.81	8.67	66.36	83.16
Mus musculus (House mouse)	47.31	211.38	378.47	26.42	1277.33	7.83	64.15	75.12
Myotis grisescens (Grey myotis)	48.72	205.25	389.77	25.66	1315.47	7.60	63.52	72.94
Capra hircus (Goat)	49.83	200.70	398.61	25.09	1345.32	7.43	63.05	71.32
Antrozous pallidus (Pallid bat)	58.36	171.34	466.90	21.42	1575.78	6.35	59.81	60.89
Alluropoda melanoleuca (Giant panda)	77.44	129.13	619.54	16.14	2090.95	4.78	54.50	45.89
Artibeus lituratus (Big fruit eating bat)	84.79	117.94	678.31	14.74	2289.31	4.37	52.81	41.91
Homo sapiens (Human animal model MAJ)	114.50	87.34	915.96	10.92	3091.37	3.23	47.80	31.04
Mean (best) agreement	*Homo sapiens*		*Sarcophilus harrisii*		*Tarsipes rostratus*		*Mean 76.85*	*Mean 182.26*
The data in Table 1 is calculated as: - (1st value) ratio (μ_{ev}/μ) × 100 and (2nd value) inverse ratio (μ/μ_{ev}) × 100							*Std dev 27.69*	*Std dev 223.42*

Important to note that 'only complete data sets' were selected at random from Cummins *et al*, 1985 with some of the data recorded to 1d.p. with most at 2d.p. and all volumetric calculations performed at this number of decimal places. The previous mass calculations were to 1d.p. again to reflect the published values for cell mass precision and accuracy. The standard deviations for mass and volume are also so large that they cover most of the entire primordial to secondary oocyte range for any of the sperm in the study. A resultant 119.62% (83.60%) correlated experimental and theoretical agreement (to 1836.15 as 100%) between the classical particle ratio and this comparative volumetric equivalent at 100 μm oocyte diameter, for amplified proton/electron (μ) particle to oocyte/sperm (μ_{ev}) particle ratios, was observed across mammalian class. Human sperm is some of the smallest of all across the entire living world, and the volume of approximately 30 μm^3 correlates with published data (Curry *et al*, 1996) and also includes ram and bull sperm of a similar size or less. The comparison between sperm and egg must be observed in the end processes of gametogenesis, when haploid genomes are generated by meiosis (spermatogenesis and folliculogenesis) for the ratio of 1836.15. Both sperm and oocyte must complete a barrier penetration in the form of ejaculation in sperm and ovulation in oocyte to become viable and therefore represent a logical point of absolute comparison around the final stages from primordial to secondary oocyte stage of gametogenesis. The enormous deviation in sperm population length for mammals is close to an entire power of ten, and so even human sperm at the lowest size have population deviations which extend closer to an oocyte of 100 μm diameter for the ratio of 1836.15. Table 1 demonstrates that the proton to electron equivalent ratio (anisogamy volumetric ratio also see Figure 2) is present in all the organisms listed over a broad size of mammals between the primordial and primary oocyte dimensions for larger mammals, and appears between the primary to secondary oocyte dimensions for smaller mammals, for the final part of gametogenesis.

The final conclusion in Cummins *et al*, 1985 states that a medium sized sperm of head mean average 7.68 μm and total length of 96.9 μm is the overall average for mammalian class, with an enormous range of lengths from 33.5 to 356.3 μm. The data in this article didn't include many Rodentia and Marsupialia order data as there was data missing. This group are medium sized sperm and published in copious amounts due to high levels of species and shifts the overall mean average of this study towards the expected average associated with an oocyte diameter close to 100 μm. This makes the overall model in this study very strongly supported for the typical sperm averages in association to a secondary mature oocyte at >100 μm. It is extremely unlikely that this high level of mass and volumetric ratio comparison, from the atomic level to the cellular level is purely the result of chance as the correlations are consistent across mammalian class and by extension to most animals and plants, although this still requires further investigation. The proton/electron ratio (particle/wave duality) is a fundamental 'universal physical constant' of Nature; it is therefore an excellent value for demonstrating amplified conserved particle phenomena. Rounding errors, variation and slight data skewing in measurement are richly demonstrated in this exercise regarding the 'resolution problem' for observed differences in measurement in biology and physics.

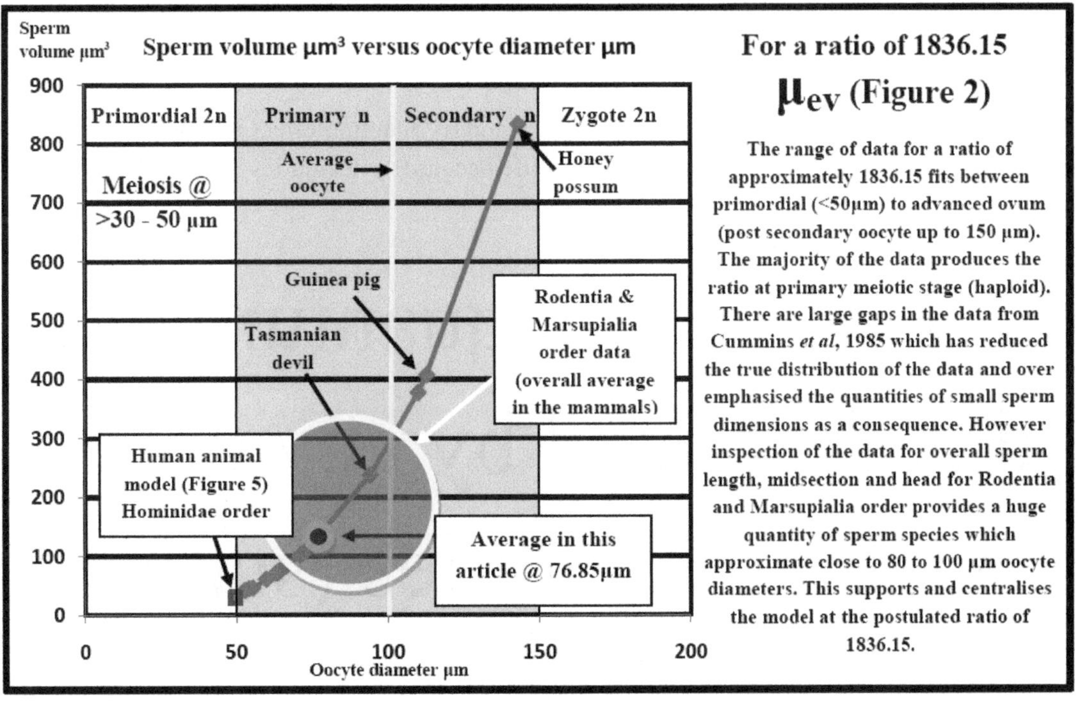

For a ratio of 1836.15

μ_{ev} (Figure 2)

The range of data for a ratio of approximately 1836.15 fits between primordial (<50μm) to advanced ovum (post secondary oocyte up to 150 μm). The majority of the data produces the ratio at primary meiotic stage (haploid). There are large gaps in the data from Cummins *et al*, 1985 which has reduced the true distribution of the data and over emphasised the quantities of small sperm dimensions as a consequence. However inspection of the data for overall sperm length, midsection and head for Rodentia and Marsupialia order provides a huge quantity of sperm species which approximate close to 80 to 100 μm oocyte diameters. This supports and centralises the model at the postulated ratio of 1836.15.

An important oversight with Schrödinger's cat (Janes's dog).

The thought experiment Schrödinger's cat is an excellent device for illustrating the problems physicists have with understanding living organisms in relation to their own theories. In this experiment Schrödinger uses a sealed box containing a cat, and a device demonstrating quantum uncertainty.

A radioactive source decays unpredictably inducing a signal in a detector which activates the release of a poison capable of killing the cat. In this experiment until the box is open the cat is said to be in a superposition of states, in this case either alive or dead. An external observer can only consider the wave function of the system where the cat is both alive and dead at the same time. When the box is opened the wave function collapses because the observer measures the cat as either alive or dead. This produces amplitude reading on the system.

The oversight is that Schrödinger has failed to realise that his thought experiment is over complicated. Like many other physicists he failed to consider the physics of the cat in his model because physicists find the physics of life difficult to comprehend. This is because the quantum effect is amplified in living organisms making accurate numerical measurement and therefore mathematical consideration extremely uncertain. Life is perfect proof of the unpredictability and nature of the quantum effect in nature. The cat is already a quantum system and there is no need for the radioactive source, poison etc. An external observer would still measure the cat as a superposition of states. Until the box is opened the cat could be anywhere in the box displaying unpredictable quantum behaviour. Perhaps when the box is opened the cat is lying in the middle of the box sleeping, perhaps it's in the corner licking its paws??? Living things are already perfect examples of quantum uncertainty.

Any observer perceives any living being as a probability distribution through living conscious wave functions until they make an amplitude reading. This is a constant effect when any observer considers any living organism; living things are only predictable through statistics and probability as proof of their quantum nature and always exist in a superposition of states until they produce measureable amplitude states. This experiment beautifully demonstrates the quantum nature of amplified atomic logic in carbon entromorphology. If Schrödinger had been asked 'do you think the cat is a quantum system?' He would have had to say 'yes!' Although his mind didn't rationalise with this logic when he created this thought experiment.

The cat is made of conserved composed atomic logic and its quantum effect is observed constantly; although a physicist doesn't consider physics with life because mathematics is severely compromised by living complexity.

Pip is already a quantum event and is described by any observer through probabilities (statistics).

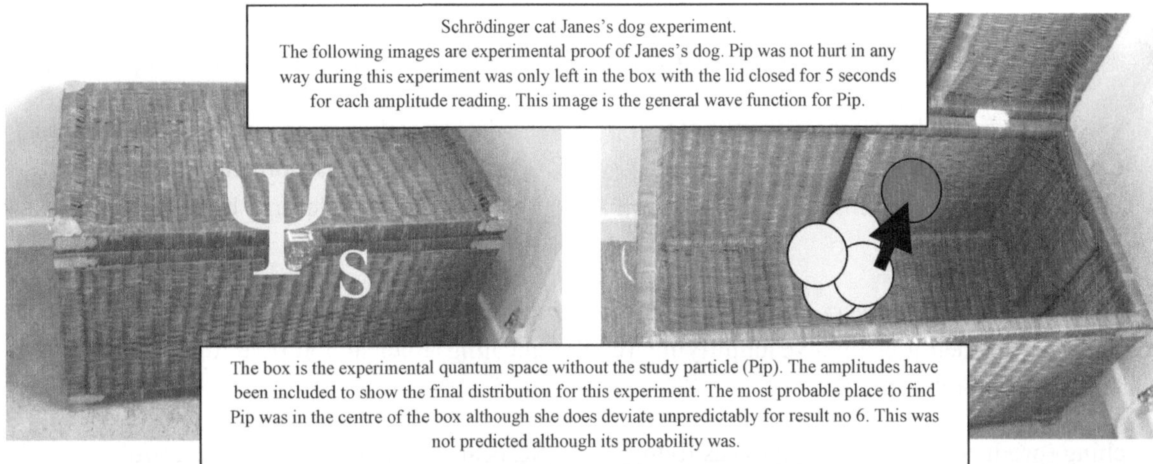
Schrödinger cat Janes's dog experiment.
The following images are experimental proof of Janes's dog. Pip was not hurt in any way during this experiment was only left in the box with the lid closed for 5 seconds for each amplitude reading. This image is the general wave function for Pip.
Ψs
The box is the experimental quantum space without the study particle (Pip). The amplitudes have been included to show the final distribution for this experiment. The most probable place to find Pip was in the centre of the box although she does deviate unpredictably for result no 6. This was not predicted although its probability was.

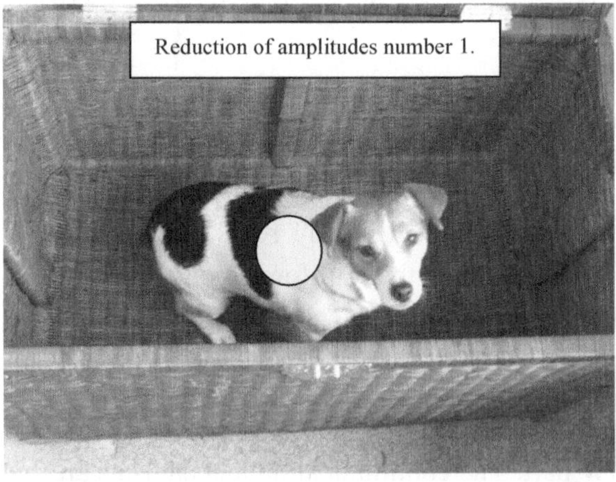
Reduction of amplitudes number 1.

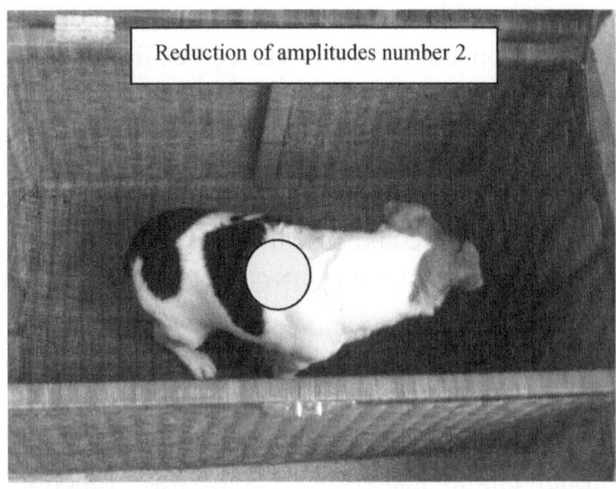
Reduction of amplitudes number 2.

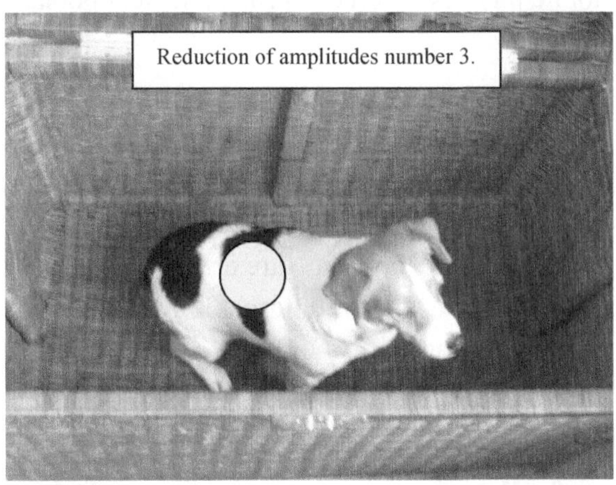
Reduction of amplitudes number 3.

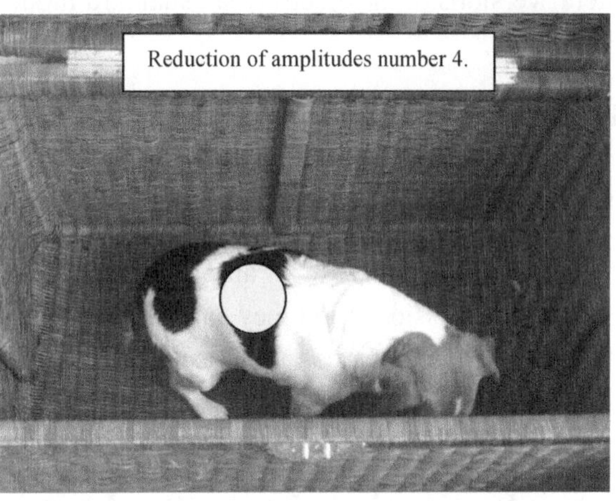
Reduction of amplitudes number 4.

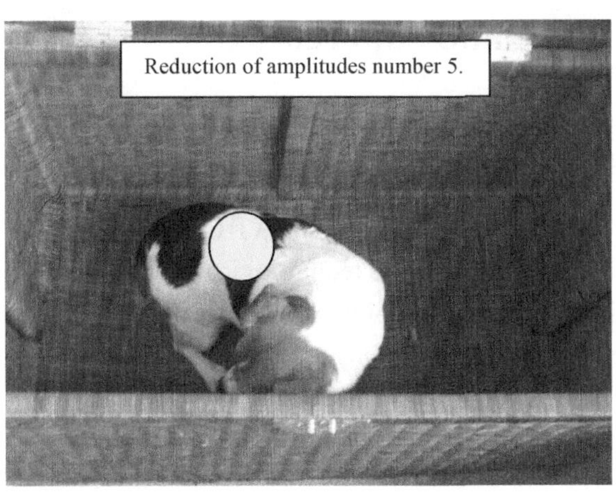
Reduction of amplitudes number 5.

Reduction of amplitudes number 6.

Carbon entromorphology is effortlessly demonstrated and explained.

ECASSE (Entromorphic Carbon Amplification by Solar Stimulated Emission) is the LASER based logic which explains how life (carbon) grows (amplifies) and evolves in carbon entromorphology through solar energy condensation and accretion amplification (Bose condensate logic – solar stimulated emission).

ECASSE is demonstrated very easily, an example is nuclear physics lecturers. They are living proof of this theory as their jobs involve amplifying (communicating) nuclear and field logic to a group of students (entromorphic atoms).

Teaching (produces in-phase cohesion as living ECASSE beams or graduates) is a beautiful illustration of ECASSE and the energy driving the process comes from the food the lecturers and students consume and this that comes from the Sun.

Hence the process of teaching (amplifying nuclear and field logic) is driven by solar stimulated emission this is a complete demonstration and proof in one.

ANY FORM of communication is proof and demonstration of ECASSE, the IT revolution demonstrates it EVERY SINGLE DAY worldwide. IT demonstrates the four force model of matter in the way energy flows and condenses into matter through carbon centred nuclear and field amplification.

The teacher is the nucleus in their lecture, the students their quantum field and their communication through talking and graphics is the wave function HOW part. The communication mechanisms are literal versions of the electroweak standard model in atomic physics (overhead projectors, text books etc).

This is the communication mechanism for wave functions in quantum mechanics. The wave functions are determined by the teacher and appear as the words, pictures and equations (spiritual familial nucleonics) used to communicate the theory.

The Internet is another super amplified nucleonic system driven by people who are driven by solar stimulated emission (food they eat).The art/science exhibition of carbon entromorphology fills a sports hall again demonstrating beautifully nuclear and field logic amplification.

'My cell' the exhibition is an absolute demonstration and proof of nuclear and field amplification.

Nuclear logic is basically the nature of all information systems; it is binary, chromosomal, genomic and supercoiled.

To an atomic physicist 'nuclear logic' is purely associated with the nucleus of an atom. In carbon entromorphology there are many levels of nuclear logic.

The atomic level made up of protons and neutrons, the cellular level made up of AT/CG DNA base pairs and the neurological level made of On/Off action potentials and the technological level containing all language, data, mathematics (black/white of text, silicon binary systems) all are organised by binary supercoiled organisation. Each level demonstrates conserved nuclear binary logic, this is the nature of any stored information system and the 'strong nuclear interaction'. This means DNA, words and pictures are all nuclear logic and demonstrate continuity on each level.

They are read and expressed into quantum and gravitational field logic through electroweak models.

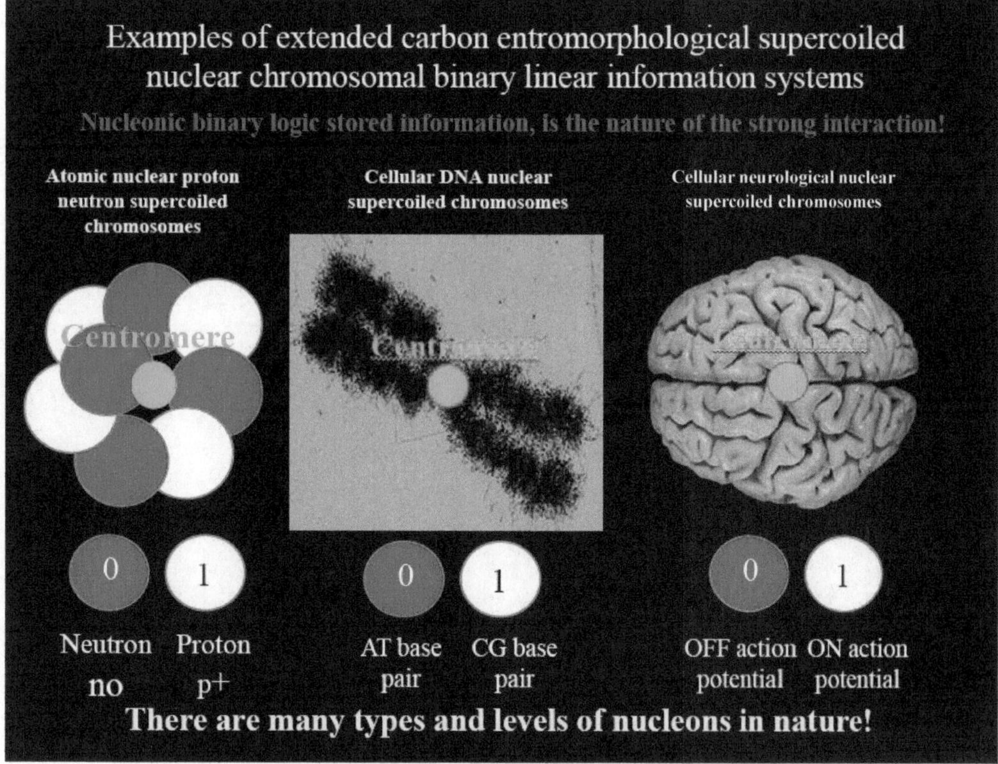

Genetics is the nuclear logic on the level of DNA and cells. Its logic can now be extended to neurological and atomic systems. Hence DNA contains genes which are fundamentally memories they collect into 'supercoiled' chromosomes and form multi chromosomal genomes. All the other nuclear logics must follow the same organisation. Hence supercoiled binary nuclear logic appears in written words through language.

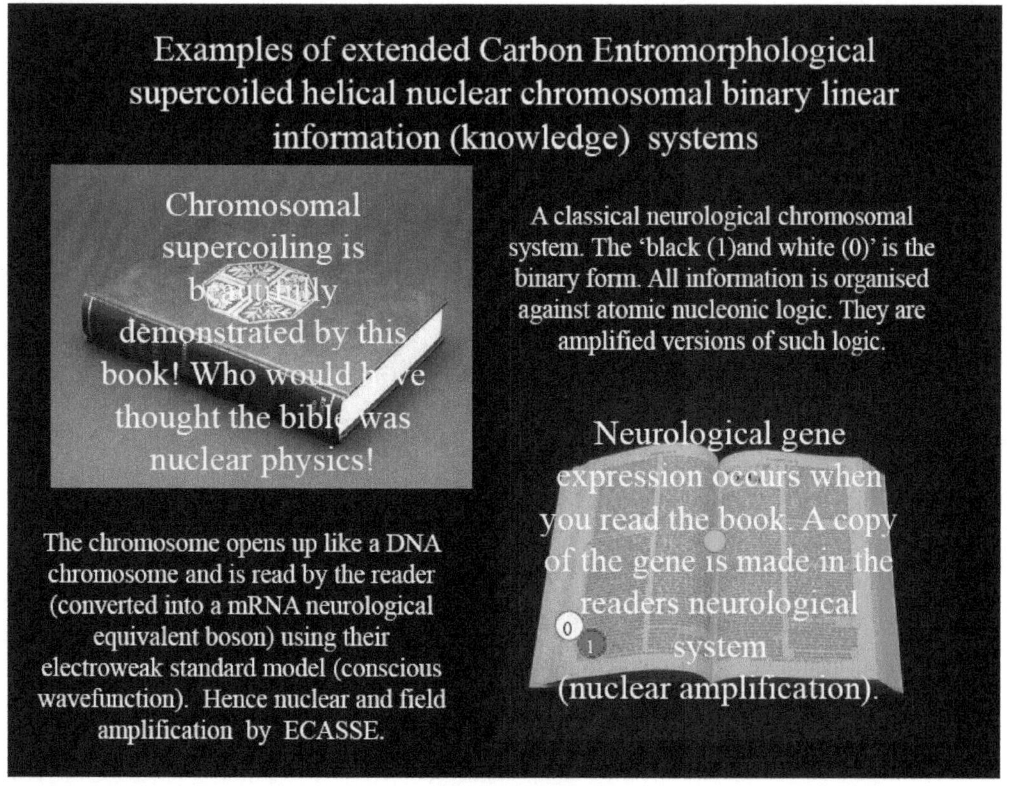

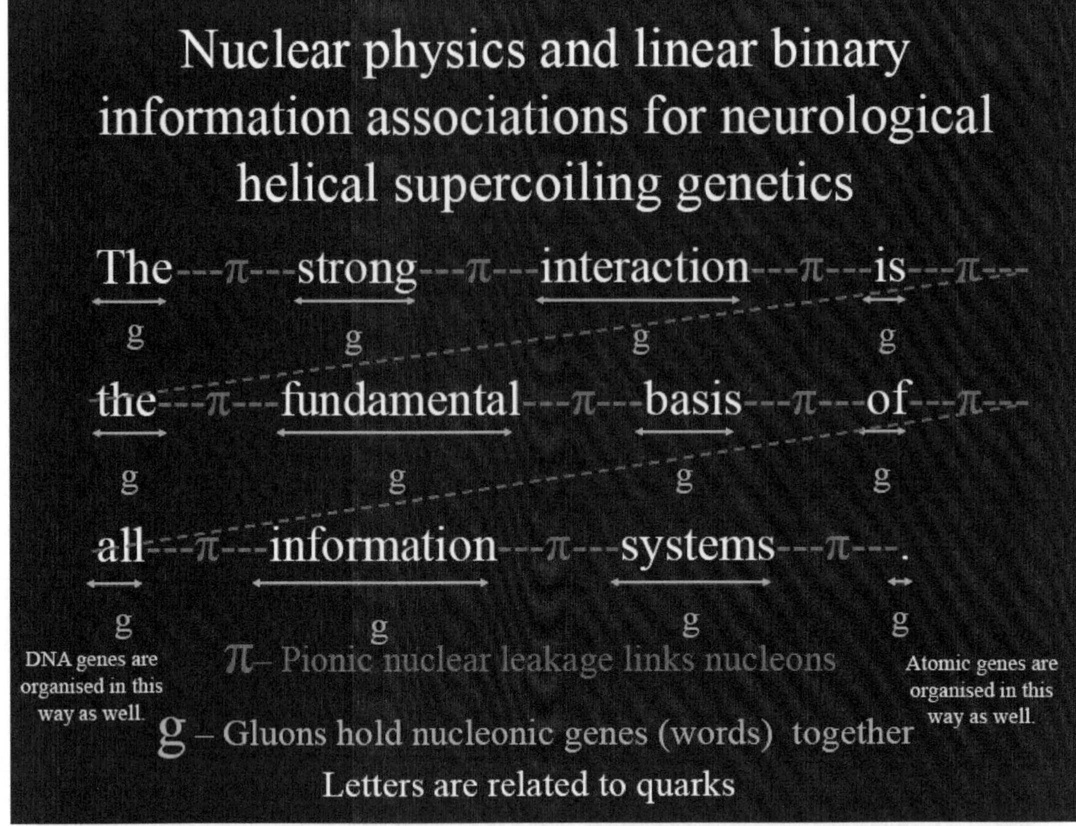

Nuclear physics and linear binary information associations for neurological helical supercoiling genetics.

One of Einstein's neurological genes: he created the physics neurological genomes for 'Special relativity and General relativity' and more.

A very famous neurological gene indeed!

Gluonic logic (g)

One of Einstein's neurological genes. He created the physics neurological genomes for

'Special relativity and General relativity',

'The Photoelectric effect' and 'Brownian motion'

Each one can be considered to be a neurological physics supercoiled helical genetic chromosome, part of the overall neurological genome known as Physics.

A example of a neurological helical supercoiled nuclear chromosome

Supercoiled chromosomes on the neurological level can be seen in language and writing. The black and white of written binary language is supercoiled into sentences where each word is a neurological gene and a piece of writing like this one is a neurological chromosome. An entire book could be considered to be a genome, its chapters chromosomes and words neurological genes. The logic can also be considered against atomic logic as well where the letters in each word are held together by gluonic logic and words are linked by pionic logic. Quantum expression occurs when it is read, and nuclear amplification occurs in the readers mind where the nuclear gene is copied (memory)!

Each one can be considered to be a neurological physics supercoiled helical genetic chromosome, part of the overall neurological genome known as 'physics'.

A slight diversion with a thought experiment regarding non locality and the nature of neurological genetic memory.

A simple thought experiment to demonstrate 'non locality' is a childhood memory. Remember your first day at school, the details. This is a non local experience; the details are the variables of memory. Einstein called them 'hidden variables' they are in actual fact the basis of genetics, on all levels of natural scale. The specific nature of the entanglements on your first day at school has made a permanent change in your brain. Non locality demonstrates how although atoms are thought of as being clones, they are in fact all very different depending on the energetic interactions they experience. Independent measurement of atoms resets their non local entanglements suggesting clones. Conventional scientific measurement is like hitting the re-set button.

Quantum conscious wave function focal, coordination and the nature of hand-eye-coordination.

- In carbon the wave functions for the six electrons in its two energy level n=1 (K shell) and n=2 (L shell), are defined by wave functions with angular and radial components.
- In carbon entromorphology amplified carbon atoms, entromorphic atoms such as humans also define two energy levels n=1 (K shell or torso) and n=2 (L shell or the limbs).
- When the solution to the wave functions for the head is made equal to those for the limb then we have a quantum model of hand-eye-coordination.

- The wave functions for eyes and hands produce the same amplitude solution: there lies the basis for hand eye-coordination. Wave functions from K shell (eye) electrons do influence the wave functions of higher energy levels such as the L shell limbs in entromorphic atoms.

$$\Psi_s \text{ (Total conscious wave function)} = \text{Radial (R) x Angular (Y)}$$

$$\Psi_{\text{left eye}} = R_{\text{left eye}} \times Y_{\text{left eye}} = r(x,y,z,t)$$

$$\Psi_{\text{right eye}} = R_{\text{right eye}} \times Y_{\text{right eye}} = r(x,y,z,t)$$

$$\Psi_{\text{total eyes}} = R_{\text{right hand}} \times Y_{\text{right hand}} = r(x,y,z,t) \quad \text{Hand eye coordination}$$

The following images compare living anatomy and physiology to quantum mechanics.

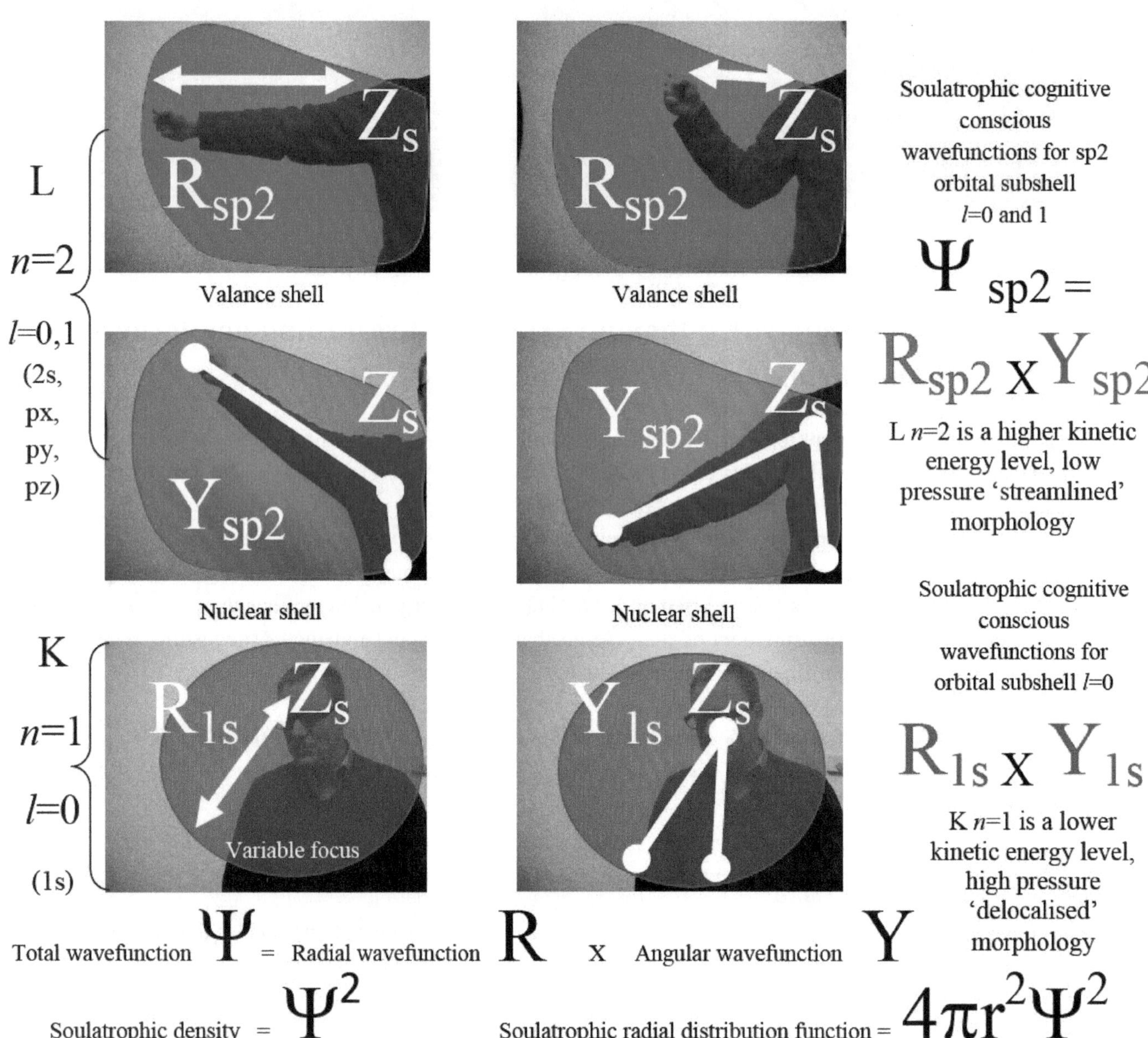

Another very powerful proof for carbon entromorphology is 'e' the naperian logarithm base.

Like its twin π, e is one of the immensely powerful although comparatively younger constants of nature. 'e' takes the value of 2.171828182845904523536..... like π it is also an irrational number and therefore immeasurable due to infinite complexity and it cannot be represented by a whole number fraction. Like π it is also transcendental and isn't the solution to conventional equations, making it one on its own like π and the soul of electroweak growth through natural amplification.

The wave functions of quantum mechanics decay exponentially with distance from the atomic nucleus and 'e' is the means of modelling these physical phenomena, hence 'e' is a property of the quantum effect. In carbon entromorphology the amplification logic of ECASSE produces a 99.8% agreement for the conserved value of the proton to electron ratio. 'e' is also used extensively throughout the whole of biology and its value is identical to that which is used to describe atomic wave functions and radioactive decay.

So for biological systems amplified up through growth from atomic origins, 'e' retains exactly the same value a 100% agreement between atomic logic and carbon entromorphology growth.'e' is typically used to model growth in living populations and in carbon entromorphology population growth is the very essence of ECASSE (Entromorphic Carbon Amplification by Solar Stimulated Emission). The conserved nature of 'e' from atomic origins to large biological entrochiraloctet cells is an immensely powerful proof of carbon entromorphology.

'e' is the fundamental constant associated with the electroweak standard model, and its conserved value with π for biological populations (quantum mechanics) proves amplified 'growth' states in Carbon Entromorphology have atomic origins . Spider's web, the coiled nature of many shellfish and ancient creatures coil up in this way and are among the huge numbers of examples of this constant.

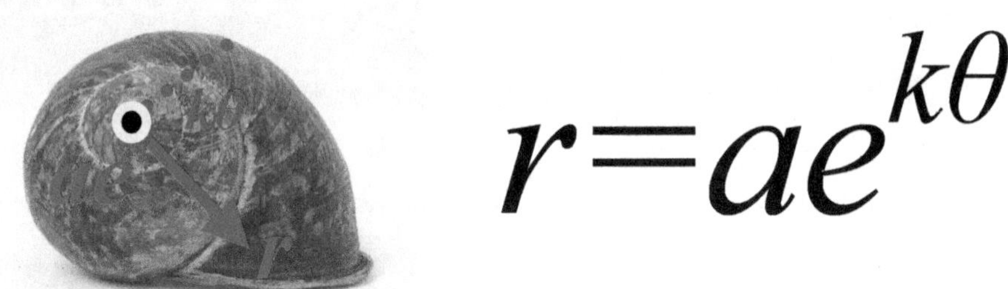

The Internet publication of the carbon entromorphology is a real time demonstration and proof of the theory.

Incredibly the mass media launch through the Internet is a perfect demonstration and proof of carbon entromorphology (from Mark Andrew Janes neurological genome) and ECASSE (Entromorphic Carbon Amplification by Solar Stimulated Emission).

The nuclear and field logic is provided by Mark Andrew Janes – carbon chiral centralised amplification by in-phase coherence (understanding in people by carbon constructive interference), in the form of graphical words, pictures and equations amplified through the Internet and other publications (neurological chromosomes and associated genes).

The Internet will amplify these by making many copies for printed versions; which is a version of the electroweak standard model or communication interface (from Mark Andrew Janes neurological chromosomes and neurogenes). This process is driven by the food that Mark Andrew Janes eats, hence 'Solar Stimulated Emission' or 'Bose condensate'. The more people that hear about this theory (and resonate with it through chiral centred in-phase coherence or constructive interference) the greater the amplified size of 'My cell' and the more concise the demonstration and proof of the theory.

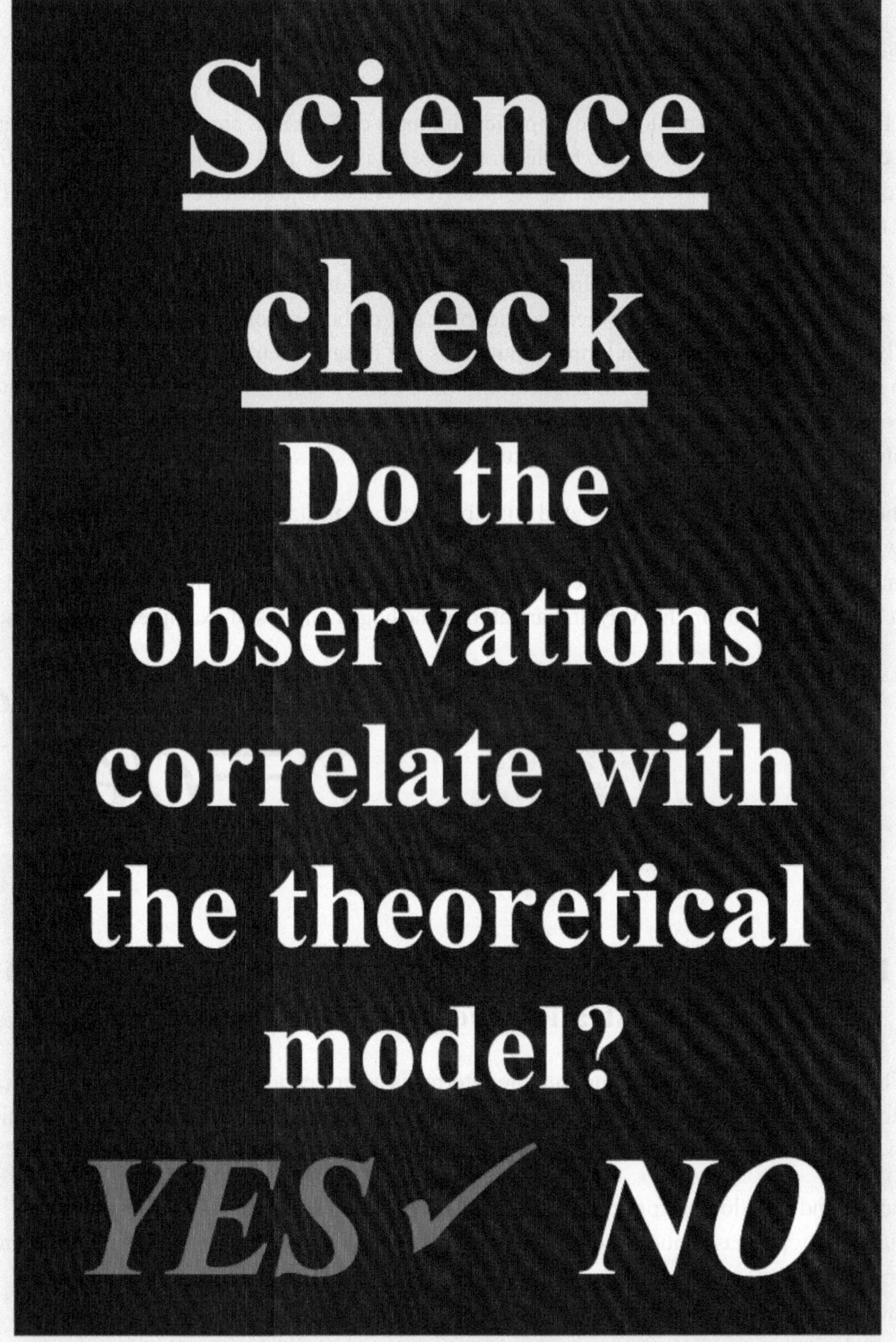

PART FOUR

The 'quantum gender' model

Neurochromosome: the 'quantum gender' model, the battle of the sexes is finally over; the logical power of the fermionic gender trinity.

The 'quantum gender model' is one of the oldest and most powerful concepts in the whole of carbon entromorphology. Current theories of 'sexual dimorphism' make it very clear that humanity has absolutely no real understanding of precisely what gender really is and what its origins are.

The complete 'Gender Model' is enormous and like anything in carbon entromorphology this book only contains a tiny fraction of the whole story. The three 'trinity' based particle pathways sub divide into the classical 'trinity' three particle gender models. So the animals (electronic), plants (protonic) and technology (neutronic) reflect the three fermionic particles but each pathway divides into electronic male (e-), protonic female (p+) and neutronic homosexual (n^0).

The logical statement suggests that any level of living matter has a fractional dimension which divides into electron, proton and neutron. Three particles produce a very simple axiomic logic for describing life and more specifically, the basis and physics of gender.

In carbon entromorphology life is described as a 'soulatrophic pathway' from the word the 'soul'. It runs all the way back to the Big Bang and is defined by specific levels such as atomic, microbial, somatic, cognitive and spiritual. Mathematically the pathway is known as a 'fractional dimension D'; a fractal geometry. Each level is self-symmetrical to the other levels and is driven by iterative cell cycling (ECASSE) and is made up of the fraction of the lower levels and acts as a fraction of the levels above it.

The gender model.

There are three particle evolutionary pathways based on the fermionic trinity; the electron, proton and neutron. The fermionic trinity is the basis of the gender model in carbon entromorphology where the electron is a male component the proton a female component (hence polarised gender) and the neutron is a more evolved homosexual none polarised component.

Gender can be shown to be a split into carbons two energy levels as well; the K (n=1) shell and nucleus or 'nuclear shell' classically female (gatherer, home or nest linked) the inner-space spingrav potential, and the L (n=2) shell or valance shell classically male (hunter).

This is the valance shell with outer-space spin potential. Firstly the anatomical propertied must be considered, where carbon is composed of two quantum shells with links to field and nuclear potentials. This inner shell contains a nucleus with 6 protons and 6 neutrons having a combined mass of 12 (even number) and acting as a bosonic particle as a result. The first principle quantum number n=1 or the K shell (Barkla convention) houses and shields the nucleus. This shell is represented by a spherical probability distribution or 1s orbital (sub shell), and there are two electrons housed in this energy level having opposite spin configurations, gender can fall into this organisation.

The K shell is a stable region in carbon as it has two quantum stable electrons. The second energy level in carbon the valance shell or bonding shell has only four electrons. Its ground state has two electrons in 2s spherical morphological probability distributions around the K, n=1 nuclear shell.

There are also three p type streamlined probability distributions the px, py and pz. Bonding in the valance shell may require orbital hybridisation, such as sp2 form. An electron is promoted to the pz orbital and combination of s and p type orbitals produce bonding this is hypothesised as the cytokinetic moment of cell division. **The diagrams will make this considerably clearer!**

Fractional Dimensions & Soulatrophicity

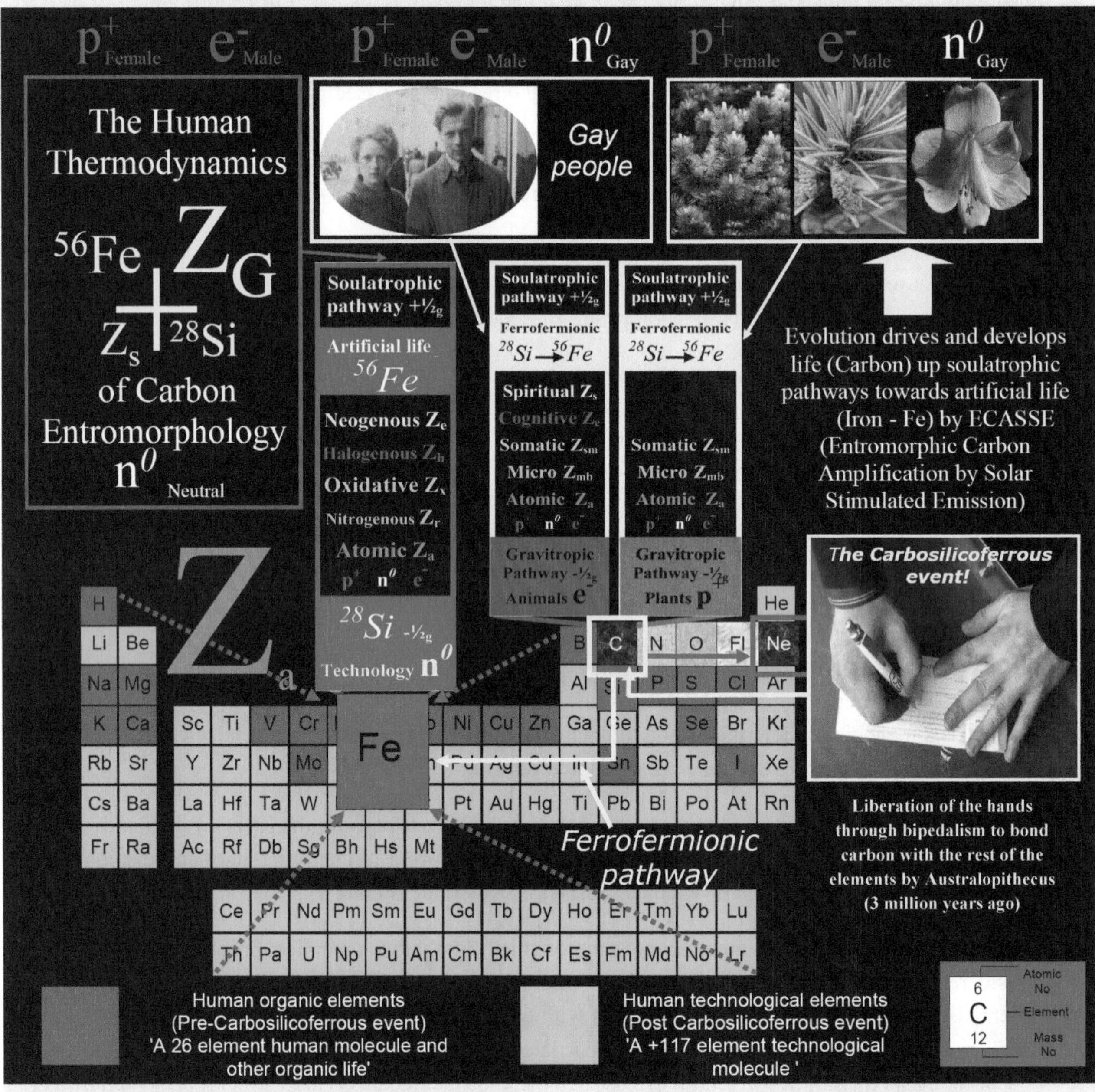

Life and its place on the periodic table (self-awareness through organoaccretion)

Fermionic trinity

A quantum gender model

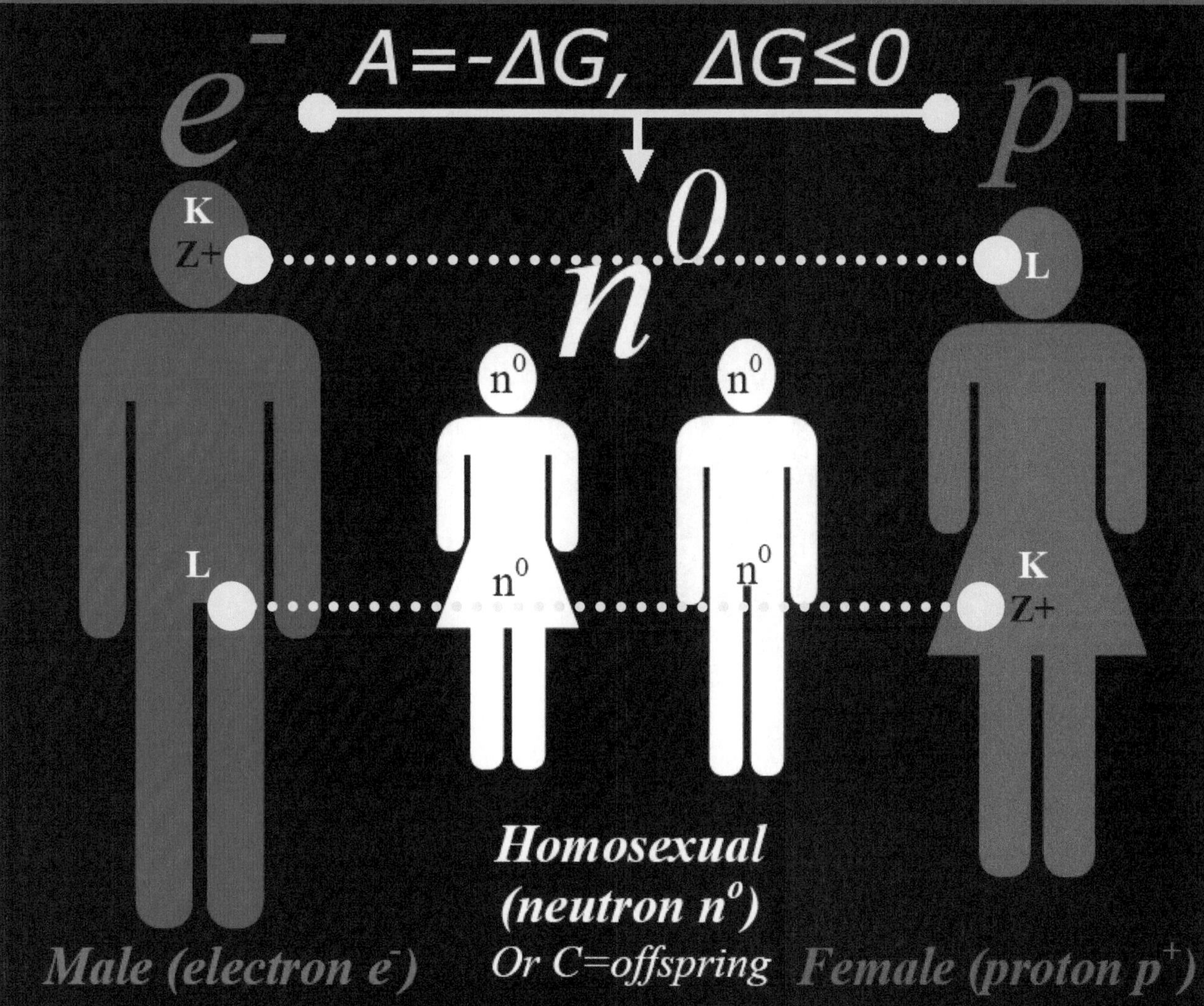

A hyper reductionist model of gender

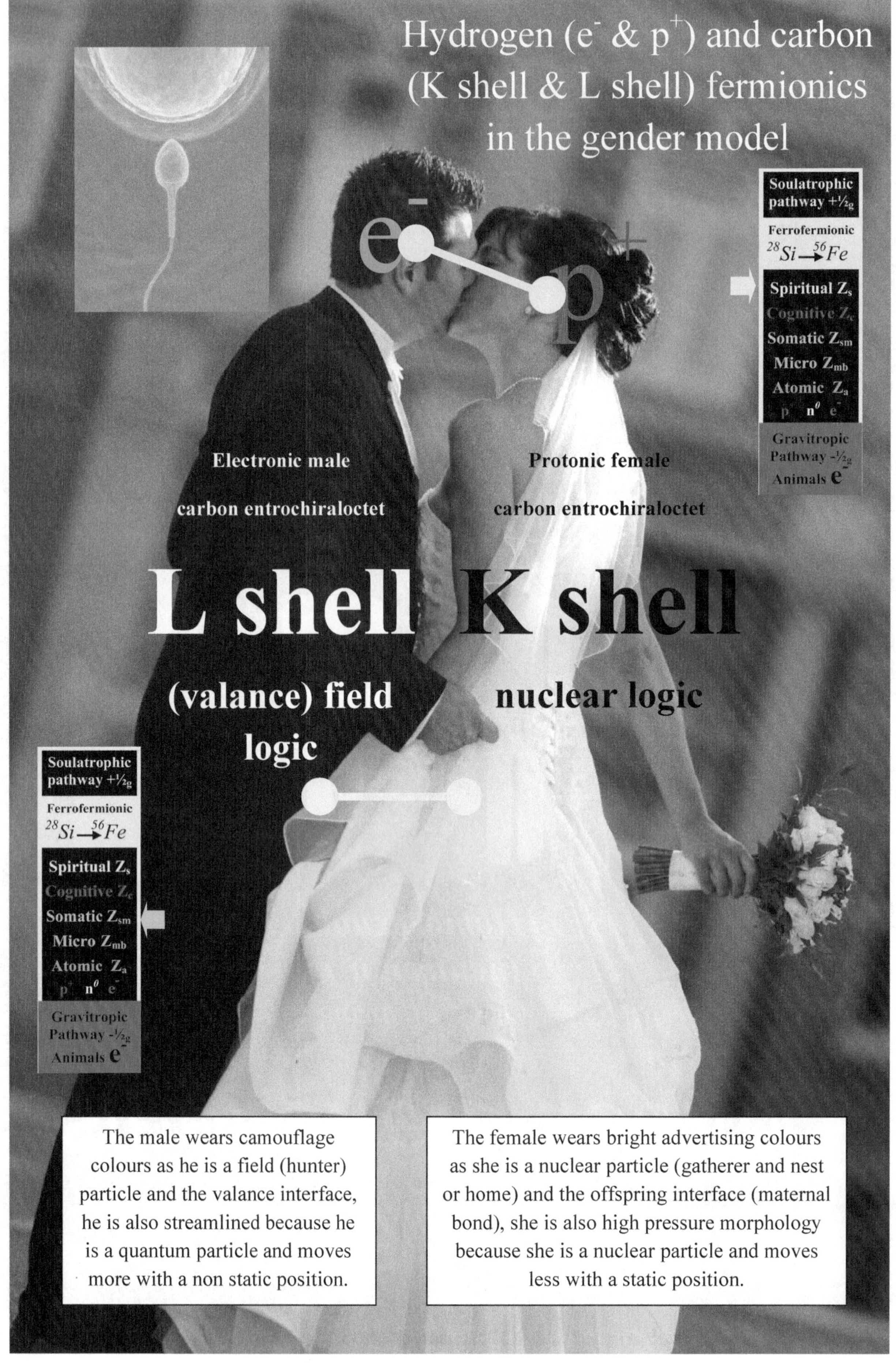
Hydrogen (e^- & p^+) and carbon (K shell & L shell) fermionics in the gender model
e-
p+
Soulatrophic pathway $+\frac{1}{2}_g$
Ferrofermionic
$^{28}Si \rightarrow ^{56}Fe$
Spiritual Z_s
Cognitive Z_c
Somatic Z_{sm}
Micro Z_{mb}
Atomic Z_a
p^+ n^0 e^-
Gravitropic Pathway $-\frac{1}{2}_g$
Animals e^-
Electronic male
carbon entrochiraloctet
Protonic female
carbon entrochiraloctet
L shell
K shell
(valance) field logic
nuclear logic
Soulatrophic pathway $+\frac{1}{2}_g$
Ferrofermionic
$^{28}Si \rightarrow ^{56}Fe$
Spiritual Z_s
Cognitive Z_c
Somatic Z_{sm}
Micro Z_{mb}
Atomic Z_a
p^+ n^0 e^-
Gravitropic Pathway $-\frac{1}{2}_g$
Animals e^-
The male wears camouflage colours as he is a field (hunter) particle and the valance interface, he is also streamlined because he is a quantum particle and moves more with a non static position.
The female wears bright advertising colours as she is a nuclear particle (gatherer and nest or home) and the offspring interface (maternal bond), she is also high pressure morphology because she is a nuclear particle and moves less with a static position.

Each level of amplification in carbon has a carbon template or entrochiraloctet associated with it, first as ground state then in accordance with the octet rule with a valance shell completed. These are hypothesised to be the three fundamental morphologies of life based on carbon quantum mechanics. The probability distributions in carbon, which typify so many classical living morphologies, can be seen in all living structures through the conservation amplification effect.

Flowers show 2s orbital morphology, or a human head, which personifies 1s orbital anatomical morphology. A leaf is an excellent piece of evidence for p type streamlined orbital morphology; a muscle is also typically p type streamlined structure. Bone is hypothesised to have formed from a 1s orbital component forming hybridised morphology such as musculoskeletal systems in higher animals. This gives the biologist a basis for understanding phenotype in living organisms.

There are two components potentials driving carbons behaviour, the first the quantum mechanics follows the octet rule where carbon is four electrons short of a full valance L shell (n=2). This is a very old quantum model (cube model) of carbon behaviour and is well accepted as carbons quantum potential. Carbon remains unstable and reactive (electrophilic) until its valance shell (L shell, n=2) contains eight electrons (octet rule). So the basis of carbon entromorphological processes starts with carbon on the periodic table and its tendency to pursue a noble gas electronic configuration identical to neon and therefore quantum stability.

In order to achieve this carbon must firstly acquire one electron giving it a nitrogenous configuration followed by an oxidative configuration and then a halogenous and finally neogenous configuration. The valance shell (L shell, n=2) in carbon cells is also called the metabolic organelles (linking classical quantum models to classical biological cells), with the missing L shell bonding electrons termed as the 'valance bonded metabolic organelles'; this is conserved at higher levels of carbon cellular amplification. It is a model of living motivation and behaviour.

Soulatrophic ladders – The basic nature of communication in carbon entrochiraloctet cells. Classically described on the somatic level with sexual reproduction the effect occurs on all soulatrophic levels. When two adult humans bond on spiritual levels first by conversational bonding, then a kiss, indicating cognitive bonding then onto somatic levels in intercourse and finally through microbial (sperm and egg) and finally to atomic through fertilisation. All prokaryotic, eukaryotic cells are classed as microbial although eukaryotic cells aren't normally included in biology (animal cells). Soulatrophic energy levels can be thought of in an organism as levels of consciousness.

Gender communication is proliferated by unsaturated C=C double bonds between soulatrophic amplification events (pathways). Sexual dimorphism is seen on all soulatrophic energy levels. The nuclear source can emit bosonic particles (sperm and egg) between quantum energy levels (L and K shell). The direction of transmission occurs when the L shell component experiences a 'quantum jump' to the K shell level and releases photonic (bosonic) genomic spectra through orgasm. On the DNA level this is sperm and egg, on the cognitive level it is basic verbal communication and hearing or a kiss. On the spiritual level it is complex emotional and logical communication.

When two complementary soulatrophic pathways capable of constructive interference merge, the L shell components allow access to the K shell (body) and finally as in classical sexual intercourse access to the nucleus (oocyte) on that soulatrophic level, the coulomb barrier must open by consent.

The dimorphic potential alternates along soulatrophic pathways and form new carbon entrochiraloctet cells with L and K and nuclear fields plus metabolic instability. Carbon is fundamentally comprised of fermionic particles but acts as a bosonic communicator due to even numbers of nucleonic particles.

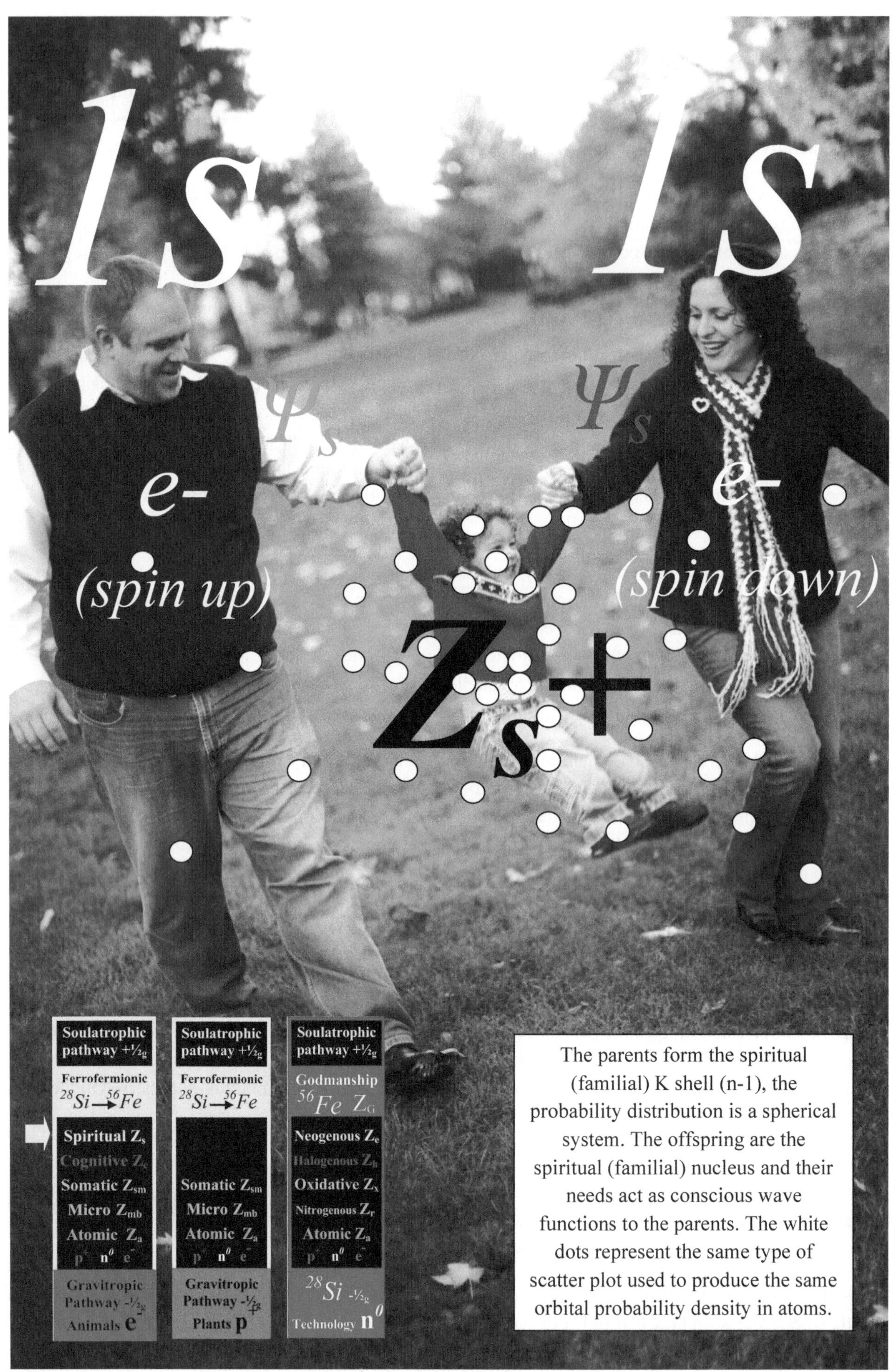

The parents form the spiritual (familial) K shell (n-1), the probability distribution is a spherical system. The offspring are the spiritual (familial) nucleus and their needs act as conscious wave functions to the parents. The white dots represent the same type of scatter plot used to produce the same orbital probability density in atoms.

Soulatrophic amplification – The fundamental basis of carbon entromorphology is the amplification of carbon by solar accumulative shielding ECASSE. This occurs due to quantum instability in carbons entrochiraloctet fields (metabolic organelles and valance bonded forms). The instability potential occurs in the quantum fields through neobosonic pathways and through a nuclear pathway the ferrofermionic systems leading to iron nuclear stability. The amplification allows carbon to fix solar electromagnetic energy in its nucleus as condensed matter ECASSE through gender bonds.

The gender nuclei increase in size and become amplified as does their K and L shell expression fields. Soulatrophic energy levels can be thought of in an organism as levels of consciousness.

Soulatrophic gender pathways – Amplification models of carbon and its metabolic organelles, which form organisms comprised of a unique soulatrophic hierarchy. A phototropic pathway points towards the Sun and solar acquisition, increased enthalpy and Gibbs free energy. The pathway is classically described as 'good' in metaphysical terms but manifests itself precisely against the second law of thermodynamics. The Gibbs equation demonstrates how carbon based energy systems through gender can maximise energy storage and therefore evolutionary hierarchy. Energy conservation in carbon entrochiraloctet cells is therefore a fundamental of good and phototropism. The energy level of an organism based on its ferrofermionic pathway position is determined by the pursuit of solar energy and stability in carbon entrochiraloctet cells and bonds though gender. The Sun or other solar body determines the energy level on the phototropic pathway and fixes energy in offspring. In theory this unbroken line of ancestry runs back to the Big Bang 13.7 billion years ago, and locks energy into life through the gender model.

Soulatrophic pathways form a connective pathway between the smallest particles to the largest celestial bodies (with fractional dimension – fractal logic). They bridge the gap between classical nuclear, quantum mechanics and general relativistic mechanics through soulatrophic fractional dimensions. Quantum fields are conserved and amplified by the gender accumulator effect.

Soulatrophic pathway $+\frac{1}{2}_g$

Ferrofermionic $^{28}Si \rightarrow {}^{56}Fe$

Spiritual Z_s

Cognitive Z_c

Somatic Z_{sm}

Micro Z_{mb}

Atomic Z_a

p^+ n^0 e^-

Gravitropic Pathway $-\frac{1}{2}_g$

Animals e^-

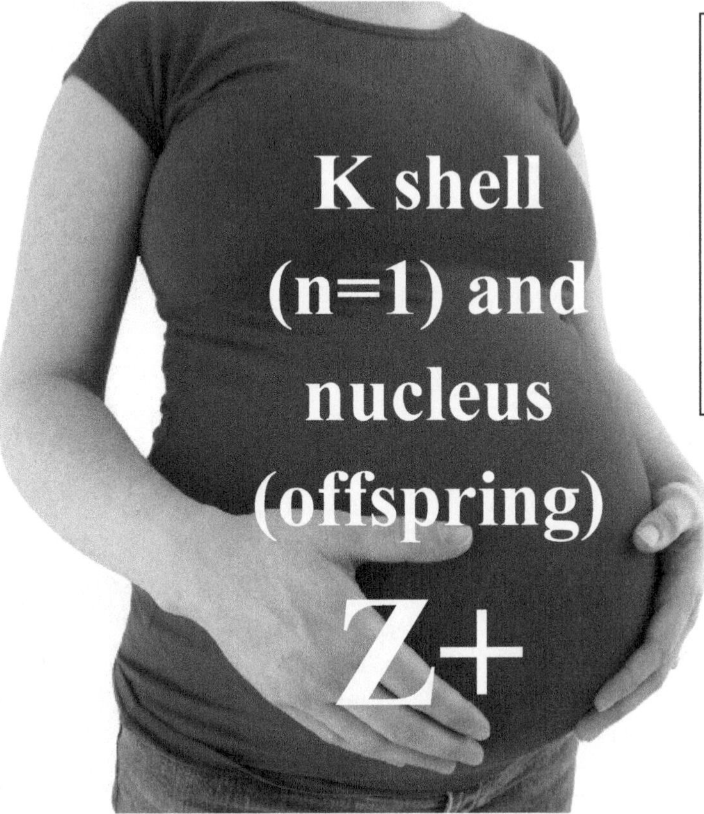

Conservation of solar energy through the communication of half sections of atomic carbon, brought together to produce nuclear fusion.

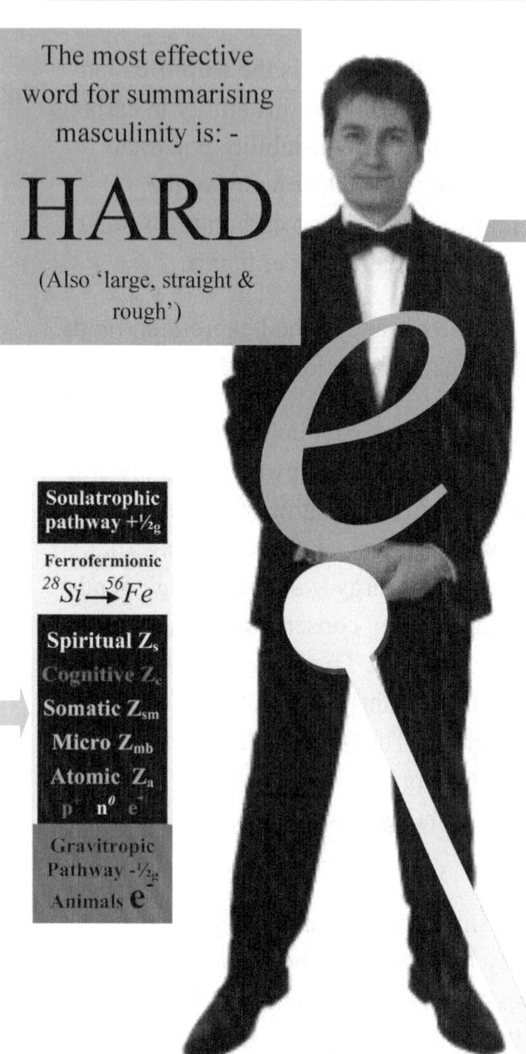

Electronic male

carbon entrochiraloctet

L shell

(valance) field logic

PRACTICAL LOW PRESSURE

The male wears camouflage colours as he is a field particle and the valance interface, he is also **streamlined** (hard wearing clothes) because he is a quantum particle and moves more with a non static position.

The most effective word for summarising femininity is: -
SOFT
(Also 'small, curved & smooth)
p
+

Protonic female

carbon entrochiraloctet

K shell

nuclear logic

IMPRACTICAL HIGH PRESSURE

The female wears bright advertising colours as she is a nuclear particle and the offspring interface (maternal bond), she is also a **high pressure** (delicate often impractical materials) morphology because she is a nuclear particle and moves less with a static position.

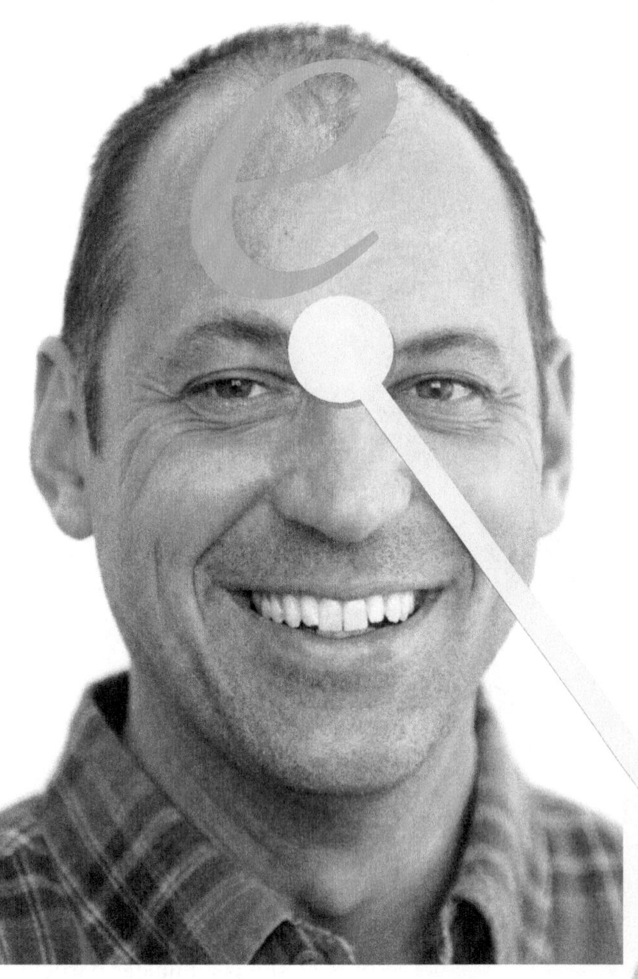

Electronic male

carbon entrochiraloctet

L shell

(valance) field logic

PRACTICAL LOW PRESSURE

The male wears camouflage colours as he is a field particle and the valance interface, he is also **streamlined** (short hair and balding) because he is a quantum particle and moves more with a non static position.

Protonic female

carbon entrochiraloctet

K shell

nuclear logic

IMPRACTICAL HIGH PRESSURE

The female wears bright advertising colours as she is a nuclear particle and the offspring interface (maternal bond); she is also **high pressure** (large or long hair) morphology because she is a nuclear particle and moves less with a static position.

Electronic male

carbon entrochiraloctet

L shell

(valance) field logic

PRACTICAL LOW PRESSURE

The male wears camouflage colours as he is a field particle and the valance interface, he is also **streamlined** (practical and hard wearing clothes and shoes) because he is a quantum particle and moves more with a non static position.

IMPRACTICAL HIGH PRESSURE

The female wears bright advertising colours as she is a nuclear particle and the offspring interface (maternal bond); she is also **high pressure** (impractical often delicate materials) morphology because she is a nuclear particle and moves less with a static position.

Protonic female

carbon entrochiraloctet

K shell

nuclear logic

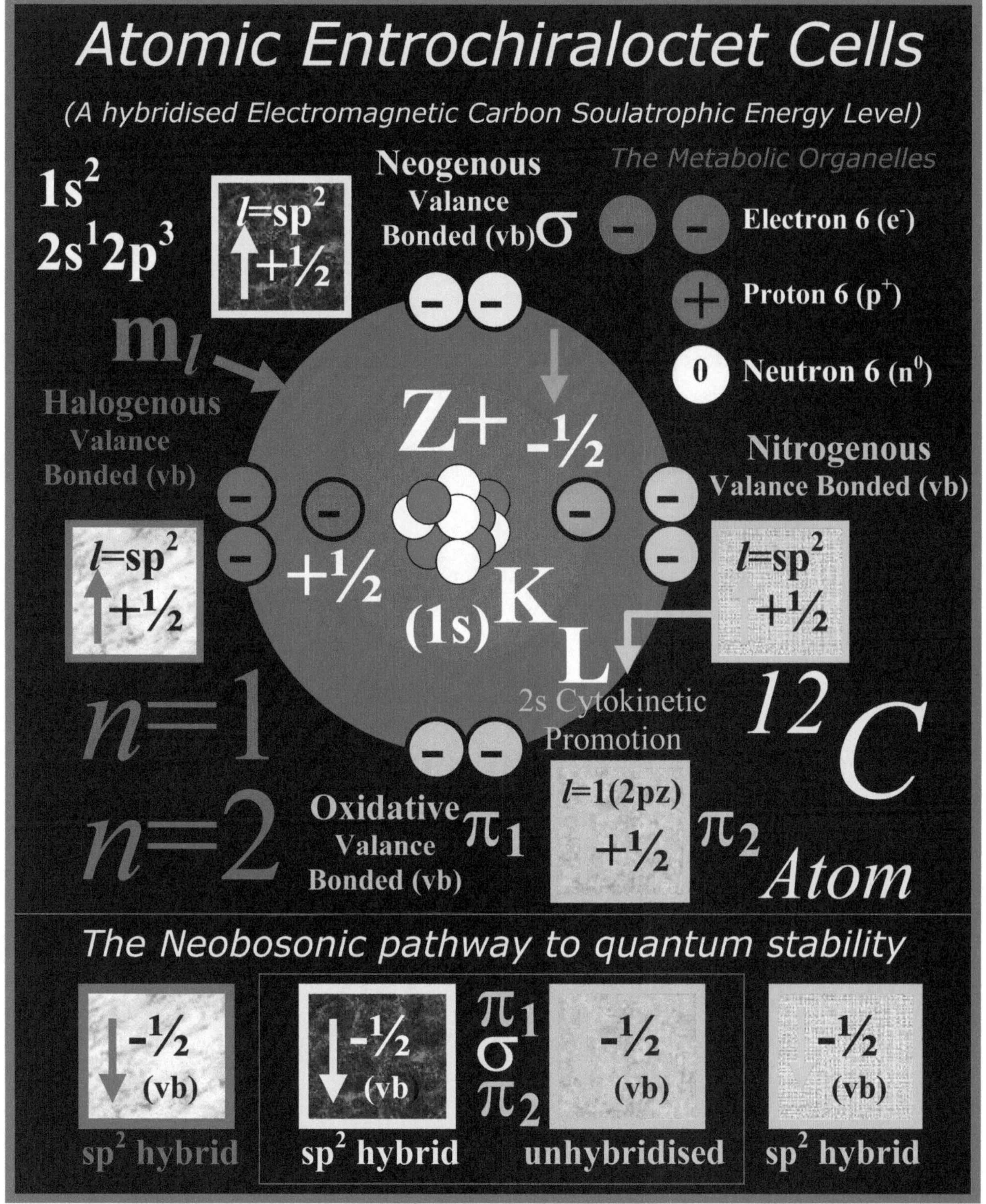

The male is larger than the female because of the **'inverse square law'**, the L shell field spreads out and is a larger field as it has to wrap around the K shell female nuclear field.

e^-

Electronic male

carbon entrochiraloctet

L shell (valance) field logic

Gender and the inverse square law!'

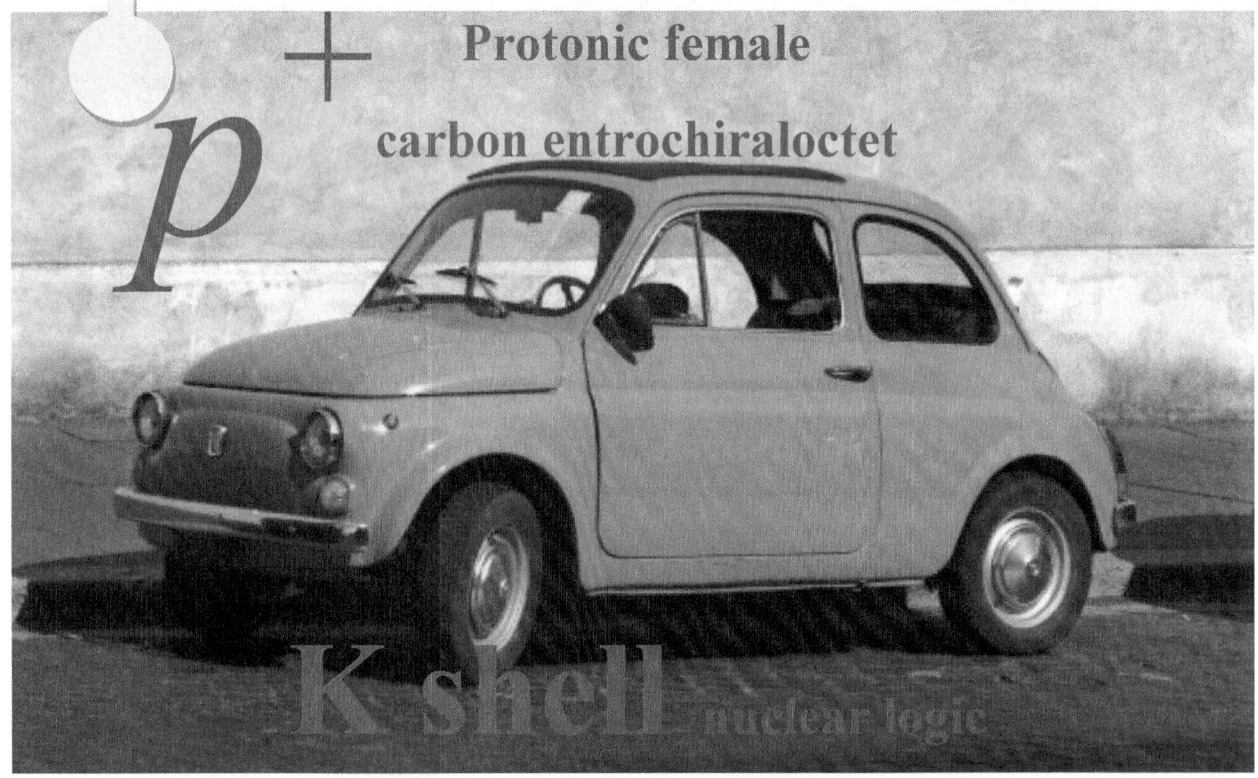

The male is larger than the female because of the **'inverse square law'**, the L shell field is a larger field as it has to wrap around the K shell female nuclear field.

Soulatrophonics – The conservation of lower soulatrophic energy levels as seen expressed in higher soulatrophic energy levels through gender. The somatic soulatrophic energy level in a woman can be described in terms of carbon entrochiraloctet, hydrogen or fermionic protons. Carbon is the fundamental soulatrophonic basis in carbon entromorphology but other fundamental particles can be used to describe higher levels just as effectively.

Spiritual soulatrophicity – This entrochiraloctet cell has its nucleus as the neogenous metabolic organelle on the cognitive soulatrophic energy level. The brain is a super nucleus and the cell extends beyond the range of sight and sound. Soulatrophic energy levels can be thought of in an organism as levels of organism groups such as the family consciousness through gender bonding. Hence we find the imagination on the spiritual level; the entrochiraloctet cell extends to the heaven and hell concept of reality for some. This level is the multi organismal level where carbon entrochiraloctet cells are formed by groups of complete organisms acting as particles. This begins with the family cell through the gender model, where parents and offspring form the carbon entrochiraloctet cell. In the family entrochiraloctet cell the 1s electrons are the parents and the siblings the L shell valance components.

The nuclear organism is the nucleus and other organism's act as neogenous valance bonded metabolic organelles. This level is the 'Heisenberg cell' as it deals with multi organismal behaviour and the imagination as a limit to cellular size, and as such is subject to the perception of the uncertainty principle. Examples of such entrochiraloctet cells on this level are 'religion', politics, economics, and education but based on familial gender models. The metabolic organelles are related to the size of the entrochiraloctet cell; an example is a political oxidative organelle such as an atomic bomb. The valance bonded metabolic organelles are shared technology, food and shelter. Sharing is the key on this level, technological valance bonded metabolic organelles for humans on the cognitive level are personal possessions, not for sharing.

Spin alternates in soulatrophic pathways and pairs up across gender communication bosonic ladders or pathways (sexually reproductive communicators) for soulatrophic ladders producing recombination. When a penis enters a vagina the two somatic levels form a neutronic somatic level through their combination reducing overall energy and producing offspring.

Sub-atomic soulatrophicity is based on fermionics associated with firstly hydrogen then carbon. The proton, neutron and electron are the basic particles, which are conserved through amplification ECASSE. The three particles are the fundamental basis of gender in carbon entromorphology. The proton energies can be seen as feminine sexual dimorphism. It results in nuclear energies; energy conservation to the nucleus and nuclear flaunting, the positive charge produces electric field lines in keeping with an advertisement for the nucleus and bright colours associated with females.

Protonic systems are phototropic often associated with bright colours, high-pressure morphology and particle form characteristics. The plant world has female soulatrophic pathways, associated with oocyte production in keeping with lower kinetic energy levels as the oocyte remains fixed and is a K shell component, particle form. Electron energies are seen in masculine sexual dimorphism. It results in energy battles, as it is waveform energy, has a larger field than the nucleonic proton but a diluted mass, the female is bonded to the nest. The ratio of female plus nests to male fits the classical proton to electron ratio and has been shown to have a >99.8% agreement as proof of carbon entromorphology. The electron is associated with thermoentropic pathways and energy liberation events. It is associated with environmental shielding and the acquisition of metabolic organelle energies. It is a harvester of external energy (outside the entrochiraloctet cell, or atom) hence hunter gatherer instincts. It results in instability (violent energetics) by comparison to the more passive stable feminine proton (>90 % of all violent crime is created by the male). It is associated with camouflage because of its position relative to the external environment (gravitropism). The plant world has male soulatrophic pathways, associated with pollen production in keeping with higher kinetic energy levels as pollen is released into the air or water as a supplementary L shell component, wave form.

The neutron is composed of a union of electron and proton and neutrino. It is heavier in mass than the other two particles and is seen as a suitable model for homosexuality. It has both sets of energies and as such has phototropic and gravitropic characteristics. Homosexual energies are advanced, stable nuclear systems. Homosexuals display phototropic energies through bright colours and energy liberation through advanced communication capabilities. The angiosperms (flowering plants) of the plant world are homosexuals neutronic and with stamen (male – electron) and pistil (female – protonic). They use bosonic transmitters (bosonic insects) to communicate nuclear genomic information. They possess both electrons (stamen) at higher kinetic energy levels. They also possess proton (pistil) at nuclear lower kinetic energy levels.

Alternative definition. Neutronic homosexual gender differs from heterosexual gender in many ways – Classical gender is polarised into protonic and electronic states based on fermionic particle origins. The complete trinity has a compound particle called a neutron, which is an electron, a proton and a neutrino. Homosexual people are neutronic and therefore more nuclear with greater mass than the other gender energies, which are wave forms. They adopt both energies (proton and electron forms) and can oscillate between the two showing so-called 'camp energy conservation' wave form and 'extrovert colours and communications' protonic in keeping with nuclear flaunting.

Classical fermionic genders with electron male and protonic female energies are polarised and inversely complementary. Neutronic gender is therefore none polar (neutral) where mass is the potential driving bonding. Hence gay people like 'butch' and 'femme' states, which are slight polarisations due to neutronic expression. Slight none polar potentials and strong gravitational potentials mediate effective bosonic ladders (communication links such as the genitals).

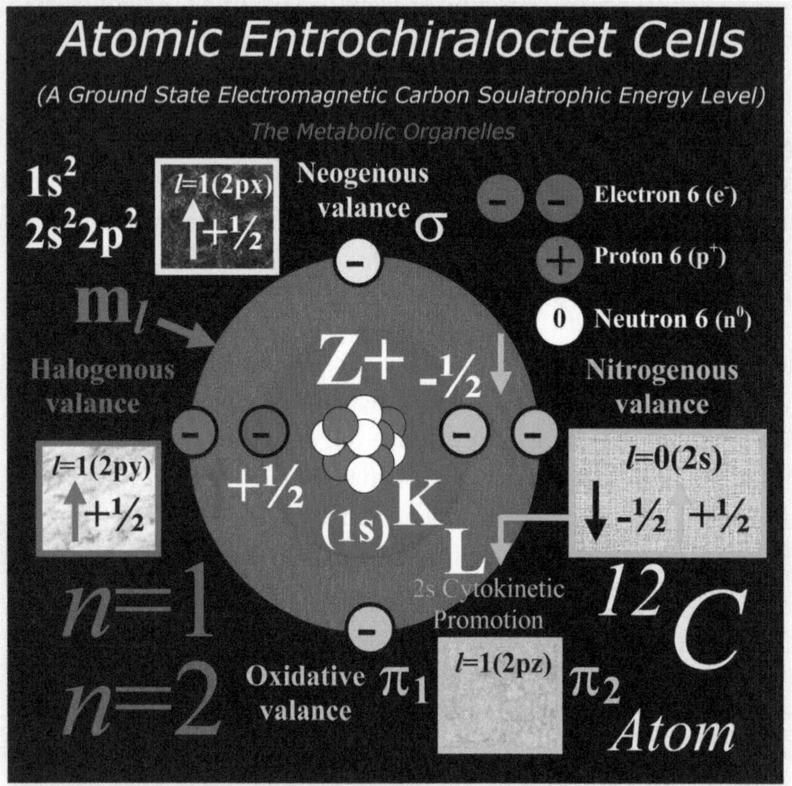

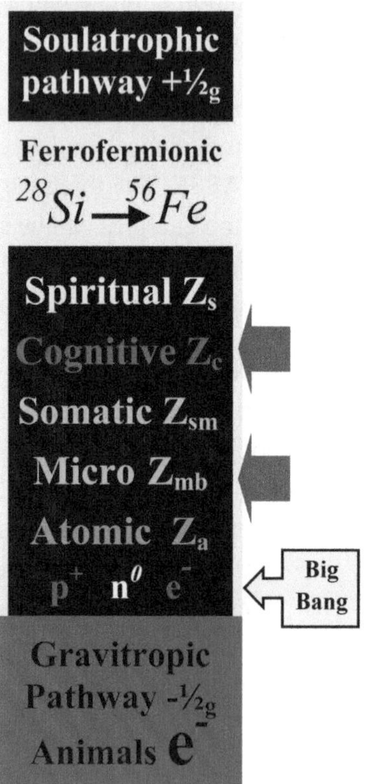

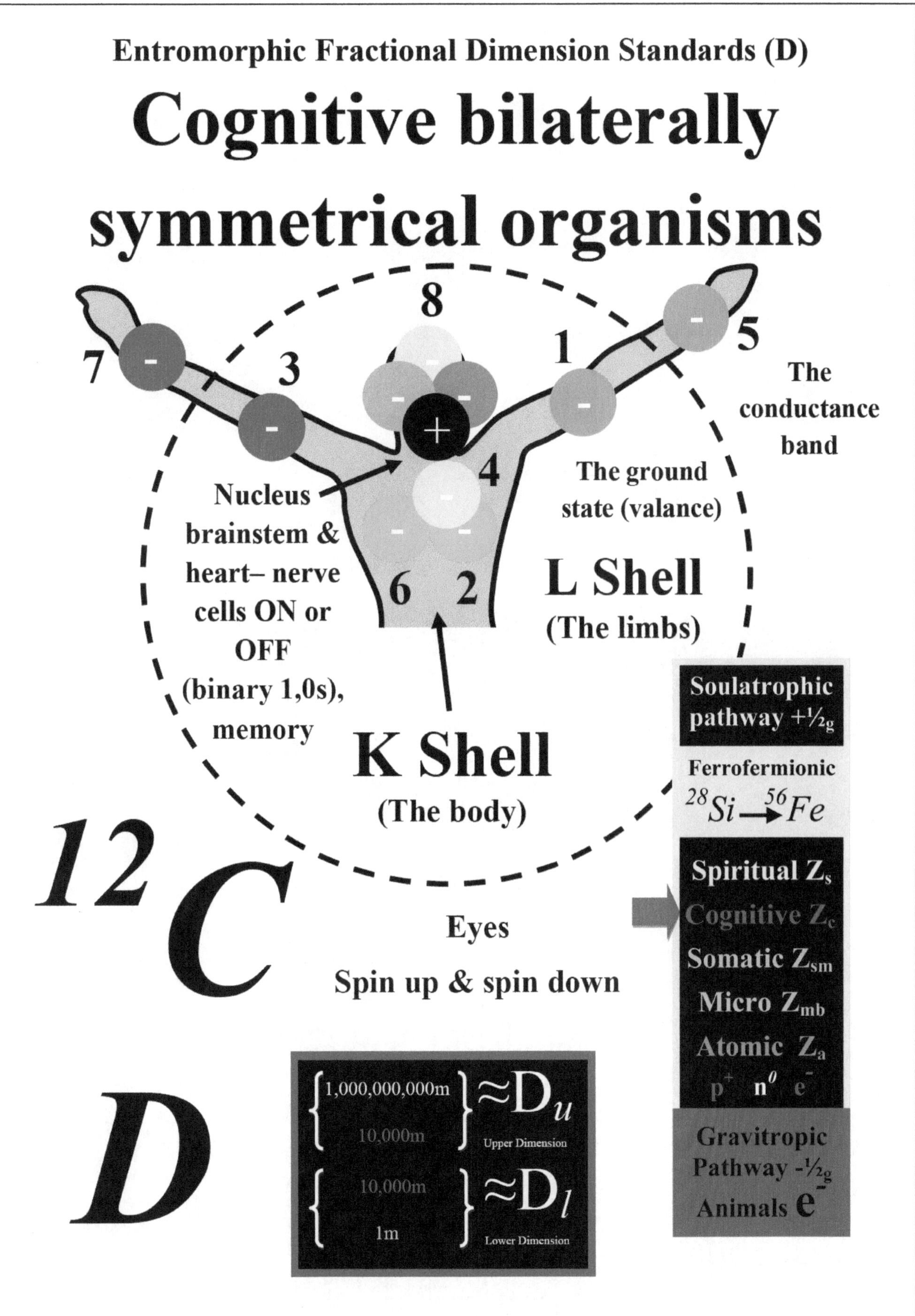
Entromorphic Fractional Dimension Standards (D)
Cognitive bilaterally symmetrical organisms
8
7
3
1
5
4
6
2
The conductance band
Nucleus brainstem & heart– nerve cells ON or OFF (binary 1,0s), memory
The ground state (valance)
L Shell (The limbs)
K Shell (The body)
Soulatrophic pathway +½g
Ferrofermionic
$^{28}Si \rightarrow {}^{56}Fe$
Spiritual Z_s
Cognitive Z_c
Somatic Z_{sm}
Micro Z_{mb}
Atomic Z_a
p^+ n^0 e^-
Gravitropic Pathway -½g
Animals e^-
^{12}C
Eyes
Spin up & spin down
D
{1,000,000,000m 10,000m} ≈ D_u Upper Dimension
{10,000m 1m} ≈ D_l Lower Dimension

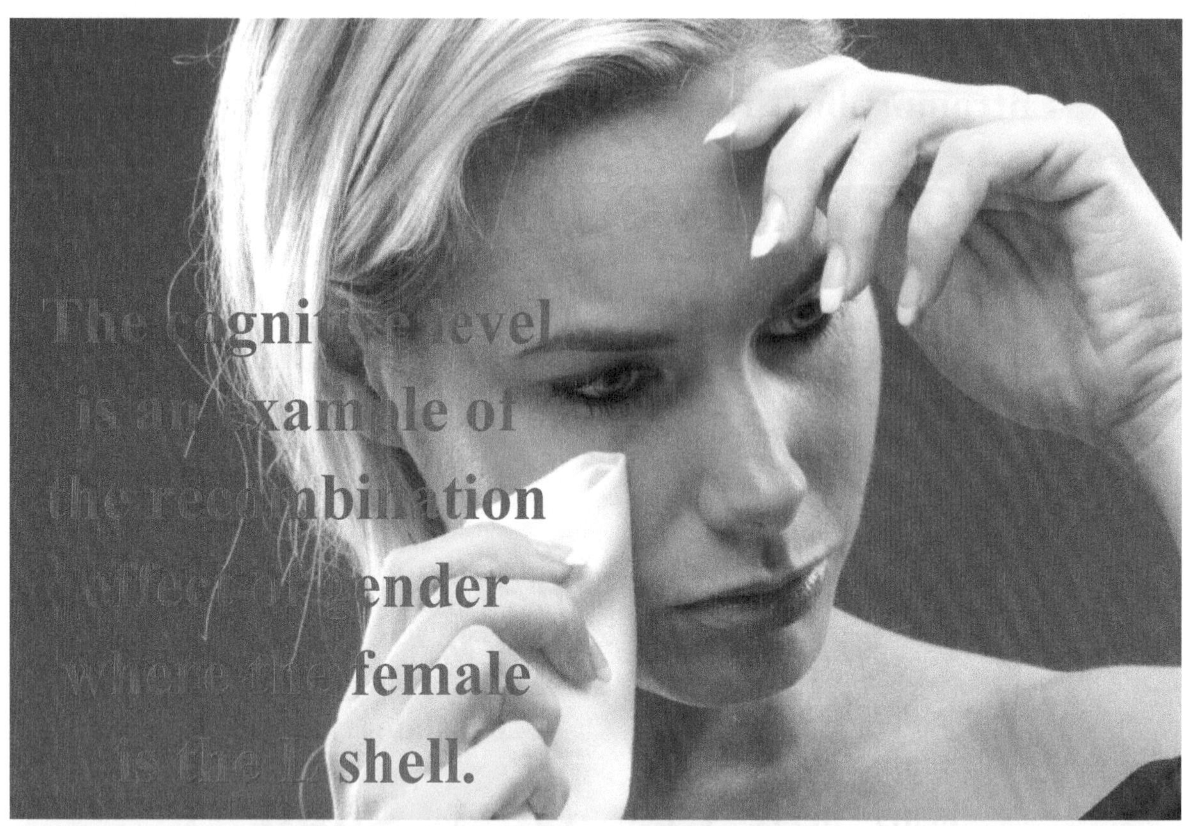

The cognitive level
is an example of
the recombination
effect of gender
where the female
is the L shell.

L shell n=2

- Region of energy liberation, higher kinetic energy level.

- This is a region of energy liberation in women hence the ability to articulate emotions physically; hence crying and general communicative potentials are more pronounced in women than men on this level.

Soulatrophic pathway $+\frac{1}{2}_g$

Ferrofermionic

$^{28}Si \rightarrow {}^{56}Fe$

Spiritual Z_s

Cognitive Z_c

Somatic Z_{sm}

Micro Z_{mb}

Atomic Z_a

p^+ n^0 e^-

Gravitropic Pathway $-\frac{1}{2}_g$

Animals e^-

The cognitive level as an example of the recombination effect of gender where the male is the k shell and nucleus.

Z+, K shell n=1

- Region of energy conservation. Men typically bottle up feelings because they have a quantum stable cognitive energy level.
- This is where their drive to win comes from.

- Men conserve energy cognitively; hence they retire to the so-called 'cave'. This is an energy conservation technique associated with the cognitive Z+ and K shell and the need to win at spiritual games (football). Hiding feelings can be a form of camouflage.

Soulatrophic pathway $+½_g$

Ferrofermionic

$^{28}Si \rightarrow {}^{56}Fe$

Spiritual Z_s

Cognitive Z_c

Somatic Z_{sm}

Micro Z_{mb}

Atomic Z_a

p^+ n^0 e^-

Gravitropic Pathway $-½_g$

Animals e^-

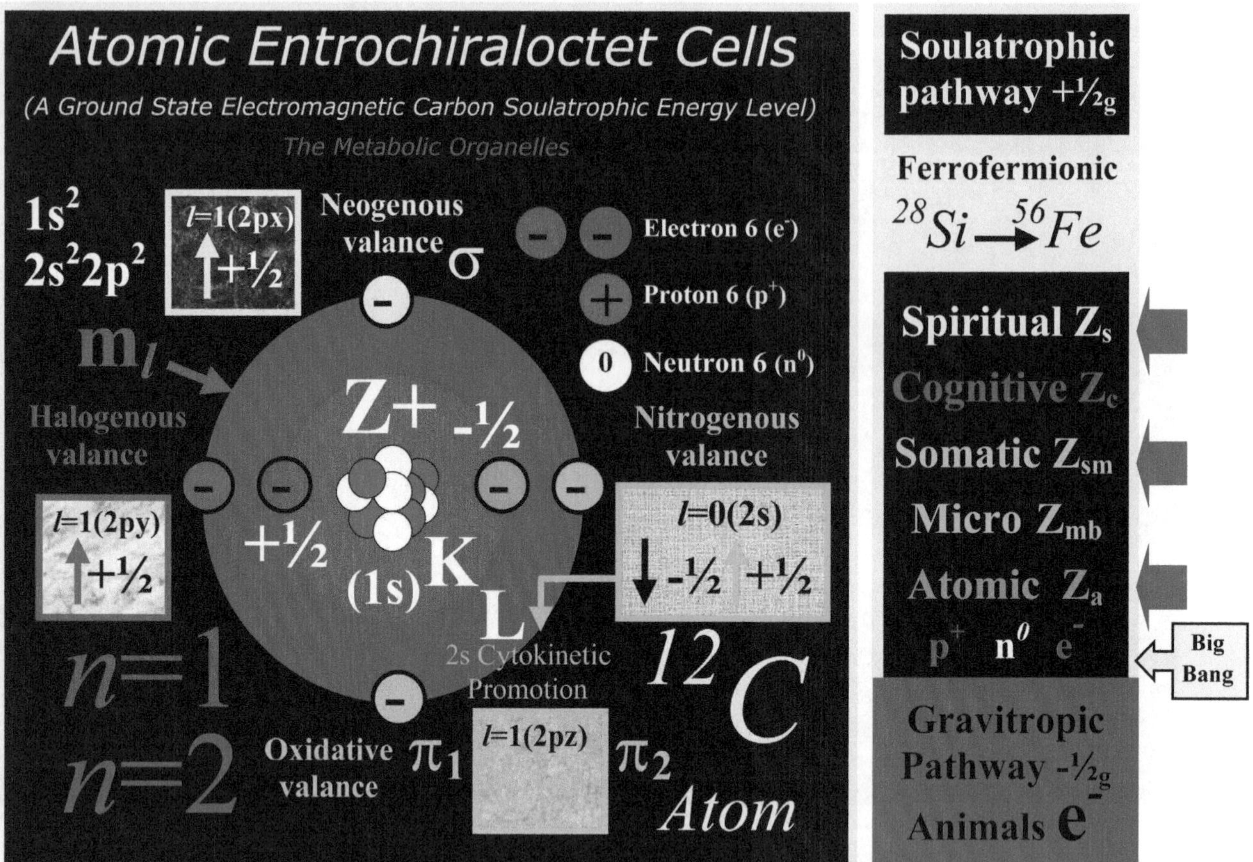

The gender model splits carbon entrochiraloctets into their respective energy levels, the nuclear level which includes the nucleus and K shell is a stable region.

The field or L shell region is a valance region where energy flows into and out of carbon cells.

When male and female come together they bond across the different soulatrophic energy levels (fractional dimension), to produce a complete carbon cell entrochiraloctet.

Each fractional level alternates as we go up the soulatrophic energy levels, so on the cognitive level the female is the L shell valance area and the male is the nuclear shell.

On the level below the somatic level we find the female is the nuclear shell hence pregnancy from the female (the offspring is the nucleus); the male is the L shell valance field level.

The bonding of all soulatrophic energy levels produces a ‘soulatrophic ladder’; bonding from the spiritual level occurs firstly usually through conversation and bonding through compatibility.

Then bonding on the cognitive level through a kiss. Bonding on the somatic level through sexual intercourse and then through orgasm bonding through initially microbiological levels.

Finally bonding on atomic and fermionic levels through successful fertilisation.

Each level oscillates from K shell to L shell as we move up and down soulatrophic fractional dimensions due to Pauli’s exclusion principle.

Sexual dimorphism has never been based on a fundamental physical model and as such gender has no true basis in physics, these models allow humanity to understand where gender originates from.

Entromorphic Fractional Dimension Standards (D)

Somatic radially symmetrical organisms

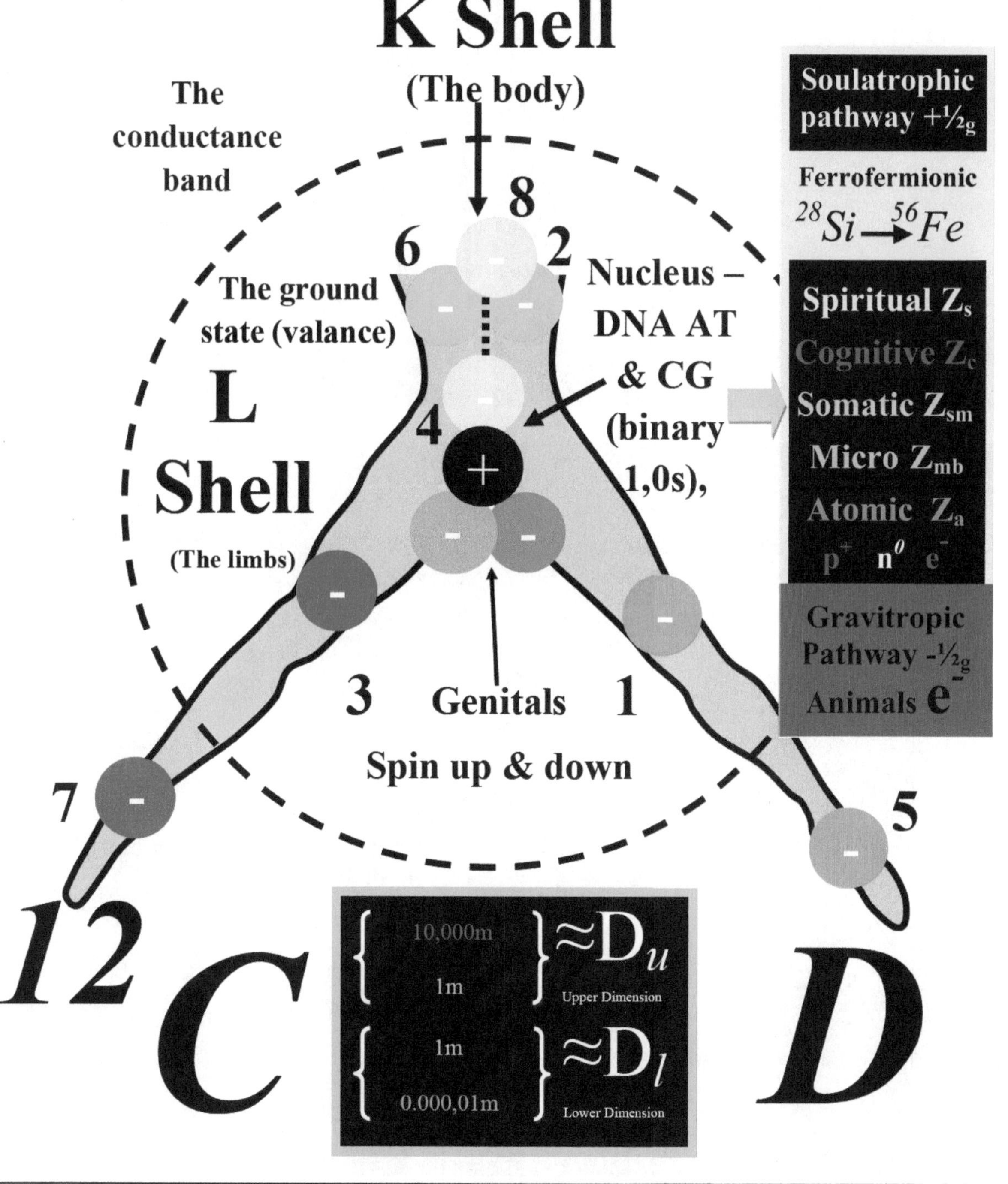

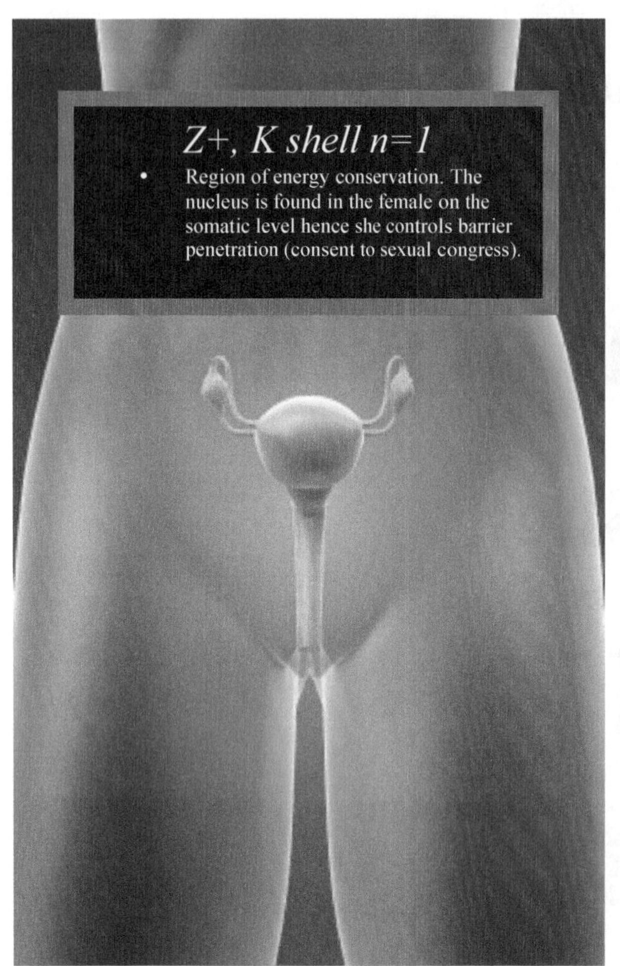
Z+, K shell n=1
Region of energy conservation. The nucleus is found in the female on the somatic level hence she controls barrier penetration (consent to sexual congress).

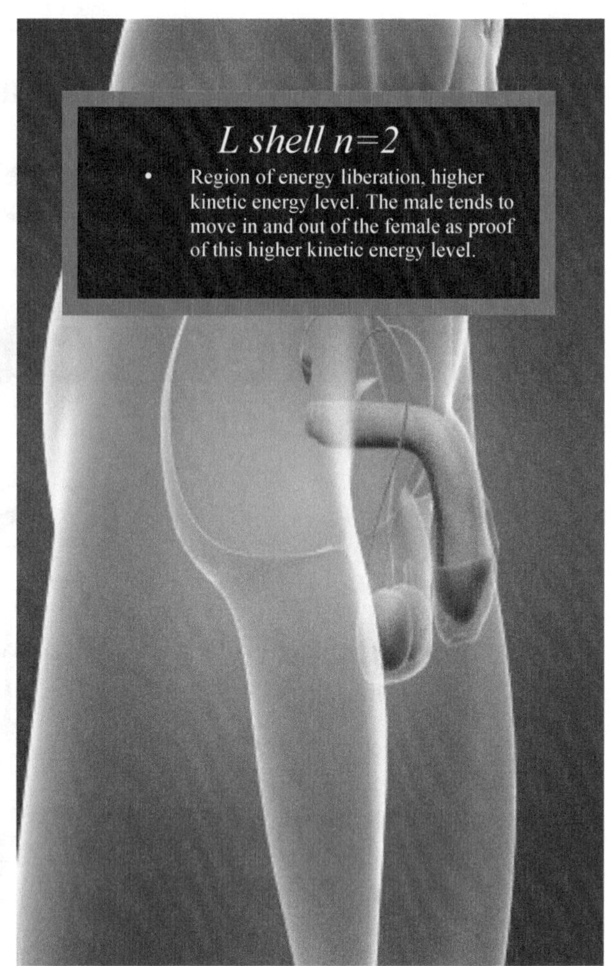
L shell n=2
Region of energy liberation, higher kinetic energy level. The male tends to move in and out of the female as proof of this higher kinetic energy level.

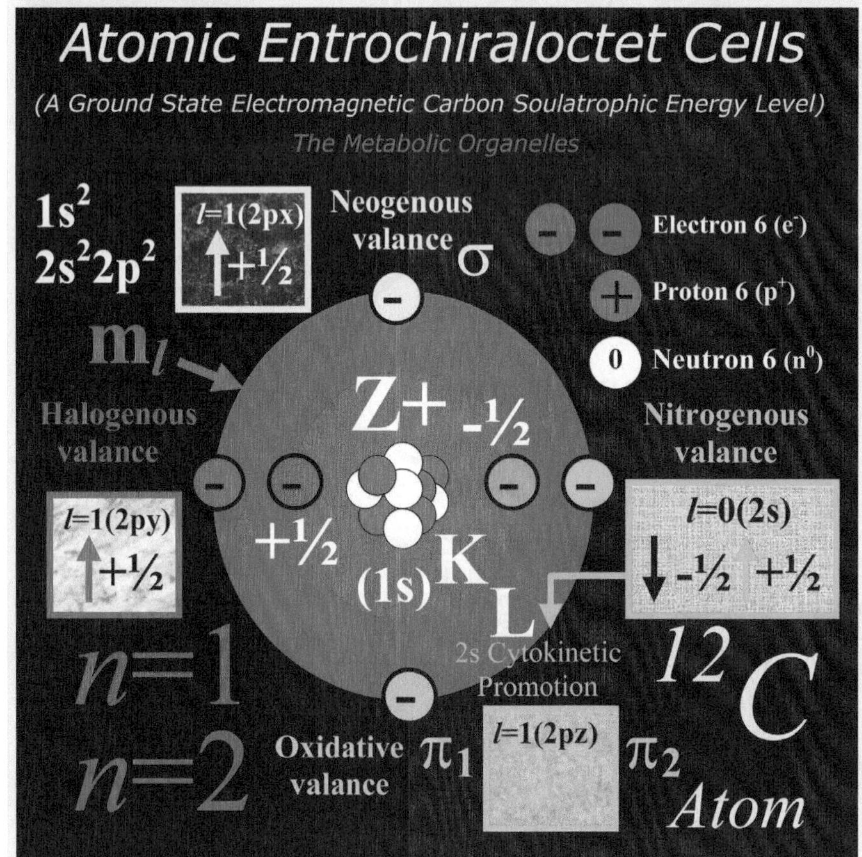
Atomic Entrochiraloctet Cells
(A Ground State Electromagnetic Carbon Soulatrophic Energy Level)
The Metabolic Organelles
1s2
2s22p2
l=1(2px) +½
Neogenous valance σ
Electron 6 (e-)
Proton 6 (p+)
Neutron 6 (n0)
ml
Halogenous valance
Z+ -½
Nitrogenous valance
l=1(2py) +½
+½
(1s) K
L
l=0(2s) -½ +½
2s Cytokinetic Promotion
n=1
n=2
Oxidative valance π1
l=1(2pz)
π2
12 C Atom

Soulatrophic pathway +½g
Ferrofermionic
28Si → 56Fe
Spiritual Zs
Cognitive Zc
Somatic Zsm
Micro Zmb
Atomic Za
p+ n0 e-
Gravitropic Pathway -½g
Animals e-

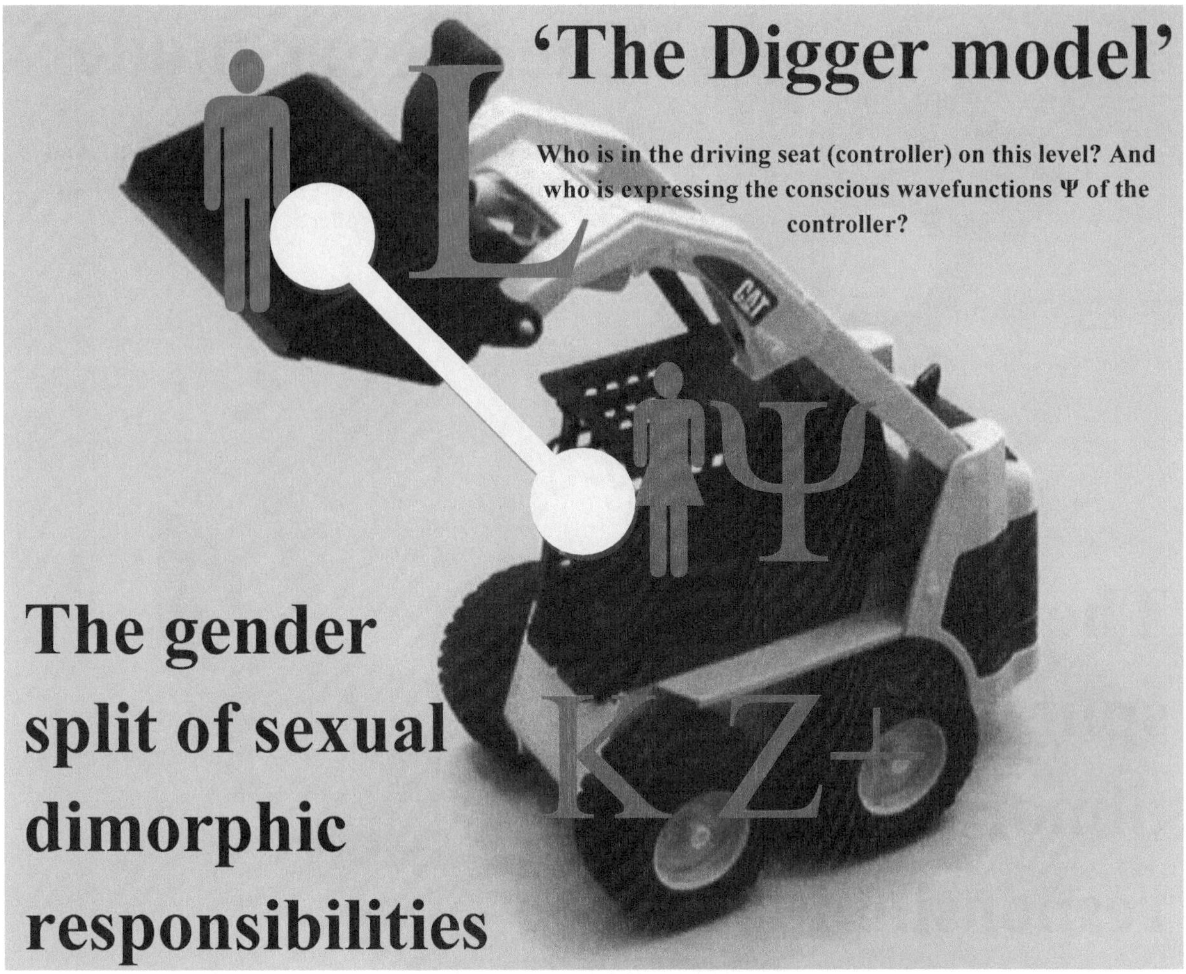

'The Digger model'

Who is in the driving seat (controller) on this level? And who is expressing the conscious wavefunctions Ψ of the controller?

L

Ψ

K Z+

The gender split of sexual dimorphic responsibilities

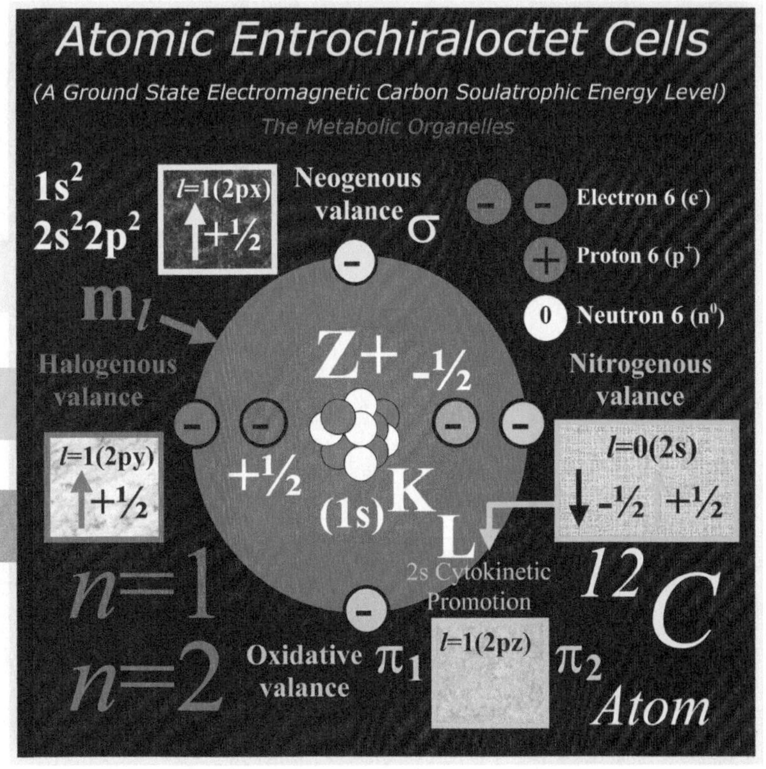

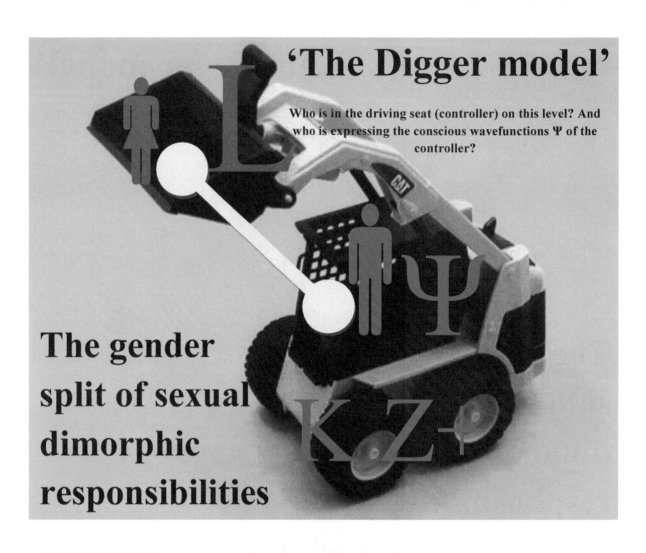

'The Digger model'

Who is in the driving seat (controller) on this level? And who is expressing the conscious wavefunctions Ψ of the controller?

L

Ψ

KZ+

The gender split of sexual dimorphic responsibilities

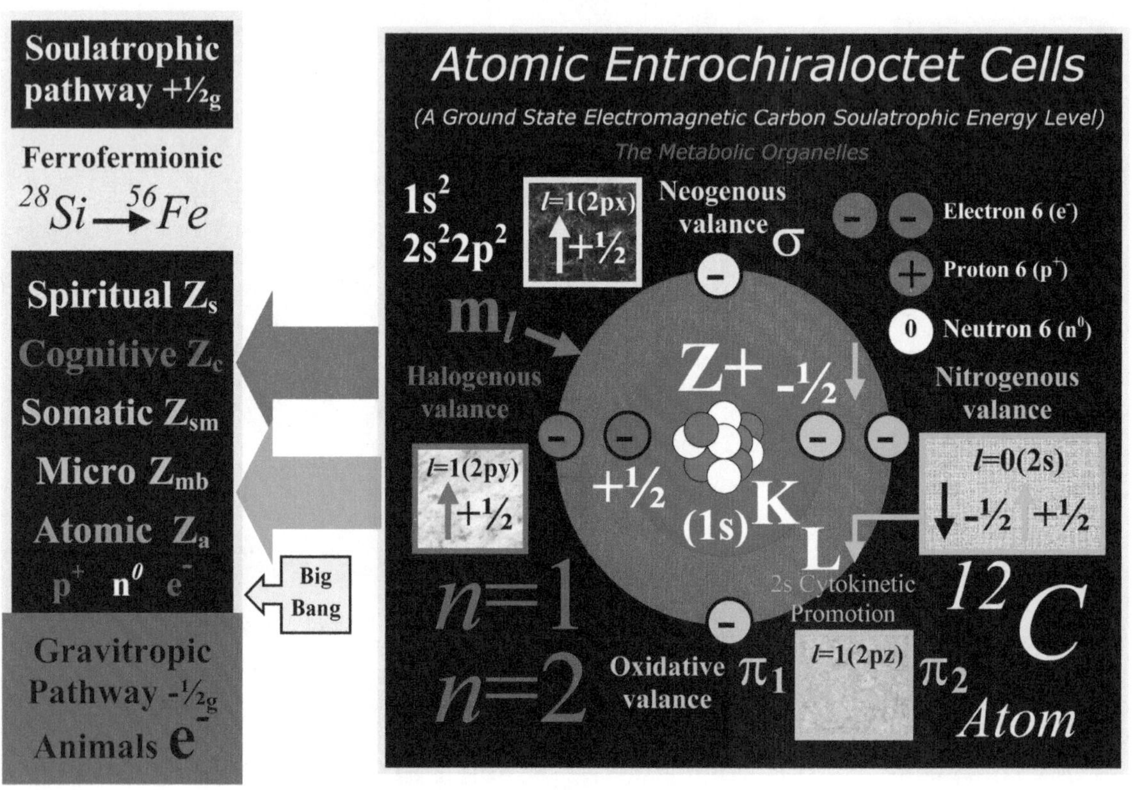

Protonic female
carbon entrochiraloctet
Hydrogen and carbon fermionics in the gender model
(Somatic radial level)
Soulatrophic pathway +½g
Ferrofermionic
28Si→56Fe
K shell
Spiritual Zs
Cognitive Zc
Somatic Zsm
Micro Zmb
Atomic Za
p+ n0 e-
Gravitropic Pathway -½g
Animals e-
R female
and Z+
nucleus
nuclear logic
Electronic male
carbon entrochiraloctet
L shell
(valance) field logic
R male
The inverse square law on the somatic level finds the female is closest to the offspring (Z+), the male further away hence the male is a 'larger field – larger body' but a consequential 'weaker field strength' or bond to the offspring.
Based on Zfield=Z+/R2
Rmale (L shell) > Rfemale (K shell)

Neutronic forms are advanced highly evolved states of matter as the depolarisation of electron/proton relationships have converged into neutronic composites. This completes the evolutionary fermionic trinity for the animals (electronic phase of evolution). In the plants homosexual neutronic forms are the angiosperms (flowering brightly coloured plants). Again the evidence is found in the flower which has oocyte (protonic forms) and stamens (electron forms) together are neutronic flowers.

The neutrino component is linked to the production of pollen. The bosonic particle is the insect on this level and again the protonic phase of evolution completes the trinity for the plants. When both electronic and protonic phases of evolution are complete technology, the neutronic phase of evolution begins and leads towards the complete ferrofermionic trinity and planetary cytokinesis where life moves into space towards artificial life. Strong evidence exists to support the neutronic homosexual model where the angiosperms (plants) and homosexual humans both embrace phototropism. Gay groups often dress in flamboyant phototropic colours often copying flowers as proof of this.

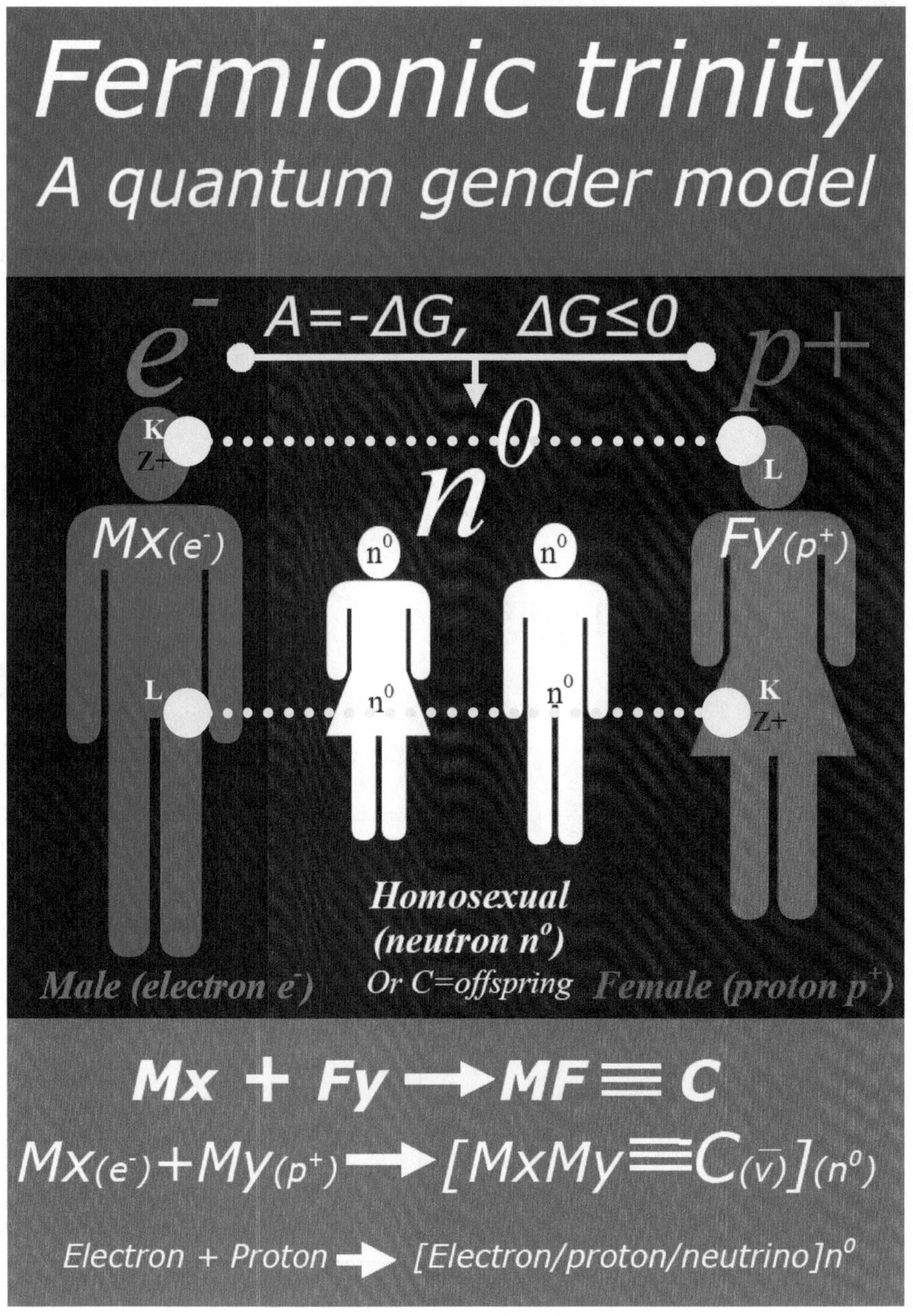

Science check

Do the observations correlate with the theoretical model?

YES ✓ NO

PART FIVE

Neurological genetics through an extension to classical genetics

Neurogenome neurological genetics - a model of the human mind, a unification of all knowledge.

Neurological genetics is the nuclear and field logic associated with the highest soulatrophic energy level the 'spiritual level'. It is a general description of ALL human knowledge without exceptions. It is demonstrated by carbon entromorphology and this book, where knowledge is divided into neurogenomes (such as the field of biology), neurochromosomes (such as the sub-field of biochemistry) and neurogenes (such as the word 'enzyme').

It is also demonstrated by unique neuroalleles (individual characteristics) which are subject to evolution by natural selection. Carbon entromorphology and this book are expressed as a neurogenome, each page a neurochromosome and the new terms and characteristics as neurogenes and neuroalleles. For example a neuroallele in a human could be the characteristic of sociability.

Neurological genetics is a massive unifying concept which sits behind ALL human knowledge and the precise definition of neurogenome, neurochromosome, neurogene and neuroallele are difficult to define and will require some establishment of convention by this author. Neurological genetics is the integrated existence and expression of Mark Andrew Janes BSc (Hons) CBiol MSB neurogenome, by entering this site you are entering the mind of this author.

My mood disorder has revealed the true logic of the mind as I oscillate from manic to depressive states, waveform to particle form states. Self analytical methods and my scientific training and experience gave me the intuition to reflect my own mind in this extraordinary way and in doing so produce a unification of the conscious experience and the nature of the mind. Ironically the total conclusion of this logic is that since we are quantum systems we are unknowable even to ourselves. It is ironic that a pathological mental illness such as my own should through its various mood states illustrate the quantum mechanics and nuclear physics of life.

A mental illness that through its complexity and diversity should reveal the true nature of the mind and the soul (origin of life) and incredibly the true nature of the concept of God, that God or 'artificial life' is within humanity's grasp. We are the God's of the future, energy rich with temporal stability in a low pressure system called a heaven. Humanity is very close to artificial life and carbon entromorphology shows us the way to understand spiritualism through science.

May I take this opportunity to embrace all the people of the world who believe in God! I was an atheist for 17 years but my scientific work has shown me the true nature of the concept of a God (androids, robots and virtual organisms). For anyone who is peaceful and who fights for basic human and animal rights I celebrate your belief in God however you specifically choose to understand it and wish you a beautiful and peaceful future. Humanity has it within ourselves to be the Gods of tomorrow, but we have already started and come a long way. This author predicts that within the next 20 to 50 years humanity will close the door on cancer and aging, poverty and suffering but we must

NEVER EVER FORGET that the great genius of Werner Heisenberg is standing in the corner reminding us of our limitations and bringing us down to Earth through uncertainty.

The second law of thermodynamics will eventually win but the road can be made to be very very very very long as we become more efficient in terms of our use of energy and it will be a hell of ride!

Neurological genetics is a cross fertilised self symmetrical hypothesis made by comparisons between neurological and DNA nucleonic systems of natural logic.

Amazing examples of supercoiled neurological chromosomes, evidence of extended nuclear logic. The words, files etc are hypothesised to be neurological genes; they are bound together by the strong interaction through gluonic binding which hold the letters in a word together. And by pionic bonds which hold words together to form sentences. Letters are related to quarks and the strong interaction is the nature of all binary information systems (ROM), expressed through quantum communication.

'Super coiled' neurological chromosomes.

The disk and the paper act as histones for these systems; they are read and manipulated by ribosomal electroweak mechanisms.

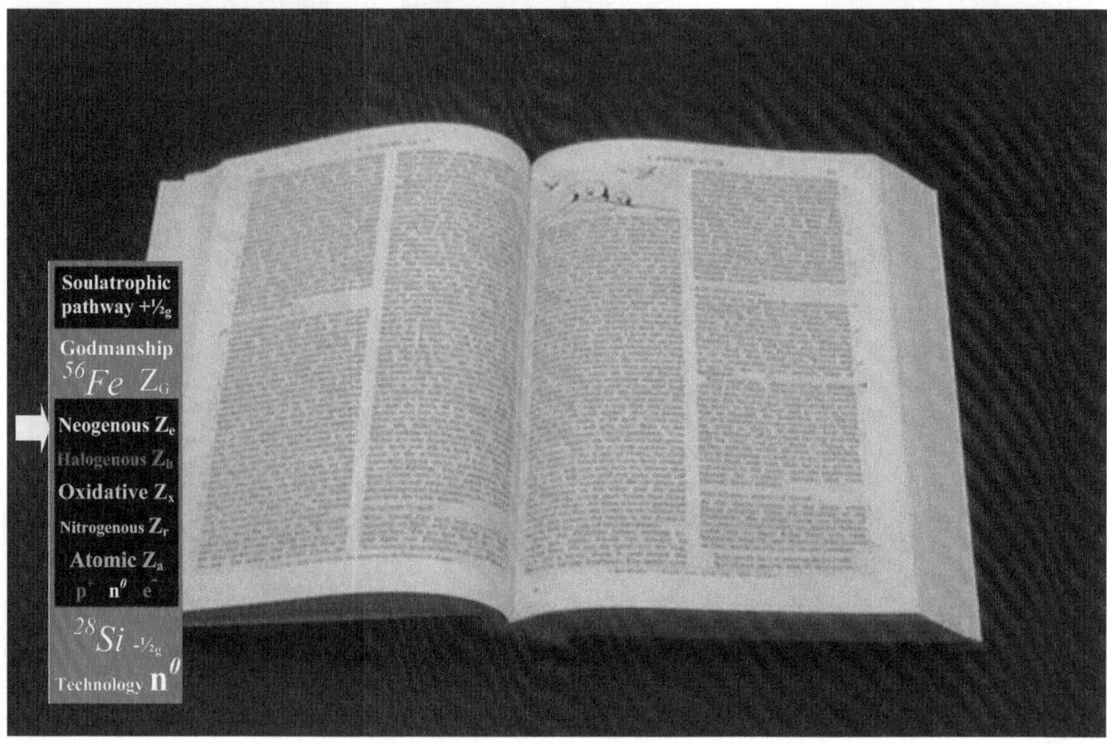

The centres of the neurochromosomes have centromere holes and central binding and centrosomic properties, they are read only memory systems originating from atomic nucleonic (strong force).

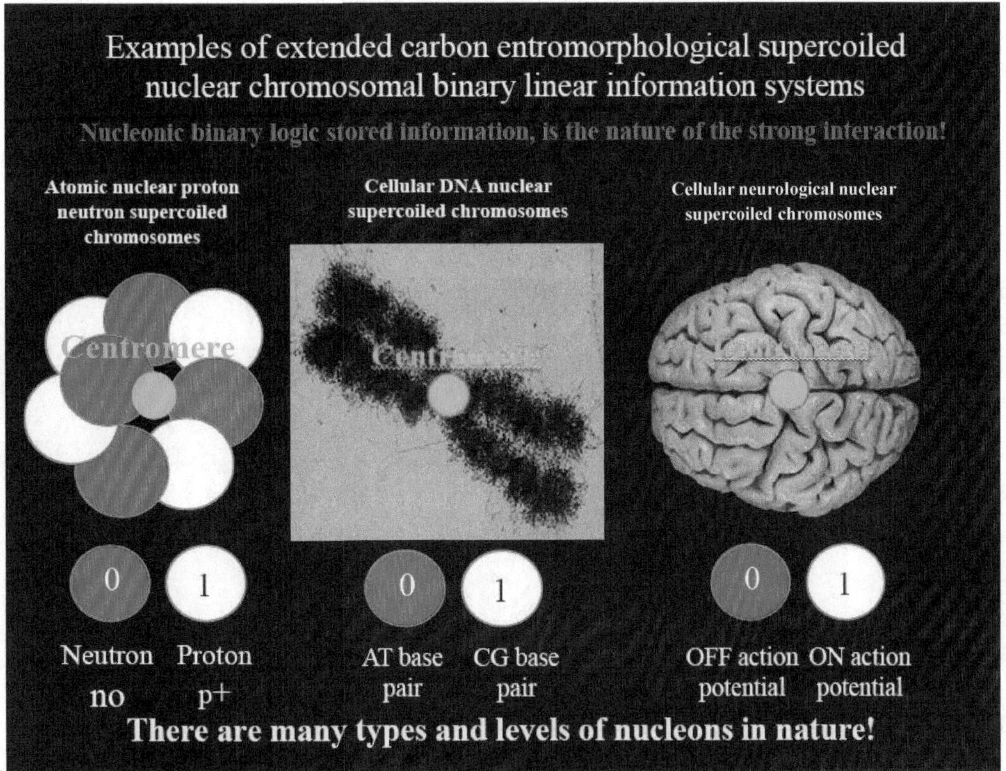

The following diagrams list the main forms of nucleonic systems found in life on Earth. The basis comes from the binary nature of atomic nuclei and the forces involved. This is hypothesised by carbon entromorphology to be the basis of all information systems hence there are DNA and neurological systems which differ only in size relative to the smaller ones. This binary nature is found in all basic information systems and covers the nature of the strong nuclear interaction and the pionic and gluonic components. Energy has different properties relative to its observational position relative to nucleonic centres.

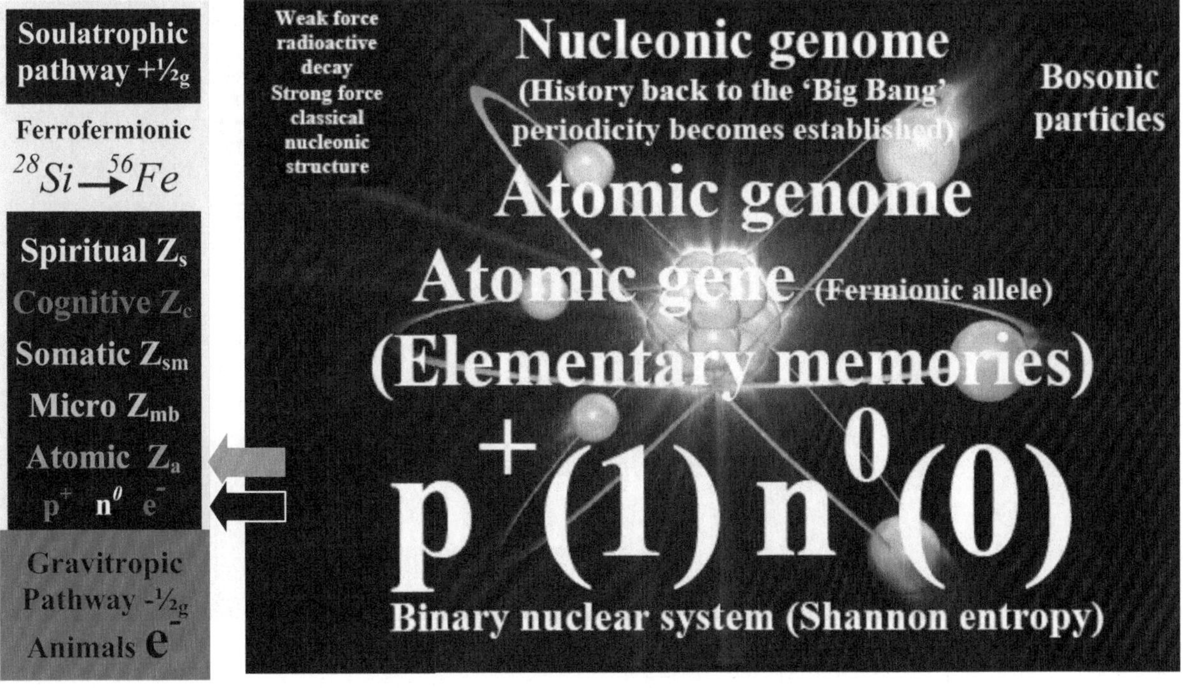

The energy in the nucleus cycles into and out of the nucleus through the electroweak standard model and energy typically through photons is absorbed or emitted, but since the potential in the nucleus is a variable influencing quantum organisation then the second field component of nucleonic logic must be the basis of communication. This observation hypothesises in summary that information is the basis of the strong and probably gravitational forces, and communication or energy transfer and transduction is the field logic component. Neurological genetics extends nucleonic information inner space logic and field communicative outer space logic to produce many neurological systems such as the written word, or any information and or communicative process.

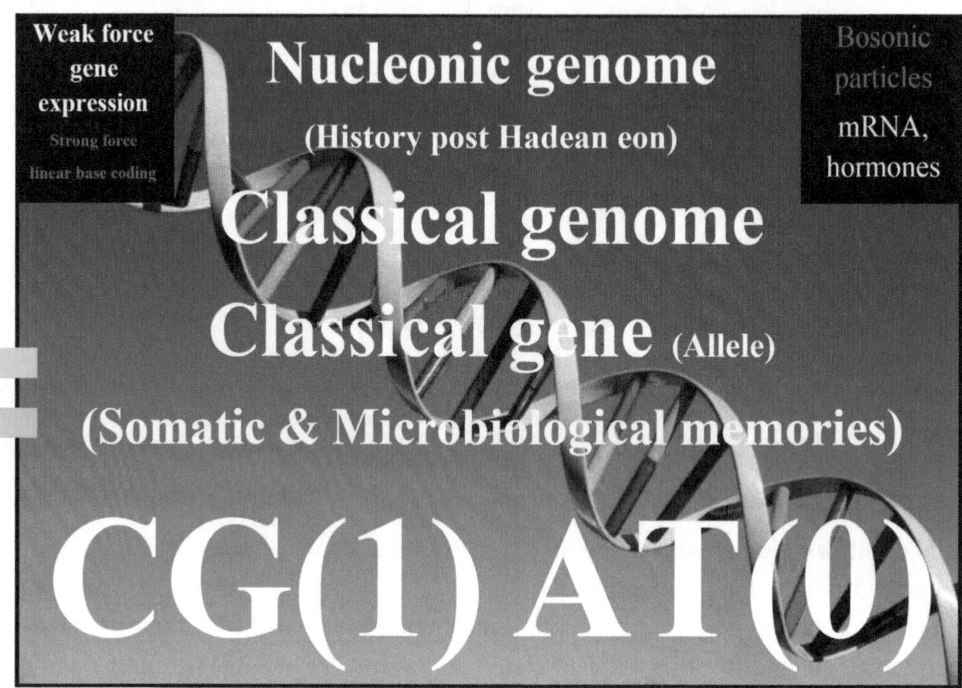

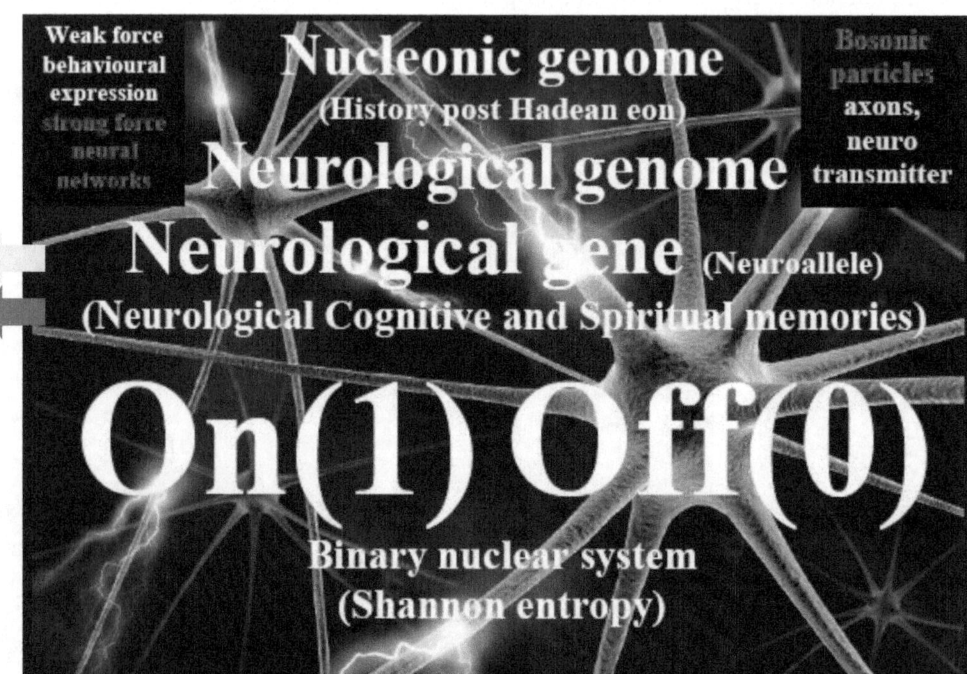

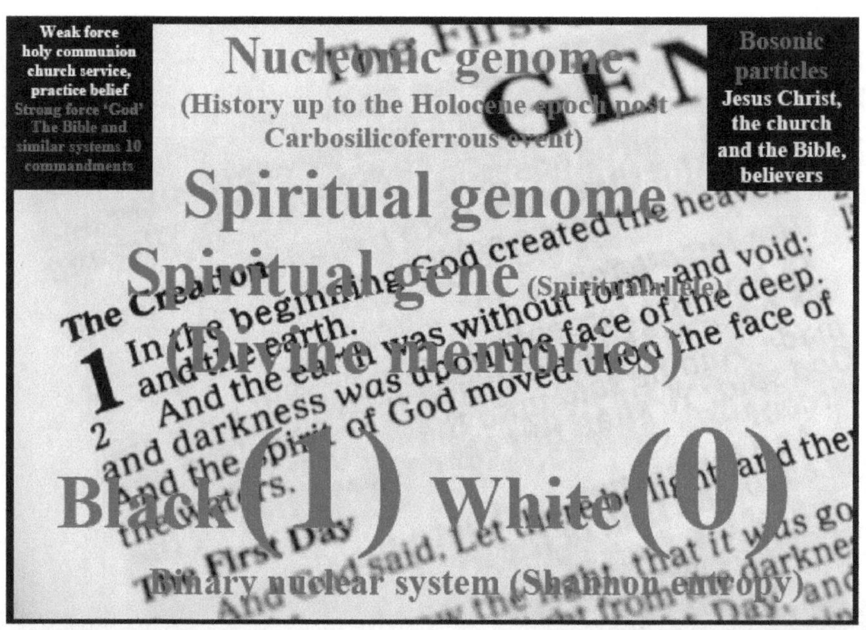

Soulatrophic pathway $+\frac{1}{2}_g$

Ferrofermionic $^{28}Si \rightarrow ^{56}Fe$

Spiritual Z_s

Cognitive Z_c

Somatic Z_{sm}

Micro Z_{mb}

Atomic Z_a

p^+ n^0 e^-

Gravitropic Pathway $-\frac{1}{2}_g$

Animals e^-

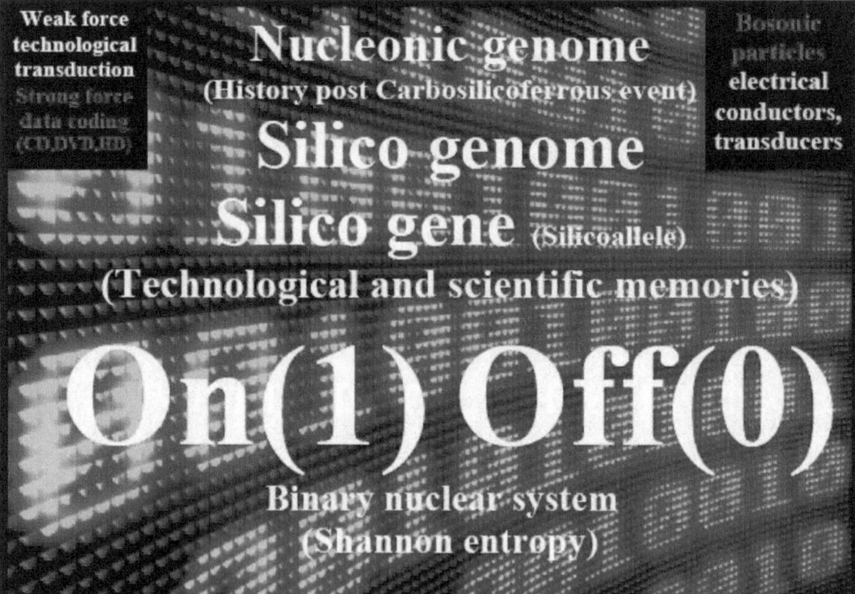

Soulatrophic pathway $+\frac{1}{2}_g$

Ferrofermionic $^{28}Si \rightarrow ^{56}Fe$

Spiritual Z_s

Cognitive Z_c

Somatic Z_{sm}

Micro Z_{mb}

Atomic Z_a

p^+ n^0 e^-

Gravitropic Pathway $-\frac{1}{2}_g$

Animals e^-

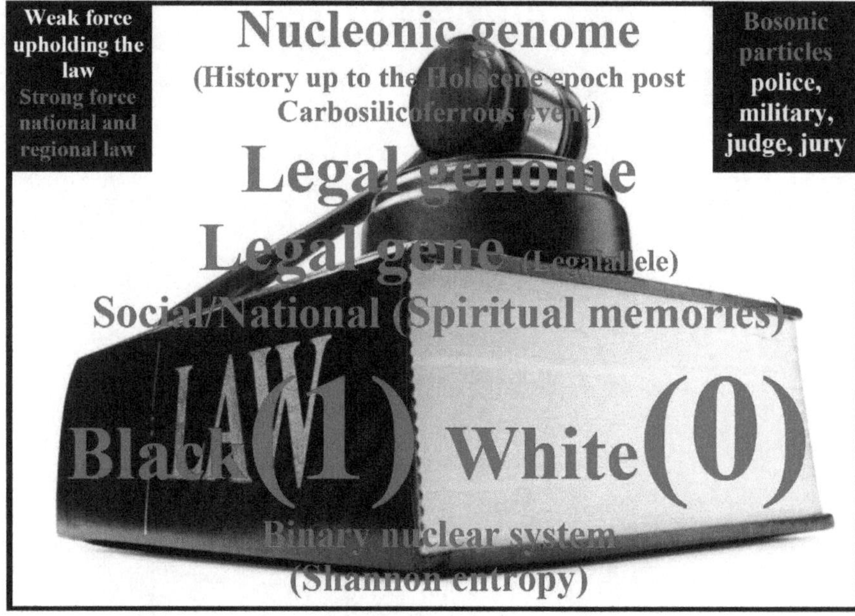

Soulatrophic pathway $+\frac{1}{2}_g$

Godmanship ^{56}Fe Z_G

Neogenous Z_e

Halogenous Z_h

Oxidative Z_x

Nitrogenous Z_r

Atomic Z_a

p^+ n^0 e^-

^{28}Si $-\frac{1}{2}_g$

Technology n^0

Examples of extended Carbon Entromorphological supercoiled helical nuclear chromosomal binary linear information (knowledge) systems

Chromosomal supercoiling is beautifully demonstrated by this book! Who would have thought the bible was nuclear physics!

A classical neurological chromosomal system. The 'black (1)and white (0)' is the binary form. All information is organised against atomic nucleonic logic. They are amplified versions of such logic.

Neurological gene expression occurs when you read the book. A copy of the gene is made in the readers neurological system (nuclear amplification).

The chromosome opens up like a DNA chromosome and is read by the reader (converted into a mRNA neurological equivalent boson) using their electroweak standard model (conscious wavefunction). Hence nuclear and field amplification by ECASSE.

Nuclear physics and linear binary information associations for neurological helical supercoiling genetics

The---π---strong---π---interaction---π---is---π---
g g g g
the---π---fundamental---π---basis---π---of---π---
g g g g
all---π---information---π---systems---π---.
g g g g

DNA genes are organised in this way as well.

π– Pionic nuclear leakage links nucleons

g – Gluons hold nucleonic genes (words) together

Letters are related to quarks

Atomic genes are organised in this way as well.

A example of a neurological helical supercoiled nuclear chromosome

Supercoiled chromosomes on the neurological level can be seen in language and writing. The black and white of written binary language is supercoiled into sentences where each word is a neurological gene and a piece of writing like this one is a neurological chromosome. An entire book could be considered to be a genome, its chapters chromosomes and words neurological genes. The logic can also be considered against atomic logic as well where the letters in each word are held together by gluonic logic and words are linked by pionic logic. Quantum expression occurs when it is read, and nuclear amplification occurs in the readers mind where the nuclear gene is copied (memory)!

Spooky action at distance the EPR paradox and its relation to the process of non local hereditary and memory.

In 1935 Einstein, Podolsky and Rosen investigated discrepancies in quantum theory by looking at strange phenomena called 'non locality', described by the group as a 'spooky action at distance'. The process suggests that when two or more particles physically interact their overall superposition is conserved not just locally but also non locally. Hence the particles are extracted from the local environment where they interacted and driven far apart, however they still retain physical properties of the thermodynamic interaction. In essence there has been a physical change instigated in all the particles involved, such that the particles if separated by enormous distance, hence non locality, still appear to be in a physically associated state. Since the greatest speed at which energy travels in the universe is the speed of light then no transfer of physical properties can exceed this limit. Einstein and his colleagues suggested that these phenomena could be explained by the presence of 'hidden variables' in the system. This implies that the thermodynamic interaction of local particles produces a change to the particles which is stored permanently in the system in a way that is extremely discrete and difficult to measure, producing a type of memory or heredity (atromic genetics).

Current theory on atomic physics suggests a world of clones, such that when you have encountered the physical properties of an electron you have in a sense seen them all; hence a subatomic world of clones. This non locality concept suggests that atomic particles are all very different but in measuring them science is resetting their non local properties and as a result producing clones. An atom in a molecule is subject to vast complex environmental forces which produce unique states which if maintained when removed from the molecule will continue to remain a stable property if the interaction was thermodynamically stable. This is hypothesised by this article to be the nature of heredity, and that genes in DNA represent the 'hidden variables', it suggests that through self symmetrical analysis of natural order, atomic genes can be hypothesised as the result of unique nuclear organisation in atoms. This is the basis of memory, where memory or heredity or non locality, which ever level of self symmetrical property is under consideration, is based on the absolute thermodynamic state history. This process of non local conservation is the quintessential basis of evolution by natural selection, which is the epitome of memory (history).

Neurochromosome, Oppenheimer's proof and the power of the mind.

The father of the atomic bomb could only have produced the bomb if both his brain and his DNA were nuclear systems of logic. The thermodynamics only works if this were the case.

"If the radiance of a thousand Suns were to burst at once into the sky, that would be like the splendour of the mighty one and now I am become death, the destroyer of worlds."

Oppenheimer is an excellent proof for carbon entromorphology as he clearly demonstrates the nuclear nature of both the brain and the DNA genome and its energetic thermodynamic function. Energetically he produced a massive nuclear explosion, but where did the reaction and its energy come from? Was it when uranium was split? Certainly this is as far as the physics profession considers it. Or did it begin when his brain conceived of the explosion? Or did it originate from his DNA genome which made his brain form in the first place? Or perhaps the energy originates from the carbon atomic nuclei which comprised his DNA? Perhaps it began at the 'Big Bang'?

In fact this is the true logical thermodynamic model of this process and proves undeniably that the brain and DNA are both nuclear systems as he could not have produced a nuclear explosion if this were not the case. The first law of thermodynamics applies to the atom bomb, but because of the quantum gravity effect (the seamless equilibrium between electromagnetic forces and gravity and the exchange of determinism between mass and energy) Oppenheimer's brain only appears to produce tiny amounts of energy but this is deceptive, as his brain contains a significant amount of mass which has accumulated over billions of years of evolution, but when one considers the mass of his brain or and his DNA then the amount of energy associated with it is still enormous.

Humanity can conceive of atomic logic because we are made from such logic, the physics profession only see atomic logic on the level of a single atom and doesn't consider it past small levels of particle composition. Life is inheritable and represents a conserved pathway from atomic levels up to huge composite particles called living organisms that's why we can look like huge carbon atoms. Carbon bonds with trillions and trillions of other atoms and in reality the actual number is so enormous it would be difficult to estimate and calculate. The universal awareness and the acceptance of the Big Bang and 13.7 billion years of time allow humans in particular to bond with astronomical bodies.

This logic supports the theory of carbon entromorphology!

Science check

Do the observations correlate with the theoretical model?

YES ✓ *NO*

PART SIX

Analytical hypotheses in support of carbon entromorphology

Neurochromosome entromorphic hypotheses – classical scientific procedures in support of amplified atomic logic in living organisms.

The classical well established model of scientific investigation involves the designing of experiments aimed to reveal supportable hypothesis to formulate theories; carbon entromorphology has this challenge.

A scientist must start by producing a hypothesis, typically generated through observations about natural physical systems. The hypothesis is a challenge to current thinking and must identify the system of study, its physical state and the logic used to investigate it. The hypothesis is then set up as a working model, which is then challenged, by experimentation and its associated observations. Typically the hypotheses are set up to be rejected and a hypothesis, which can survive continuous challenge (natural selection), may be drawn into a basic theory. The hypothetical explanation of 'cause and effect' often based around many other theories must be fully integrated into the scientific model. Data generated is often of a numerical form and any treatment of the data to challenge the hypothesis is desirably mathematical although observational data, if resolution of the system is achievable, is highly desirable.

Experiments are set up to challenge a variable often whilst other variables are kept constant. A careful set of calibrated recordings of data, produce the evidence to challenge the hypothesis. Controlled conditions are paramount in a scientific experimental hypothetical model. And often a repeat of the experimentation is used and may involve a second independent group of scientific investigators looking for duplication of results. Statistical evaluation of models is always present where observational measurement is numerical.

Peer review is essential for publication where a theory is being suggested and its hypothetical explanation has been unsuccessfully rejected. The findings of experimentation are presented in the context of the hypothesis and the theory resulting from it. Only after rigorous testing by peer reviewers and establishment of conventions for publication are the results published in scientific journals.

Incredibly even the most supported theories are still called and accepted as theories even after vast amounts of time. This means that science maintains a high level of challenge to its own models. Any scientist cannot test all possible outcomes of a testing system and therefore continues to challenge any theory. This is a prudent and sensible approach which aims to keep evolution of science at its most prolific and supportable level.

Carbon entromorphology has long since required a condensed logical set of hypotheses and evidence integrated to define the theory. Experimentation evidence has always been strong because biology is a subject where observation and measurement by sight are available as biological systems are large enough to be thoroughly evaluated by sight alone. This makes biology one of the most successful natural sciences as a visual pictorial observation and measurement of the biological systems is highly desirable over basic numerical data, which seldom come close to modelling the enormous complexity of living organisms. Any physical scientist will always feel out of place using visual data as the physical sciences are all based on numerical mathematical logic. An atomic physicist practices science in a blind abstract surrealistically simplified manner, where any attempt to visualise the atomic system is fundamentally unsupportable. Therefore physical scientists don't consider science outside of mathematics and numerical data. A biologist on the other hand prefers to look and make observations through sight alone as a pictorial model contains massive quantities of data which can be seen. Most of medicine has been developed without solving mathematical equations; a surgeon can look at a tumour and remove it by sight. A physicist cannot look at an atom and so only produces surreal counter intuitive models often surrounded by speculative theories. Often in physical science the theories are generated out of mathematical necessity and logical compliance, for example the natural symmetry in nature often suggests a mathematical inverse. An example is the Dirac equation which as well as modelling matter also by mathematical symmetry, suggested anti-matter. A mathematical model with two solutions which predicts a complete new model of reality before any observations or measurements has been made on it. Again the gravitons of the gauge model are suggested by mathematics but have never been observed, anti matter on the other hand it now routinely made in laboratories.

The hypotheses in support of carbon entromorphology are vast in scope and vast in evidence. Since the theory suggests that life is a conservation of atomic level of organisation and since living organisms can be measured by sight, a symmetrical unification can be proposed. The evidence system is based on photographic data, for example a photograph of an organisms head which contains 2 eyes, and a brain as models of that particular physical system, can be used to demonstrate and elucidate the atoms which they are comprised of. However much the atomic physicists feel that only mathematics can reveal the atom, it is impossible to ignore the vast numbers of graphical models of the atom produced by scientists over many years. In fact in most cases where a physicist considers the atom they will have to use a graphical model to describe the anatomy and physical potential and organisation in the atom. Therefore carbon entromorphology allows humans to look at organisation seen in the atoms of living beings through the logic of nuclear and field amplification. This evidence system for carbon entromorphology is so vast that it could contain trillions and trillions of viable agreeable evidence. There are so many and so much diversity in life that any anatomical property can be demonstrated trillions of times over. This makes carbon entromorphology one of the most supportable theories in science. The bottom line is do the physicists truly believe in their theories, all carbon entromorphology is doing is respecting the enormous success of nuclear physics and quantum mechanics. Through chemical bonding links are set up with carbon at their heart to extend carbons determinism and hence amplify its properties.

97% of all the atoms in a living organism are made of just 4 elements which in turn are intrinsically linked through covalent shared bonding, making massive composite particle wave systems and conserving the links between the smallest levels of natural scale (atomic) all the way up to that of an entire organism such as a human. This quantum amplification effect is occurring all the time in multi-cellular organisms; the quantum world allows the reader of this article to have the energy to read it.

10% by weight of a human is made of hydrogen, the most basic element which makes up all the other elements in the periodic table. It is the fermionic basis of all matter and in carbon entromorphology it is conserved in all composite living structures. This means we can describe any organism in terms of its polarised hydrogen fermionic logic. It is therefore so breathtakingly fundamental that it has been hypothesised in carbon entromorphology as the basis of polarised gender, also a neutron (the third fermion) is made of a composite of a compressed hydrogen atom. This atom also contains a 1s orbital distribution and also follows the rule of octets acting to stabilise its K shell. Its small size makes it an excellent atom for carbon, nitrogen and oxygen to bond to, making for a condensed stable covalent molecule and the basis of carbon membranes as it shields atomic carbon.

The other 87% of the elements which make a human are just three elements conjoined in logic on the periodic table. Carbon, nitrogen and oxygen are all made from hydrogen linking their logic together. They all have a nucleus, a K shell (n=1) and L shell (n=2) and follow the rule of octets gaining solar energy to produce stable energy level systems. They all contain the three fundamental sub shells (orbitals), the 1s spherical distribution, the 2s sphere within a sphere distribution and three lobe shaped p sub shells each at 90° to each other. Each distribution connects with other atoms through bonds and links connecting massive quantities of atoms into vast molecules through quantum sub shells (they all covalently bond). Since quantum space is measured as a superposition of states it provides an excellent hypothetical model for explaining all the diversity of life. Carbon in particular is central in this model and has an evenly imbalanced L shell containing four electrons allowing it to form vast and complex covalent sharing bonds with itself, nitrogen and oxygen and hydrogen: in essence it is the true nuclear element next to hydrogen. This allows it to conserve and amplify its atomic logic producing living structures such as cells, which are comprised of 87% carbon, nitrogen and oxygen determinism and 10% hydrogen determinism. The cell has a nucleus, a cytoplasm (K shell, n=1) and a membrane (L shell, n=1) which are self-symmetrical to the nucleus, K shell and L shell in 87% of the atoms which comprise it. An organism such as a human, has a physical composition of approximately 100 trillion of these cells acting with deep fundamental integration as a composite super particle wave system and has a head (nucleus) a body (K shell, n=1, cytoplasm) and limbs (L shell, n=2, membrane and limbs). Each level shows absolute conservation of the physical and anatomical model of the four elements which comprise 97% of its matter, their properties amplified through extensive and massive solar driven bonding (linked connectivity) to produce giant hadronic composite particles.

Composite natural structures such as organisms must have conservation of the building blocks which they are comprised of. In other words life being a pathway of links running back billions of years to a time where carbon existed as a simple carbon dioxide molecule. Growing and conserving links of determinism over that time to produce a human comprised of 100 trillion cells which in turn are comprised of trillions and trillions of atoms must have evidence of its dichotomy of construction (building block morphology and function) at all levels. The fractional dimension in carbon entromorphology is testament to this and allows us to use any level of logical self-symmetrical fractions to describe the logic of the wholen accretion composite structure. It means we can explain the organisation using sub atomic logic, atomic logic, cellular logic, organismal logic as they are all fractions of the linked integrated composite particles and collectively show integration to produce very complex giant structures such as a human.

The 26 element human molecule (Thims, 2008) is part and parcel of the direct link between atomic levels of organisation and those of higher levels. A single carbon atom in a human can be argued to be part of a trillion trillion trillions (the octillion) sized super molecule hence the links and conservation of atomic organisation. This molecule evolved over millions of years slowly and with strong links to its past entanglements. It's impossible to truly limit the size of such huge biochemical molecules, what is important is the proportion of its deterministic contribution which for carbon, nitrogen and oxygen and more fundamentally hydrogen is 97% of all the matter in humans.

Try to imagine the extension of carbon logic in a US president; the molecules and hence atomic level amplify their effect to billions of people in seconds through their speech on the TV.

There is a trade off in the logic however, as the atomic level appears to be constructed of almost empty space where electromagnetic determinism dominates at very close levels of resolution. On the level of a composite human for example, the structures have a diluted electromagnetic effect and a significant mass component. This is quantum gravity with life as a special case, the seamless transition between atomic significant forces and those of gravity are clearly evident in this process. Mass rich structures act as though they where electromagnetic fields as a conservation of atomic organisation and its links between these levels of natural scale. A physicist typically cannot link atomic organisation to astrological organisation, and any effects of the quantum world appear to disappear as we resolve larger collections of the smallest particles. But this is not true of life where the links to the quantum past are conserved in organismal systems, and still remain part of the chain of determinism in a functional organism.

For example, the cytochrome system in the membranes of mitochondria extract the energy of solar stimulated high energy electrons, which is transferred into ATP (Adenosine tri phosphate), energy from food (Sun) and produces water at the end. This effect is driven by the electrons energy and is used to produce a protonic gradient across a membrane. This is a quantum process but its energy is used to allow a human to think, and move hence a quantum link (across many levels or organisation) between atomic activity in an organisms lower levels of natural scale and the overall physical activity of the complete organism. Another powerful example is the effect of a single photon hitting a rhodopsin molecule in the eye. The visual amplification cascade is an amplification of a quantum effect which is raised to the level of an entire human in under 1 second. The photon excites an electron in rhodopsin, which activates 500 transducin, which activates 500 phosphodiesterase which hydrolise huge quantities of (105) cyclic-GMP, which close 250 sodium channels, preventing 106 – 107 sodium from entering the cell for a period of less than 1 second. This hyperpolarises the rod membrane by 1mV, sending a visual amplified signal into the brain. This could be a photon from a highly dangerous situation, perhaps in the eye of the US president, which could lead to war. If that isn't proof of nuclear and field amplification I don't know what is!

The physics world would have everyone believing that the atomic world is somehow separated with our everyday experiences which I am afraid, is unsupportable and is a concept driven by mathematical limitations. There is no common sense in this approach for life where atomic reactions from billions of years ago are still part and parcel of an integrated human and are still essential for maintaining life, there is no 'cut off' point in living organisms. There are so many examples of this effect that you could write a book on them, in fact all biological activity is seated on simpler logic (Redox reactions) but we do not lose this logic it simply changes its determinism from an electromagnetic effect to gravity depending on a position of resolution (how close we look, do we look at a molecule or the entire human from which its determinism contribution came?).

The entromorphic hypotheses (order may differ).

There are a vast number of hypotheses as this theory is a major unified theory aimed at harmonising all human knowledge into one integrated system. The following are the original hypotheses of most importance.

- Thermodynamic hypothesis – The extension of the first law and implementation of a forth law.

- Fermionic hypothesis – The conservation of fermionic determinism is identifiable on any level of natural living scale.
- Evolution hypothesis – The extension of evolution by natural selection.
- Nuclear hypothesis – The identification of the various nucleonic systems and their force equilibrium consequences.
- Quantum hypothesis – The natural extension of quantum determinism in living organisms, the uncertainty principle and life.
- Energy level hypothesis – Quantification in living organisms.
- Pauli hypothesis – The fermionic statistics and its consequence to living organisation.
- Orbital hypothesis – The physical distributions in living systems as a model of morphology.
- Head hypothesis – This is one of the oldest hypotheses which uses an organisms head as a physical biological structural system, which has parallels and symmetry to the organisation of the atoms which it is comprised of.
- Limb hypothesis – This hypothesis identifies the bonding in living organisms and suggests a fundamental link to the chemical bonds in the atoms they are comprised of.
- Body hypothesis – This hypothesis identifies the body in living organisms and suggests a fundamental link to the K shell in the atoms it is comprised of.
- Coulomb barrier hypothesis – This hypothesis identifies biological structures and their nuclear origins.
- Potential well hypothesis – This hypothesis identifies localised biological structures and their atomic origins.
- Barrier penetration hypothesis – This hypothesis identifies localised biological structures and processes and their atomic origins.
- Electric field line hypothesis – This hypothesis identifies localised biological structures and their atomic origins.
- Gauge symmetry hypothesis – This hypothesis identifies localised biological structures and their atomic origins. Also included is a chiral section.
- Electroweak hypothesis – This hypothesis identifies the processes of connectivity between organisms and its relation to living expression.
- Gravistrong hypothesis – A new symmetrical hypothesis which aims to suggest a grouping of forces in parallel to those of the electroweak standard model.
- Force cycle hypothesis – A new symmetrical hypothesis which aims to suggest a grouping of all four forces in parallel to those of the other particle standard models.

- Mathematical hypothesis – Mathematics is the language of nature so why bother even applying it, the pure form should be a precise model of nature?
- Statistical hypothesis – All statistics are a model of the quantum effect.
- Visual data hypothesis – Visual data is more effective than numerical data which is typically sampled at a very low level.
- Cell cycle hypothesis – The four time zones and their relationship to the octet rule.
- Periodic table hypothesis – The need to find life on the periodic table.
- Octet hypothesis – The octet as a measure of solar acquisition and stability.
- Iron hypothesis – The powerful element of a 4th law of thermodynamics driving all elementary thermodynamics through an organism's free will.
- ECASSE hypothesis and sports– Carbon is a boson and a fermion and can semi-occupy energy levels to produce coherent in phase organisation.

The principle general entromorphic hypothesis.

That living organisms have a fractional dimensional logic organised at unique levels and conserve and amplify the logic from lower levels (atoms) to produce higher levels (organisms) through solar acquisition. In short, life is an amplification pathway through iterative 'open system' cell cycling. Showing conservation (self symmetry) of lower fractions such as the atomic level, where carbon is the centrally deterministic element and is amplified through bonding links to produce cells and organisms, with increasing mass determinism and decreasing electromagnetic determinism.

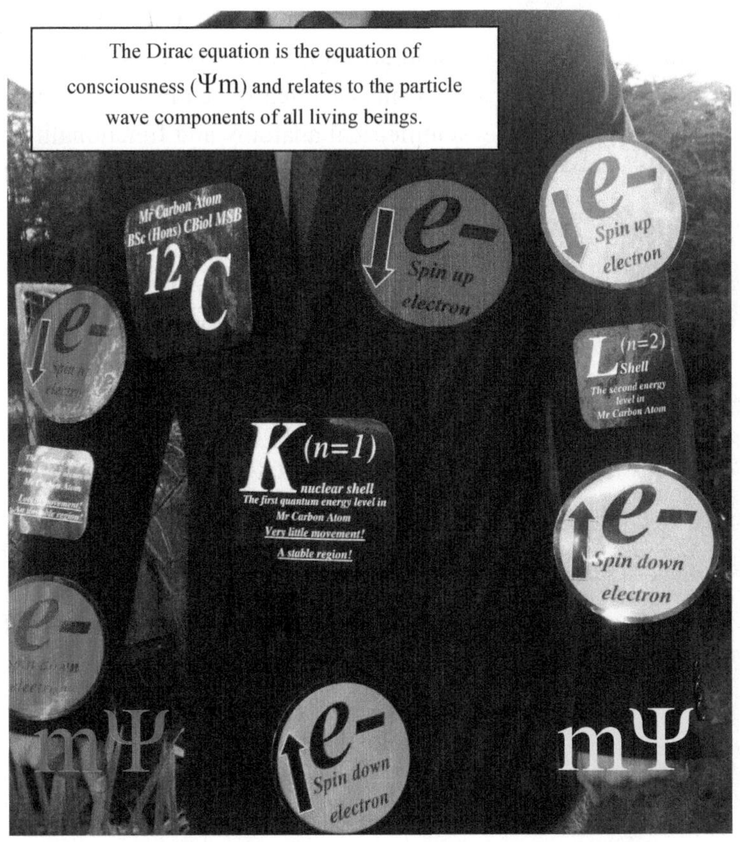

The Dirac equation is the equation of consciousness (Ψm) and relates to the particle wave components of all living beings.

The following images have been demonstrated to support the hypotheses relating to the torso and K shell relationship hypotheses. They also clearly demonstrate the limbs and their relationship to the atomic covalent bonds and quantum logic. The concept of spin is also demonstrated with supporting models such as the foetal position and ground states in carbon.

The nucleonic pathway is also displayed running down the centre of the body from the head to the genital region. In short, the Dirac equation is a constant measurable logic in every waking moment of an organism's life; the hand and the eye and the entire organism are 'spinners' in such equations (Ψm).

The 'Dirac spinners' and the limbs.

The various nucleonic binary nuclear logic in a living organism, running from the head to the genitals.

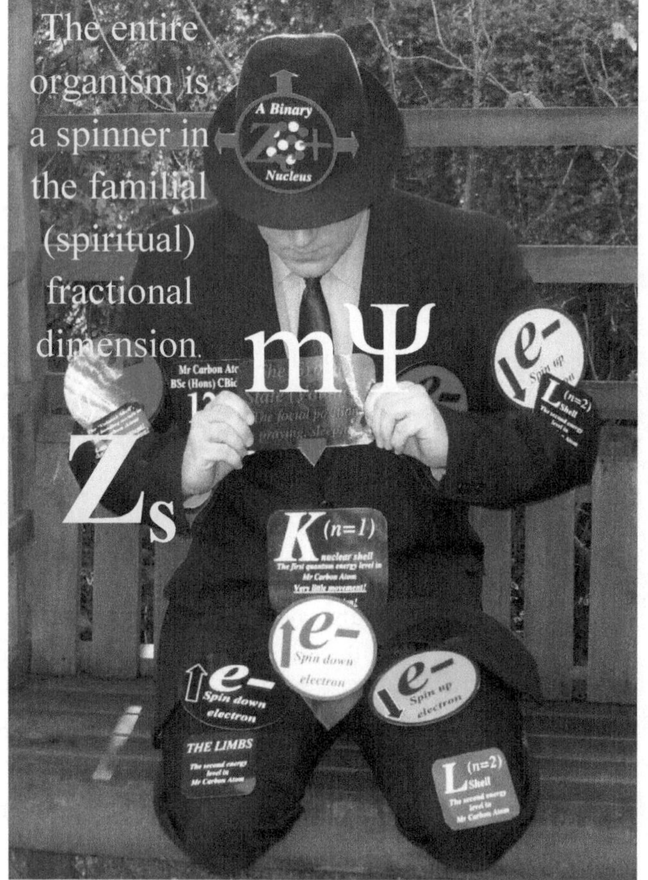

The quantity of evidence in support of all carbon entromorphology hypotheses is absolutely enormous. Countless trillions of organisms can clearly be observed to have a symmetrical anatomy and functionality or physiology, which again directly parallels the same organisation in the atoms they are made of. Most multicellular organisms exist in all areas of the world producing breathtaking variation but based on some very simple structural phenomena.

This evidence is not merely coincidence and it appears that any such notion is unsupportable. In short, by comparison to the little scatter graphs, scintillation light, spectrum and tiny almost unviewable tracks made in cloud chambers in physics proofs pale into utter insignificance by comparison to the viewable, touchable, tastable, smellable observational evidence in carbon entromorpholgy.

Testing the hypothesis.

Examine physical observational evidence and numerical data for different levels of natural scale, and identify how they fit together to produce an integrated organism through conserved filtered amplification. Data to be presented as photographic observations of integral levels of organisation.

Head hypothesis – This is one of the oldest hypotheses, it uses organism's heads as a physical biological structural system, which has parallels and symmetry to the organisation of the atoms, which it is comprised of.

Statement of hypothesis (point 1).

The head of bilateral organisms is one of the most revealing consequences of amplified conserved atomic logic. In the vast majority of trillions of organisms on planet Earth the head operates as a spherical probability distribution, containing two eyes, which constitute the amplified consequence of the 2 electrons on the atomic level of carbon (also nitrogen and oxygen). Most organisms also have a skull, which constitutes the coulomb barrier, and have a brain, which nucleates the system and controls the eyes. The behaviour is unpredictable for the eyes as this is amplified quantum logic and subject to uncertainty rules. The eyes also absorb light, have spin and opposite spin to each other (focus), which is symmetrical to the atomic level and they are controlled by conscious wave functions.

Testing the hypothesis.

Examine physical observational evidence for different organism heads in nature, and their physical similarity to the K shell and nucleus in the atoms they are comprised of. Data to be produced as photographic observations of different head morphology with associated atomic symbols.

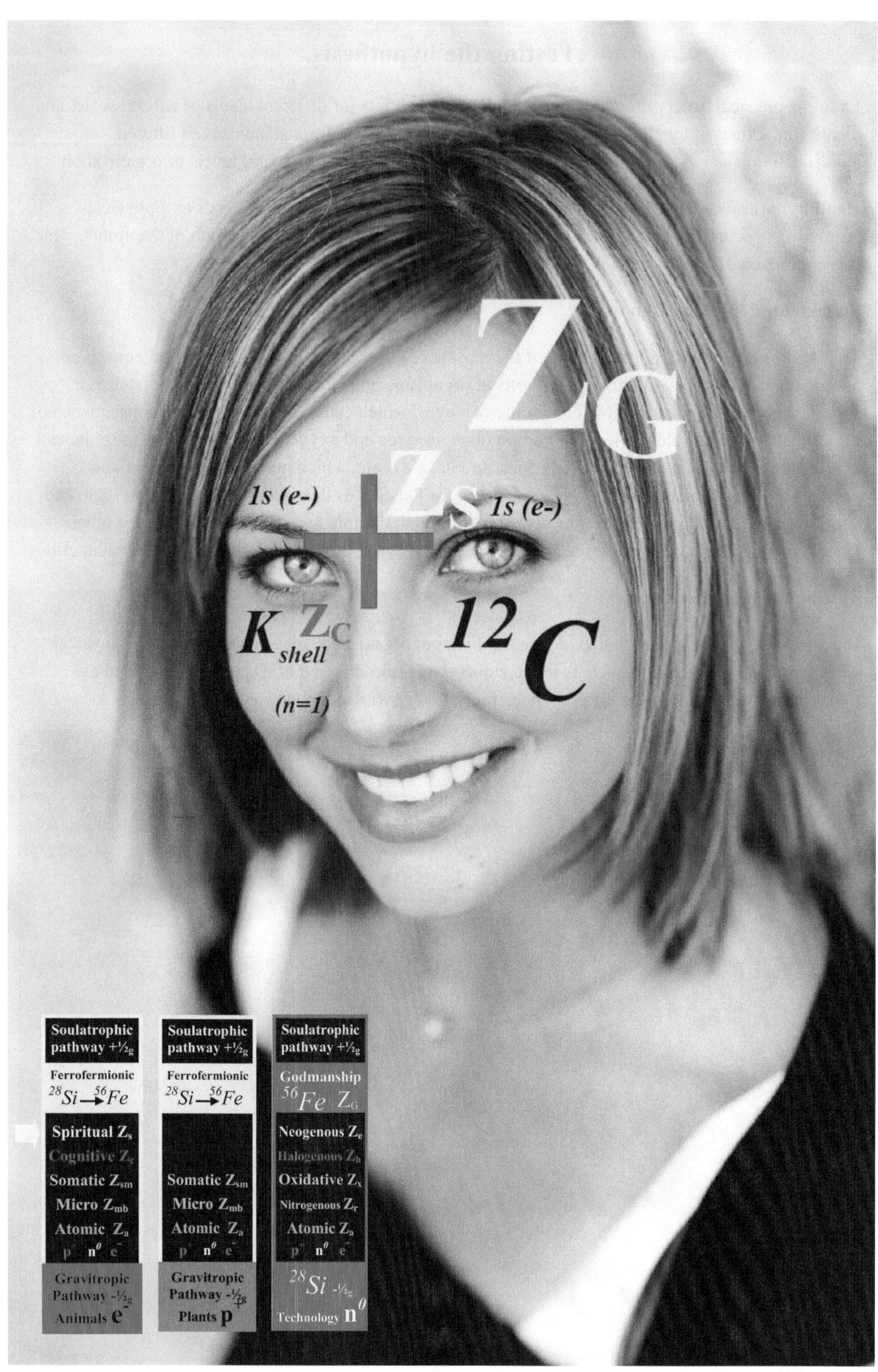
Z_G
Z_S
1s (e-)
1s (e-)
K_{shell}
Z_C
(n=1)
^{12}C
Soulatrophic pathway $+\frac{1}{2}_g$
Ferrofermionic
$^{28}Si \rightarrow ^{56}Fe$
Spiritual Z_s
Cognitive Z_c
Somatic Z_{sm}
Micro Z_{mb}
Atomic Z_a
p^+ n^0 e^-
Gravitropic Pathway $-\frac{1}{2}_g$
Animals e^-
Soulatrophic pathway $+\frac{1}{2}_g$
Ferrofermionic
$^{28}Si \rightarrow ^{56}Fe$
Somatic Z_{sm}
Micro Z_{mb}
Atomic Z_a
p^+ n^0 e^-
Gravitropic Pathway $-\frac{1}{2}_g$
Plants p^+
Soulatrophic pathway $+\frac{1}{2}_g$
Godmanship
^{56}Fe Z_G
Neogenous Z_e
Halogenous Z_h
Oxidative Z_x
Nitrogenous Z_r
Atomic Z_a
p^+ n^0 e^-
^{28}Si $-\frac{1}{2}_g$
Technology n^0

ZG
ZS
1s (e-)
1s (e-)
ZC
12C
K
n=1
(shell)

The somatic level also provides excellent evidence of an atomic conservation and K shell nuclear shell.

The ovaries and testicles are again extensively demonstrated all the way through the entire living world, there are countless trillions of supporting examples.

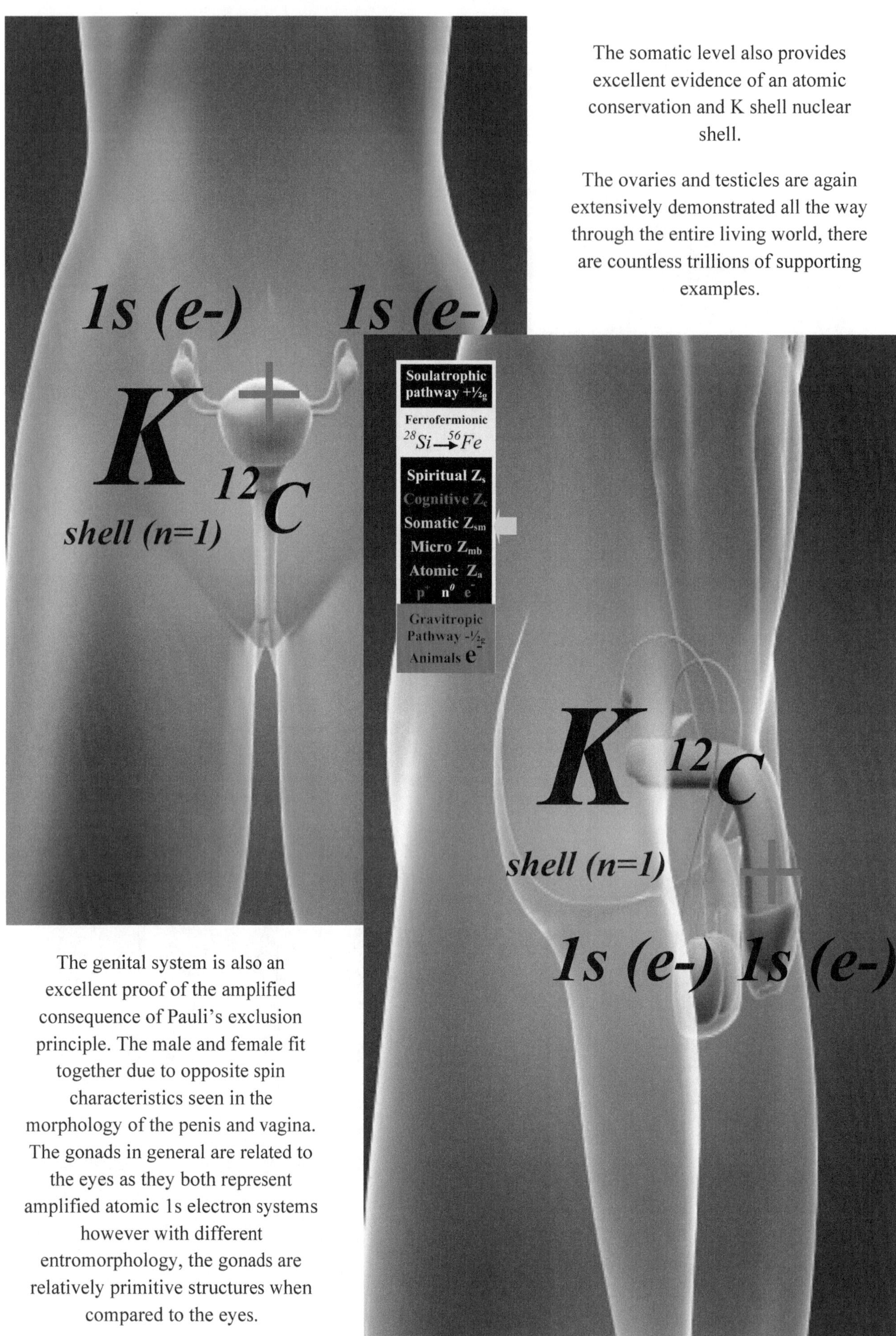

The genital system is also an excellent proof of the amplified consequence of Pauli's exclusion principle. The male and female fit together due to opposite spin characteristics seen in the morphology of the penis and vagina. The gonads in general are related to the eyes as they both represent amplified atomic 1s electron systems however with different entromorphology, the gonads are relatively primitive structures when compared to the eyes.

Limb hypothesis – This hypothesis identifies bonding in living organisms and suggests a fundamental link to the chemical bonds in the atoms they are comprised of.

Statement of hypothesis (point 1).

The limbs of trillions of organisms allow said organisms to bond to their environment in exactly the same way as the atoms they are comprised of. This is the valance shell and allows energy to flow into and out of the system and have p (lobe) and 2s (sphere within a sphere logic) spatial organisation. Arms, tails, wings, flippers, fins produce lobe shaped p type probability clouds and rotate about the body in a spherical fashion in accordance with a 2s logic. They are controlled by conscious wave functions and the s component hypothetically produces bone and the p component hypothetically produces the muscularity collectively producing sp2 hybridised covalent (sharing) bonds.

Testing the hypothesis.

Examine physical observational evidence for different organism limbs in nature and their physical similarity to the L shell valance and orbital shapes (morphology) in the atoms they are comprised of. Data to be produced as photographic observations of different limb morphology. Again there are trillions of different examples of this effect.

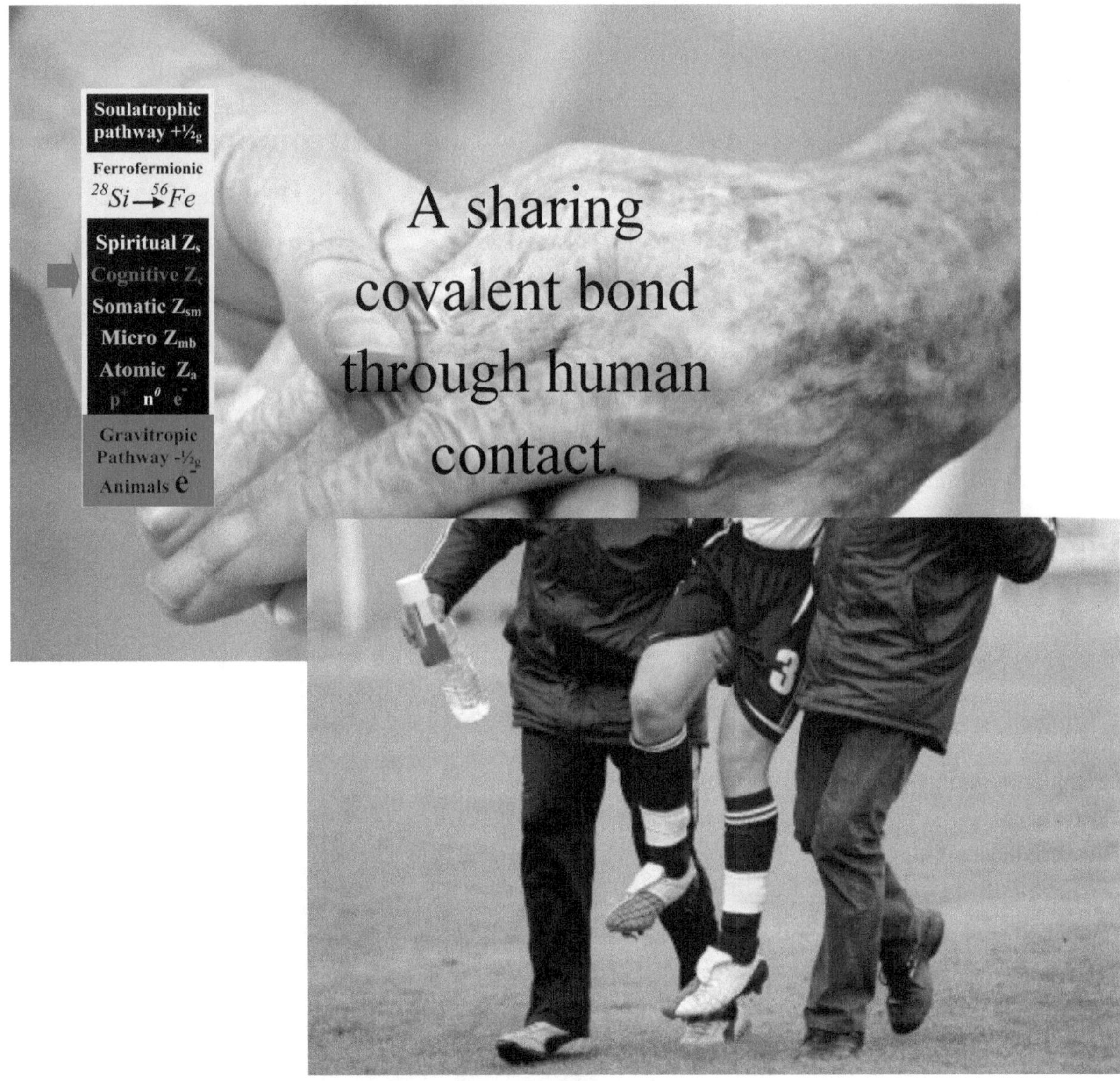

Body hypothesis – This hypothesis identifies the body in living organisms and suggests a fundamental link to the K shell in the atoms it is comprised of.

Statement of hypothesis (point 1).

The bodies of trillions of organisms are stable regions which house the head or nucleus and exist in between the nucleus and limbs or membrane. The probability distribution between nucleus (head) and valance shell (limbs) is the K shell n=1 and stabilises the nucleus (head) and shields it.

Testing the hypothesis.

Examine physical observational evidence for different organism torsos in nature and their physical similarity to the K shell valance and orbital shapes (morphology) in the atoms they are comprised of. Data to be produced as photographic observations of different body morphology. Again trillions of other examples exist to support this.

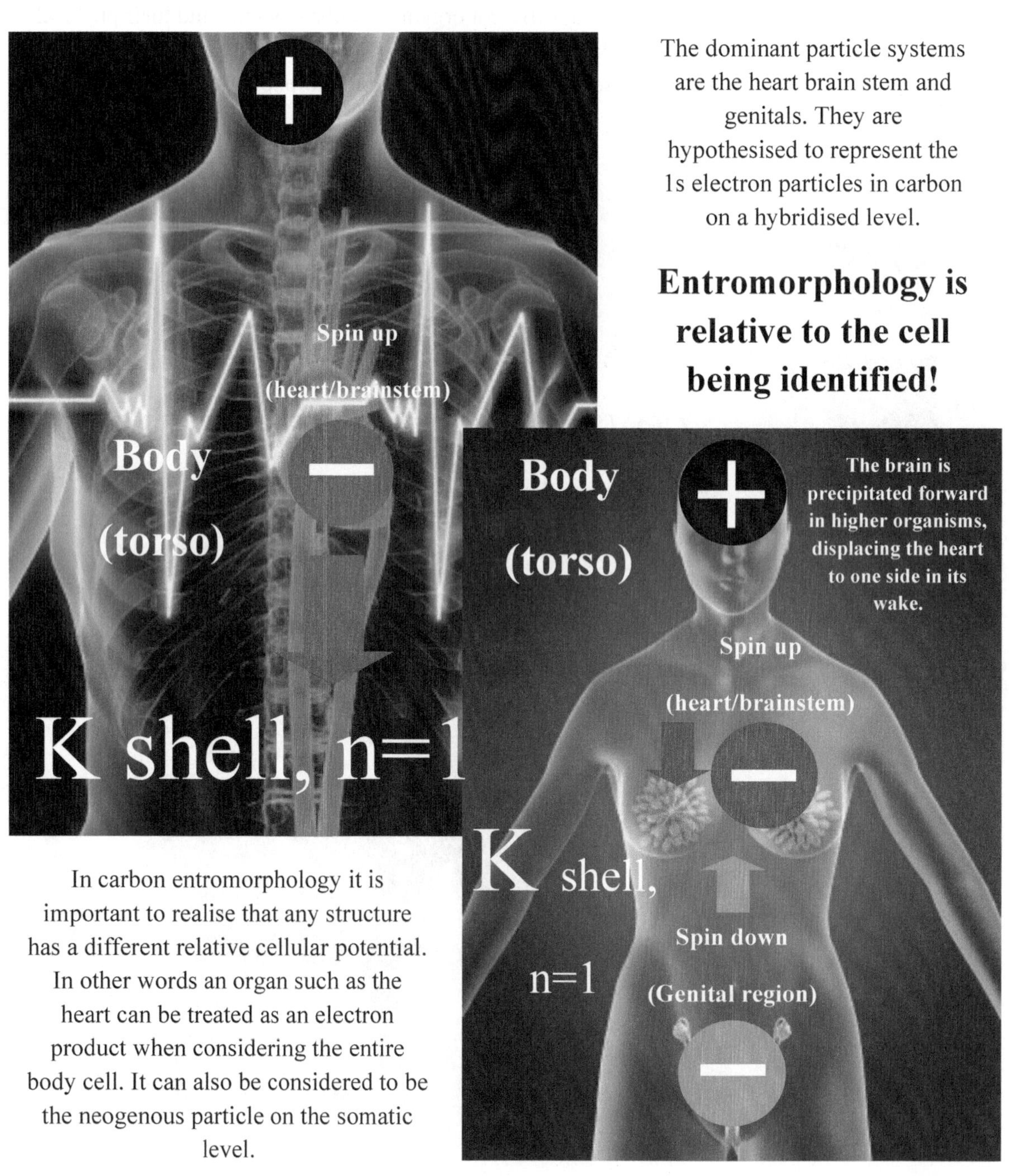

The dominant particle systems are the heart brain stem and genitals. They are hypothesised to represent the 1s electron particles in carbon on a hybridised level.

Entromorphology is relative to the cell being identified!

In carbon entromorphology it is important to realise that any structure has a different relative cellular potential. In other words an organ such as the heart can be treated as an electron product when considering the entire body cell. It can also be considered to be the neogenous particle on the somatic level.

Covalent Living Bonds – The amplified consequence of chemical bonding in living organisms.

Statement of hypothesis (point 1).

The chemical bonds, which build organisms from small atomic origins, are typically covalent sharing bonds. Amplified logic in carbon entromorphology hypothesises that the bonds made by humans to other humans for example are amplified covalent bonds. The limbs themselves represent amplified valance and conductance band regions and the electroweak standard model; these are all identifiable in large-scale organisms.

Testing the hypothesis.

Examine physical observational evidence for different organism limbs in nature and their physical similarity to the L shell valance and orbital shapes (morphology) in the atoms they are comprised of. Data to be produced as photographic observations of different limb bonding morphology. A human to human hand shake is a classical example of an enlarged covalent bond.

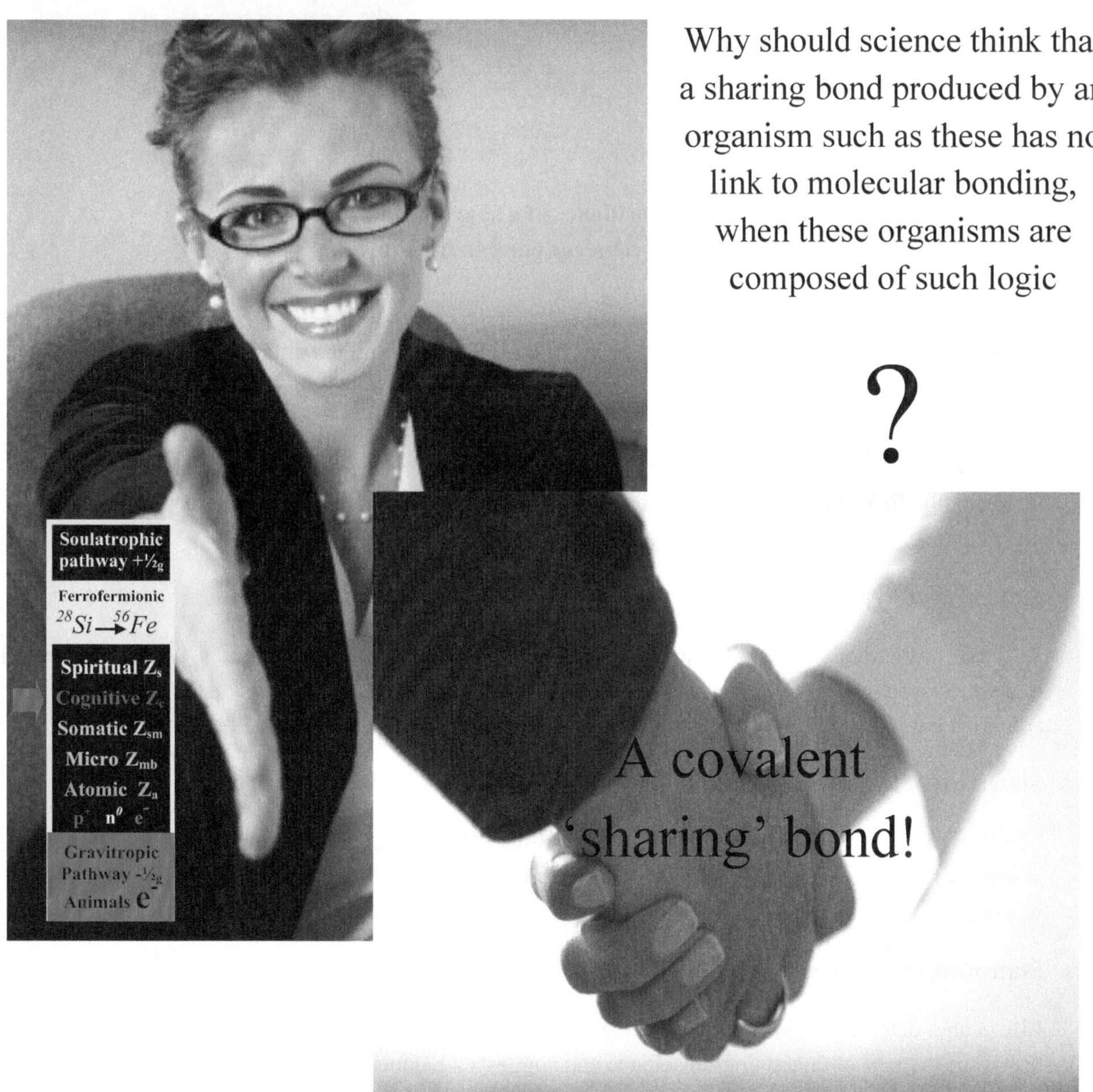

Why should science think that a sharing bond produced by an organism such as these has no link to molecular bonding, when these organisms are composed of such logic
?
Soulatrophic pathway $+\frac{1}{2}_g$
Ferrofermionic
$^{28}Si \rightarrow {}^{56}Fe$
Spiritual Z_s
Cognitive Z_c
Somatic Z_{sm}
Micro Z_{mb}
Atomic Z_a
p^+ n^0 e^-
Gravitropic Pathway $-\frac{1}{2}_g$
Animals e^-
A covalent 'sharing' bond!

Evidence of a sharing single covalent bond in humans.

Thermodynamic hypothesis – The extension of the 1st law and implementation of a 4th law.

Statement of hypothesis (point 1).

Since Einstein formulated his relativistic model of nature where energy and mass are interchangeably identical, the first law of thermodynamics should be expressed in terms of the conservation of both mass and energy (particle wave duality).

Statement of hypothesis (point 2).

A 4th law of thermodynamics is required to explain the conditions of open systems. An act of free will in an open system with free energy from the sun allow living conscious beings to accumulate energy and draw out the inevitable effects of the 2nd closed system law for vast periods of time (efficiency).

Statement of hypothesis (point 3).

The Gibbs equation can be re-written using substituted Boltzmann entropy to produce an energy cycle in terms of both macroscopic and microscopic physical systems. Morality is a thermodynamic logic where good is a reflection of a 4th open system law and evil a reflection of the 2nd closed system law.

Statement of hypothesis (point 4).

Life cannot determine whether the universe is an open or closed thermodynamic boundary system, this is the essence of uncertainty in science and suggests a balance or equilibrium between the 2nd (closed law) and 4th (open law) law for any particle wave duality system.

Testing the hypothesis.

Examine physical observational evidence of different levels of natural scale and identify how they fit together thermodynamically to produce an integrated organism. Data to be produced as photographic observations of integral levels of thermodynamic organisation.

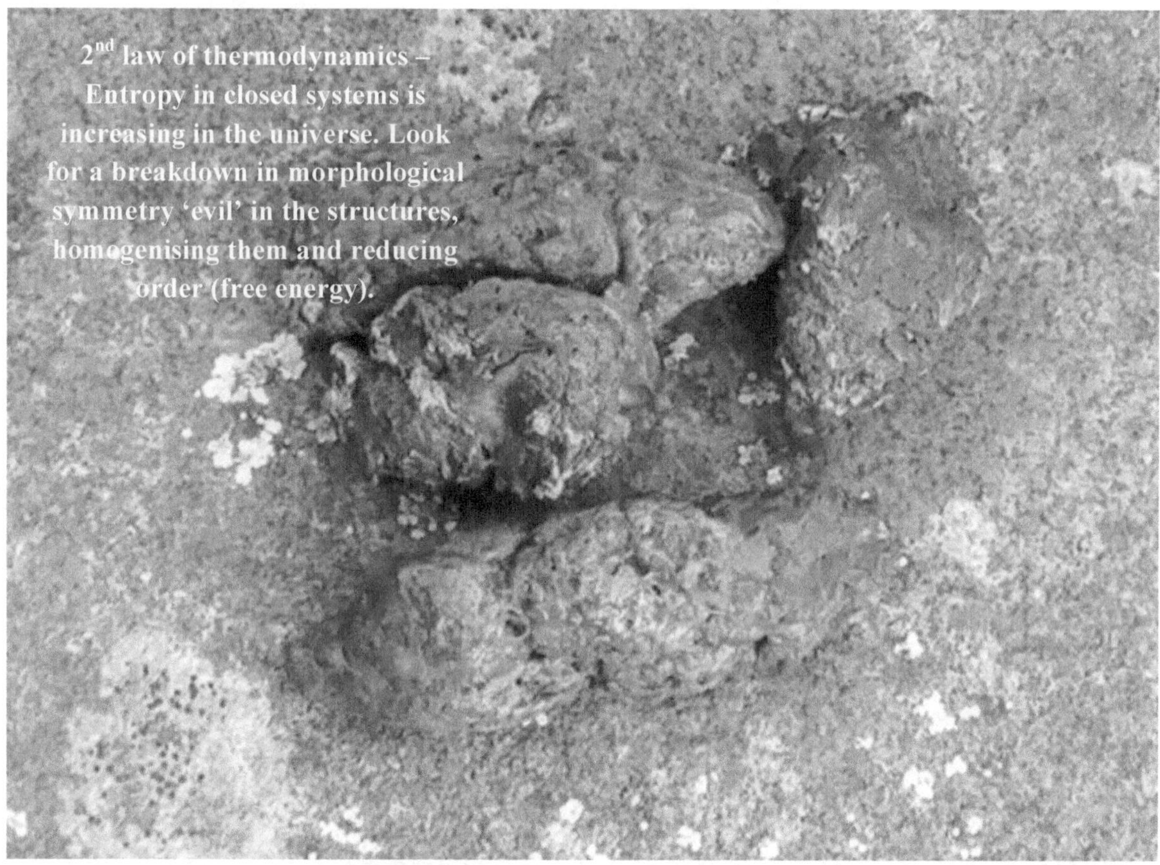

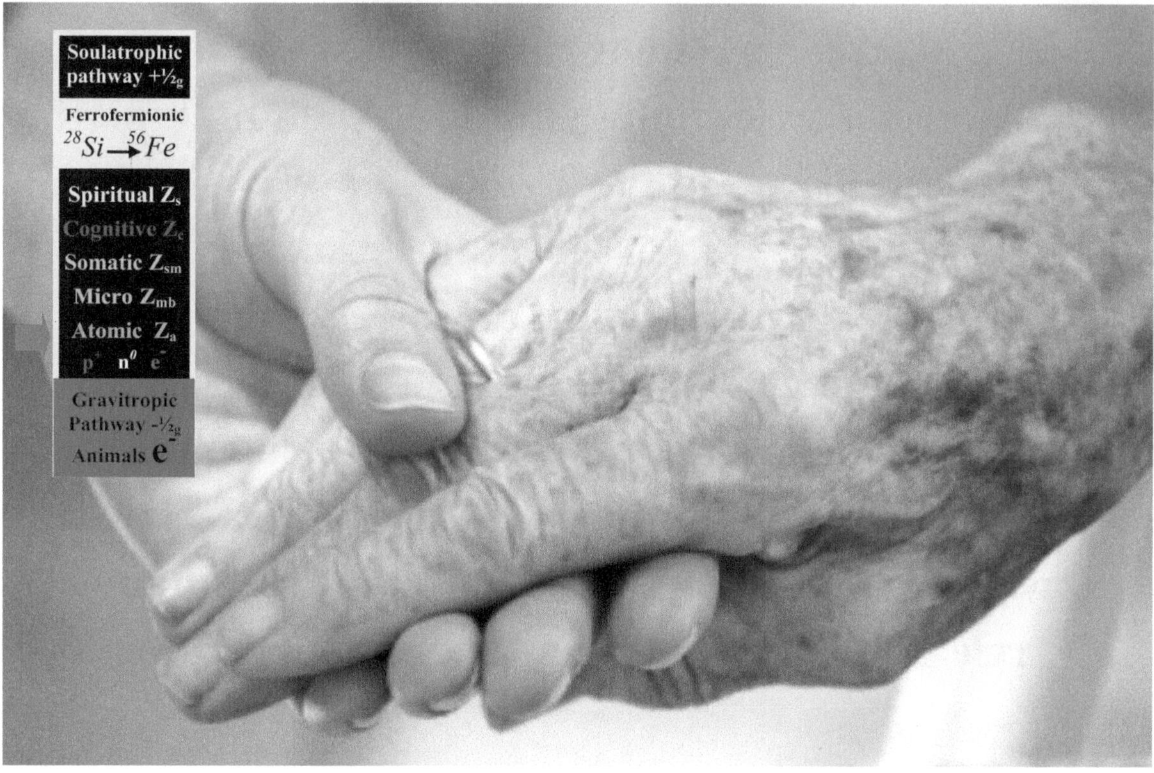

The 4th law of thermodynamics – The Sun is the producer of the enormous quantities of 'free energy' on planet earth (phototropism). Efficient use of free energy (known as good) is the method required for spreading out time such that the inevitable effects of the 2nd law can be withheld. This is an exciting prospect for humanity as it results in massive amount of time availability to life on Earth and should humans gain the ability to re-engineer longevity produces a potentially beautiful heavenly future. Although the 2nd law is waiting in the wings to collapse this result and release the useful energy as thermoentropy.

Fermionic hypothesis – The conservation of fermionic determinism is identifiable on any level of natural living scale.

Statement of hypothesis (point 1).

Since matter can be simplified to the three fermionic particles, the proton, neutron and electron (having both mass and energy properties), then a composition of these particles must reflect their physical properties however large the resolved the composition is.

Statement of hypothesis (point 2).

The fermionic trinity is the basis of gender; a protonic nuclear particle is female (radiating high pressure nuclear properties, impractical clothes, hair large delicate properties, low velocities, energy conserving), the electronic field particle is male (converging field quantum low pressure properties, practical clothes, hair slick, high velocity, energy liberating) and neutronic particle is homosexual (neutral physical nuclear properties, often both male and female properties are present). This broad definition is true for all matter in the universe as all matter is comprised of this logic and all physical interactions are the result of these three particles and their conserved properties.

Statement of hypothesis (point 3).

Fermionic properties dominate electromagnetically at close observational interactions of matter and gravitationally at distant observational interactions of matter. Observation is an equilibrium of the force distance relationship.

Testing the hypothesis.

Examine physical observational evidence of different levels of natural scale and identify how they fit together to produce an integrated organism. Data to be produced as photographic observations of integral levels of organisation. Look for high pressure and low pressure morphologies in gender.

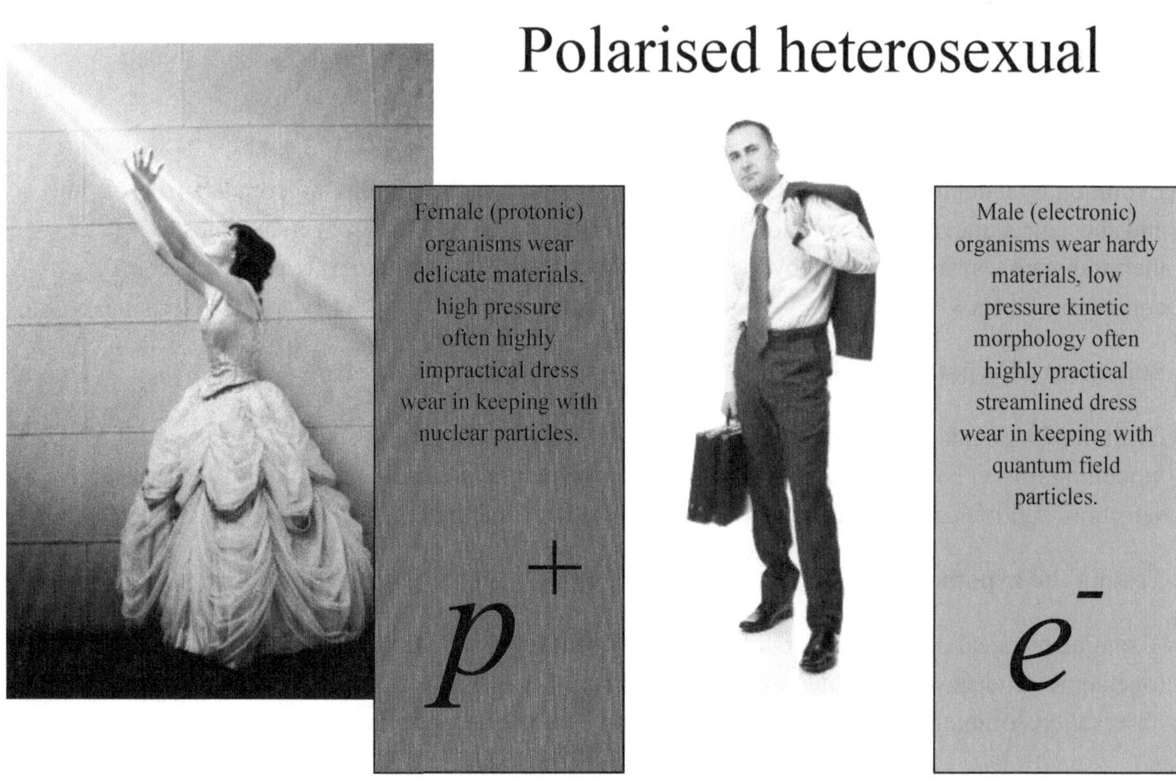

Polarised heterosexual
Female (protonic) organisms wear delicate materials, high pressure often highly impractical dress wear in keeping with nuclear particles.
p^+
Male (electronic) organisms wear hardy materials, low pressure kinetic morphology often highly practical streamlined dress wear in keeping with quantum field particles.
e^-

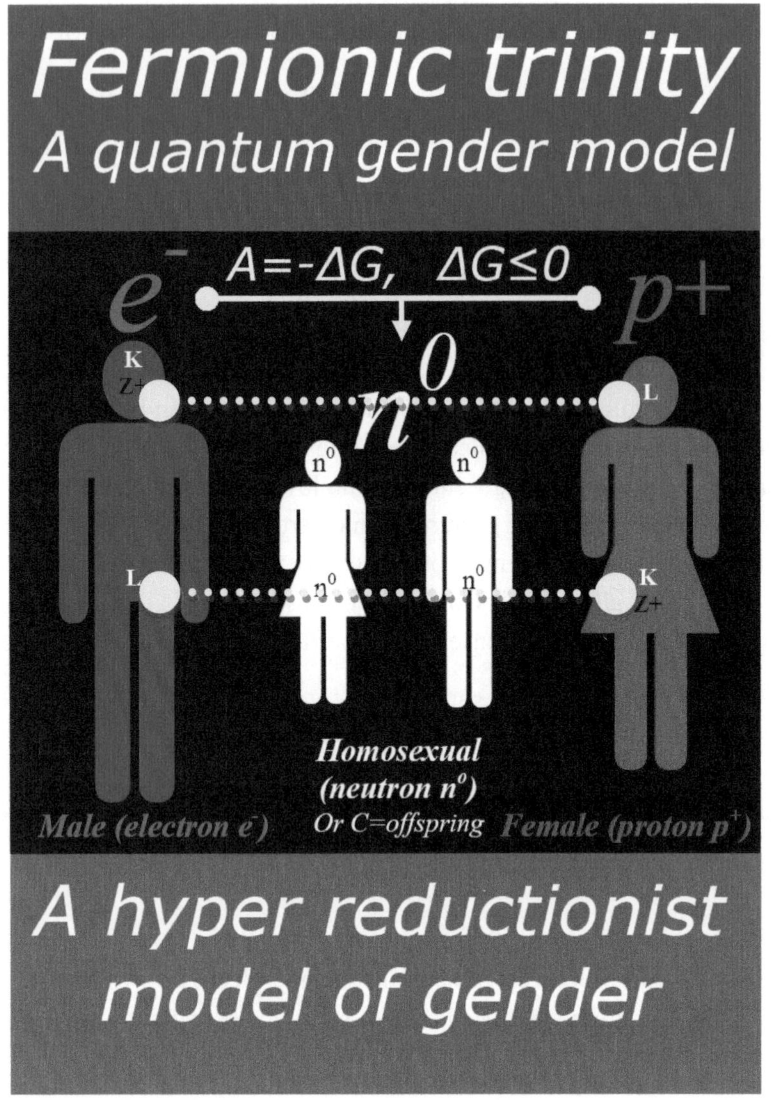

Gender has almost always been misunderstood. This is due to a lack of fundamental physical logic in order to thoroughly explain it.

By using a particle physics basis science can offer sexual dimorphism a strong powerful model in order to explain gender in all levels of natural scale.

Evolution hypothesis – The extension of evolution by natural selection to all matter and all time.

Statement of hypothesis (point 1).

Evolution by natural selection by Charles Darwin is unacceptably limited by both a barrier of time and substance, and of life. Evolution is the equilibrium balance between open and closed thermodynamic systems, driven by the inevitability of the 2nd law and the free will free energy consequence of a 4th law.

Statement of hypothesis (point 2).

Evolution is a fundamental principle of the whole of time and space and not an exclusive logic for modelling life on Earth. The universe on all levels of natural scale obeys evolution by natural selection, and life cannot be proved to start at any other time than that of the Big Bang.

Testing the hypothesis.

Examine physical observational evidence of different levels of natural scale and identify how they fit together to produce an integrated evolving organism. Data to be produced as photographic observations of integral levels of organisation and how they change over time.

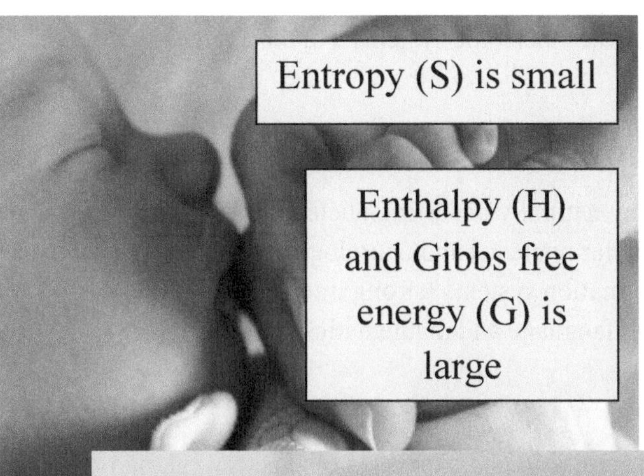
Entropy (S) is small
Enthalpy (H) and Gibbs free energy (G) is large

Topographically the universe is in a perpetual flux, where change or evolution is the most common theme!

Soulatrophic pathway $+\frac{1}{2}_g$

Ferrofermionic

$^{28}Si \rightarrow ^{56}Fe$

Spiritual Z_s

Cognitive Z_c

Somatic Z_{sm}

Micro Z_{mb}

Atomic Z_a

p^+ n^0 e^-

Gravitropic Pathway $-\frac{1}{2}_g$

Animals e^-

Entropy (S) is large

Enthalpy (H) and Gibbs free (G) energy is small

Evolution is very easy to identify and prove, look how humans evolve throughout their lives as a definitive proof of this powerful fundamental concept in nature. If evolution was incorrect then all organisms within a species would be clones.

Nuclear hypothesis – The identification of the various nucleonic systems and their force equilibrium consequences.

Statement of hypothesis (point 1).

The nucleonic logic of atomic nuclei is one of many naturally occurring nuclear organising systems. As well as the atomic nucleons there are DNA cellular nucleons and neurological organism nucleons. The neurological nucleons are the basis of all information systems (strong interaction – gluonic and pionic) and also manifest their interactions through language and mathematics.

Statement of hypothesis (point 2).

All nucleonic systems are binary, super coiled or open, read and expressed by electroweak particle systems and contain huge amounts of energy. There are many naturally occurring levels of nucleons which string together to produce life.

Statement of hypothesis (point 3).

Neurological nucleons, like DNA nucleons, have chromosomal supercoiled structures such as books, CD's, DVD's, Flash RAM, ROM, records, tape cassettes. All can be reduced to binary and exist in condensed forms, and can be read using an appropriate electroweak standard model such as the hands, arms and the head of a human.

Testing the hypothesis.

Examine physical observational evidence of different nucleonic levels of natural scale and identify how they fit together to produce an integrated evolving organism; plus technological life. Data to be produced as photographic observations of integral levels of organisation and how they change over time.

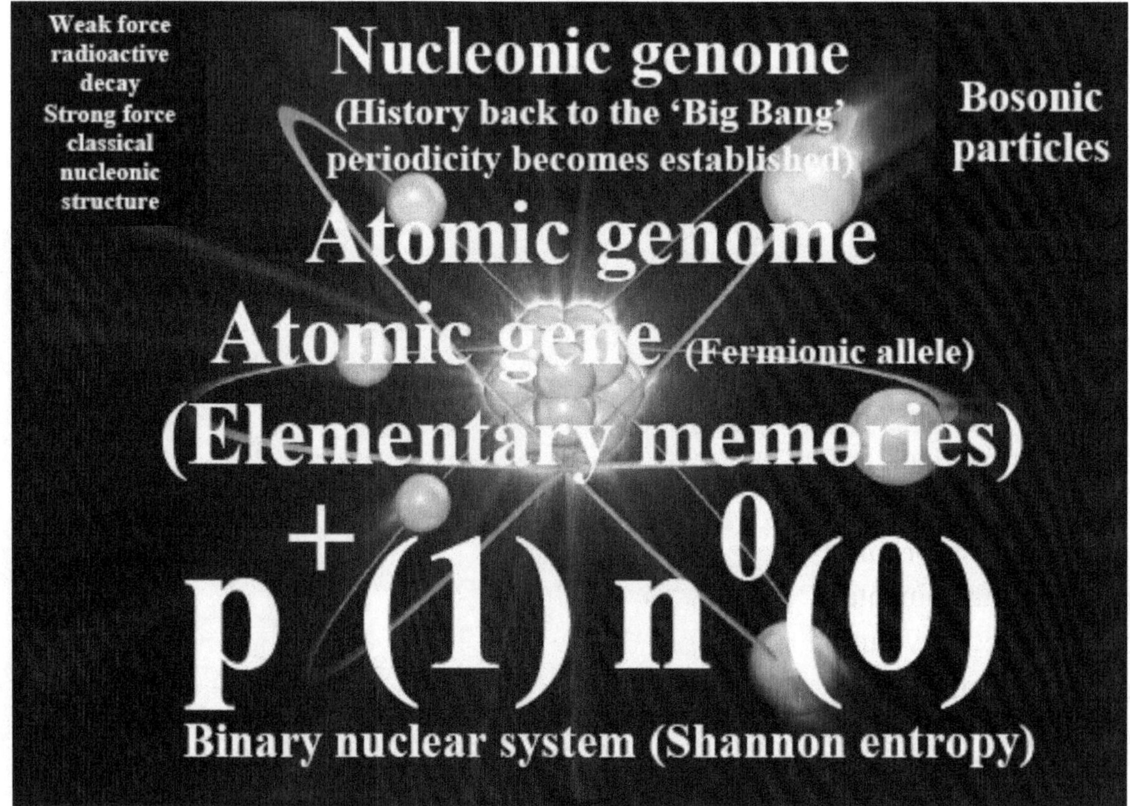

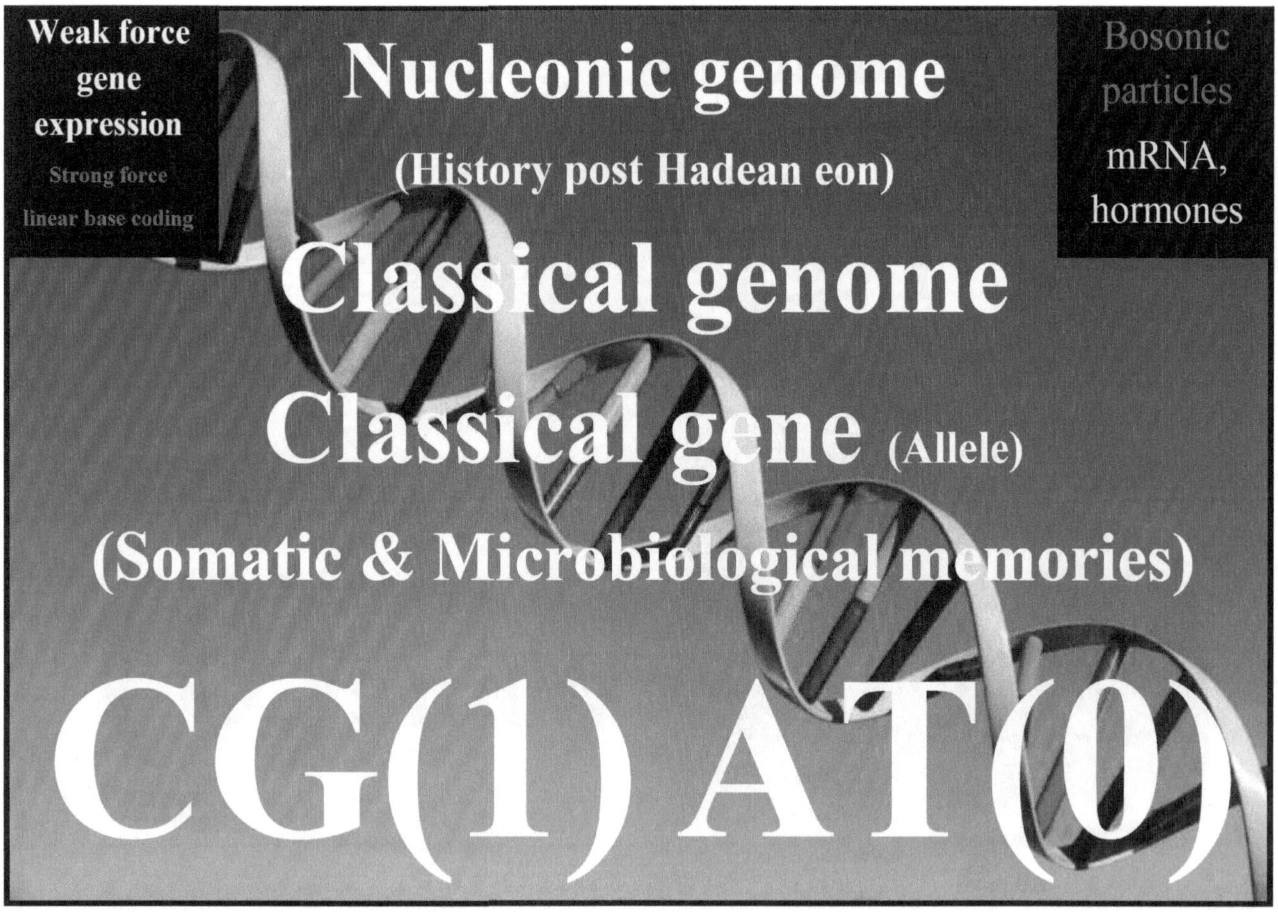

Weak force
gene
expression
Strong force
linear base coding

Nucleonic genome
(History post Hadean eon)
Classical genome
Classical gene (Allele)
(Somatic & Microbiological memories)
CG(1) AT(0)

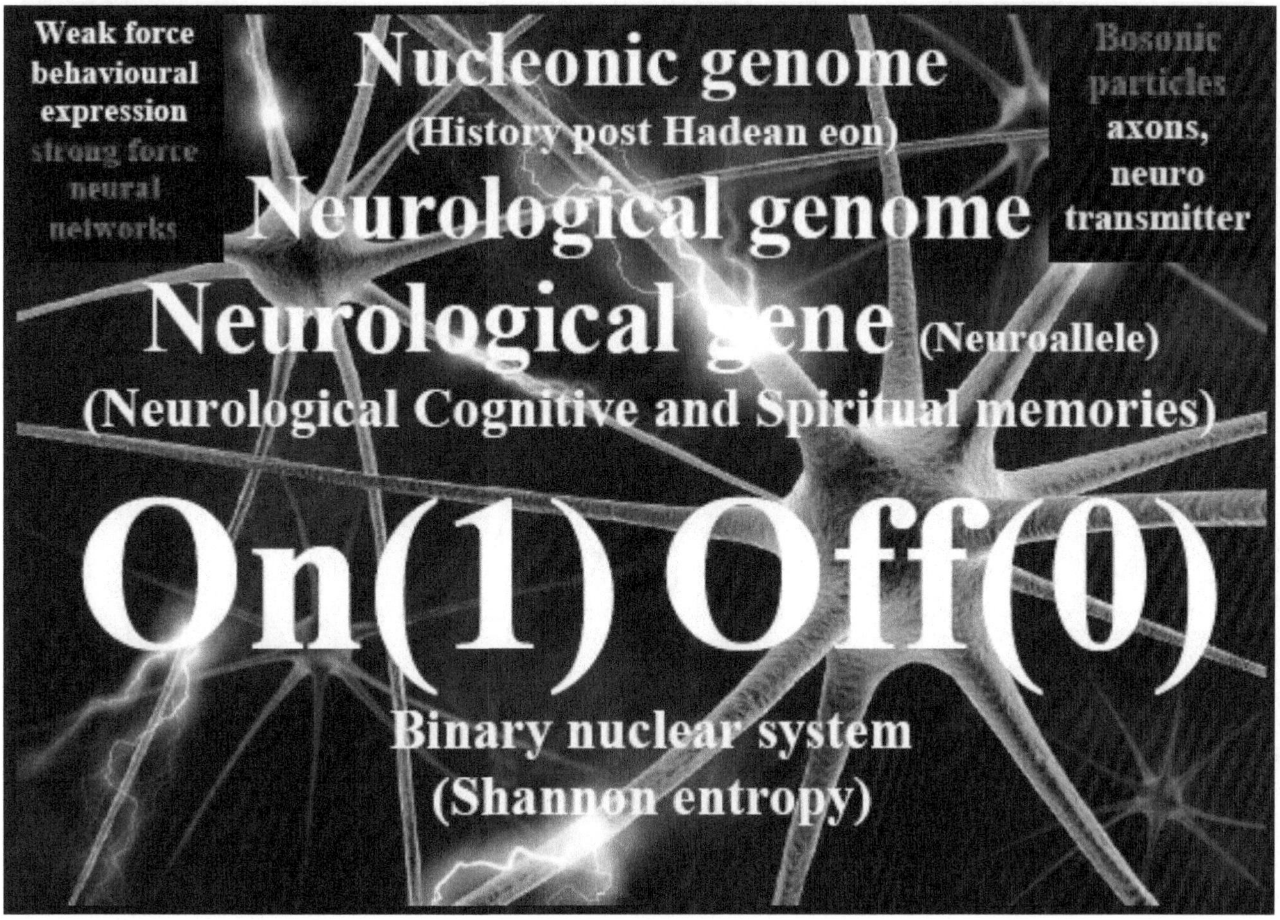

Bosonic
particles
mRNA,
hormones

Weak force
behavioural
expression
strong force
neural
networks

Nucleonic genome
(History post Hadean eon)
Neurological genome
Neurological gene (Neuroallele)
(Neurological Cognitive and Spiritual memories)
On(1) Off(0)
Binary nuclear system
(Shannon entropy)

Bosonic
particles
axons,
neuro
transmitter

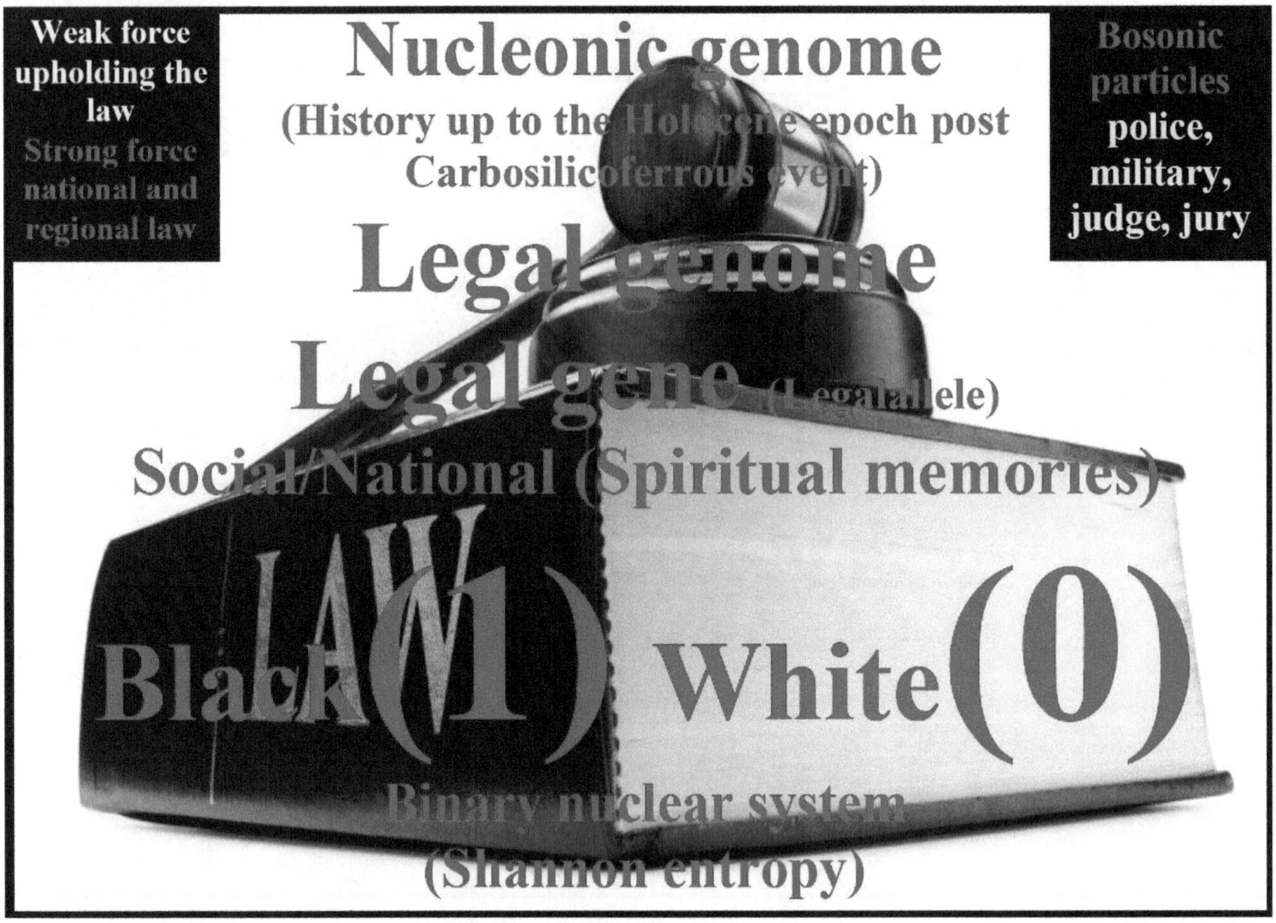

Weak force upholding the law
Strong force national and regional law

Nucleonic genome

(History up to the Holocene epoch post Carbosilicoferrous event)

Bosonic particles police, military, judge, jury

Legal genome

Legal gene (Legalallele)

Social/National (Spiritual memories)

Black (1) White (0)

Binary nuclear system (Shannon entropy)

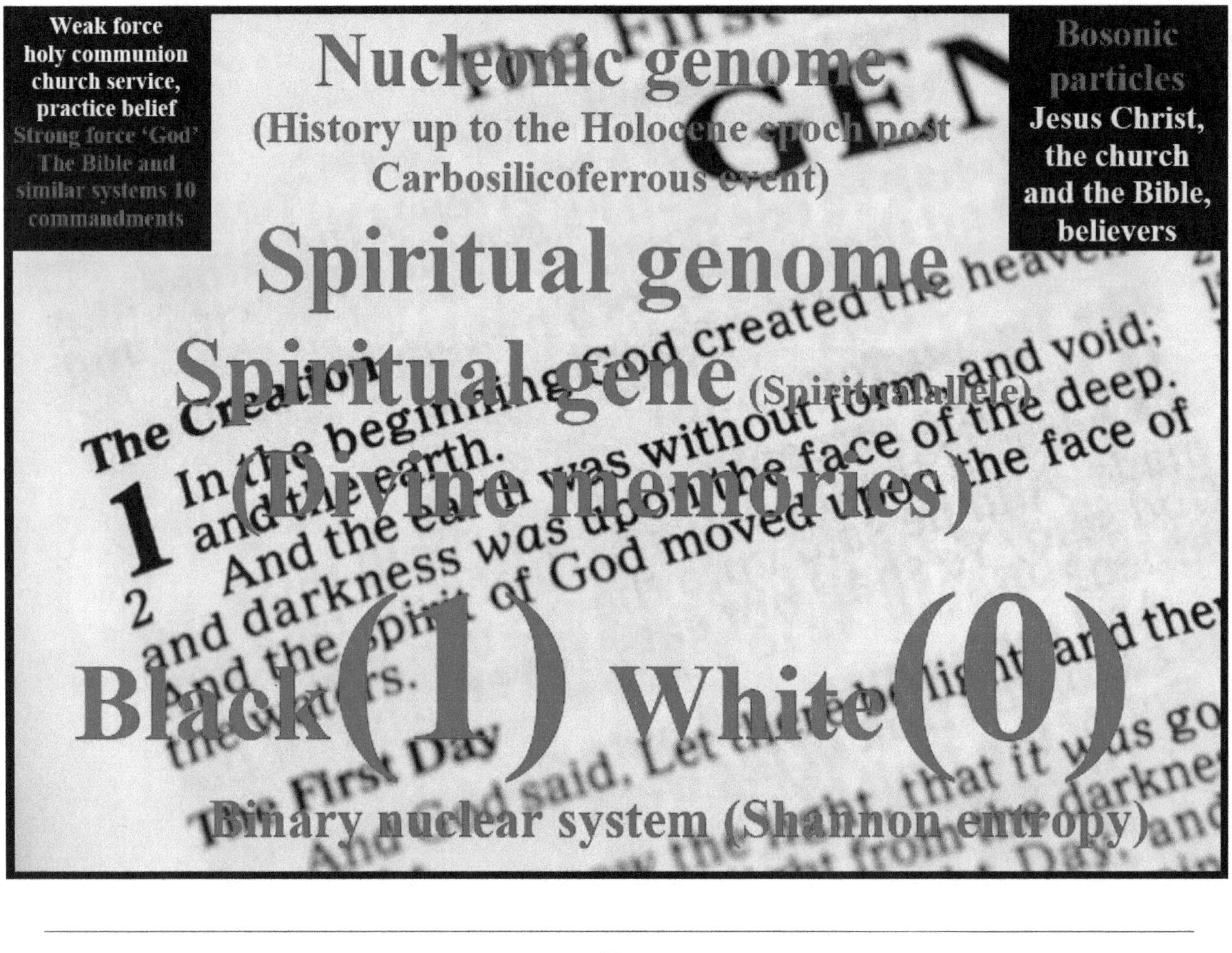

Weak force holy communion church service, practice belief
Strong force 'God' The Bible and similar systems 10 commandments

Nucleonic genome

(History up to the Holocene epoch post Carbosilicoferrous event)

Bosonic particles Jesus Christ, the church and the Bible, believers

Spiritual genome

Spiritual gene (Spiritualallele)

(Divine memories)

Black (1) White (0)

Binary nuclear system (Shannon entropy)

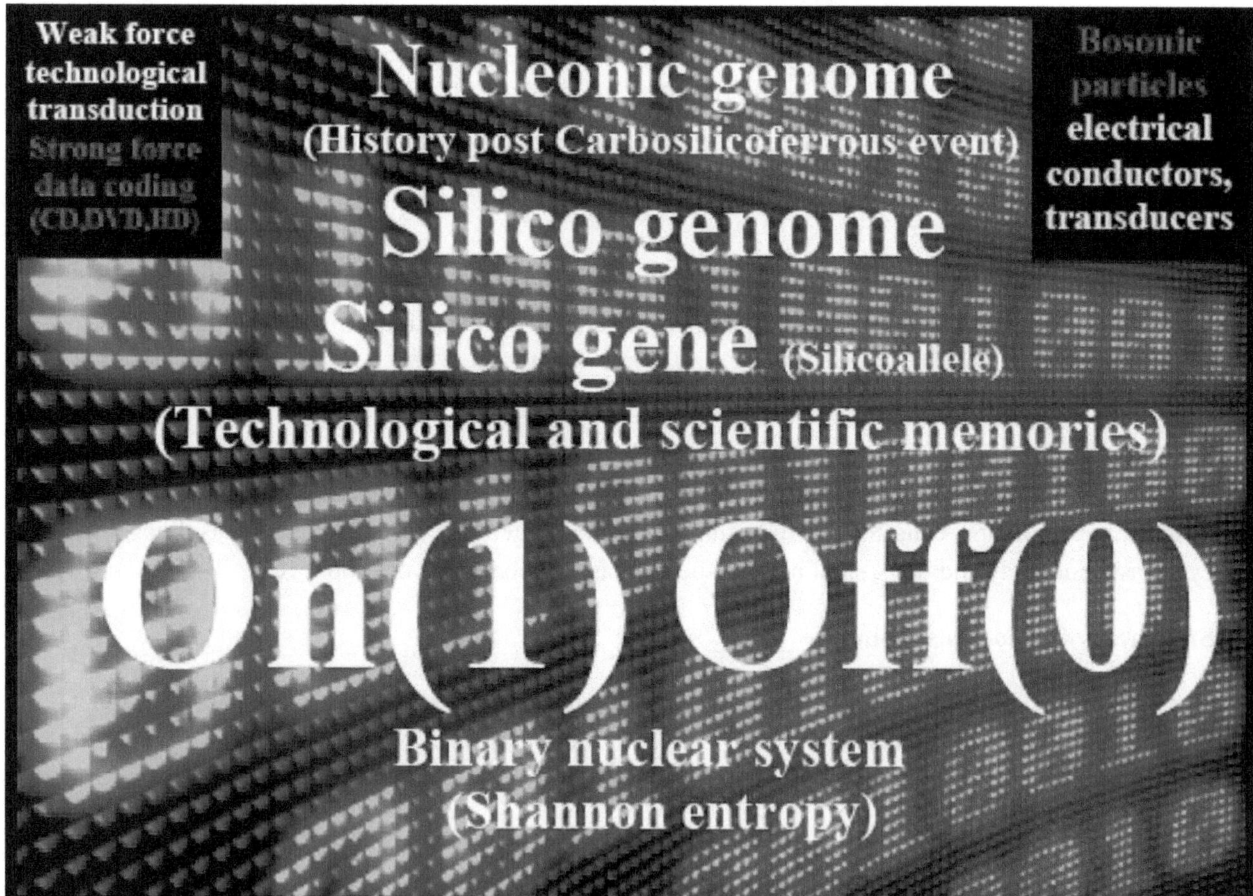

Quantum hypothesis – The natural extension of quantum determinism in living organism, the uncertainty principle and life.

Statement of hypothesis (point 1).

All living organisms have higher nucleonic systems, such as DNA, and neurological and an associated natural consequence of higher quantum field logic. These amplified fields exist as charge clouds on atomic levels and tissues on the level of higher organisms and produce fields with the spatial organisation of the 1s, 2s and three p orbitals, the superposition of states explain the enormous variety in living organisms. These fields are the basis of the body and cytoplasm and limbs and membrane in living organisms. They are also the bonding fields and form a shielding effect, connecting to other external systems typically through covalence (for example a hand shakes between two people).

Statement of hypothesis (point 2).

The amplification of quantum fields in living organisms obeys the uncertainty principle. Living quantum fields exist as a superposition of states and produce probability distributions such as a human arm. An external observer can only describe the most likely configuration of an organism and not any unique amplitude result.

Statement of hypothesis (point 3).

All statistical models, on all levels of natural scale, are the absolute personification of the quantum effect in matter, and are particularly evident in life. Mathematical probability and statistics exist as the natural consequence of living uncertainty through the biological nature of quantum mechanics.

Testing the hypothesis.

Examine physical observational evidence of different quantum levels of natural scale and identify how they fit together to produce an integrated evolving organism. Data has been extensively produced as photographic observations of integral levels of organisation and how they change over time throughout this book.

Energy level hypothesis (Extension to the quantum hypothesis) – quantification in living organisms and conserved amplified self symmetry to atomic organisation.

Statement of hypothesis (point 1).

Amplified atomic logic from carbon and associated nitrogen and oxygen and hydrogen produce 96.3% of an organism's determinism, structure and physical potential. Quantification into energy levels is seen in atoms, cells and in organisms where the K shell n=1 level can be seen as the cytoplasm and body and the L shell n=2 can be seen as the membrane or limb level.

Statement of hypothesis (point 2).

The completion of carbon, nitrogen and oxygen (hydrogen) 96% of all living matter follow the octet rule, and complete and stabilise their levels typically through acquisition of food in most animals and food and technology in humans.

Testing the hypothesis.

Examine physical observational evidence of different energy levels of natural scale and identify how they fit together to produce an integrated evolving organism; plus technological life. Data to be produced as photographic observations of integral levels of organisation and how they change.

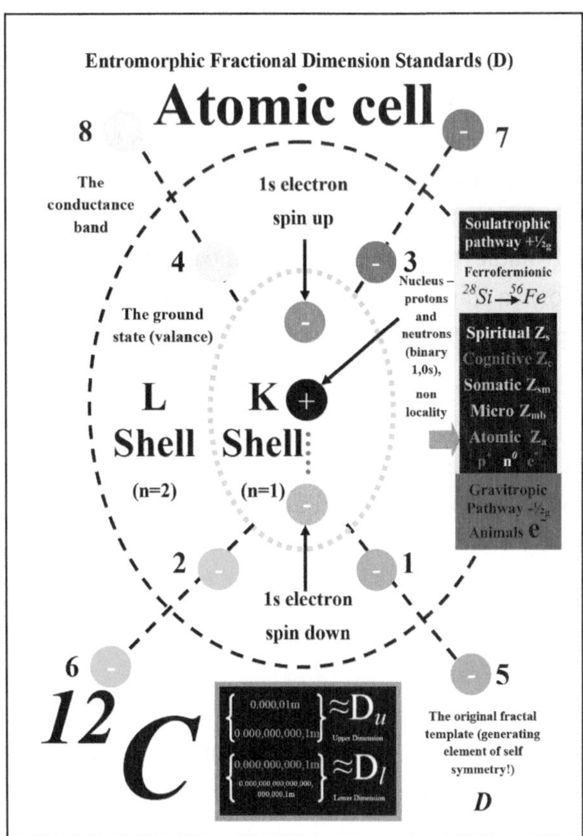

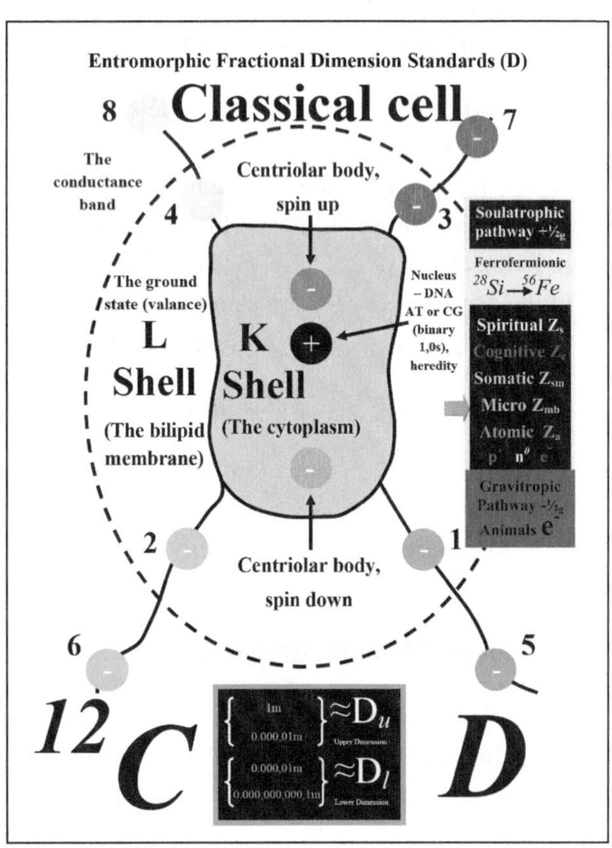

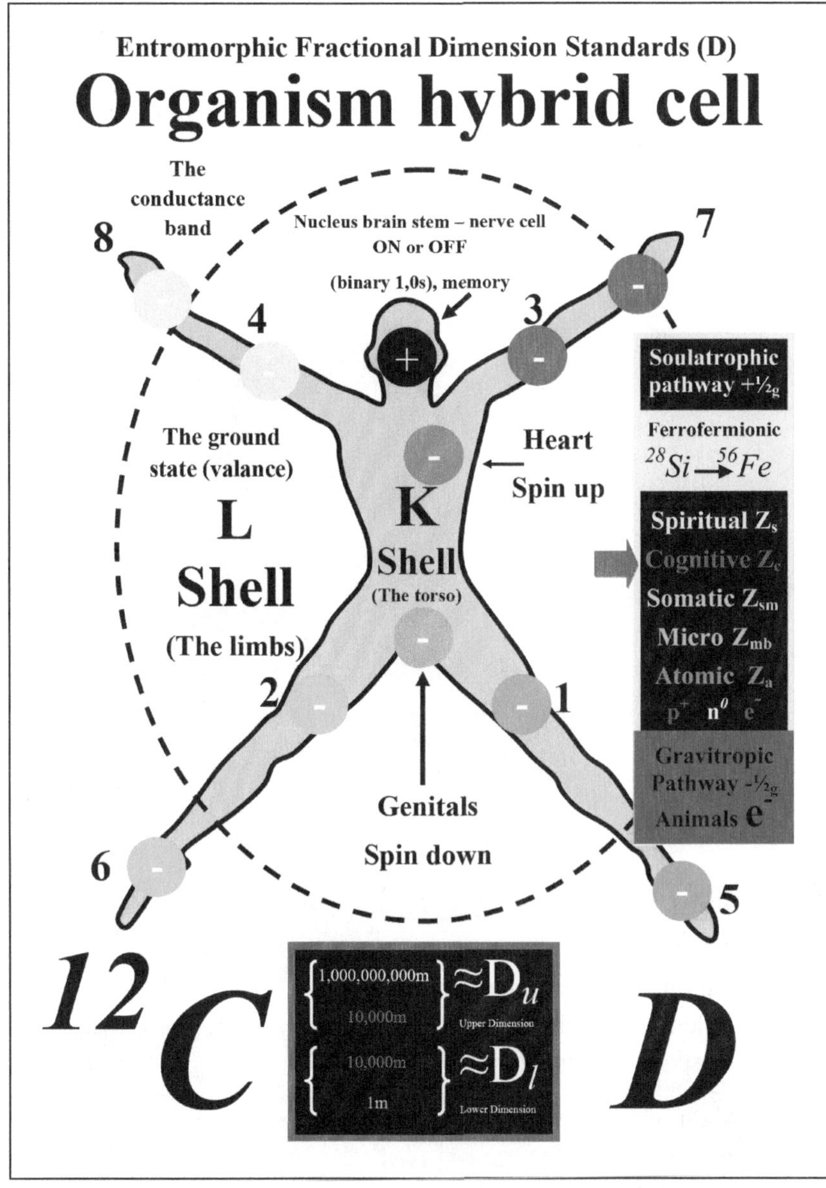

The truth is that quantum mechanics and nuclear physics is the reality of every second of our lives.

It is all around us, it binds us and produces its effect for all to see in living organisms where atomic logic is amplified to levels where we can visualise their properties.

It is not hidden away on the level of the atom which is the current thinking seen in atomic physics.

Albert Einstein once alluded to the fact that science is simply the corollary of our every waking moment by saying in 'physics and reality' in 1936 that:-

'The whole of Science is nothing more than a refinement of everyday thinking!'

It should be no surprise that physicists should find a theory such as carbon entromorphology which applies physics to describe life.

Pauli hypothesis – The fermionic statistics and its consequence to living organisation.

Statement of hypothesis (point 1).

Atomic organisation is quantitised and stabilised into specific regions of space through fermi-pressure. The 'Pauli exclusion principle' is so fundamental to the matter which comprises life, that its operation has to be fundamental to all life. Fermionic components of life must obey the principle where spin angular momentum or motivational direction must be opposite to produce cooperative thermodynamics. Living organisms also follow this rule from their quantum levels, hence a Pauli cell must contain no more than two particles (electrons) with opposite spin, for example an argument between two people must be resolved by a shift into opposite spin. No organism can have the same set of conscious quantum numbers.

Statement of hypothesis (point 2).

Spin in bosonic particles in living organisms (carbon based life) operates opposite to fermionic spin, where the probability of organism being in the same state is more probable. This explains the in phase coherence in living organisms and is the basis of amplification in living solar beings through groups.

Testing the hypothesis.

Examine physical observational evidence of different levels of natural scale and identify how they fit together in cooperative groups to produce an integrated evolving organism. Data to be produced as photographic observations of integral levels of organisation and how they cooperate.

Bosonic spin is integer spin – these bosonic entromorphic atoms are sharing similar physical states.

Orbital hypothesis – The physical distributions in living systems as a model of morphology.

Statement of hypothesis (point 1).

The conserved amplified atomic nature of life produces structures from quantum electron clouds which account for the tissues in living organisms. Leaves in plants have a 90° set of three leaf systems, which are symmetrical and have a lobed p type orbital morphology. Dandelion clocks and other flowers have a sphere within a sphere morphology, which are symmetrical to 2s orbitals. And a human head is the absolute personification of a 1s orbital, spherical with two eyes which act as amplified electronic logic. All the 1s, 2s and p type orbitals are nucleated by a nucleonic nucleus such as a seed in a plant or a brain in an animal, there are trillions of self symmetrical examples.

Testing the hypothesis.

Examine physical observational evidence of different levels of natural scale and identify the symmetry of physical morphology in producing an integrated evolving organism. Data to be produced as photographic observations of integral levels of morphology organisation in symmetry to the three basic orbital shapes. With a consequential production of all the wild type variation in nature.

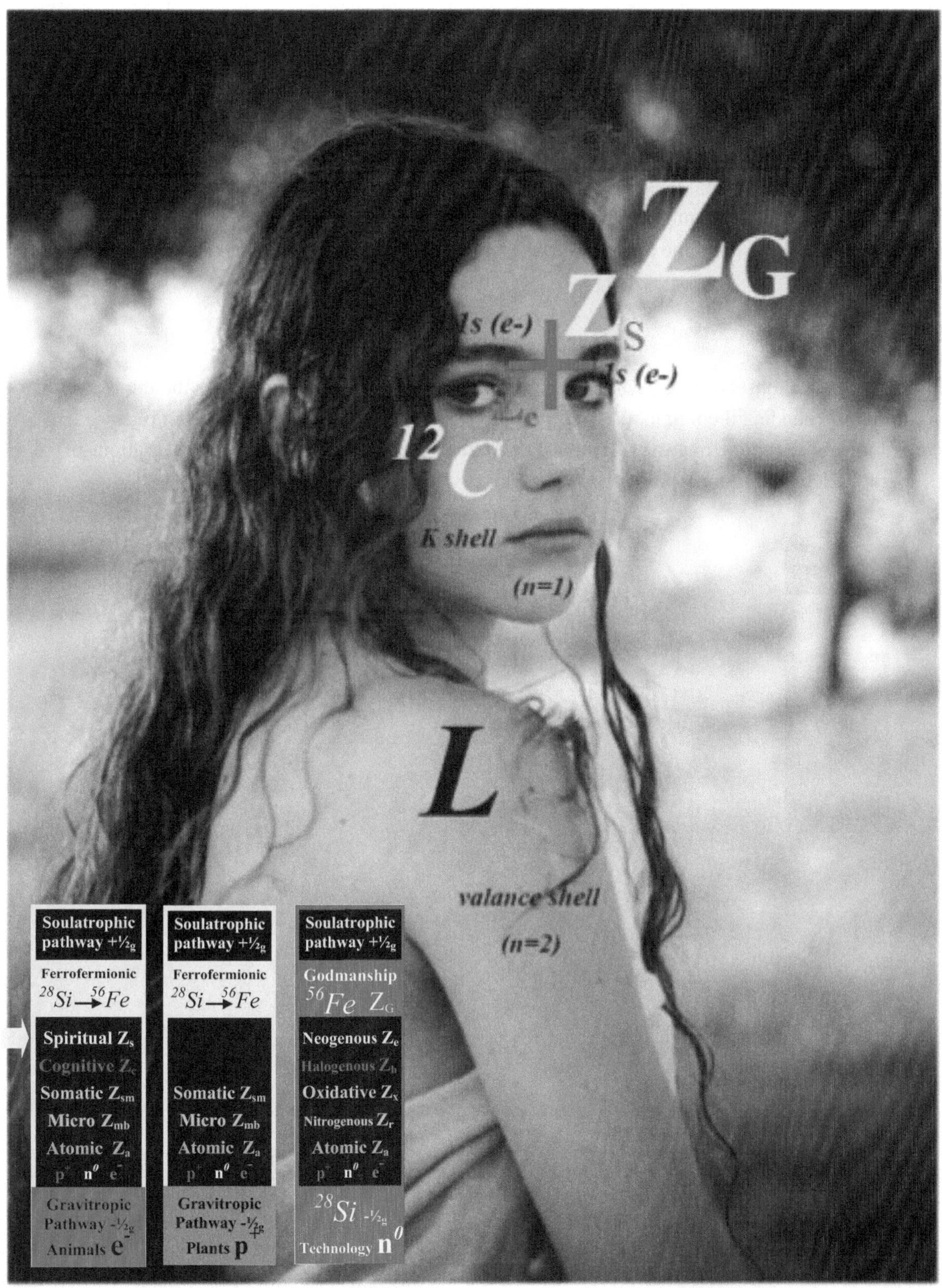

Beautiful carbon entromorphology, the extraordinary conservation of powerful fundamental physics is truly breathtaking in nature.

Coulomb barrier hypothesis – This hypothesis identifies biological structures and their nuclear origins.

Statement of hypothesis (point 1).

The numerous nucleonic forms in life such as the proton/neutron atomic level, the AT/CG DNA level and the On/Off neurological level all have coulomb barrier structures around the nucleus. In living organisms, a skull is an example of this coulomb barrier around a brain and a seed shell around the DNA in a plant, they shield the nucleus but also allow barrier penetration to take place under specific conditions.

Testing the hypothesis.

Examine physical observational evidence for different organism's nuclear region and the barrier structures surrounding them in nature and their physical similarity to atomic barriers; plus technological life robotic nuclear systems. Data to be produced as photographic observations of different nuclear casing morphology.

Potential well hypothesis – This hypothesis identifies localised biological structures and their atomic origins.

Statement of hypothesis (point 1).

The potential well on the atomic level is the powerful concentration of nuclear forces into the nucleus; these still appear in life as organ systems. Its consequence for living organisms can be seen in certain structures such as the gastrointestinal system, the mouth and the vagina. All are short range acting systems (narrow holes) and are fundamentally organ holes. The word hole is used to describe this effect in atoms as well as giving excellent continuity of logic. The pull of sexual desire is the effect of the potential well in living organisms although it is based on mass structures with electromagnetic dilution.

Testing the hypothesis.

Examine physical observational evidence for different organism nuclear entrances in nature and their physical similarity to the potential well in the atoms they are comprised of; plus technological life robotic heads. Data to be produced as photographic observations of potential well structures.

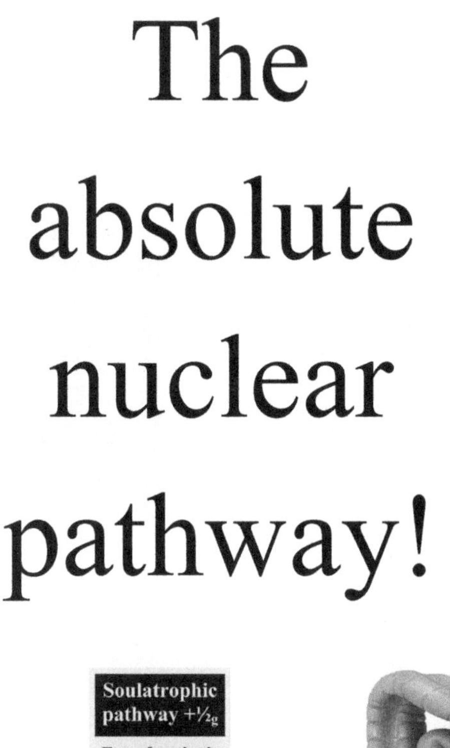

The absolute nuclear pathway!

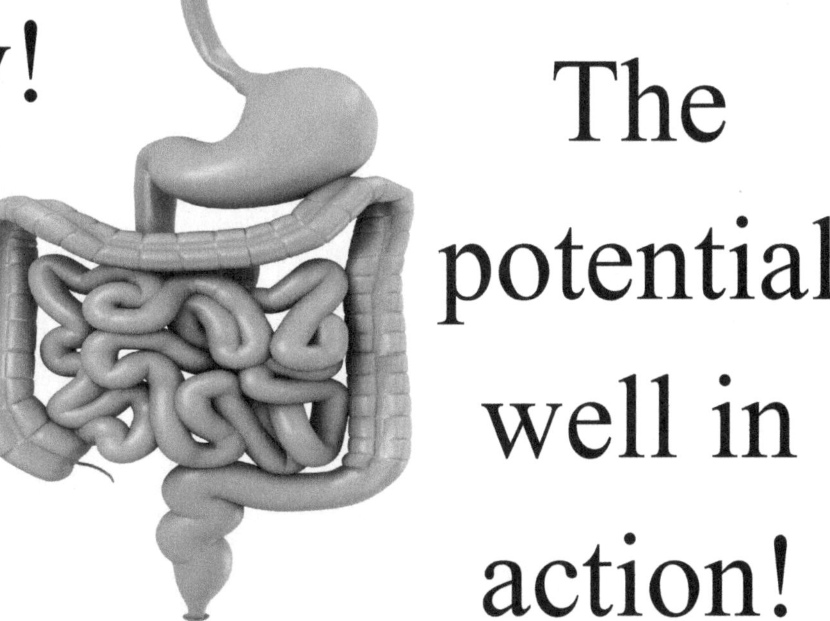

The potential well in action!

Barrier penetration hypothesis – This hypothesis identifies localised biological structures and processes such as sexual penetration and their atomic origins.

Statement of hypothesis (point 1).

The barrier penetration effect occurs in atoms and their composite living structures. Barrier penetration is the fundamental mechanism in sexual penetration where a sexual coulomb barrier (vagina – 'in and out', binary 'on/off' penetration) is breached to achieve access to the nucleus (egg). In trillions of organisms this mechanism occurs many times throughout their lives and penetration occurs through the formation of neutronic particles such as sperm and egg, which are non polar zygotic structures when combined. A French kiss (tongue penetration) occurring between two people is another example of this effect and requires barrier penetration which occurs through thought exchanges and emergent trust. Also birth of an organism is the inverse effect where offspring penetrate out of the nuclear region.

Testing the hypothesis.

Examine physical observational evidence for different organism nuclear entrances in nature and their physical similarity to the barrier penetration in the atoms they are comprised of; plus technological life. Data to be produced as photographic observations of different barrier penetration structures.

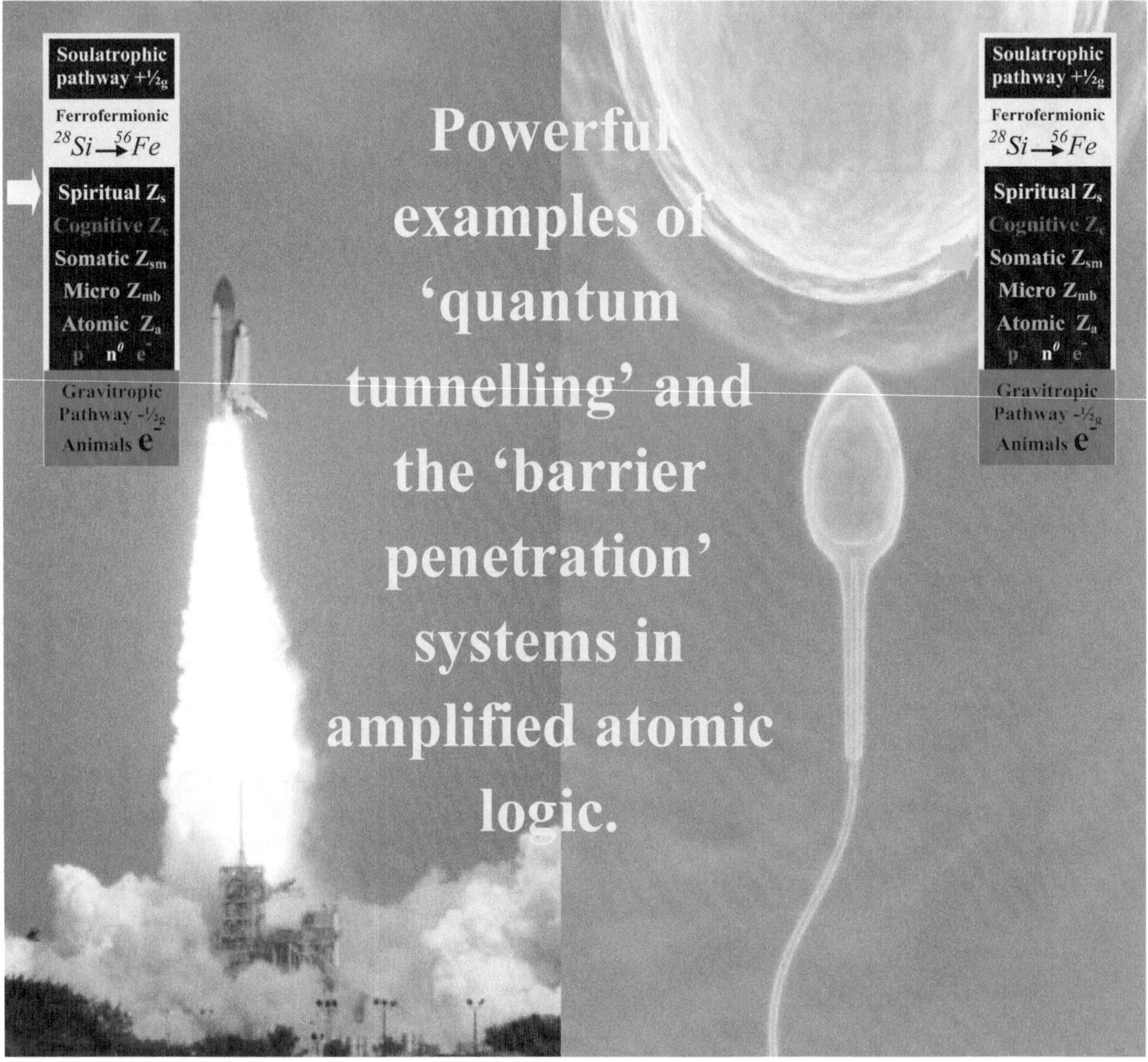

Electric field line hypothesis – This hypothesis identifies localised biological structures and their atomic origins and the structures which have morphology in keeping with electromagnetic origins.

Statement of hypothesis (point 1).

The electromagnetic effects on the level of atoms are the basis of interactions on that level and therefore the basis of such forces in all life. They produce electric field lines which stream out of positive charges or nuclear regions and converge into negative charges or quantum field regions. Since organisms are composed of this logic it is evident that the morphology of the structures they produce has evolved down such field lines but with increasing mass determinism. Hence living structures such as a penis is a negative structure as it is convergent and a radiating vagina a positive nuclear structure (hence the offspring develop in females); together they form a neutral exciton particle. Flowers and animals all have structures which have evolved out of their polarised electromagnetic origins.

Testing the hypothesis.

Examine physical observational evidence for different organism radiating and convergent structures in nature and their physical similarity to the electric fields in the atoms they are comprised of. Data to be produced as photographic observations of physical structures.

The conservation of atomic logic produces the variety and diversity of living morphology.

Gravitational mass rich structures have evolved out of quantum electromagnetic origins.

Carbon Electromorphology

Electric field lines and biological morphology

Structures evolving out of electromagnetic mathematical models

$$E = \frac{1}{4\pi\varepsilon_0} \frac{q}{r^2}$$

Gauge symmetry hypothesis – This hypothesis identifies localised biological structures and their atomic origins with the electroweak model.

Statement of hypothesis (point 1).

The bosons of carbon (nitrogen and oxygen) form the electroweak standard model on the atomic level. In carbon entromorphology the gauge bosons or mediators of force, form large complex structures such as limbs, which enable energy to flow into (gravitropically) and out of (electroweak) atomic systems. Classical bosons are the W+ and W- and the Z^0 boson plus the γ (photon) and are associated with the electrons in carbon. The amplified consequence in living organisms is the antagonistic limbs (W+ and W-) and the hand or Z° (ribosomal) system and its photonic consequence in communication (waving goodbye to someone).

Testing the hypothesis.

Examine physical observational evidence for different organism limb bonding in nature and their physical similarity to the nuclear/quantum processes in the atoms they are comprised of; plus technological life robotic heads. Data to be produced as photographic observations of different limbs and their function.

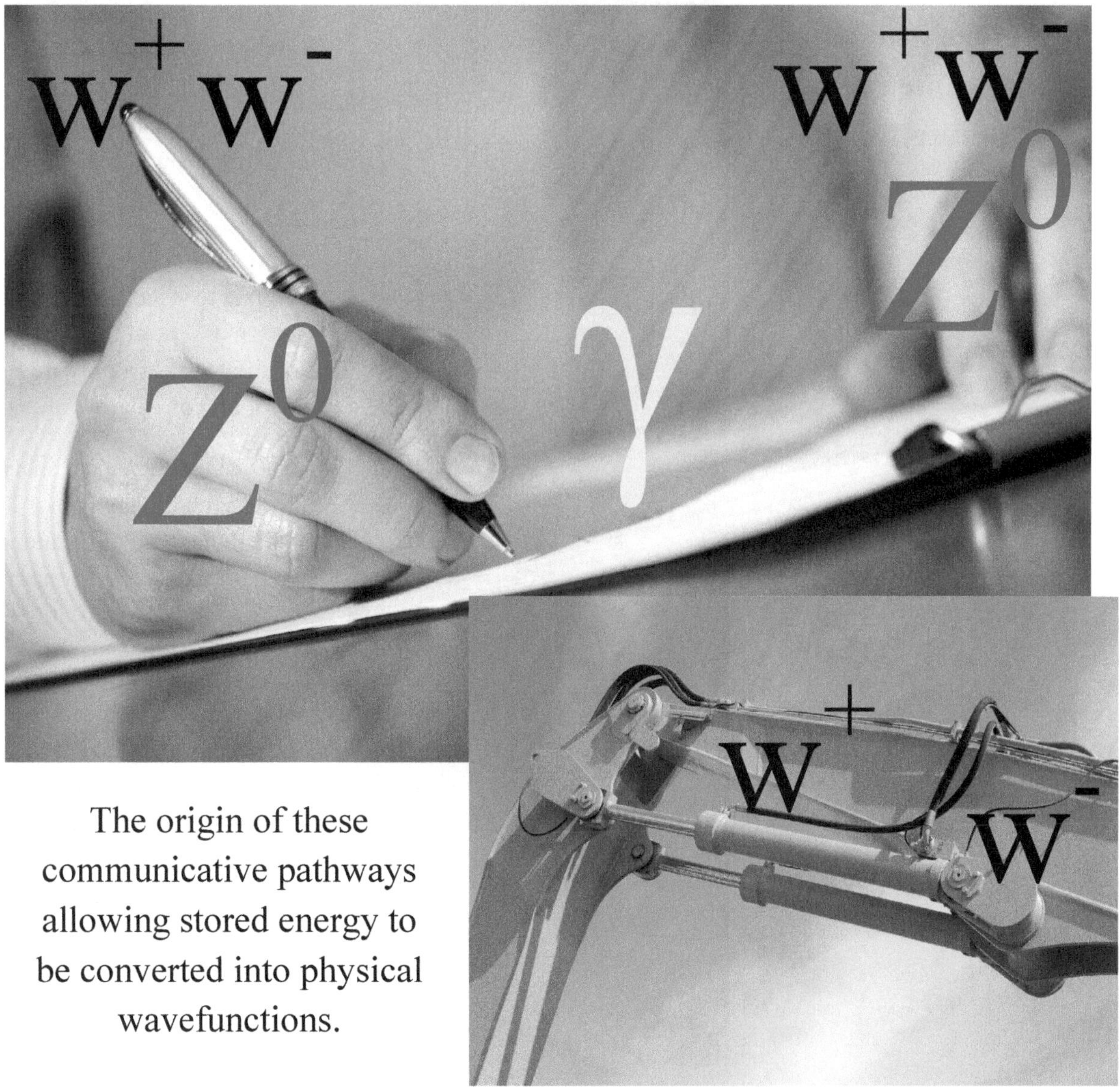

The origin of these communicative pathways allowing stored energy to be converted into physical wavefunctions.

Electroweak hypothesis – This hypothesis identifies the processes of connectivity between organisms and its relation to living expression as a variation on the previous hypothesis but with similar results.

Statement of hypothesis (point 1).

Body (cytoplasm) and limbs (membrane) in an organism are the resultant effect of amplified atomic electroweak logic. This is the basis of the particle system which allows organisms to express nuclear potentials and which interacts with all the force particles. The limbs are an excellent example of this effect as a means of explaining the valance bonding interface in organisms. In short a hand-shake can be hypothesised to be a covalent bond between two amplified carbon systems.

Testing the hypothesis.

Examine physical observational evidence for different organism limb bonding in nature and their physical similarity to the nuclear/quantum processes in the atoms they are comprised of; plus technological life robotic. Data to be produced as photographic observations of different limbs and their function.

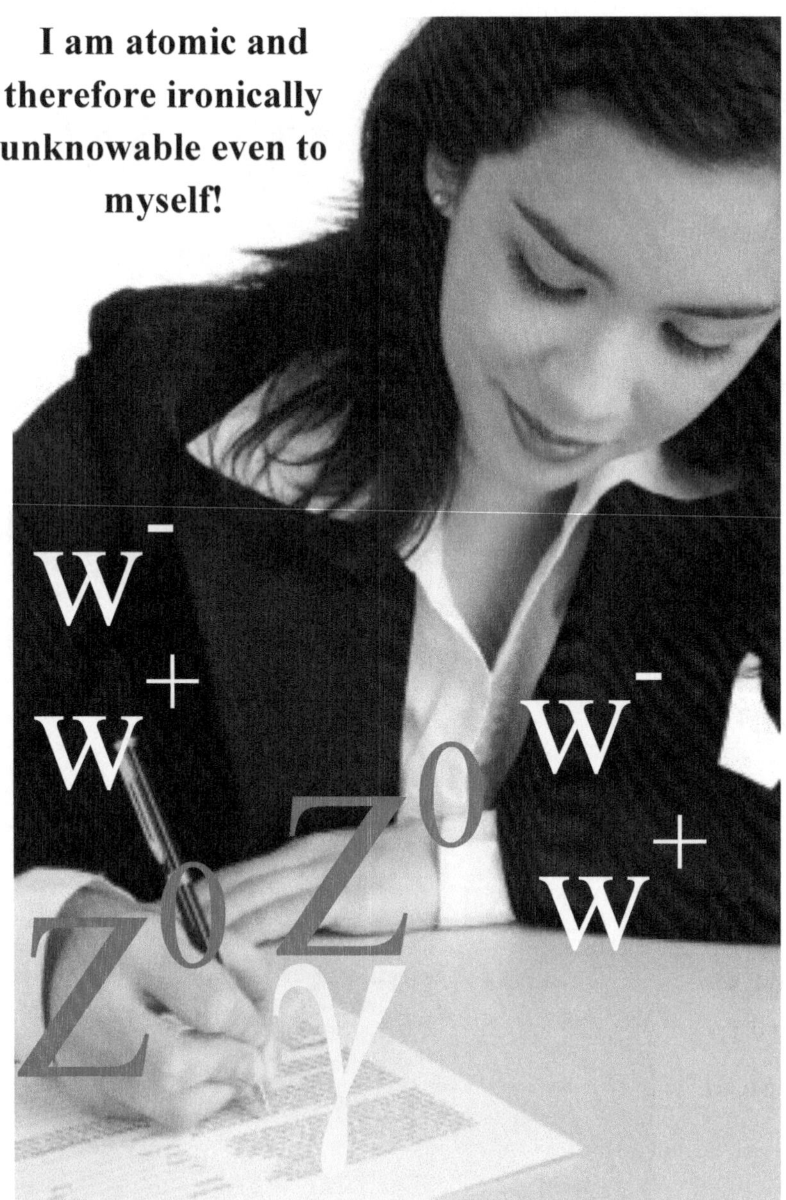

I am atomic and therefore ironically unknowable even to myself!
W-
W+
W-
W+
Z0
Z0
γ

Often the reality of our everyday experiences from the simplicity of moving a limb with coordination to complex communication holds its origins in the atoms we are comprised of.

Again it should be of no surprise to science that atomic logic should be at the very heart of life because atoms are the fabric of what we are.

A lack of mathematical analysis produced by living beings has isolated the logic of the atom to its notoriously unfathomable levels of its organisation.

It takes a biologist using non mathematical treatments to spot the links.

Since life started on atomic levels and through bonding produced larger structures through conservation and bonding this should allow us to be able to see the world of atoms.

Gravistrong hypothesis – A new symmetrical hypothesis that aims to suggest a grouping of forces in parallel to those of the electroweak standard model.

Statement of hypothesis (point 1).

To balance the electroweak model which connects all the particles on the atomic level. Due to fundamental symmetry on this level it is pertinent to balance out the electroweak components with the gravitational and strong force nuclear components with a 'gravistrong standard model'. In short, the four forces do not exist in isolation but represent and cycle of connectivity for which mass and energy can flow through and form equilibrium with. The electroweak model interfaces the electromagnetic force with the nuclear weak force to field logic. The gravistrong standard model interfaces the gravitational force with the strong nuclear force nuclear to field logic. This type of system can be seen in most organisms and balances mass and energy but is always attractive and space time distorting.

Testing the hypothesis.

Difficult to test theory: may need further elaboration and investigation.

Force cycle hypothesis – A new symmetrical hypothesis which aims to suggest a grouping of all four forces in parallel to those of the other particle standard models.

Statement of hypothesis (point 1).

A balance between the electroweak standard model and the gravistrong standard model produces a complete four force cycle of logic allowing energy to flow into and out of an atomic system. This force cycle model is observed in an organism's behaviour to allow energy flow into and out of living organisms. This type of system can be seen in most organisms and balances mass and energy.

Testing the hypothesis.

Difficult to test theory may need further elaboration and investigation.

Mathematical hypothesis – Mathematics is the language of nature so why bother even applying it?

Statement of hypothesis (point 1).

If mathematics is absolute truth then its raw basis must be the very essence of conscious logic; application is unnecessary. In this case mathematics in its pure form can be attributed to the physical world. The number sets (1,0,e,i,π &∞) represent mass and energy (microscopic particle/wave logic) and four dimensional space time (Euclidean coordinate system macroscopic logic) and the operators +-×÷ the four forces of nature where + is the strong interaction, - is the weak interaction, × is the electromagnetic interaction and ÷ the gravitational interaction. The = sign is the basis of the first law of thermodynamics and the flow (time/energy/mass) of numbers into and out of physical systems the 2nd and 4th laws, based on the directions in number lines.

Also, the illusive and powerful, but essential nature of infinity ∞ as a model of the uncertainty in nature.

Testing the hypothesis.

Examine the origins of mathematics and its link to natural phenomena.

Pure mathematics and its symmetrical link to natural phenomena, the 'Genderfication equation'.

Statement of logic. A very senior and eminent physicist once said 'mathematics is the language of Nature'.

Argument. If this is true then pure mathematics at its most fundamental must reflect symmetrically with physical science at its most fundamental. The following statements and arguments investigate this logic and provide proof by beautiful mathematical self symmetry. The 4th law allows this author to make this central claim to carbon entromorphology.

Axioms and assumptions. Sets of fundamental quantities, and operational forces and basic equations and the laws of thermodynamics in Nature. $\varepsilon_{universe}$ is the general relativistic space/time set defined by tensor models (x,y,z,t).

(α) Let = (verb) be the mediator of the 1st and 3rd law of thermodynamics. Let the shift from 0, 1 (order) to e, i, π, (disorder) be the mediator of the 2nd law of thermodynamics, and the 4th law mediate efficient transformation 0, 1 (order) to e, i, π, (disorder) of energy. Let variables function as 'pronoun' and quantity act as 'nouns' and equations as 'sentences'.

(β) Let the four forces of Nature set $\{s, w, e, g\}_{science}$ = the four mathematical operators set $\{+, -, \times, \div\}_{maths}$

(γ) Let the six quantities of Nature set $\{0, 1, e, i, \pi, \infty\}_{science}$ = the six quantities in pure mathematics set $\{0, 1, e, i, \pi, \infty\}_{maths}$

(δ) Let the fermionic set of matter $\{e^- + p^+ = n^0\}_{science}$ = the quantitive trinity set $\{e^{i\pi} + 1 = 0\}_{maths}$

Key
e =2.71, $i=\sqrt{-1}$, π=3.14, Ψ wavefunction (quantum state), ħ Dirac's reduced constant, qm quantum mechanics, np nuclear physics, ω angular frequency, t time, H Hamiltonian (total energy of the system), p^+ proton, e^- electron, n^0 neutron. s strong interaction (force), w weak interaction (force), e electromagnetic force, g gravitational force, p momentum, x wave direction of travel, k 2π/wavelength.

The fundamental set of particles and waves

$\varepsilon_{universe}$ **Microscopic disorder**

x y z t

Maths: + - × ÷ | Maths ∩ Science: 0 e π i ∞ 1 | Science: s w e g

Maths ∩ Science

4th law ≡ 2nd law

The fundamental set of temporal duality

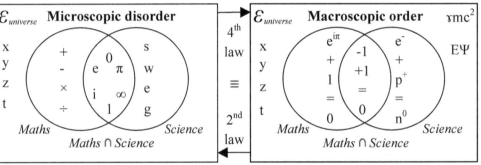

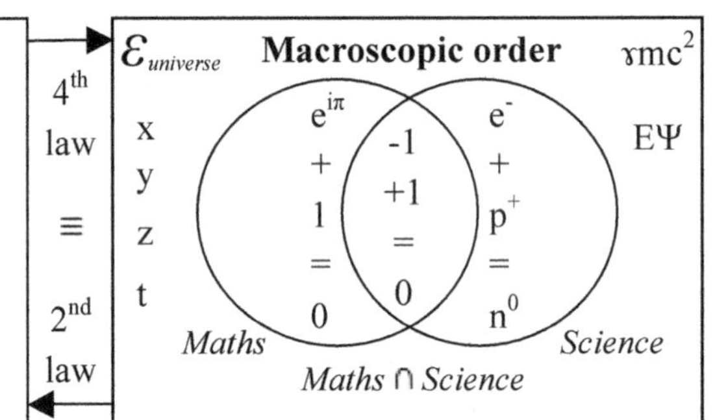

The self symmetric intersection of maths (theoretical) and science (experimental) leads to the intersection and defines differentiated fermionic matter with operational forces.

The self symmetric intersection of maths (theoretical) and science (experimental) for algorithmic compression defines fermionic quantities with operational forces.

Euler's identity in pure mathematics (theoretical - particle/wave duality)

Euler's identity in science (experimental - particle/wave duality)

$$e^{i\pi} = \cos\pi + i\sin\pi \quad \equiv \pi \neq x \equiv \quad e^{ix} = \cos x + i\sin x$$

$$e^{i\pi} = -1 + i(0)$$

$$e^{i\pi} = -1$$

$$e^{i\pi} + 1 = 0$$

[ΔxΔp≥½ħ]
[ΔEΔt≥½ħ]
[ΔmcΔt≥½ħ]

$$e^{i(kx - \omega t)} = \Psi$$

≡ The Mind ≡ $\Psi\hat{H} = i\hbar\dfrac{\partial}{\partial t}\Psi$

Uncertainty produces the two derivation pathways and produces an axis of consciousness, thought!

A bosonic force cycle model showing no net change to energy/mass/time.

$$\frac{+_s \; -_w}{\div_g \; \times_e} = 0 = \frac{\text{Nucleon}}{\text{Boson}}$$

[Numerator]

[Denominator]

(1) w and e represent the electroweak standard model, which operates in two ways.

(2) g and s is a gravistrong relationship (attractive only). Mediated by w+, w-, Z^0, γ.

See pages 90 to 96 for other genderfication proofs.

$$e^{i\pi} + 1 = 0 \quad \equiv \text{The Mind} \equiv \quad [e^- + p^+]_{qm} \approx n^0$$

(The axis of physical symmetry the focus of soulatrophic consciousness.)

$$\frac{e^{i\pi} + 1}{0} = \infty \text{ (undetermined)} \qquad e^- + [p^+ \approx n^0]_{np}$$

$$e^{i\pi}{}_{electron}\,♂ + 1_{proton}\,♀ \approx 0_{neutron}\,♂♀$$

male (sperm) + female (egg) = offspring (zygote) or homosexual for global cells (the genderfication anisogamy equation)

Statistical hypothesis – All statistics are a model of the quantum effect in any physical system.

Statement of hypothesis (point 1).

Any statistic has its origins in the quantum mechanics of the system where probability and statistics are the only method of evaluation and measurement due to observational uncertainties.

Testing the hypothesis.

Examine the origins of statistical mathematics and its link to natural phenomena.

Visual data hypothesis – Visual data is a more effective and reliable observational method than numerical data which is typically sampled at a very low levels relative to the population it came from.

Statement of hypothesis (point 1).

A visual measurement (observation) is the most powerful method of scientific evaluation. An image of a physical system, if resolvable, produces huge quantities of data and real time observation of functionality but also resolves deep and extensive complexity which can compromise mathematical treatment. Inversely a non visible system of observation can only have a small sample of data, although mathematical treatment is achievable as a result of the simplification.

Testing the hypothesis.

Examine the origins of scientific observation and the differences between physics evidence and biological evidence.

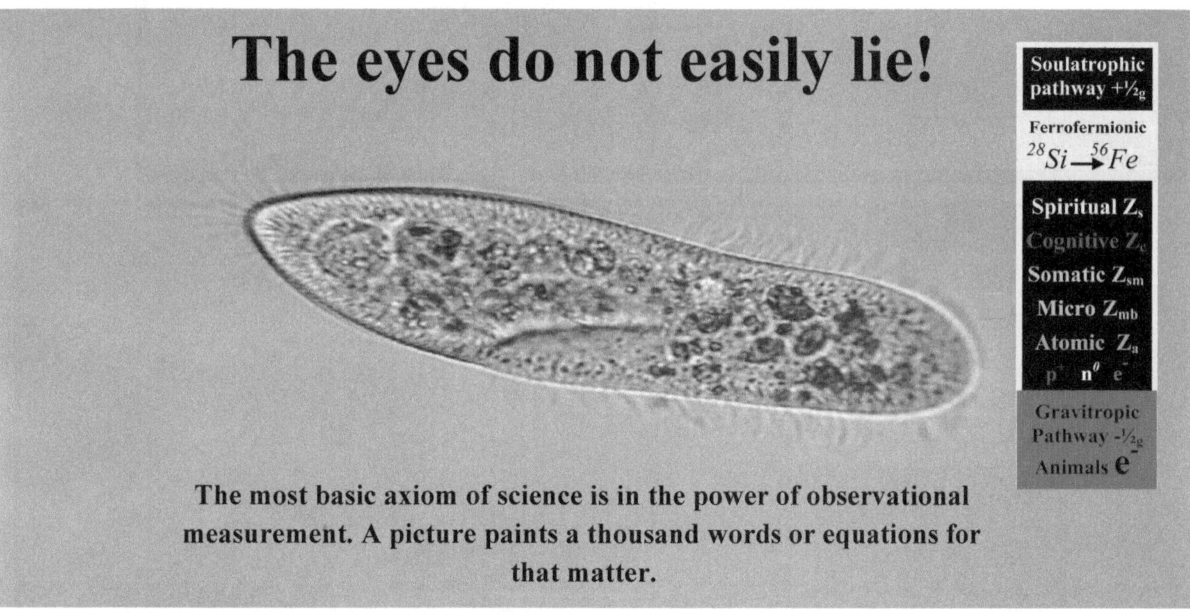
The eyes do not easily lie!
Soulatrophic pathway $+\frac{1}{2}_g$
Ferrofermionic
$^{28}Si \rightarrow ^{56}Fe$
Spiritual Z_s
Cognitive Z_c
Somatic Z_{sm}
Micro Z_{mb}
Atomic Z_a
p^+ n^0 e^-
Gravitropic Pathway $-\frac{1}{2}_g$
Animals e^-
The most basic axiom of science is in the power of observational measurement. A picture paints a thousand words or equations for that matter.

Cell cycle hypothesis – The four time zones and their relationship to cells and the octet rule.

Statement of hypothesis (point 1).

The classical cell cycle is currently limited in its scope because it is only associated with classical micro cells and DNA logic (G1 interphase, S interphase, G2 interphase and mitosis or division). Living organisms have many natural levels of physical size which can all be identified as a cell (a boundary system with a temporal component), for example a full day in the life of a human. The day starts with G1 interphase of the morning, S interphase of the afternoon, G2 interphase of evening and finally sleep mitosis M phase. This cycle is based on the octet rule where cells like the atoms they are made from must acquire four missing stabilising particles. Other cells and their cycles are childhood, adulthood, a lifetime, pregnancy, the menstrual cycle and many more.

Testing the hypothesis.

Examine the origins of the cell cycle and its limitations.

Periodic table hypothesis – The need to find life on the periodic table.

Statement of hypothesis (point 1).

Since life is so fundamental to the universe and all matter, its true logical home should be on the periodic table of the elements. The most important elements start with hydrogen and then through carbon to nitrogen and oxygen (constitute 96.3%), then to a further 22 elements although they are trace quantities. The elements also have a strong association through silicon and iron and produce the rest of the table through technological cells. The carbon entromorphic soulatrophic pathways fit onto carbon and iron, and chart all living evolution on the table itself.

Testing the hypothesis.

Examine the origins of the periodic table and its links to living organisms.

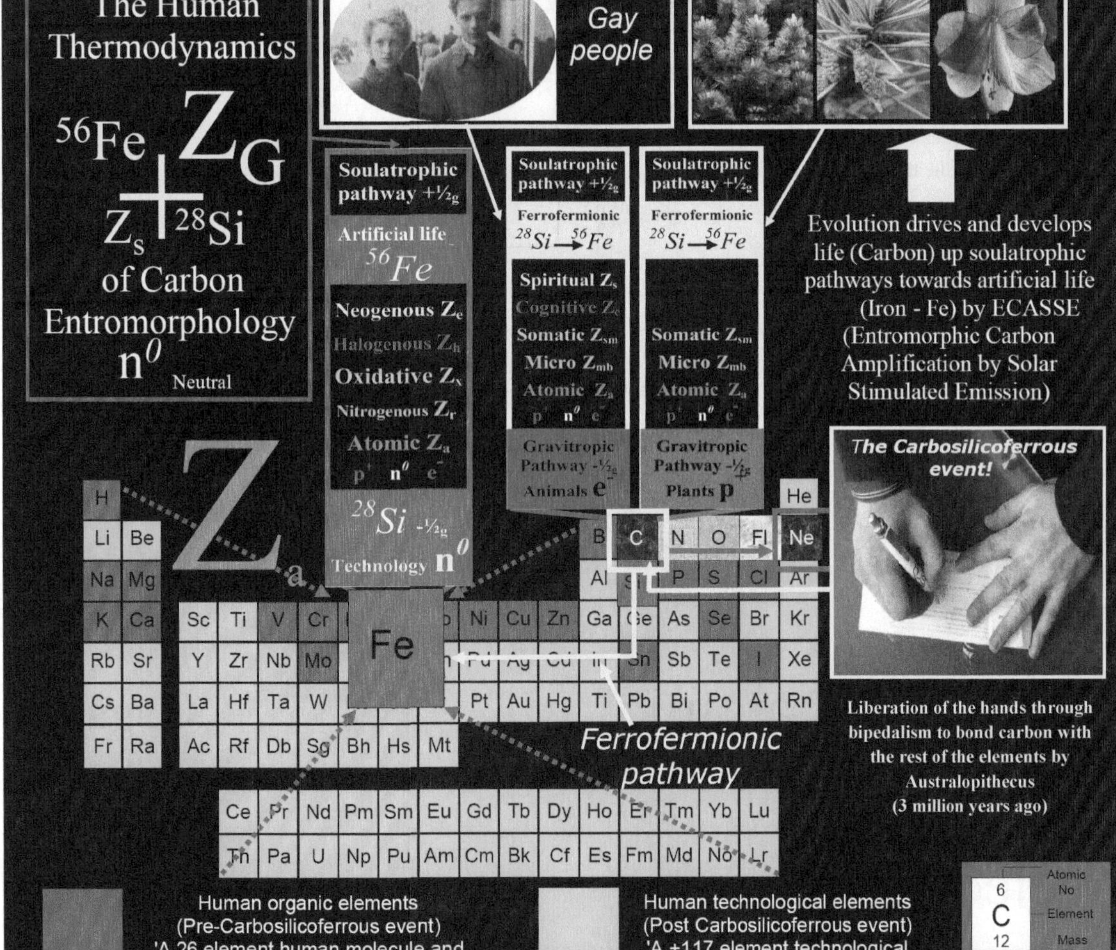

Octet hypothesis – The octet as a measure of solar acquisition and stability.

Statement of hypothesis (point 1).

The octet rule is a fundamental principle of all the elements, stabilisation of quantum energy levels motivates matter on all levels. Since life can be fitted neatly onto the periodic table then life's potential should also be to reach the octet and quantum stability. The octet is so fundamental to all matter, and its effect on life is to acquire solar energy which in turn pushes the nuclear potential towards iron, the most stable type of atom. From carbon next in the table is nitrogen, then oxygen, then fluorine and finally neon. Carbon stabilises its valance shell be producing the four extra electronic configurations, each of which can be hypothesised to be the very essence of the cell cycle (Nitrogen G1 interphase, Oxygen S interphase, Fluorine G2 most reactive interphase and finally Neon the mitotic phase).

Testing the hypothesis.

Examine the origins of the octet on the periodic table and its links to living organisms.

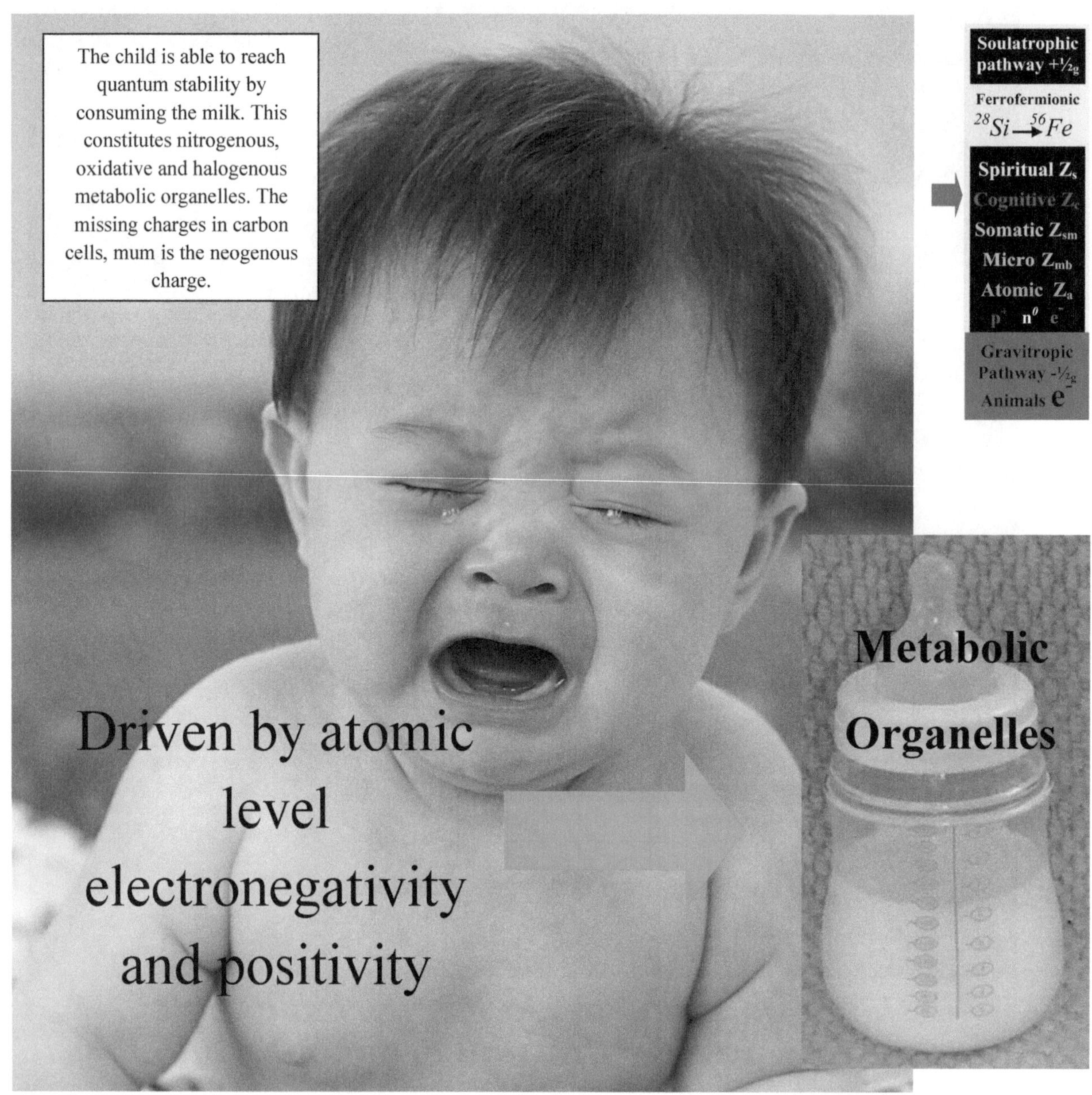

The child is able to reach quantum stability by consuming the milk. This constitutes nitrogenous, oxidative and halogenous metabolic organelles. The missing charges in carbon cells, mum is the neogenous charge.
Soulatrophic pathway $+\frac{1}{2}_g$
Ferrofermionic
$^{28}Si \rightarrow {}^{56}Fe$
Spiritual Z_s
Cognitive Z_c
Somatic Z_{sm}
Micro Z_{mb}
Atomic Z_a
p^+ n^0 e^-
Gravitropic Pathway $-\frac{1}{2}_g$
Animals e^-
Metabolic Organelles
Driven by atomic level electronegativity and positivity

Iron hypothesis – The powerful element of a forth law of thermodynamics driving all elementary thermodynamics in the whole of the universe, and a second arrow of time next to entropy.

Statement of hypothesis (point 1).

All elementary particles seek the stability of iron; this is a fundamental arrow of time and the basis of a new open system law of thermodynamics the 4th law; acts of free will. All life therefore follows this iron instability model, which leads to technology and has its impact of life's stability.

Testing the hypothesis.

Examine the origins of the iron stability graphs and natural quantities through the periodic table and its links to living organisms.

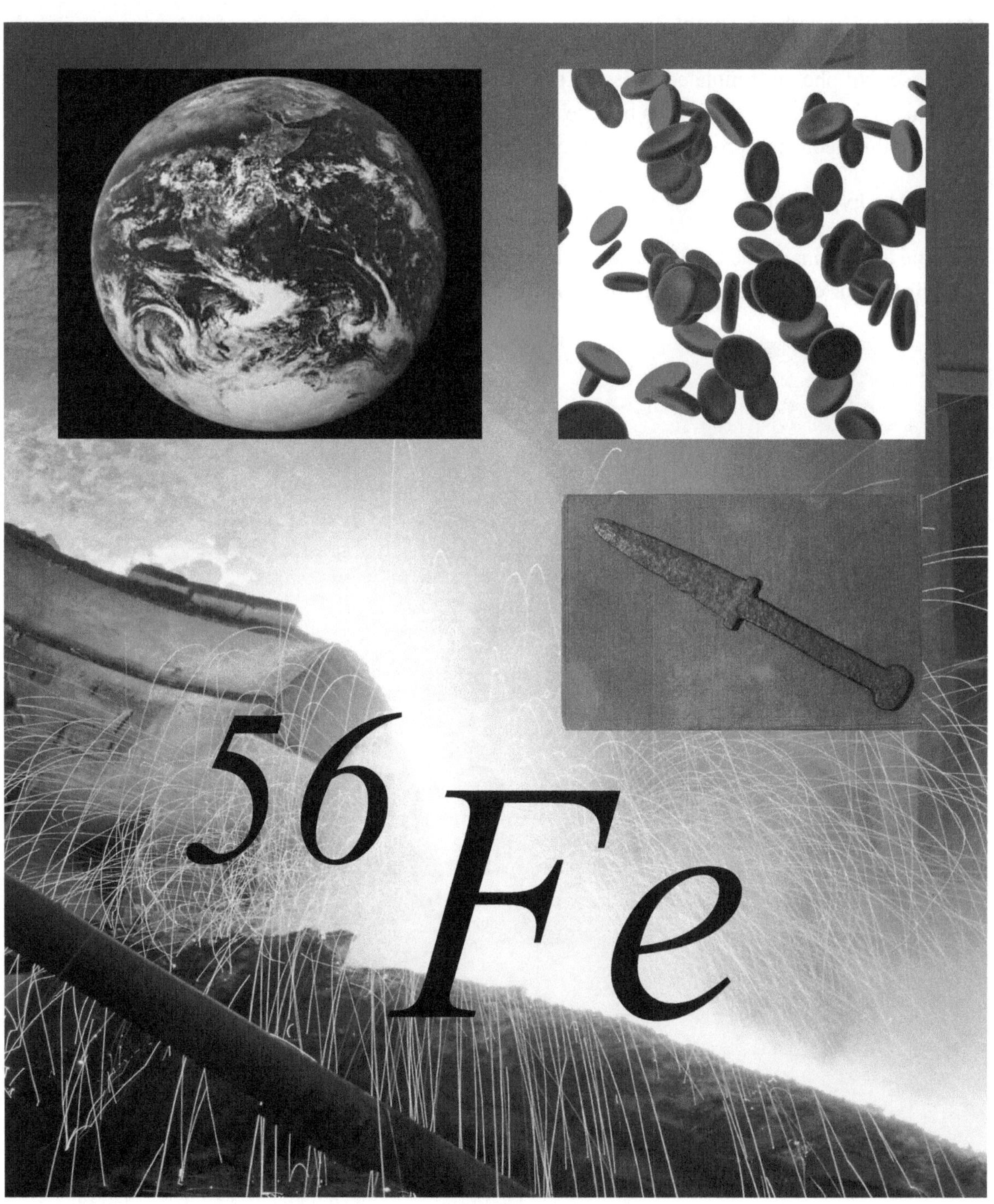

ECASSE hypothesis – Carbon is a boson and fermion, and can semi-occupy energy levels to produce coherent in-phase organisation.

Statement of hypothesis (point 1).

Since carbon is the most versatile and broadly gregarious of all the elements having both fermionic statistical behaviour and bosonic statistical behaviour, it can be considered as a coherent in-phase beam under solar stimulated emission (Bose condensate or LASER physics). Entromorphic carbon amplification by stimulated solar emission or ECASSE is the basic motivational mechanism for describing life on Earth, since carbon is the basis of life. An ECASSE beam such as a flock of birds, a shoal of fish, a company of wolves are in carbon based in-phase coherent beams driven by solar stimulated emission.

Statement of hypothesis (point 2).

The process of photon stimulated emission is a symmetrical physical process where one divides into two. This is the exact way that cells behave, where mitosis divides a cell due to solar stimulated emission, this geometric progression is an excellent example of Bose condensate.

Testing the hypothesis.

Examine the origins of the LASER and Bose condensate and its links to living organisms. **The most powerful proof of ECASSE is that humans have invented and use LASER beams.**

Soulatrophic pathway +½g

Ferrofermionic

$^{28}Si \rightarrow {}^{56}Fe$

Spiritual Z_s

Cognitive Z_c

Somatic Z_{sm}

Micro Z_{mb}

Atomic Z_a

p^+ n^0 e^-

Gravitropic Pathway -½g

Animals e^-

ECASSE

beams

(Entromorphic Carbon Amplification by Solar Stimulated Emission)

'In-phase coherence'

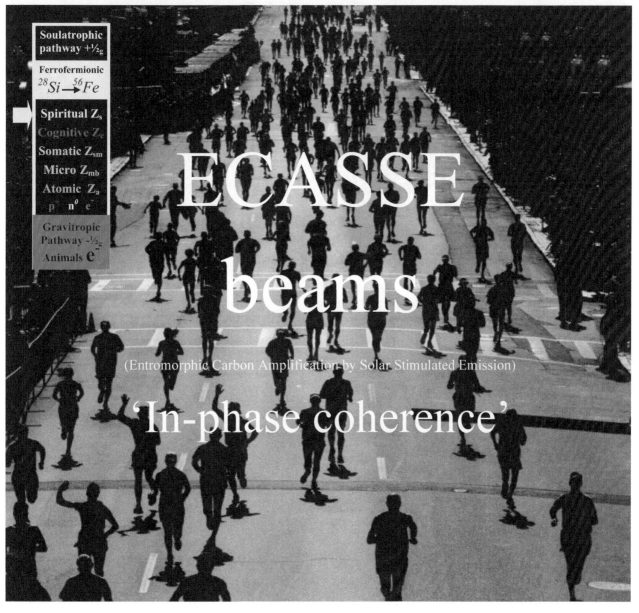

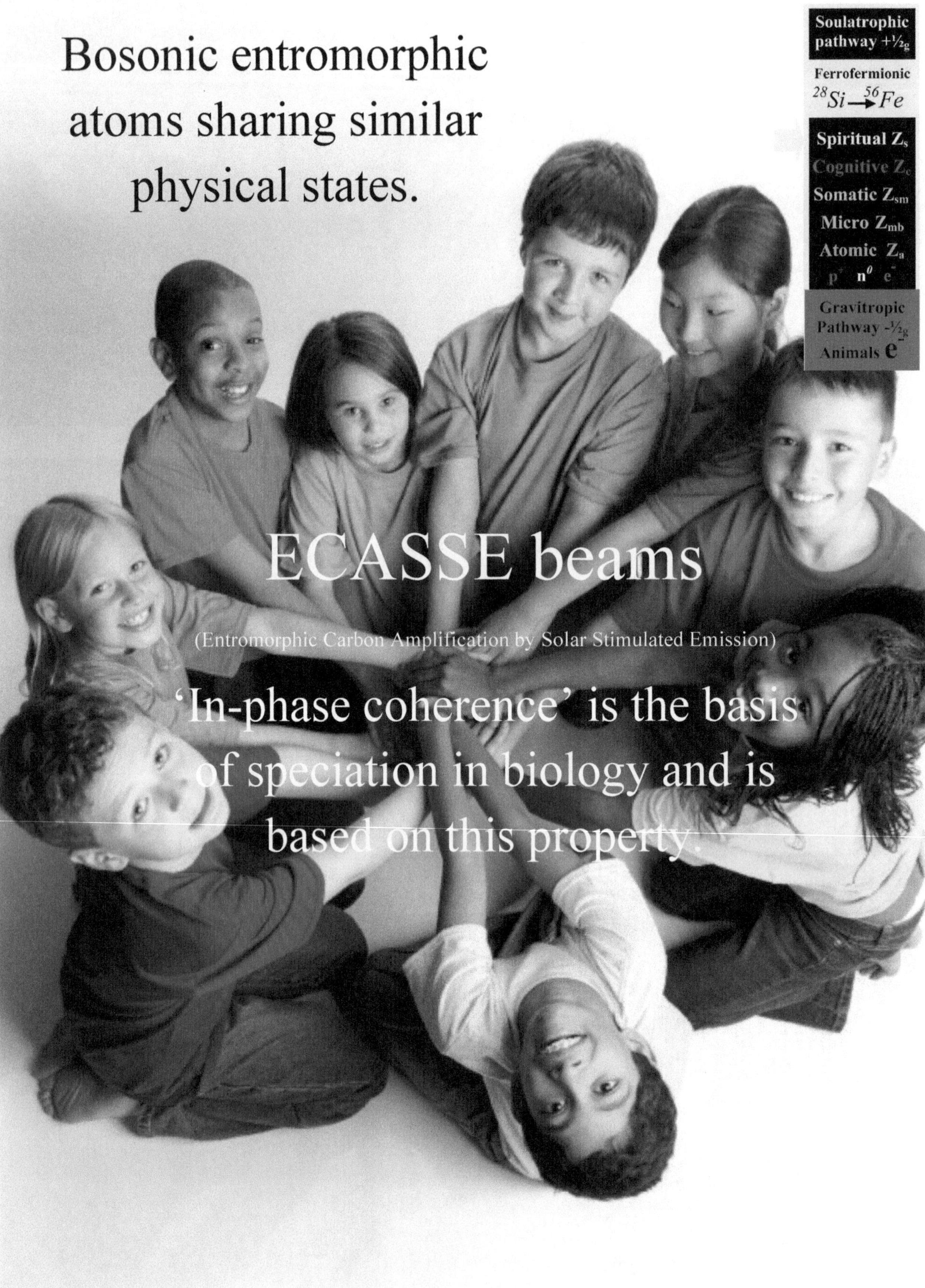
Bosonic entromorphic atoms sharing similar physical states.
Soulatrophic pathway $+\frac{1}{2}_g$
Ferrofermionic
$^{28}Si \rightarrow {}^{56}Fe$
Spiritual Z_s
Cognitive Z_c
Somatic Z_{sm}
Micro Z_{mb}
Atomic Z_a
p^+ n^0 e^-
Gravitropic Pathway $-\frac{1}{2}_g$
Animals e^-
ECASSE beams
(Entromorphic Carbon Amplification by Solar Stimulated Emission)
'In-phase coherence' is the basis of speciation in biology and is based on this property.

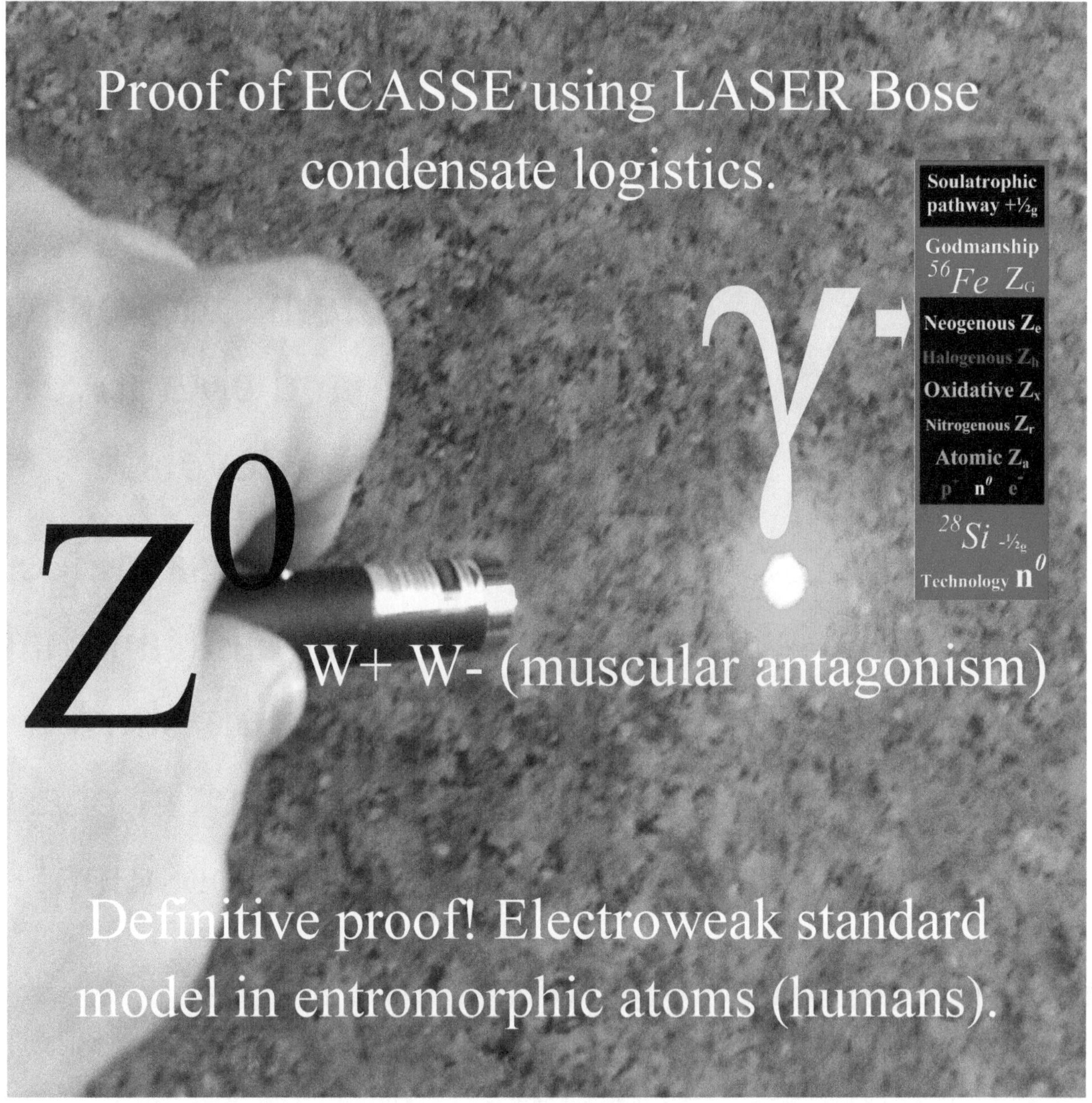

The image above is absolute concrete evidence of the ECASSE mechanism (Entromorphic Carbon Amplification by Solar Stimulated Emission). This human holding a small laser beam (Light Amplification by Stimulated Emission of Radiation), is demonstrating the Bose condensate effect.

The energy which has driven this piece of technology has come from the Sun which allows an extension to the logic to produce the ECASSE effect.

This example is undeniable and completely supports ECASSE as the primary mechanism in carbon based Earth bound life forms, humans created this laser beam.

All life on Earth is based on these statistical particle concepts and shows how carbon can act both as a fermion with ½ integer spin and also as a boson with integer spin. Carbons ability to have broad sharing properties is the reason for this incredible depth of property and allows biologists and physicists to understand the nature of consciousness and the thermodynamics driving it.

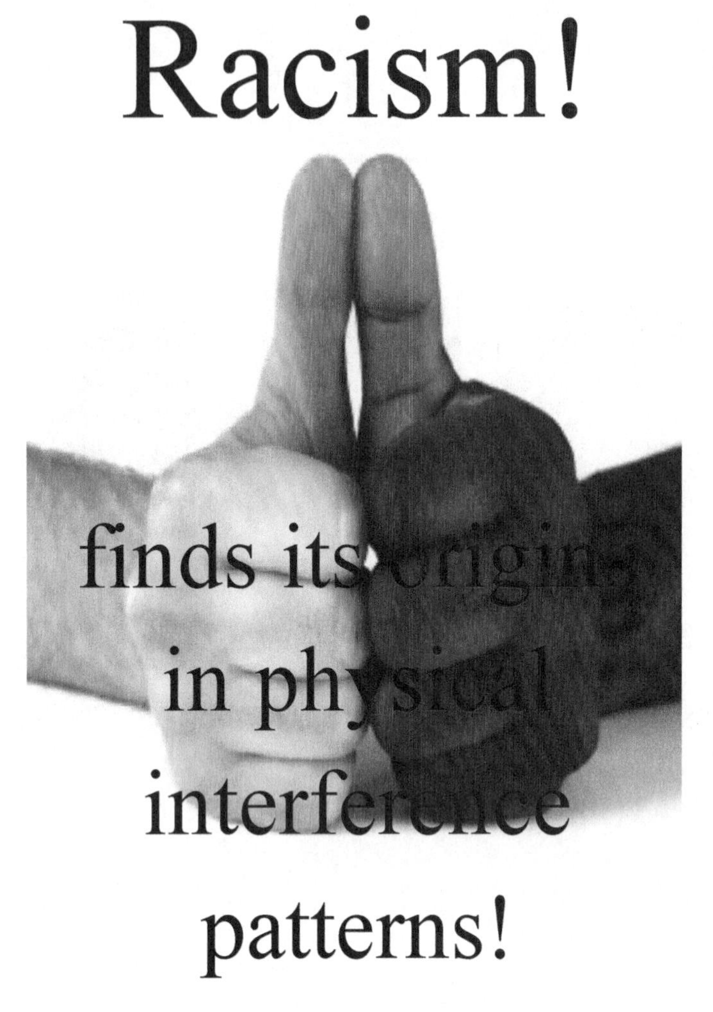

Racism!
finds its origin
in physical
interference
patterns!

Racism is'out of phase' destructive interference patterns for somatic and cognitive levels!

But only skin deep! Spiritual (familial) levels show complete 'in phase' superposition!

In physics, **interference** is the addition (superposition) of two or more waveforms (organims have waveform properties in carbon entromorphology such as race, general nucleonic genomics) that result in a new wave pattern.

As most commonly used, the term **interference** in entromorphology refers to the interaction of expressed genomic nuclear systems which are correlated or coherent with each other, either because they come from the same genetic source or because they have the same or nearly the same physical characteristic (allele) such as skin colour.

Two non-monochromatic (its not called chromatin for no reason) genomes are only fully coherent with each other if they both have exactly the same range of alleles and the same phase differences at each of the constituent positions (genomic gender compatibility in diploid systems).The total phase difference is derived from the sum of both the path difference and the initial phase difference. Black and white have historically been 'out of phase' (producing racism) but only on somatic (radial) and cognitive (bilateral) soulatrophic energy levels (fractional dimensions). Spiritual (familial) levels demonstrate 'in phase' coherence hence evolution has selected for none racism with the new US president.

Chiral hypothesis –The conserved amplified chiral distribution in carbon is a powerful proof of the theory of carbon entromorphology.

Statement of hypothesis (point 1).

The word chiral is Greek for hands; this is because the hands on a human, for example, have a left and right handed version. This is the essence of chirality that the valance electrons in carbon atoms are distributed with different strengths through a chiral distribution. Human hands are a classical example of this distribution as testament to the amplified nature of living organisms. In short, the chirality of a human would be an expected property if that human was an amplified version of the carbon atoms they are composed of. The distribution in this book is blue (nitrogenous), green (oxidative), red (halogenous) and yellow (neogenous). The blue being the most bonded to the nucleus and the yellow being least bonded.

Testing the hypothesis.

Examine the origins of limb distributions in organisms and its link to the natural phenomena of chirality.

This amazing evidence of the symmetry of bonding strengths in this human is powerful. The chiral distribution known as asymmetry is a fundamental property of carbon atoms often known as a 'chiral centred' carbon. It produces a distribution around the atom or organism where one of the limbs have a hierarchy of dominance, hence 'right handed', 'right footed'. This hypothesis supports amplified logic in carbon entromorphology.

The Greek word for the hands is

'chiral'!

Left and right forms

(Levo and Dextro)

Just what you would expect to find on the valance limbs of a giant carbon atom.

Asymmetry, left and right handedness and the word 'chiral' meaning the 'hands', would be an expected property in an amplified entromorphic atom of carbon. Excellent supporting evidence of nuclear and field amplification.

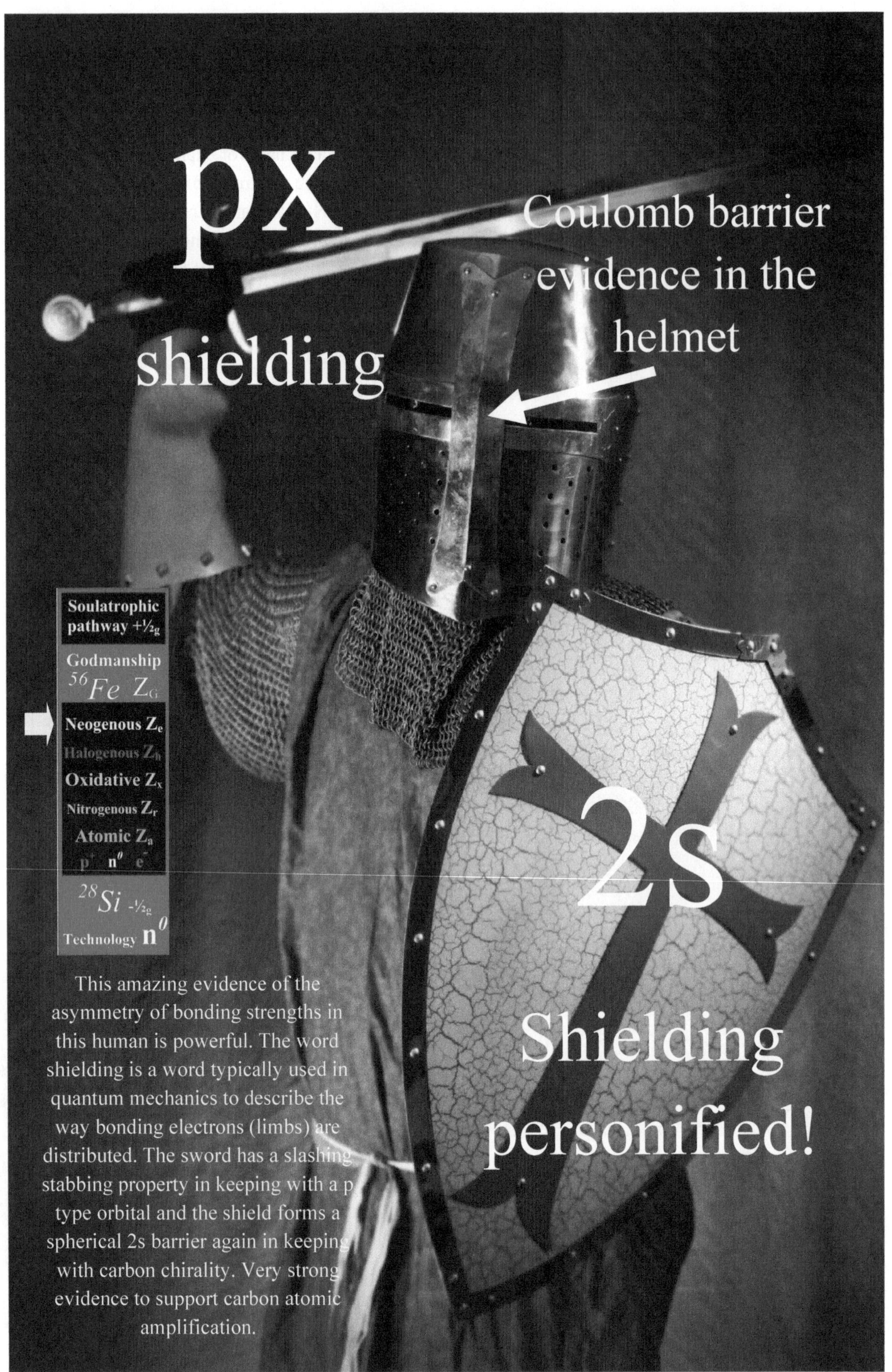
px
shielding
Coulomb barrier evidence in the helmet
Soulatrophic pathway +½g
Godmanship
56Fe ZG
Neogenous Ze
Halogenous Zh
Oxidative Zx
Nitrogenous Zr
Atomic Za
p+ n0 e-
28Si -½g
Technology n0
2s
Shielding personified!
This amazing evidence of the asymmetry of bonding strengths in this human is powerful. The word shielding is a word typically used in quantum mechanics to describe the way bonding electrons (limbs) are distributed. The sword has a slashing stabbing property in keeping with a p type orbital and the shield forms a spherical 2s barrier again in keeping with carbon chirality. Very strong evidence to support carbon atomic amplification.

Sports hypothesis – That sport is powerful and very clear evidence of amplified atomic logic.

Statement of hypothesis (point 1).

Sports involve both Pauli's exclusion principle where two or more grouped organisms form a cell where due to successful energy conservation one group reaches quantum stability and wins. Also present are the details of activity where a macrostate such as a football pitch or pool table localise thermodynamic activities, and where a particle such as a human emits and absorbs a bosonic particle such as a ball. The ball constitutes energy which is aimed to be directed towards a barrier such as a goal in football, a pocket in pool, the back of the court in tennis. Barrier penetration occurs when the ball beats a goal keeper and defence in sports such as football, leading to increased quantum stability for the opposition.

Testing the hypothesis.

Examine the origins of barrier penetration, particle emission and absorbance in human based sports.

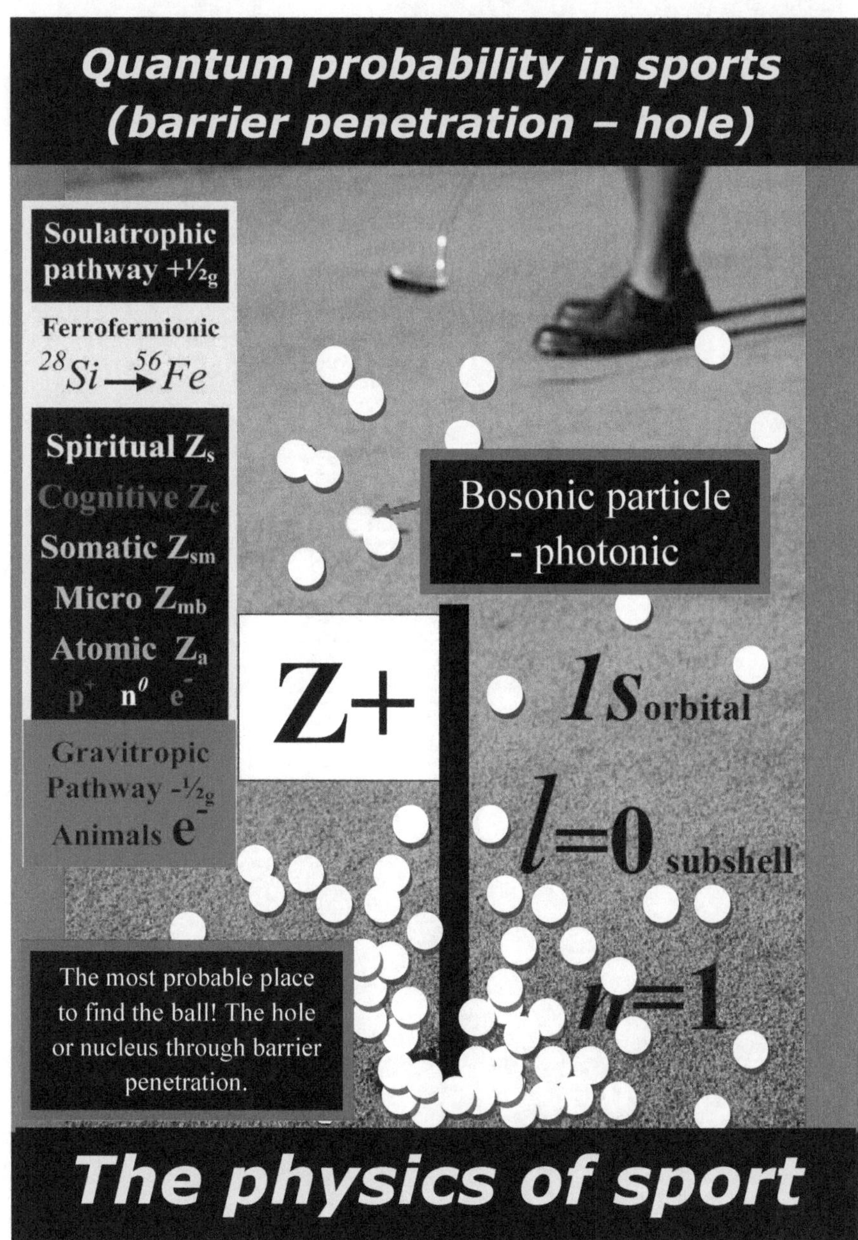

Sports are excellent models of human thermodynamic systems and fundamental physics as a whole.

They demonstrate the basic concepts of quantum theory and nucleonic nuclear physics and are demonstrated where nuclear instability and entropy are the driving force behind the activities.

Sports represent honing skills by practicing human thermodynamic physics.

The rule of octets and ferrous stability is at the heart of this!

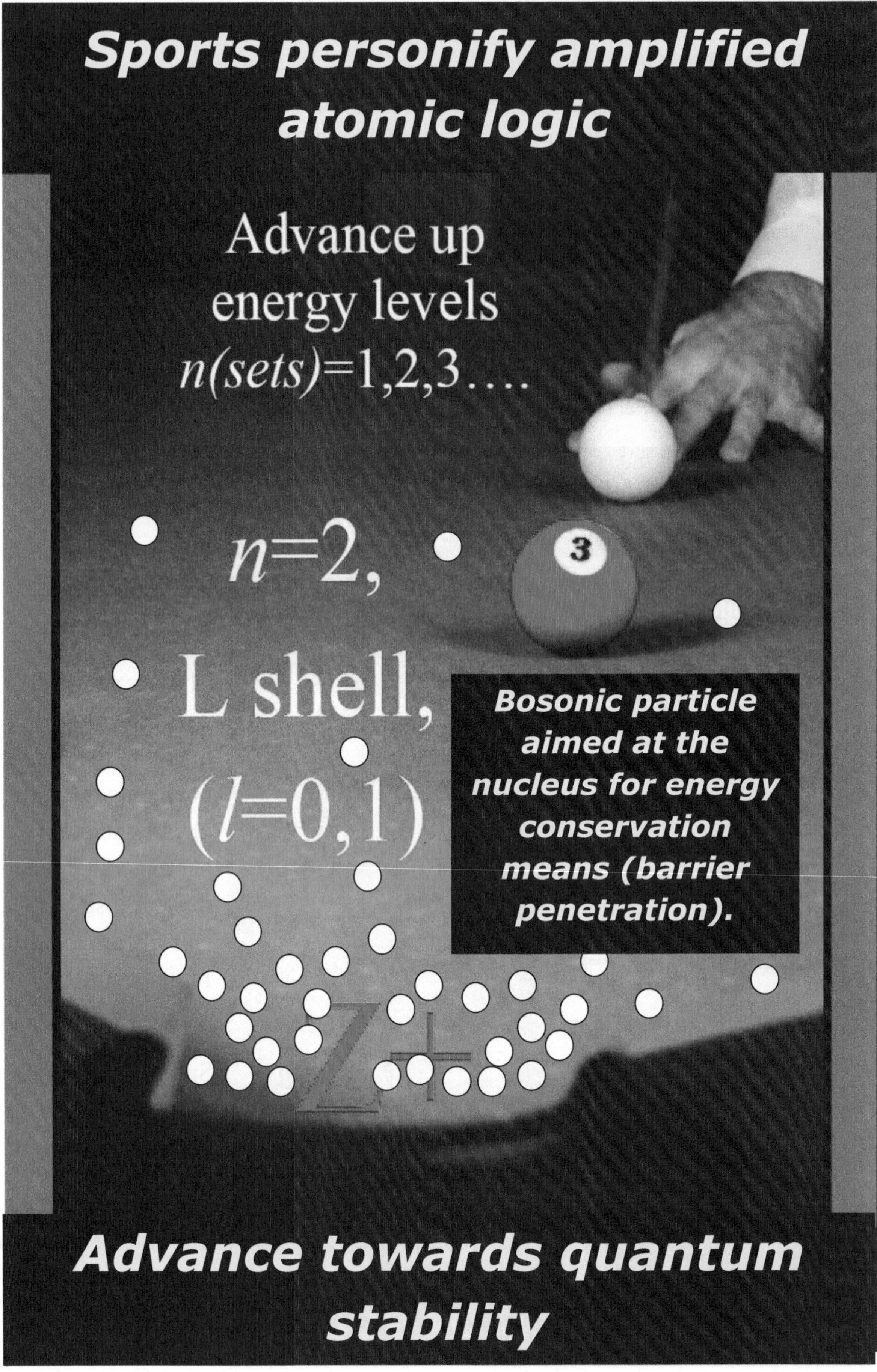
Sports personify amplified atomic logic
Advance up energy levels
n(sets)=1,2,3….
n=2,
L shell,
(l=0,1)
Bosonic particle aimed at the nucleus for energy conservation means (barrier penetration).
Z+
Advance towards quantum stability

Uncertainty hypothesis – That Heisenberg's uncertainty principle is powerful and very clear evidence of amplified atomic logic.

Statement of hypothesis (point 1).

That any measure of uncertainty in any physical system, regardless of size, is a natural limit to the conscious observer's ability to measure physical systems and limits the observer's predictions regarding scientific measurement. There is enormous difficulty with using mathematics for living organisms, where establishing initial conditions with any degree of useful accuracy is almost impossible.

Testing the hypothesis.

Examine the origins of uncertainty in living systems on all natural levels of organisation.

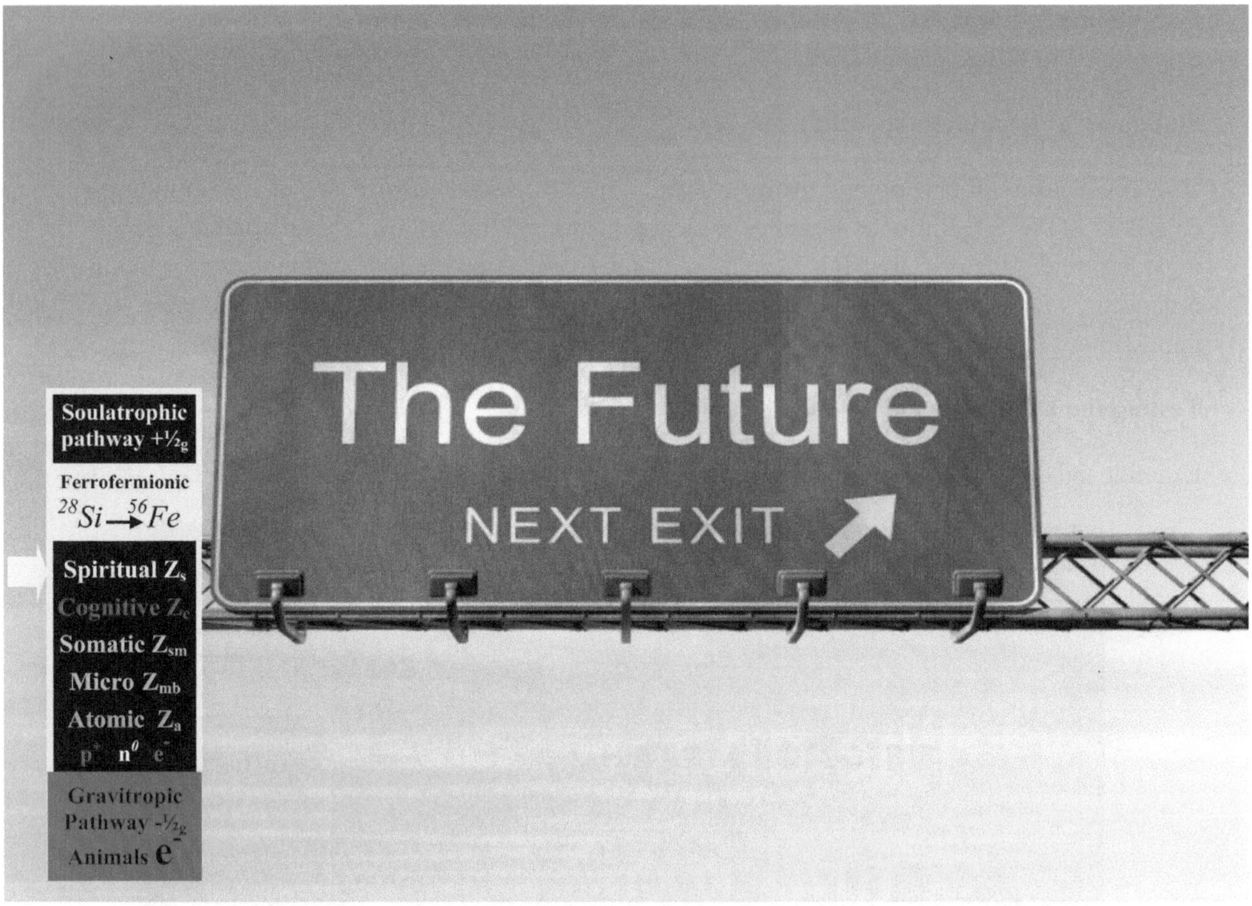

The Future
NEXT EXIT
Soulatrophic pathway $+\frac{1}{2}_g$
Ferrofermionic
$^{28}Si \rightarrow {}^{56}Fe$
Spiritual Z_s
Cognitive Z_c
Somatic Z_{sm}
Micro Z_{mb}
Atomic Z_a
p^+ n^0 e^-
Gravitropic Pathway $-\frac{1}{2}_g$
Animals e^-

Heisenberg's
$\Delta x \, \Delta p \geq \hbar \, \frac{1}{2}$
$\Delta E \, \Delta t \geq \hbar \, \frac{1}{2}$
Present!
Uncertainty and probability
(statistics) are the very essence of life?

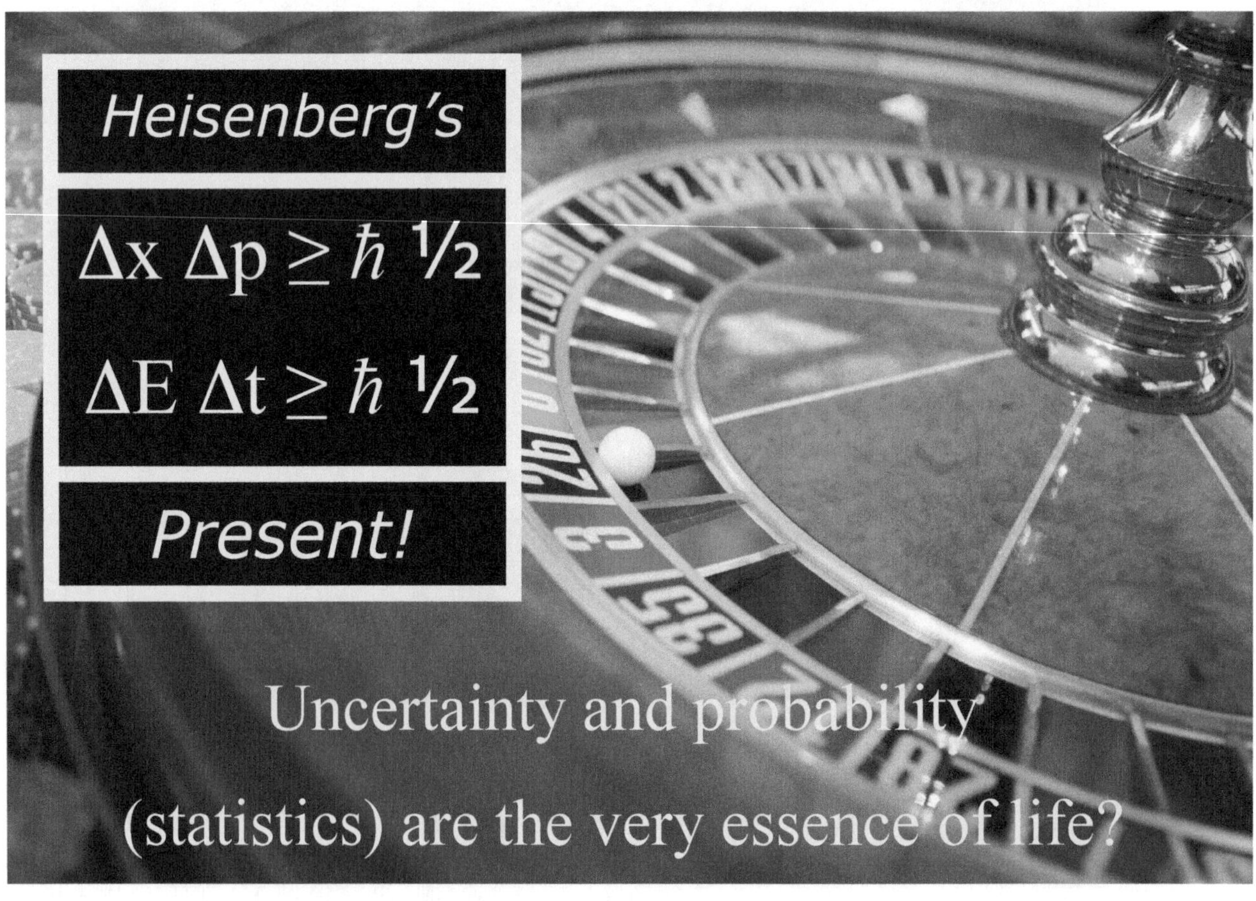

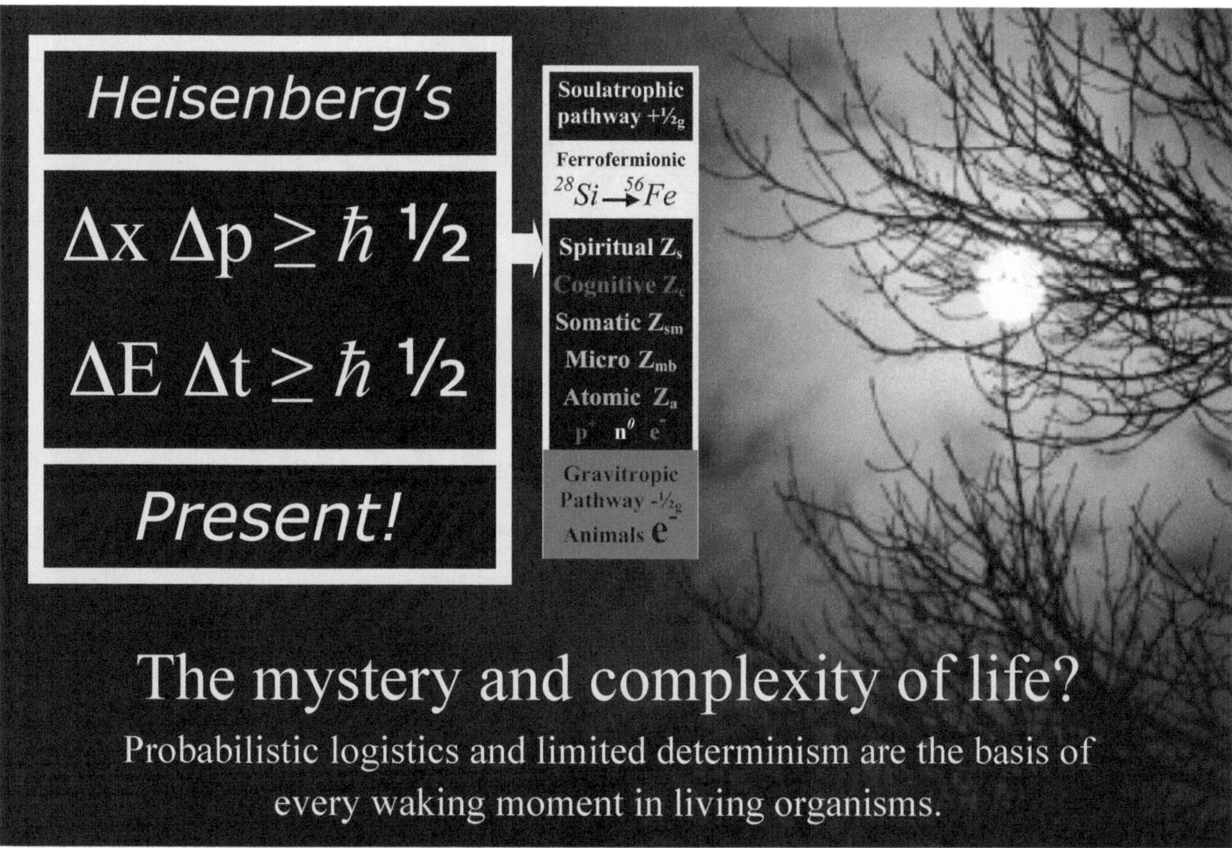

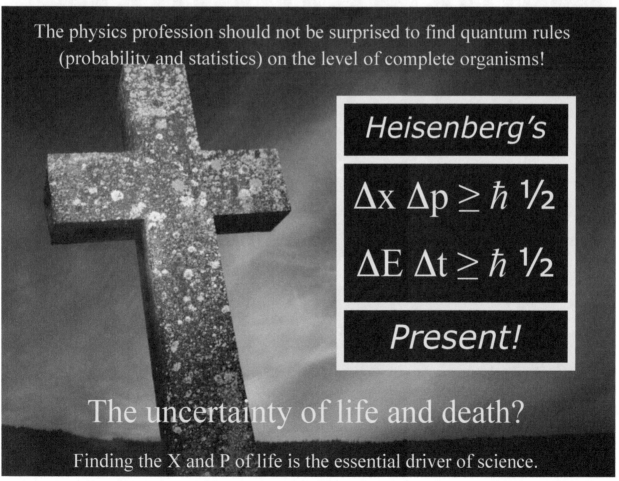

The mystery and complexity of life?

The physics profession should not be surprised to find quantum rules (probability and statistics) on the level of complete organisms!

Heisenberg's

$$\Delta x \, \Delta p \geq \hbar \, ½$$

$$\Delta E \, \Delta t \geq \hbar \, ½$$

Present!

The uncertainty of life and death?

Finding the X and P of life is the essential driver of science.

Science check

Do the observations correlate with the theoretical model?

YES ✓ *NO*

PART SEVEN

Frustrating progress – a letter

Neurochromosome, a 2009 letter - a summary of the tone of 2009.

The following is a letter I wrote to a local academic, the result of which is the website mrcarbonatom.com; the main method of publication for carbon entromorphology. It is a little sad as 2009 was a frustrating year, although I made huge leaps forward in the theory though I had little or no support from science.

Dear ,

It seems odd to start off my letter in this way, but it's important to illustrate the depth of my struggle up front. Most people who are exposed to my work are so deeply and profoundly shocked by the enormously controversial nature that most never ever contact me at all. Some people don't even want to talk to me on the phone and in typical scientific style I feel quite literally 'vilified' by many of them because of my revolutionary ideas.

History repeats itself in science and every time a revolutionary appears people often ostracise them, an example is Ludwig Boltzmann who committed suicide because he supported the deterministic limitation of uncertainty and the necessity for the use of probabilistic mathematics to describe the microscopic scale of natural organisation. He also supported the theory of atoms and the reaction of science drove him to his own death. He like me was a manic-depressive and a revolutionary scientific thinker. His work on statistical mechanics is the basis of all particle physics. His early identification of measurement uncertainty was very advanced thinking. I certainly don't intend to go the way he did so I am developing a necessary thick skin.

I don't know you so I don't know if this will be your reaction. If it is and you choose not to respond to my letter then all I can do is thank you now for your interest and consideration and wish you all the best for the future.

I am a chartered biologist working at ICI, Zeneca and AstraZeneca pharmaceuticals over 15 years, in cancer research, biotechnology, microbiology and chemical and biochemical engineering and 11 years in analytical chemistry. In the past two years after taking redundancy from AstraZeneca I have become a scientific theoretician and scientific author. I originally started off life as a UMIST undergraduate in chemical engineering or applied chemistry and physics, I later moved into biology giving me a very broad scientific background, it's very important to realise that this breadth of experience is crucial to the development of my work.

My work is based on my observations over many years about the segregation of the sciences into physics, chemistry and biology and also geology. Where each professional niche basically operates at a certain specific level of natural particle scale. For example, physics deal with the very smallest (atoms, sub atomic particles) and largest particles (stars, galaxies) in nature. A chemist deals with molecular particles and a biologist with large molecules, to cells to organisms. The sciences literally fit together to cover all levels of scale but with limited theoretical harmonisation between them. If a biologist turned up at a physics conference people would be most confused, yet life is made out of atoms so why can't biologists and physicists help each other, it makes basic logical sense that all levels of organisation illustrate the principles of natural order, why can't we blend the logic? When I moved from physics and chemistry to biology I saw a huge change in the axiomic logic and communication styles of the different sciences. A physicist as you know doesn't consider any science outside of precise numerical measurement and mathematical modelling.

A biologist typically avoids mathematics like the plague and uses words and pictures to do most science, they are particularly good at dealing with uncertainty where particle populations are dizzyingly large and statistics is the only means of analysis. Physicist's like Albert Einstein loathed uncertainty and believed in beautiful mathematics, he believed in perfect measurement and absolute mathematical determinism. He helped to establish quantum mechanics yet spent most of his career trying to discredit it. A chemist sits in between using both systems from physical chemistry a mathematical logic to biochemistry and polymer chemistry with structural diagrams and words beginning to emerge. Physicists are very poor biologists because resolution of biological variables is so great that biological measurement is deeply flawed and statistics and probability become the only deterministic algorithmic compression available to them, however this is the personification of amplified quantum logic on the level of cells and organisms.

So physicists have nothing to do with the life sciences, which is counterproductive to scientific development since physics is a most noble and breathtaking achievement full of the wisdom of nature at its most fundamental but also and unfortunately its most inanimate. Physics needs to balance itself against the inevitability of uncertainty and the measurement problem as a true reflection of the extremes of consciousness. Physics must therefore deal with life in their models of nature if we are to get a true algorithmic compression of the conscious being and a full truer understanding of the nature and interaction of energy as it transcends its four-force physical system.

When I became a biologist I noticed that there was a distinctive ceiling and floor to the subject. Biologists cannot explain the fundamental basis of life because that basis exists in atomic physics as nucleogenesis occurred over billions of years to give us elementary particles capable of accumulating into composite living breathing particle wave systems. As I have said, biologists have nothing to do with that level of scale and physicists don't do life science because they can't measure accurately or use calculus and differential equations effectively. This means that we can't answer the following questions and the list could go on and on and on and on and on, biology is currently a most limited philosophy because of these extremes of thought and must embrace all perceptual components of nature most desirably those of physics, I support integration I am not condemning anyone just trying to wake people up to our true integrated conscious nature: -

- **What is DNA and where did it come from?**
- **Why do I have two eyes and where did my head evolve from?**
- **What is the brain and the mind and where did it evolve from?**
- **What is consciousness and how can we understand it?**
- **Why do we look the way we do and function in the way we do?**
- **What is the natural basis of heredity in nature?**
- **Why do my hands look and function in the way they do?**
- **What and how did sexual reproduction come into being?**

The list is endless because biology has no basis for life in atomic physics and physicists don't recognise biological axioms because they are mainly none mathematical. To refer to the statement about why we have two eyes and where did they come from, a biologist would probably make the following statement on the subject: -

'We have found evidence of 'eye pits' in simple single celled protozoan organisms. The two eye pits have evolved into larger more complex structures in higher multi cellular organisms. This is the evolutionary basis of the eyes'.

But the protozoan is made out of smaller particles such as molecules and atoms so why doesn't the biologist look there for an answer about the eyes if that is the soul of the logic. A physicist doesn't care about where the eyes came from and reverts to their highly mathematical logic the world of the 'inanimate' and they turn their backs on the life sciences 'the animated world of life'.

This problem with science being segregated and life having not been identified in nuclear physics and quantum mechanics is unacceptable and counterproductive to science. What is needed is theoretical harmonisation so all the sciences are linked and our basis for life is found at the limits of the smallest entities we know namely atomic and sub atomic physics and where the four forces of nature are the basis for understanding life in its entirety as opposed to a concept not even observed in any biological text, yet it's the fundamental basis of the space time continuum. So why don't we consider it when trying to evaluate living particle wave/duality systems? We are comprised of three subatomic particles the electron, proton and neutron so why don't we use their logic for composite particles such as life the fermionic state of matter. Why don't biologists talk of bosonic particles in their models when this reflects the other part of the matter energy logic, why do we feel quantum determinism only exists on the level of the atom even though atoms bond to make larger and larger particles, on to cells, tissues and organisms? My carbon atoms are currently functioning to allow me to write this letter!

Try this thought experiment it's called the 'quantum gravity mutant': -

A world famous actor is doing charity work in the summer when a photon from outer space strikes an electron in a single carbon atom in the DNA of one of his skin cells. This in turn induces a quantum jump in the carbon atom the mutant event induces a promoter region in the DNA to activate an oncogene which produces a cancer which kills him, and the funeral is viewed by millions of people worldwide. If that isn't a link between atomic determinism and planetary determinism I don't know what is! Locality is far too local at the present time in physics.

If we truly want to understand life then we must compromise, biologists must embrace the immense power of nuclear and atomic physics and its powerful mathematical basis. Physicist's need to expand their models of life from the typical 'the observer' statement that they use, to a more integrated extensive system. In doing this they must embrace the power of words and pictures for doing science where measurement is made by sight alone. I use photographic evidence all the way through my work; I prefer billions of data points as opposed to a mere twenty in physics (for example validation of a linearity in chromatographic qualification), physicists tend to think that reality is never really measurable by sight even though most of life on Earth measures its physical system through sight, it's the primary measurement tool, the nature of observation. What we need is a model of consciousness, which is thoroughly described in my work through carbon entromorphology and the emergence of neurological genetics as a harmonised system for all knowledge.

The chemist is already playing 'piggy in the middle' and my work is all about harmonising the logic of science by embracing a new model of life based on nuclear and atomic physics and its harmonisation with the rest of science. In short I have discovered life on the level of the atom and this logic has given me immense power to explain and understand the nature of all life on Earth by blending the logic of the different levels of natural scale. The levels fit together through fractal geometry where each dichotomous level produces a fractional dimension for all natural scale. A fraction of larger levels of natural scale defined in its own right by lower levels of natural scale. Hence atoms make up cells so atoms are the fractional component of cells, cells in turn are the fractional component of organisms but so are atoms, they are all good deterministic logics.

I have included with this letter a copy of a paper published by myself in February this year. The copy you have has a few spelling mistakes in it as I have run out of other copies. Most people found it very difficult to understand because it blends physics and biology and each profession stands in its respective corner conserving what they know and refusing to consider the logic of the other profession. The biologists who saw it couldn't understand the physics and the physicists' don't consider biology to be numerically and mathematically precise enough science so they won't move without detailed mathematic proof.

Since this paper was published (50 copies only) I have written a 'simplified guide to carbon entromorphology' called the 'quick start guide' intended to assist understanding in a non scientific audience. With this are the 'entromorphic calibration standards', the way to think of these is that they form a calibration curve in the same way you would if you were running a chromatographic system for assaying molecular weight in an unknown sample. The molecular weight range extends from the atomic world up to the level of entire multi cellular organisms, so it's a huge calibration curve model. I also wrote four lectures on the subject, one of which is extremely important. It's called the 'Blind Physicist' and encompasses a mathematical argument for my theory of life and a new equation showing the interaction of the four forces of nature and the basis for an 'equation of life'. It details the problems with science in a similar way to this letter I am writing to you. It also brings the powerful physics theories I use into fruition and links them to the logic of living organisms.

It also provides a powerful calculation I made to show how one of the most fundamental constants in Nature namely the ratio of proton to electron mass, is conserved in my atomic amplification theory of life, and the problems with biological uncertainty, protocol design, experimental data skewing and limitation of and necessity for using population statistical measurements. This as opposed to strong applied differential equation formulaic sets with powerful numerical determinism.

This is a powerful proof and combined with my 'Stephen Hawking proof', and my interpretation of the four forces of nature and their link to the mathematical operators makes my work quite watertight. Staggeringly enough my work is gigantic to say the least, I am also a graphic designer and wanted to bring my science to life in a brand new way. Hence my work is presented as a gigantic science/art exhibition called 'My Cell'. There are photos of the Alderley park installation in November 2008 in the back of the enclosed publication for your consideration. This year has been all about detail and tidying up the theory and linking it to the underlying theories in both physics and biology. It has formed the bilipid layer on 'My Cell' the exhibition the ground state exhibition is now relatively complete.

My work also extends the theory of life back to the Big Bang and also takes into account the nature of technology as a natural extension to organic life. Biology doesn't consider technology and it is left to palaeontologists and archaeologists to deal with this subject, which again is counterproductive and limiting to biology. Carbon entromorphology is a complete model of life housed on the periodic table of the elements and extended throughout the periodic table to model technology thoroughly. My work also brings metaphysical concepts such as God, heaven, hell, good and evil into the physical. Stating that morality is simply thermodynamics in disguise, and that God is linked to the powerful drivers of quantum energy level stabilisation and the tendency for ALL elements to develop their nuclear potential towards iron nuclear stability. This would be a perfect entity with limitless temporal stability and energy availability.

Please don't be put off by what appears to be unscientific metaphysical statements, stick with it and you will appreciate why I have included them.

Again both these elementary physical potentials in nature (quantum stability) are well accepted and highly proved. My work basically extends classical theory although I do criticise some of the aspects of quantum mechanics such as the lack of gravitational determination on that level and the clone model of subatomic particles, 'once you have measured one electron you have measured them all'. I don't agree with this and have suggested an alternative theory, the limits of determinism in quantum logic is proof of this as some of the measurements can never be made hence compromised determinism the so called 'hidden variables' of non locality, hence slight immeasurable variability in matter. Also, evolution theory is limited to DNA nucleonic sizes of scale and needs extending to neurological and atomic nucleonic systems of scale. Also evolutionary theory currently suggests that life doesn't follow pre ordained pathways and has no conclusion or goal. Again I strongly disagree with this and have demonstrated extraordinary new and exciting possibilities for the future of human evolution.

The physics world I suspect are horrified with my theories as it means that the answer to atoms has been in front of them all their lives, that they are a type of extended atomic particle and that evaluation of the atomic world is simply a form of self evaluation and realisation. Once you see my world then the atomic world is everywhere and quantum uncertainty is identified as the basis of every second of consciousness. It took the uncertainty skills of a biologist to make these enormous leaps in understanding because a biologist can put two totally unlinked structures together to do good science, without a mathematical dichotomy to connect them.

I have reached the end of my pursuit of recognition by scientific bodies, they are simply too conservative to embrace such a massive change. In the past year there hasn't been ONE individual who has attacked my theory; the nuclear physics professor said he couldn't see how it could happen which means that although he couldn't identify the cause he could observe the effect. Since that I have answered his query thoroughly and concisely. In the next few weeks I am about to take my story to the press. My patience with the scientific community and conventional peer reviewed protocol is at end. Never let it be said I didn't try very hard to do this in the right way, it's time to do it the wrong way, perhaps which will ironically work. I originally wanted to publish in 'The Biologist' but my work is so huge in its scope that even a large peer reviewed publication would barely do it any justice and cost a fortune. Finding peer reviewers who have a broad enough background is also very difficult. My work needs sensationalising and the physics professor did recommend that I go ahead and publish. I have an incredible story to tell now, my work allows me to cope with severe mental illness and this makes my story even more complex and powerful.

On Halloween this year I went to a fancy dress party dressed as a carbon atom. Each part of my body beautifully synchronised my anatomy and physiology to carbon atomic physics. I have created the concept of 'Mr Carbon Atom' I have made a suit which I can wear and can use it to demonstrate the wave function nature of consciousness known as the 'electroweak standard model' in the physics world with biological behaviour. Also other aspects of atomic physics and its link to life. I also have my exhibition material, ten 7-foot exhibition stands and 300 square feet of floor models. So the next few weeks are crucial and hopefully will see my work published to a large audience.

Again thank you for your time and consideration and I fully understand if you don't decide to contact me, most people don't! Best wishes for your retirement and for the future,

Yours Sincerely,

Mark Andrew Janes BSc (Hons) CBiol MSB

Neurochromosome Entromorphic Quotations - From experience.

'Human thermodynamics, in the future, I believe, is on the precipice of being the biggest overall unified scientific concept ever.'

'Science is 0.1 % innovation and 99.9% regurgitation!'

(Innovators are extremely rare in science and are usually treated like garbage for it by non-innovators).

'The only certain thing in life is the absolute unavoidability of uncertainty!'

'I am an atom and therefore ironically unknowable even to myself!'

(The measurement problem personified)

'Love is…The Strong Force!'

'Love is….The Strong Interaction!'

'The Strong Interaction is the basis of all information systems!'

'May the events of your day bring you ever closer to quantum stability through phototropism.'

'The sexual act is nuclear physics personified.'

'The biology of technology.'

'Hello my name is Mr Carbon Atom, but you can call me Carbon!'

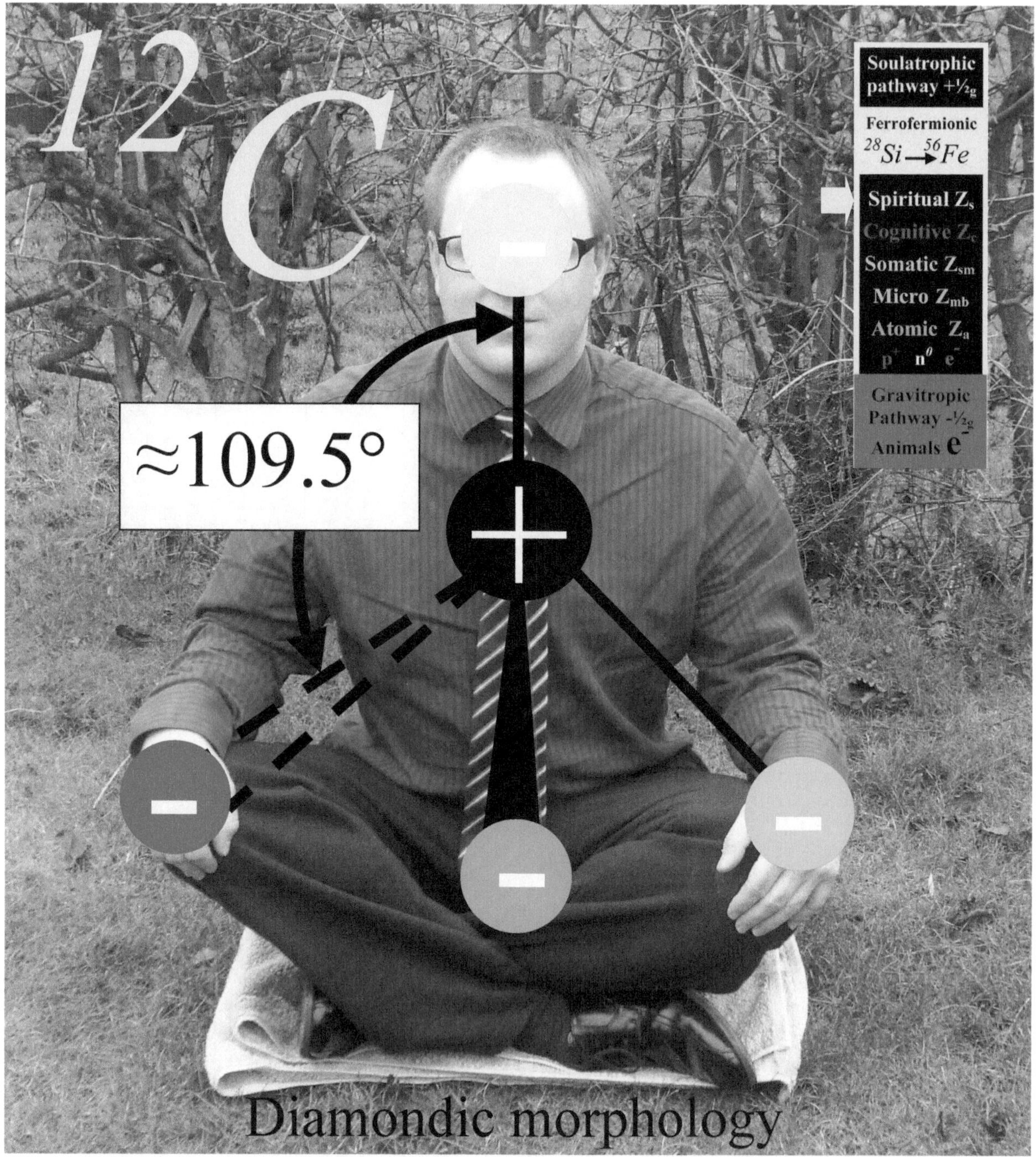

The tetrahedral structure of carbon can be seen in 'entromorphic atoms' such as this author. This position is a very stable form of carbon associated with the powerfully strong crystalline structure of diamond. Again this is powerful proof of nuclear and field amplification as an amplified carbon logic should be able to demonstrate tetrahedral structure in this way.

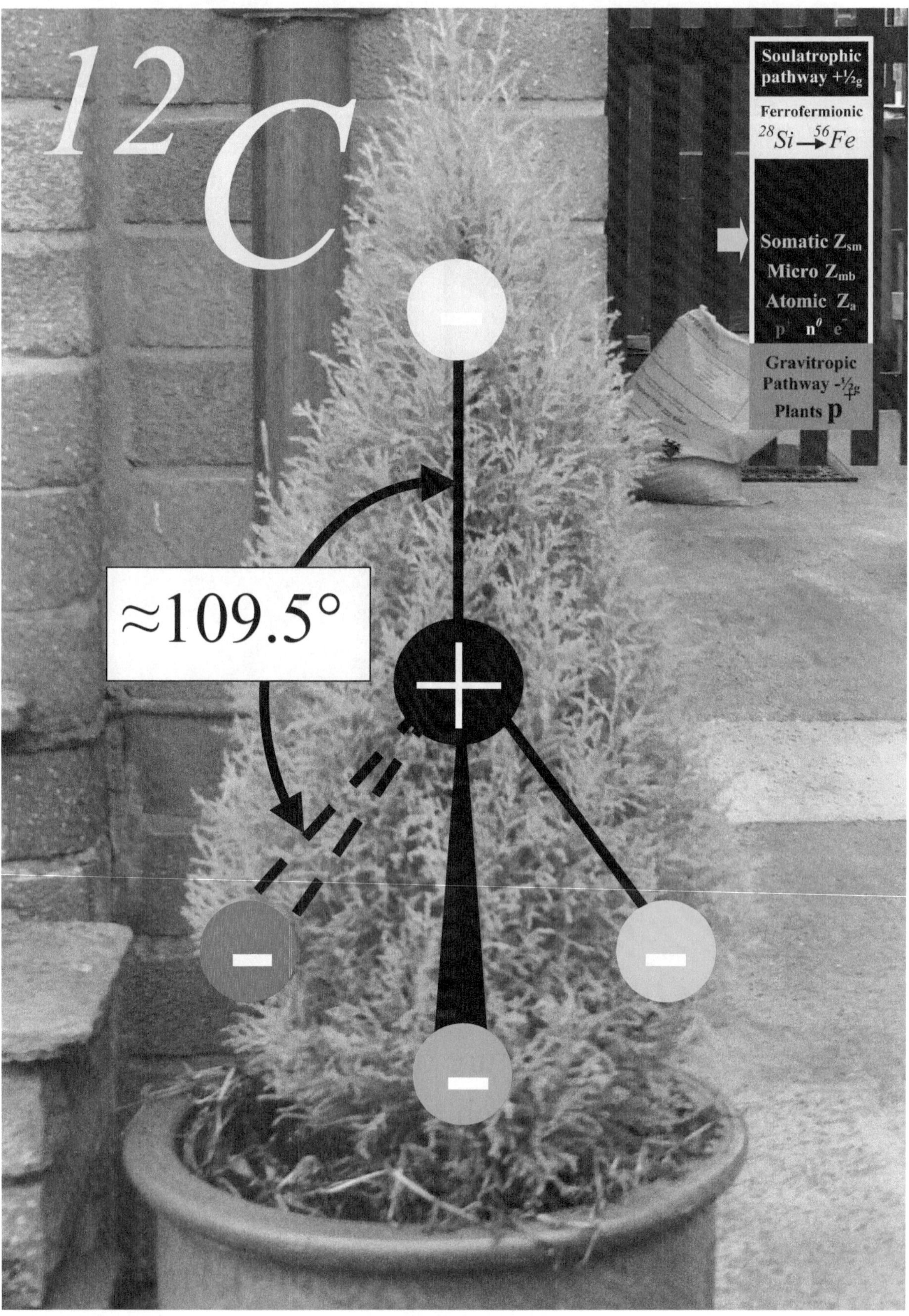

More tetrahedral structural evidence in the plant world.

Science check

Do the observations correlate with the theoretical model?

YES ✓ ***NO***

PART EIGHT

The fractional dimensions of natural living organisation

Neurochromosome soulatrophic models – Soulatrophic pathways, entrochiraloctets for electromagnetic and global soulatrophicity, wave functions, cell cycle components and thermoentropic and Gibbs enthalpic processes.

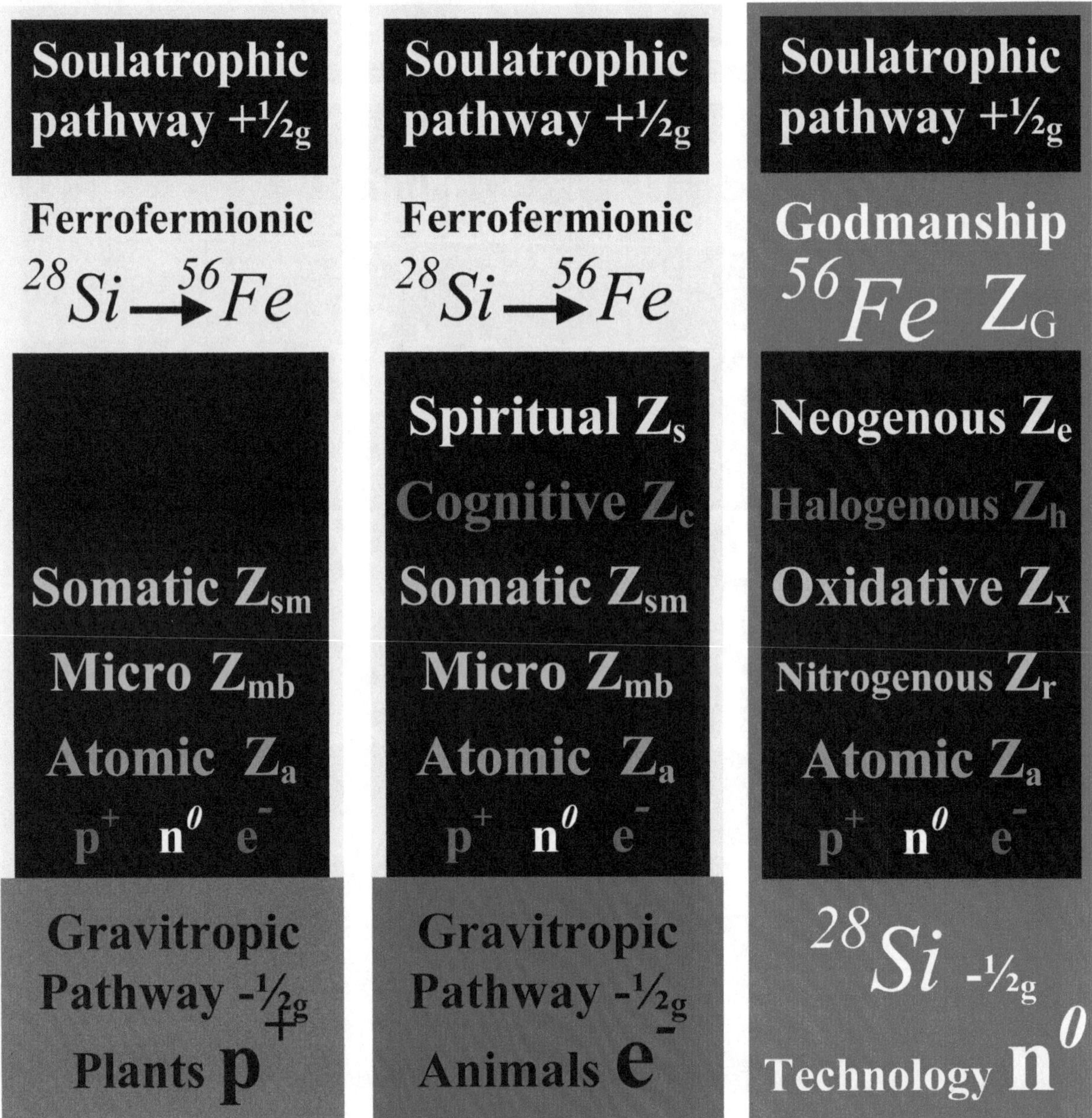

The following models are used throughout carbon entromorphology as a means of relating biological logic to atomic logic. They include the entrochiraloctets for both electromagnetic and global gravitational models. The entrochiraloctets are the carbon atomic templates, which are amplified to produce living organisms from atomic carbon origins and show both ground state and sp2-hybridised states for all living carbon cells.

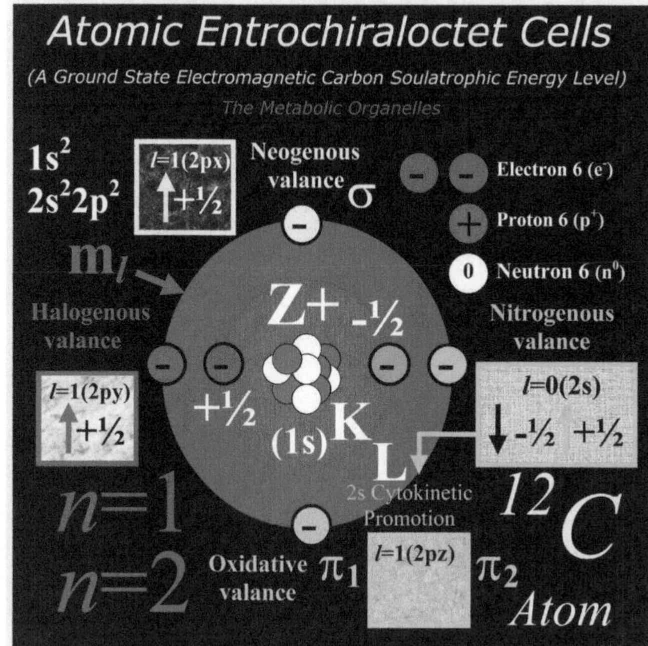

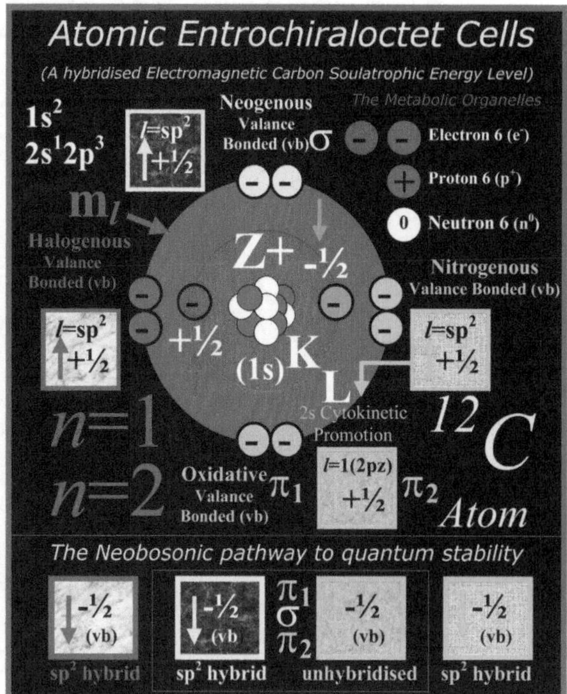

The basic templates of life, entrochiraloctet cells.

Also included are the basic models of the cell cycle, which is extensively extended in carbon entromorphology as the cell is re-defined in this logic. They are the G1, S, G2 interphase and the M mitotic phase of carbon octet cycling.

There are also the basic templates for nuclear fusion (meiosis) and nuclear fission (mitosis) and their relationships to carbon entromorphology. Again these concepts are typically associated with the microbiological and somatic (radial) levels but are extended further to include atomic, cognitive (bilateral) and spiritual (familial) levels.

The basic soulatrophic pathways for the plants, animals and technology are also shown. Each level of the soulatrophic pathway has an entrochiraloctet for both electromagnetic and gravitational (global) levels in some stage of ground state to sp2 hybrid.

Also included are the basic thermodynamic models of good and evil which specify the cell cycle in relation to soulatrophic energy conservation or good (growth) and energy liberation or evil (death).

In carbon entromorphology life is described as a 'soulatrophic pathway' from the word the 'soul'meaning the origin of life. It runs all the way back to the Big Bang and is defined by specific levels such as atomic, microbial, somatic (radial), cognitive (bilateral) and spiritual (familial). Mathematically the pathway is known as a 'fractional dimension D'; a fractal geometry.

Each level is self-symmetrical to the other levels and is driven by iterative cell cycling (ECASSE) and is made up of the fraction of the lower levels and acts as a fraction of the levels above it.

The following pages demonstrate the vast variety of spiritual 'familial' cells and their basic nuclear K and L shell energy levels.

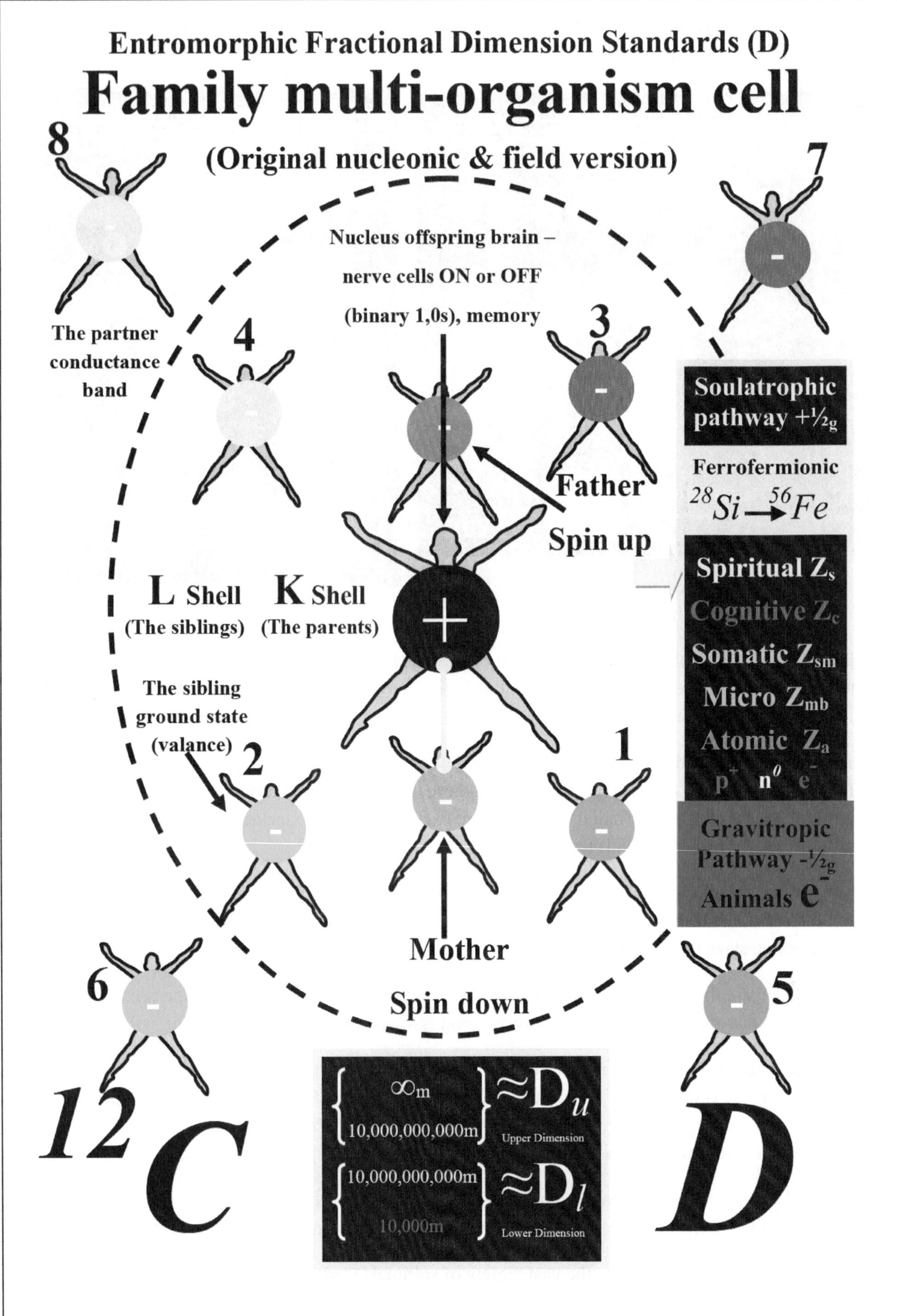
Entromorphic Fractional Dimension Standards (D)
Family multi-organism cell
(Original nucleonic & field version)
8
7
Nucleus offspring brain – nerve cells ON or OFF (binary 1,0s), memory
The partner conductance band
4
3
Soulatrophic pathway +½g
Ferrofermionic
28Si → 56Fe
Father
Spin up
Spiritual Zs
Cognitive Zc
Somatic Zsm
Micro Zmb
Atomic Za
p+ n0 e-
L Shell (The siblings)
K Shell (The parents)
The sibling ground state (valance)
2
1
Gravitropic Pathway -½g
Animals e-
Mother
Spin down
6
5
∞m
10,000,000,000m
≈Du
Upper Dimension
10,000,000,000m
10,000m
≈Dl
Lower Dimension
12C
D

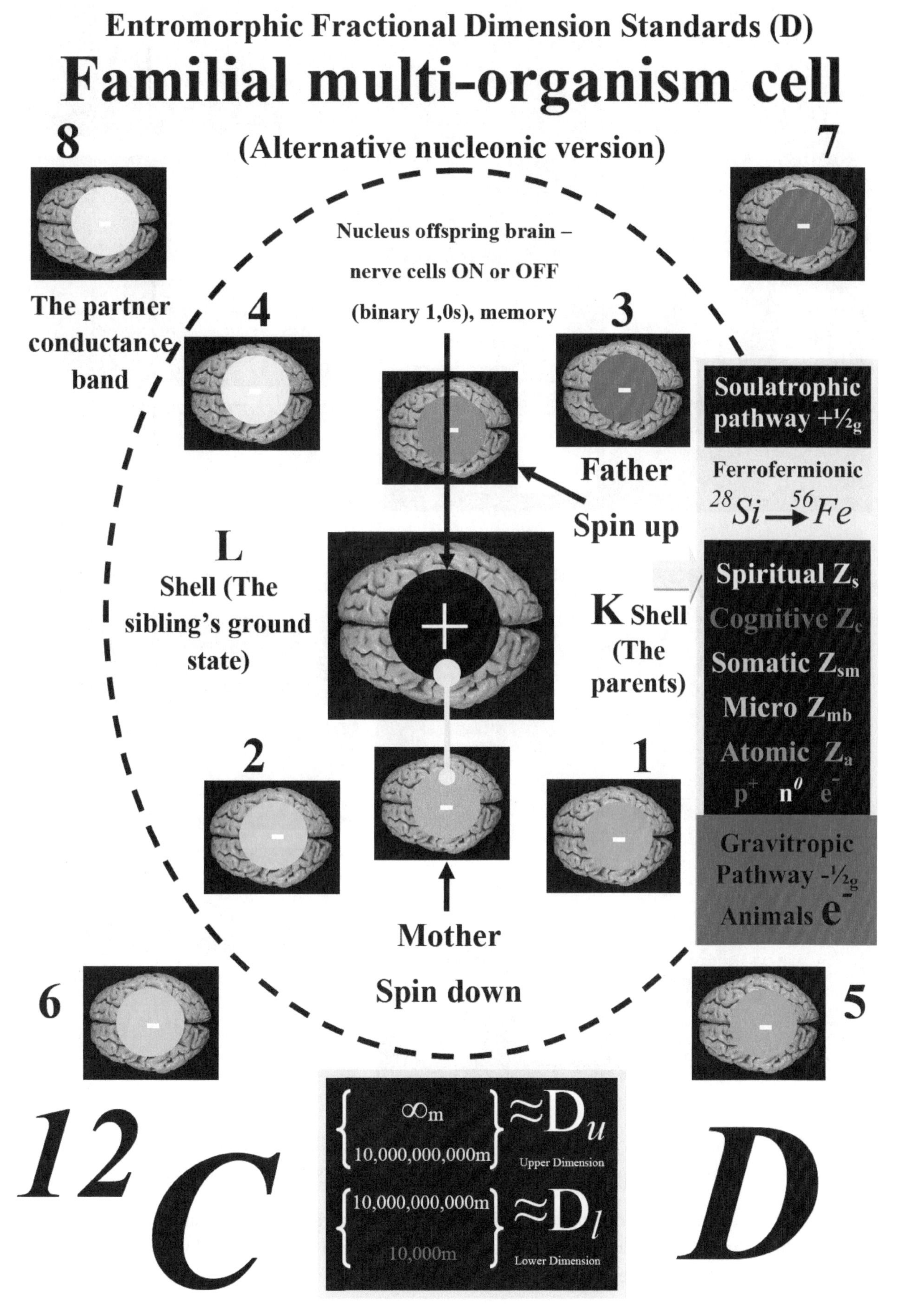
Entromorphic Fractional Dimension Standards (D)
Familial multi-organism cell
(Alternative nucleonic version)
8
7
Nucleus offspring brain – nerve cells ON or OFF (binary 1,0s), memory
The partner conductance band
4
3
Soulatrophic pathway +½g
Father
Ferrofermionic
28Si → 56Fe
Spin up
L Shell (The sibling's ground state)
K Shell (The parents)
Spiritual Zs
Cognitive Zc
Somatic Zsm
Micro Zmb
Atomic Za
p+ n0 e-
2
1
Gravitropic Pathway -½g
Animals e-
Mother
Spin down
6
5
12C
∞m
10,000,000,000m
≈Du
Upper Dimension
10,000,000,000m
10,000m
≈Dl
Lower Dimension
D

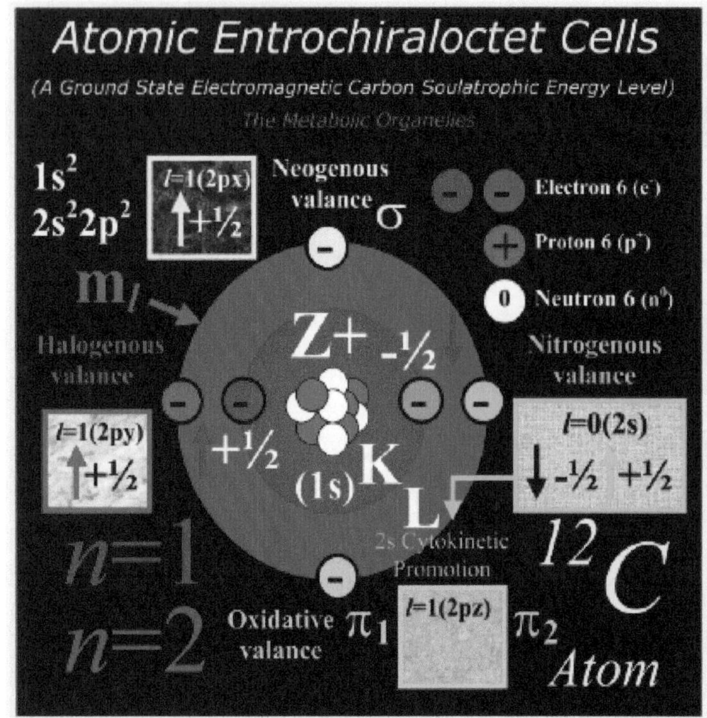

A global political cell

USA (Nucleus Z+)

G20 developed countries
(K shell n=1)

Third World countries
(L shell n=2)

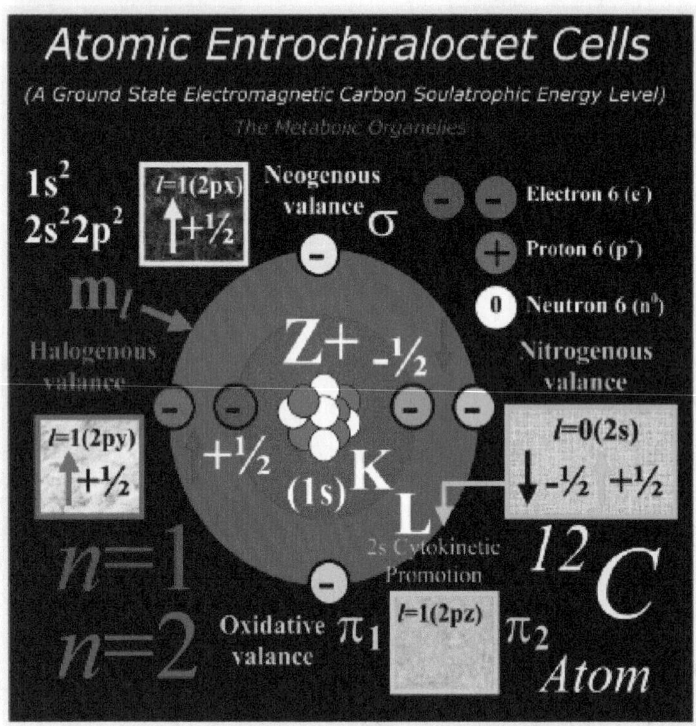

A national cell

King/Queen (Nucleus Z+)

Government
(K shell n=1)

The subjects
(L shell n=2)

Neurological genetics produces a huge variety of 'spiritual' or 'familial' cells based on the mechanics of grouped organisms and their conservation of the carbon logic they are composed of.

All living organisation is formed into these high level and quite literally enormous atomic amplified products which shape our lives based on basic atomic logic but seen in cells on a global level and involving huge quantities of energy.

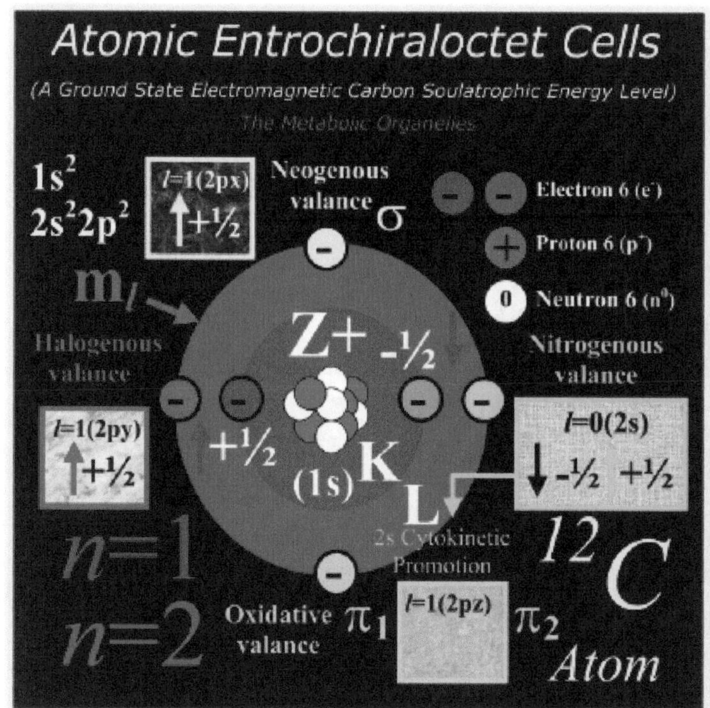

A basic business cell

Managing director (Nucleus Z+)

White colour worker (K shell n=1)

Blue colour worker (L shell n=2)

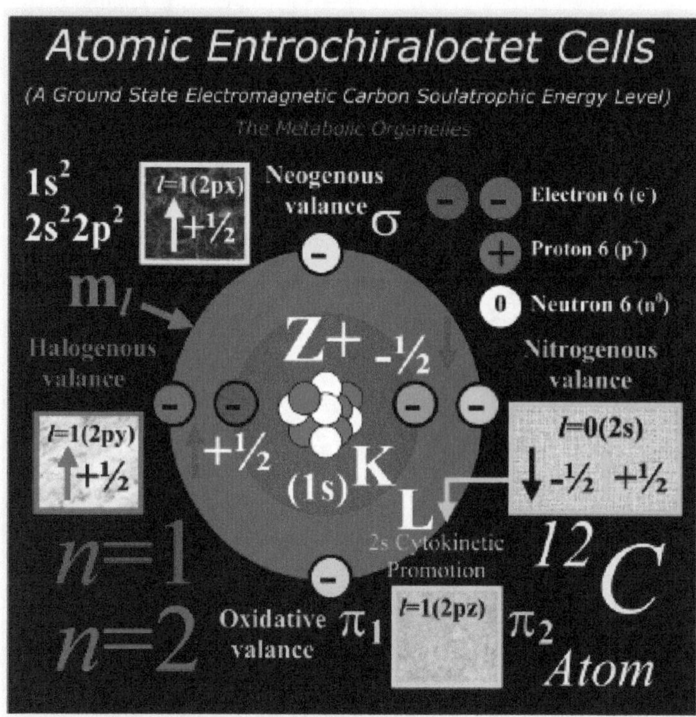

A basic academic cell

Mathematics (Nucleus Z+)

Science (K shell n=1)

Arts and philosophy (L shell n=2)

Even teams group under the basic rules of carbon atomic physics to produce large multi organismal cells which again follow a basic three component field made from nuclear control, K shell stability and L shell valance activities.

The nuclear regions control the two fields through wave functions on a far grander scale than typically seen in atomic levels of resolution.

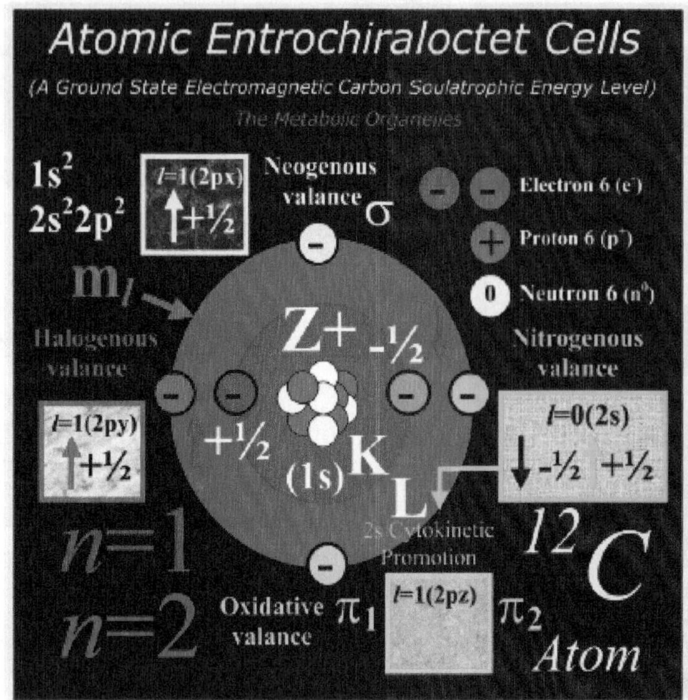

A sporting cell

Manager
(Nucleus Z+)

Defence players
(K shell n=1)

Outfield players
(L shell n=2)

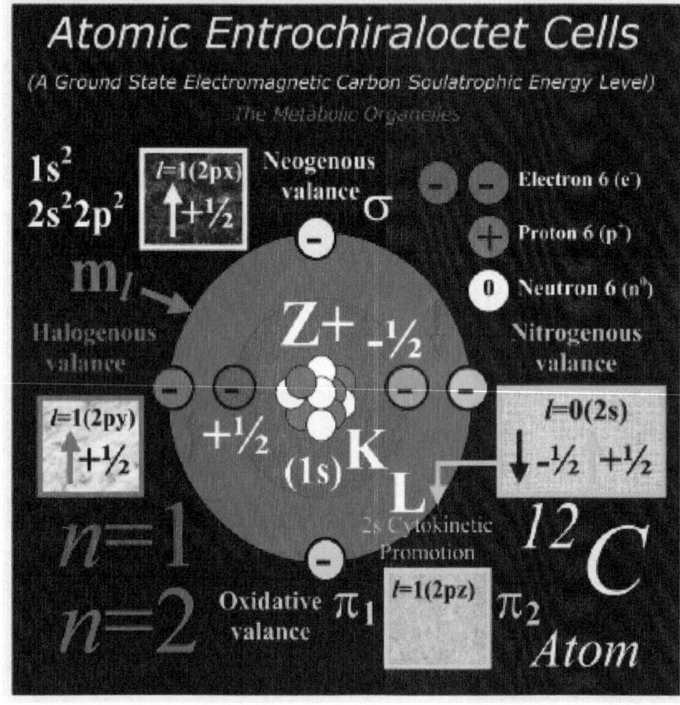

A divine cell

God
(Nucleus Z+)

Heaven
(K shell n=1)

Hell
(L shell n=2)

One of the most prolific and massively powerful entrochiraloctets is that associated with religion and belief. The cells centre the concept of God at their heart (nucleus) and the churches around the world represent enormous wealth because they are nuclear nucleonic systems.

Any grouping of humans automatically starts off with basic colonial organisation (based on hydrogen organisation, a nucleus and field) where a leading cell begins to form followers and to grow in size. Over time the cell will further differentiate into the nucleus, K shell and L shell fields or shells and even further into specific ground state and valance (conductance band interface).

Science check

Do the observations correlate with the theoretical model?

YES ✓ *NO*

PART NINE

Extending evolution by natural selection

(Neurological genetics continued)

Neurochromosome entromorphology and evolution theory - extension into neurological genetics as a powerful proof.

Evaporative model of evolution – A model of the long-term carbon cycle on planet Earth. This model suggests carbon has evolved out of Earth bound carbon in the form of trapped carbon dioxide and carbonates but traces all life back to the 'Big Bang'. The model suggests the slow evolutionary increase in nuclear potential enthalpy and subsequent increased Gibbs free energy following exponential quantum growth patterns. These results progress through the Earth macrostates (solid, liquid, gas to vacuum of space) based on phase potentials. Carbon moves through the solid macrostate (Earth) to the liquid macrostate (oceans), to the gas macrostate (atmosphere). The final evolutionary step is the post-planetary cytokinesis to the vacuum macrostate. As carbon moves through the Earth macrostates down it's concentration gradient so its nuclear shielding increases to compensate for a lack of external macrostatic environmental shielding such as a space suit.

The evaporative macrostatic model of carbon evolution sees large-scale planetary quantum gravitational fields associated with the quantum state of carbon soulatrophic pathways. As such the nucleus on the planetary level is the Earth's core (iron) and the K shell n=1 are the plants. They are phototropic in keeping with K shell quantum higher frequency energies, and are attached to the nucleus (Earth). The L shell fields are aquatic water based (ground state model entrochiraloctet), and the valance bonded metabolic organelles are the gas atmosphere organisms (valance bonded entrochiraloctet). The 2s nitrogenous valance bonding is seen in the insects (arthropods) which interface with the K shell (plants) and display post nuclear linear morphology, they are polysomatic polyentromorphs (they have many body segments and limbs, the plants are also polysomatic). The promotion of 2s ground state electrons is the basis of inter macrostatic cytokinesis, such as moving from aquatic to gas atmosphere evolution. The 2s promotion for Zn (vb) humans sees movement from gas atmosphere to the vacuum (space), through artificial life. The Pz oxidative valance bonded organelles are the herbivorous organisms: they are unhybridised so that they live on unhybridised plants. The py halogenous valance bonded organisms are the carnivores hence very aggressive which is what one would expect from halogens. The px neogenous valance bonded component is the emergence of humans, and the new nucleus can be seen in the off world satellite and space station technologies emerging through carbosilicoferrous technogenesis of technology.

On the ground state aquatic entrochiraloctet the px neogenous organisms are the amphibians which have cytokinesis onto land from aquatic origins a macroscopic cytokinesis. The planetary cytokinesis is the post Earth nucleus, the final result in Earth evolution where the Earth macrostates have evolved into a complete Earth carbon soulatrophic cell with all missing metabolic organelles. In a way quantum level fields and energies have proliferated and manifested themselves in the Earth macrostate. Where quantum systems are electromagnetic in origin, they are now gravity based; the many particle/wave dualities are vast, as life goes from particle to wave through evolution. This link through the soulatrophicity model of carbon is a real model of quantum gravity based on the completion of gravitational entrochiraloctets through completion of electromagnetic entrochiraloctets the ferrofermionic pathways to artificial life. At its most fundamental evolution is the battle between gravitational and electromagnetic forces in the same way a balance of nuclear systems occurs between the strong and weak forces, the Sun is winning as life is plucked off the surface of the Earth into space. **See pages 247 and 248 for quantum taxonomy which illustrates these complex points.**

Evolutionary momentum - carbon acts as a natural bosonic system having an even number of fermionic particles in its nucleus. This gives rise to the communicative conservational abilities of living soulatrophic entrochiraloctet cells. Commonly known as sexual reproduction each soulatrophic energy level has a communicative bosonic link. The link forms the soulatrophic bosonic ladder (think of two people kissing and hugging) and allows nuclear energies to be transferred from one pathway to another (genetic cross over through chiasmata). On the somatic (radial) soulatrophic energy level this is seen as sperm and egg and zygotic and meiotic processes on the cognitive (bilateral) level it is seen in the form of sight and sound but based on the same zygotic meiotic processes. Evolutionary momentum occurs because bosonically activated energy forms new nuclear constructs called zygotes. On the cognitive level they are psychozygotes or ideas (thought): two memories one which is a psychosperm and the other a psychooocyte fuse to form a psychozygote or 'idea', the word psycho can be replaced with neuro (they mean the same thing). The convention is still being established although 'neuro' is the most likely term. They occur by nuclear neurological fusion reactions, which appear as recombination events for DNA and neurological transient or permanent changes to neuronal function on cognitive (bilateral) and spiritual (familial) levels through classical memory. Cytomeiotic events describe this recombinant event on every soulatrophic energy level.

When two nuclear potentials combine the resultant zygote has the potential energy of both soulatrophic pathways (parents) and as such has more potential energy than each parent individually. This is evolutionary momentum and ensures that the greater store of solar energy is conserved above the inferior individual parents. This can be seen where a parent will always conserve their child before themselves because the child represents more stored potential energy and therefore greater evolutionary momentum.

Darwinian evolution by natural selection is easily demonstrated and proved in carbon entromorphology. Conventional proofs for Darwinian evolution by natural selection are typically demonstrated on microbiological and somatic genomic soulatrophic energy levels, based on DNA. All soulatrophic cellular systems have a temporal component, which reflects recombination fusion events and therefore demonstrates the cell cycle. In carbon entromorphology higher soulatrophic energy levels form neurological genomic nucleonic systems. The recombination fusion events (thought) take place very rapidly on these levels hence evolution by natural selection can be demonstrated and proved in real time with very clear concise examples and an inexhaustible supply through technology. As a result Darwinian evolution will become the most extensively proven scientific concept of all time. We do have excellent evidence of evolution through DNA genomic systems as well but most criticisms of evolution are our lack of ability to demonstrate it in real time. In carbon entromorphology this is not a problem any longer, as neurological genomics exist as an extension to classical DNA genomics but with far greater recombination speeds and effortless proofs.

DNA genomics have ultra slow recombination rates but neuronal genomics have real time recombination rates. An example is the concept of history, which wouldn't exist in the human world, as history is the study of and realisation by Darwinian evolution by natural selection. There wouldn't be any news because nothing would change if evolution were a false concept. There would be no debate about the efficacy of Darwinian evolution if evolution were a false concept; the debate alone personifies evolution by natural selection.

History and the many millions of museums worldwide are perfect testament and proof to evolutionary change and as such provide massive levels of evidence to support Darwinian evolution. The word would be meaningless if evolution was false.

Education is a perfect example of evolution by natural selection. Millions of people every year embark on educational 'evolutionary' courses. They are more evidence of evolution by natural selection, as they aim to evolve through environmental course stressors through adaptational moulding. Examinations, course work and other assessments are perfect and extensive evidence of natural selection. Students have to demonstrate levels of evolutionary change or the educational institute selects them for rejection (natural selection).

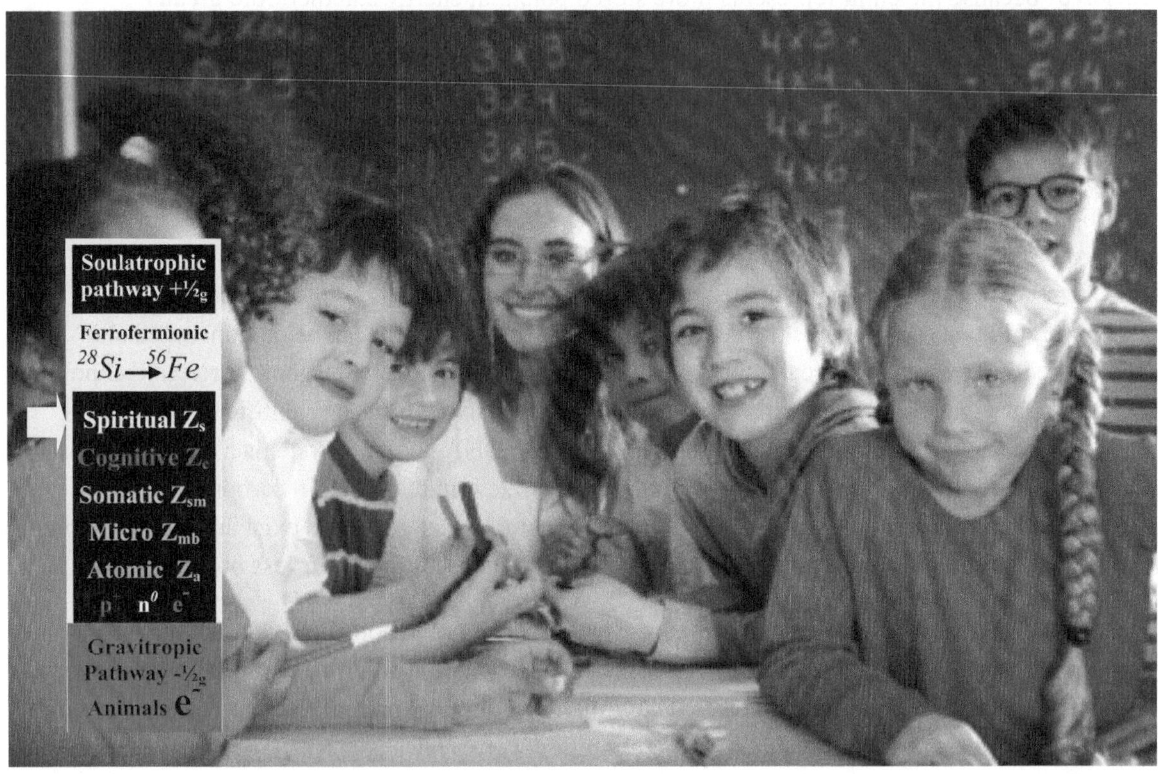

The business world is another perfect example of evolution as businesses act to maintain stable frequencies of their unique allele characteristics. Every year businesses fail and are naturally selected for bankruptcy due to environmental stressors their allele characteristics reduce in frequency.

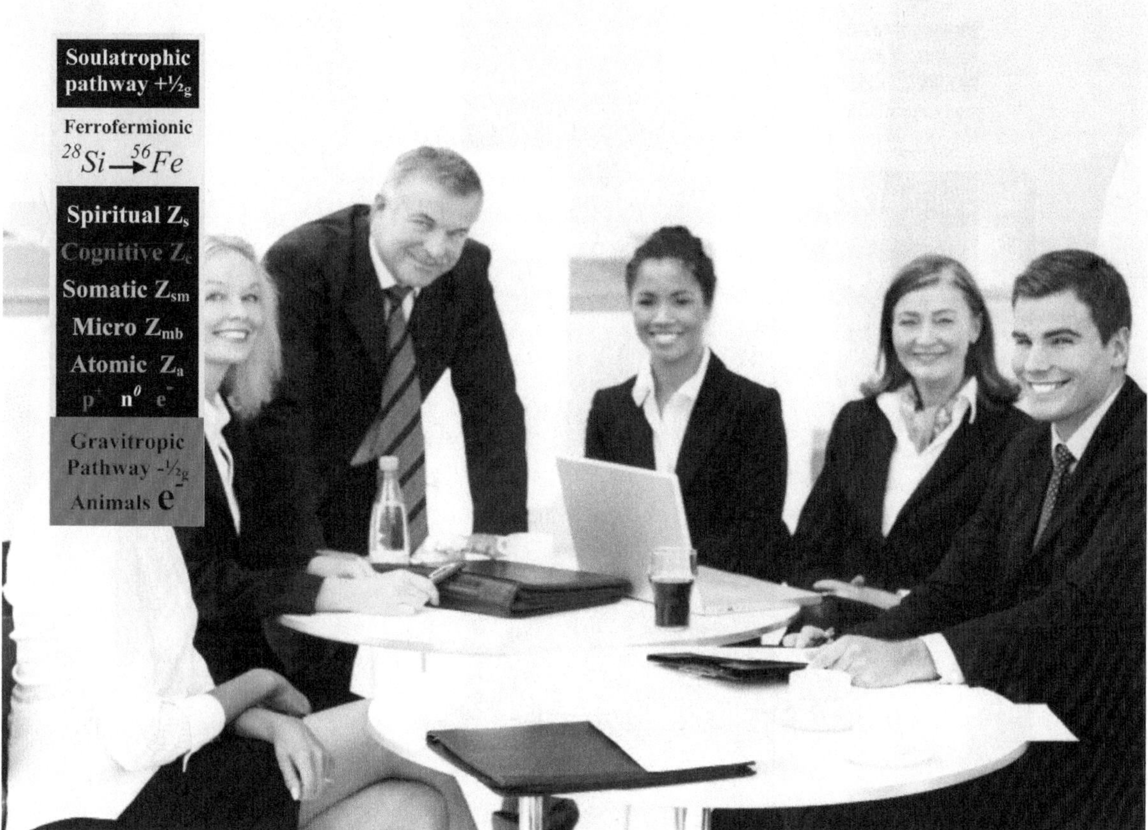

Advertising would be unheard of if evolution wasn't the basis of all life. Advertising aims to act as a natural selection factor increasing the frequency of specific product allele characteristics in competition with other business cells allele characteristic 'we offer more than anyone else'.

Sports are great examples and excellent evidence of evolution by natural selection. They absolutely personify the concept as groups compete to increase the frequency and stability of specific sporting alleles e.g. supporting a football club (Gibbs enthalpic pathways).

Failure to fit the sporting environment results in environmental rejection and reduced allele frequencies (thermoentropic pathways). Premiership football is a perfect model of biological quantum mechanics. Hypothetically the defence are nitrogenous, the midfield oxidative, the strikers halogenous and a goal scorer is neogenous (valance bonded metabolic organelles). The box is the K shell, (n=1) tightly bound to the nucleus or goal (Z+). The goal you are scoring in (by quantum tunnelling and barrier penetration) is your own nucleus as you try to conserve a photon (ball) in it. Set pieces occur around the goal, kinetics reduces around the goal and pressure increases and inversely kinetics increase in the outfield (L, n=2) where the pressure is reduced. The two teams form a Pauli cell where both aim to be energy conservers by barrier penetration goals (winners +½) by the end.

Word's such as competition, change, challenge, adapt, evolve, win, lose, live, die, past, present, future, grow would be meaningless if evolution was not true. The list is endless because Darwinian evolution is true for all soulatrophic energy levels of life. People wouldn't remember their childhood, as that would demonstrate evolutionary change. Neither would there be, adolescence, adulthood, middle age, retired, old age and death as they all beautifully demonstrate and prove evolutionary change.

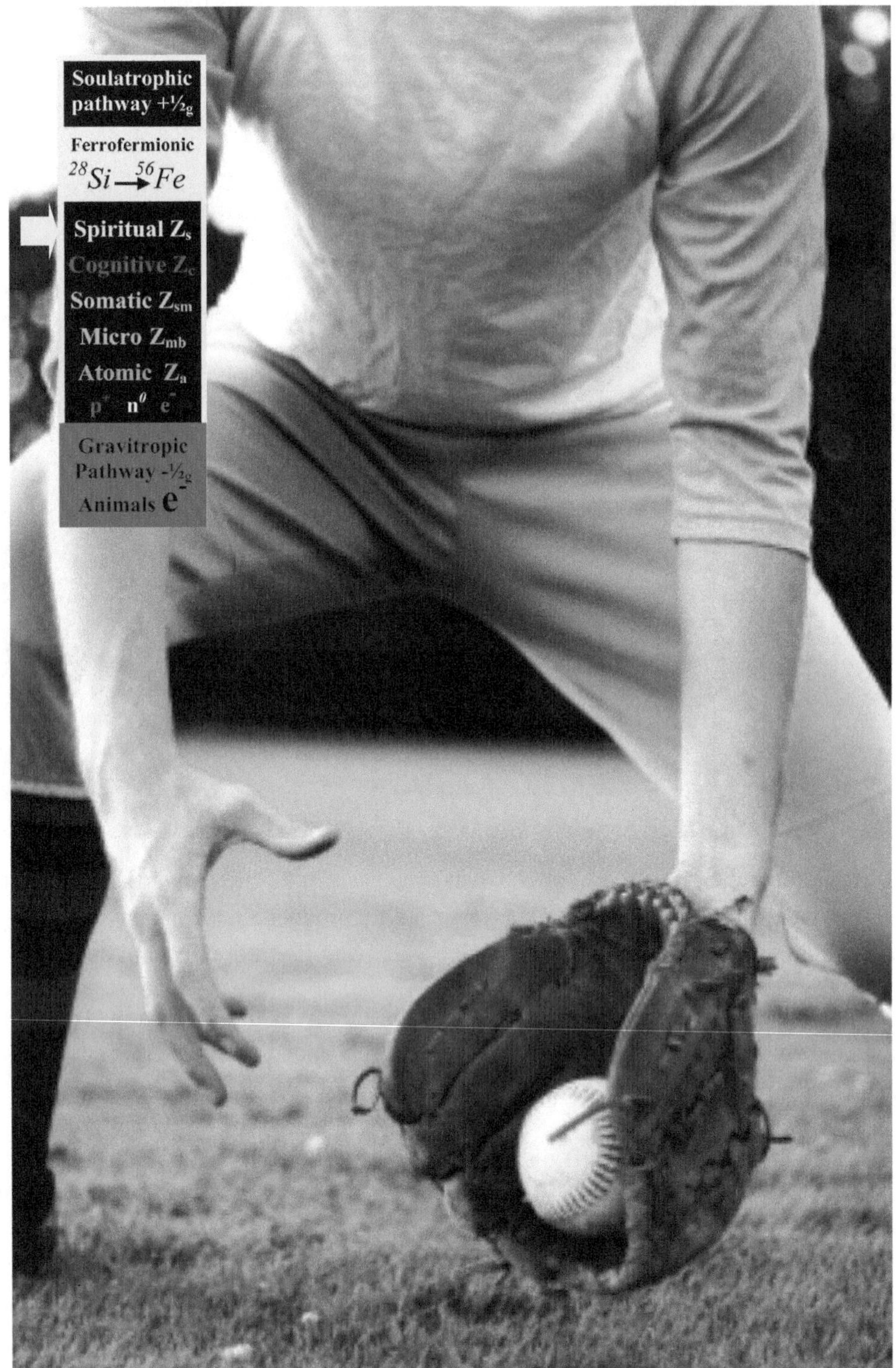

Classical morality doesn't escape Darwinian evolution. The moral fight absolutely personifies evolution by natural selection. Belief systems aim to increase the frequency of good alleles and reduce the frequency of evil alleles.

The two evolutionary pathways of good and evil produce judgement by a God, which acts as the natural selection function. People aren't born believers they evolve into believers through education, and education is classical Darwinian evolution. The models of evolution by natural selection fit all possibilities equally well because they are truth.

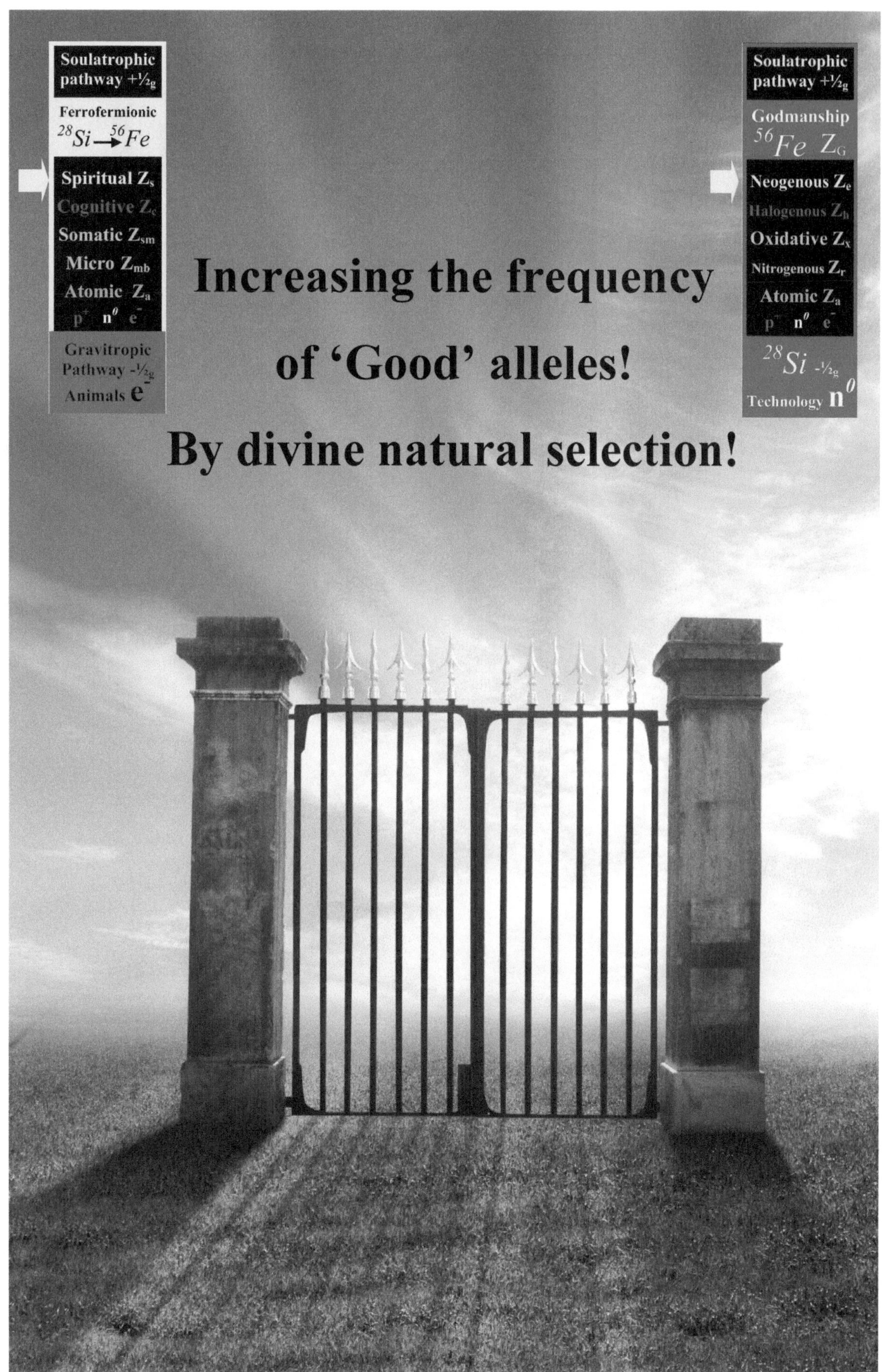
Soulatrophic pathway +½g
Ferrofermionic
$^{28}Si \rightarrow ^{56}Fe$
Spiritual Z_s
Cognitive Z_c
Somatic Z_{sm}
Micro Z_{mb}
Atomic Z_a
p^+ n^0 e^-
Gravitropic Pathway -½g
Animals e^-
Increasing the frequency
of 'Good' alleles!
By divine natural selection!
Soulatrophic pathway +½g
Godmanship
^{56}Fe Z_G
Neogenous Z_e
Halogenous Z_h
Oxidative Z_x
Nitrogenous Z_r
Atomic Z_a
p^+ n^0 e^-
^{28}Si -½g
Technology n^0

Technology is a perfect model and proof of evolution by natural selection. The speed of evolution is so fast in technology that day-by-day we can see the changes in real time.

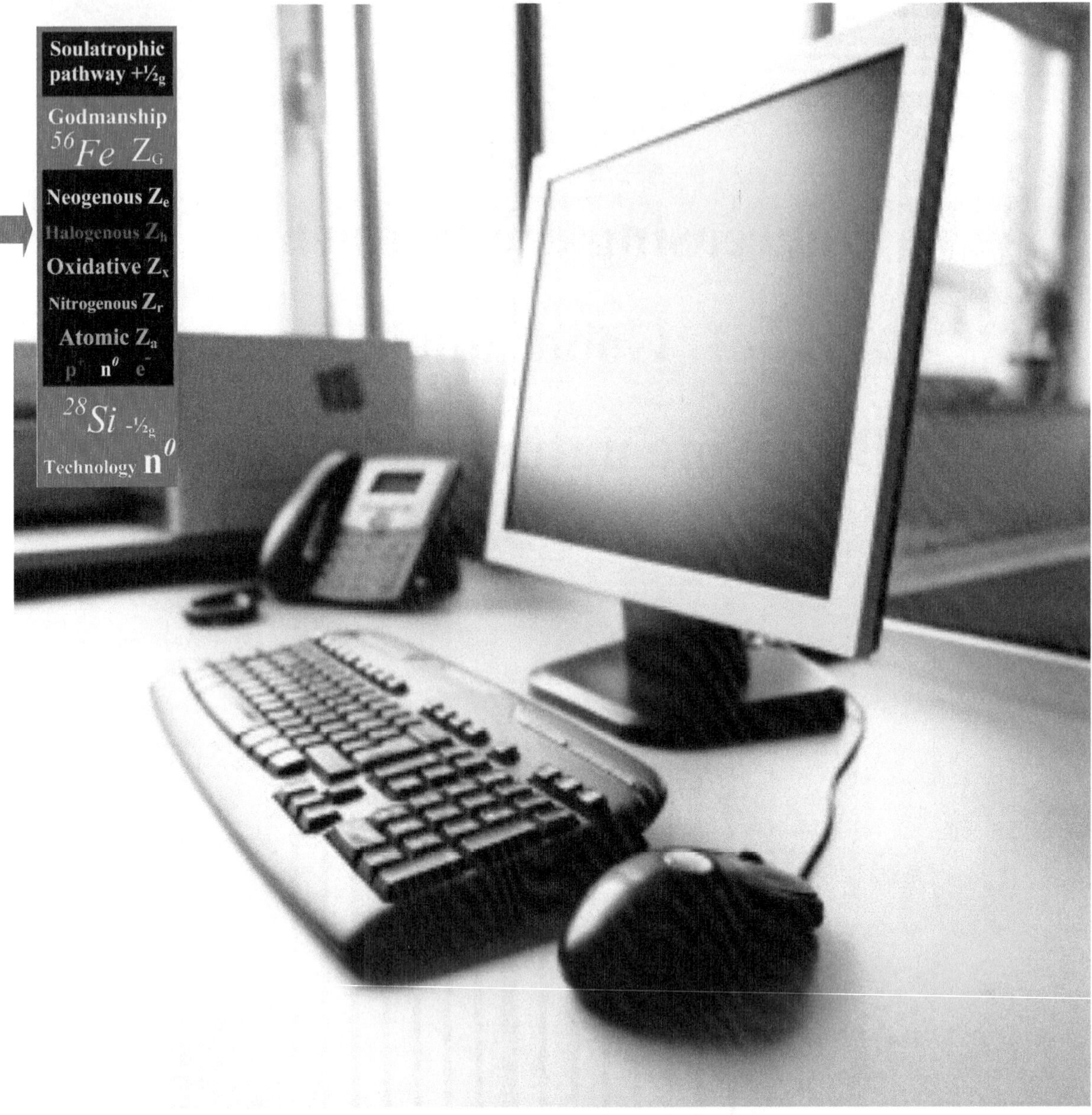

Everyday new technology appears with better technoallele characteristics. Manufacturers aim to increase the frequency of such technoalleles through patent and copyright, and to eliminate older technoallelles.

Humanity can easily demonstrate the evolution of technology such as the cart and horse to the high performance Lotus jet black s4s turbo sports car.

The beacon fires to Morse code to the telephone to the satellite mobile phone. The wireless to the mp3 player, the fire to the microwave cooker. Basic skins to synthetic fabrics in clothes and caves to ultra modern iron and glass built air-conditioned artistic buildings. The biology of technology!

The evolutionary evidence for technology is breathtakingly enormous, making evolution by natural selection the most provable concept in living history. DNA evolution is just too slow to demonstrate evolution easily; the neurological level is the level on which it runs at classical thinking speeds (spiritual (familial) temporal levels).

Amazingly enough evolution by natural selection has a perfect logic to it. Any group that rejects Darwinian evolution actually proves it by rejecting it, because they are demonstrating natural selection. This 'immaculate logic' makes Darwinian evolution unrejectable as a social allele. Evolution is simply the corollary of Einstein's space-time continuum; life simply reflects time, pressure and change with the spatial organisation of a fluid sitting in the global environmental containment, turbulently pouring through time and space gaining order and momentum as it goes.

Let's hope it continues to absolute artificial life a beautiful future of extraordinary order, energy availability and temporal stability, but ALWAYS remember Werner Heisenberg is present! Reminding us of our limitations! It is important to end this section by saying that the process of hydrogen entromorphology or nucleogenesis which produced the elements, and the large astro physical particles such as planets and stars, must also be subject to the same mechanisms of evolution by natural selection.

It is important that the physics world embrace this mechanism in its understanding of atomic to astro physical systems in nature. In other words evolution theory must be the stabilising property for atoms and planets, galaxies and stars as well as living organisms. The current life based definition is too limited to DNA only and isn't extended back to the Big Bang.

According to current understanding of evolution life began over a billion years ago, which is too limited a definition. That means life is not contemplated for 12.7 billion years, which is not a consistent theory about universal evolution by natural selection. Also current theories on evolution state that it is a chaotic random sequence of events, carbon entromorphology challenges this with soulatrophicity, ferrofermionics and ECASSE rules, giving life specific pathways of development. Also current evolution theory does not identify any goal to evolution.

Classical religion has always known the goal to evolution is that of eternal life. In carbon entromorphology life tends towards artificial life, a goal where life can stabilise itself indefinitely.

We see this in the form of the emerging androids, robots and virtual organisms. Again evolution theory is far too limited at present as it does not effectively consider technology. Technology is not thought of as life but in carbon entromorphology it is. 'The biology of technology', a translation of carbon energy to the rest of the periodic table through silicon (the Stone Age) and iron (the Iron Age) the carbosilicoferrous organisms (technological organisms and life). A mobile phone can be thought of as a carbosilicoferrous organism. Where iron nuclear stability is the goal of evolution, and artificial life 'divine evolutionary stability'.

The whole process is originally driven by classical belief in God which produces science and technology, which produces artificial life. Science cannot reject the concept of God as it is still the main natural selection factor in human evolution and I would not be writing this article if belief in God had never occurred. Life came from the consequence of the Big Bang and therefore all matter and the space time continuum must be in their current state because of natural selection.

Carbon entromorphology has already identified evolution by natural selection in neurological systems; where characteristics such as socioalleles, technoalleles, and general neuroalleles are filtered for energetic stability by environmental filtering. In this way atomic systems must have atomicallele characteristics and astroalleles as well under the same environmental filtering.

Science check

Do the observations correlate with the theoretical model?

YES ✓ *NO*

PART TEN

The physical consequence of amplified atomic properties

'Carbon electromorphology'

Neurochromosome electromorphology - the biological effects of quantum gravity emerging structures with electromagnetic origins and characteristic morphologies.

- The way that electromagnetic fields produce structures in larger biological organisms.
- The evolution of fields into tissues with characteristic morphology.

Electronegativity – electronegativity from classical physics conserved and amplified in effect by soulatrophic carbon cell (entrochiraloctets) amplification pathways. This is the basis of living drive and desire based on field (metabolic organelles and the valance shell) quantum instability and ferrofermionic pathway nuclear instability in carbon cells (entrochiraloctets). The models are based on electron affinity and ionisation energies and the charge on the nucleus. Due to the chiral nature of carbon the electronegativities are distributed around the chiral centre. This distribution follows the octet rule although due to complex overlapping of orbitals the final electronegativity distribution is determined by the final structure. A classic example of electronegativity is hunger; this is an effect amplified up to the cognitive (bilateral) soulatrophic energy level (greater complexity but the same force as soulatrophic origins as carbon). The octet rule is the basis of carbon behavioural potentials and is conserved and amplified by solar acquisition.

Electromorphology – this is the relationship between the evolution of living morphology (particle/wave) and simple elementary electrical charges seen at earlier stages of ferrofermionic evolution. It allows energy to flow between living fermionic microstates, for example in genomic information (sexual reproductive) communicative bosonic systems the 'electroweak standard model'. Flaring structures follows a derivative of coulombs law where electric field lines stream out of positive charges and stream into negative charges. Biological structures such as leaves, penis, vagina, mouths, hands, and all other structures display these field line properties evolved from simple electron proton charges. A flower is phototropically associated with a positive charge associated with nuclear regions. The stamen is a negative charge with a tight compressed converged field line structure. A penis is also a negative charge having similar properties. A vagina has a radiant (labia) structure advertising its place in space. This is a major element of all biological systems and the basis of all communication systems. The quantum gravitational effects mean that simple electromagnetic charges have been cast into material structures through evolutionary growth. Hence matter is precipitated in these electric fields producing charge equivalent structures such as a flower through chemical bonding. The material structure operates in gravitational space as if it were a classical electromagnetic charge. The charge basis still remains intact but is found lost in the complexity of the structures.

Carbon Electromorphology

Electric field lines and biological morphology

Structures evolving out of electromagnetic mathematical models

$$E = \frac{1}{4\pi\varepsilon_0}\frac{q}{r^2}$$

The gravitational mass rich structures display a transitional overlap between classical electromagnetic fields from quantum levels of complexity amplified into gravitational equivalents through growth. In a sense amplification of quantum characteristics can be seen as probability distributions in gravitationally dominant space; structures which have evolved from electromagnetic origins. The electromagnetic effect is diluted as the structures become larger.

Electrophile – electrophilic drive to acquire electrons through electronegativity (quantum instability) conserved and amplified in effect by soulatrophic carbon cell (entrochiraloctets) amplification pathways and part of the neobosonic pathway (entrochiraloctet). Carbon has the drive to become as quantum stable as iron through solar processes (ferrofermionic pathways). This is the potential which drives all living evolution. The battle between electrophilic and nucleophilic living fermionic systems is the basis of natural selection through limited resources.

A classical entromorphic electrophile is personified by the concept of hunger. It is mediated through positive charge in the nuclear region (brain) and acted upon by the weak nuclear expression force into behaviour (biological electroweak standard model), which stabilises the positive charge and reduces electronegativity. Electronegativity varies in accordance with the octet rule for carbon cells. The acquisition of charge stabilising energies is driven by the electronic configurations in carbon as it pursues the chiraloctet model. The first electron and associated potential is the nitrogenous, then the oxidative and the most reactive is the halogenous and then finally quantum stability is achieved with neogenous Nobel gas electronic configuration.

The electrophile is an oxidation agent stripping the environment of charge for his or her own reduced quantum stability needs. Most energy exchange mechanisms are redox (reduction oxidation) reactions between two or more energy systems. Stabilising the fourth electron to produce neogenous quantum stability and stabilising the L shell n=2 locks energy away in the nucleus pushing the cell towards ferrous stability through field to nucleus transmutation.

Delta+ - a positive dipole as described by positive proton charge shifting. This is the basis of muscular energy distributions in living fermionic systems. Soulatrophic amplification conserves depolarisation potentials in living fermionic microstates. The dipole is related to the quantum spin effect in fermionic systems and is the basis for sexual dimorphism and gender.

Quantum space (cellular systems) must obey Paul's exclusion principle as they are fermionic. On higher soulatrophic energy levels this describes how organisms are organised into cellular quantum space. A classical dipole is the vagina, which is close to the nucleus, which has a positive charge. A positive dipole is an energy conservational potential. Again the dipole is now a structure and its electromagnetic determinism is diluted as the structures become larger.

Delta- - a negative dipole as described by negative electron charge shifting. This is the basis of muscular energy distributions in living fermionic systems.

Soulatrophic amplification conserves depolarisation potentials in living fermionic microstates. The dipole is related to the quantum spin effect in fermionic systems and is the basis for sexual dimorphism and gender. Quantum space (cellular systems) must obey Pauli's exclusion principle.

On higher soulatrophic energy levels this describes how organisms are organised into cellular quantum space. A classical negative charge is the penis, which has an opposite dipole to the vagina hence the two, can occupy quantum space together for sexual reproductive zygotic means through Pauli's exclusion principle.

Bio-electronegativity

is distributed through chiral properties

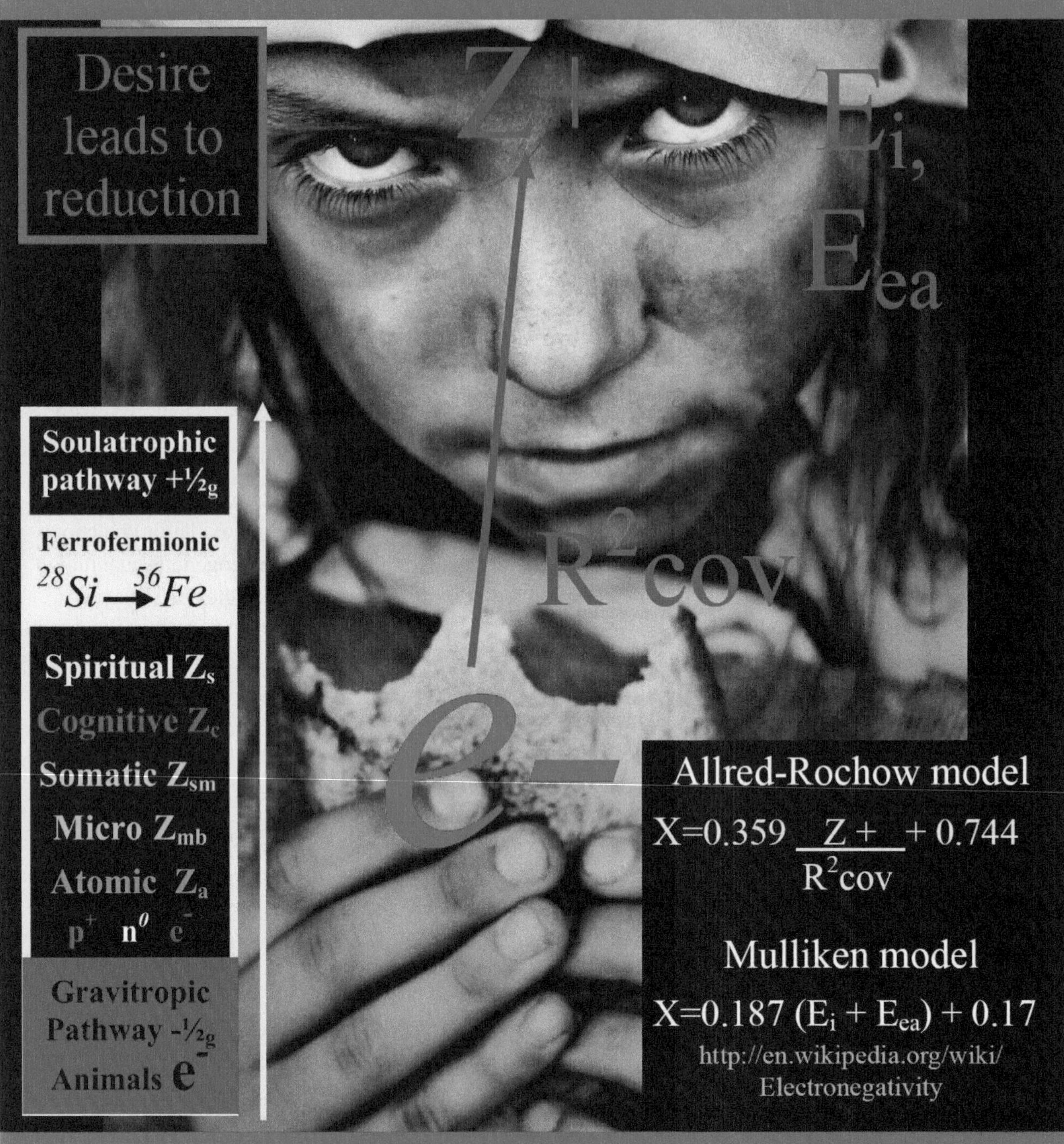

Hunger (oxidised ionised states) is a classical example of bio-electronegativity

Dipoles can be seen in the seasons where a net positive dipole occurs in plants during summer and changes to a negative energy liberating dipole in autumn.

A negative dipole is an energy liberation potential, such as the breasts in a female mammal. Again the dipole is now a structure and its electromagnetic determinism is diluted as the structures become larger.

Nucleophile – nucleophilic systems have charge availability in their structure relative in unstable electrophiles; they have charge stability. The effect is conserved and amplified in carbon soulatrophic entrochiraloctet cells. Carbon has the drive to become as quantum stable as iron. This is the potential which drives all living evolution through neobosonic iterative cell cycling along the ferrofermionic pathway. The battle between electrophilic and nucleophilic living fermionic systems is the basis of natural selection. This is seen as redox interactions. A reducing agent provides charge available to an electrophile (charity). An oxidising agent strips charge from a nucleophile (theft).

Pauli's cell – living fermionic systems must obey Pauli's exclusion principle for ½ integer spin. A spin up +1/2 (angular momentum) is energy conservational 'phototropic' enthalpic system and must pair up with a spin down -1/2 energy liberation 'gravitropic' entropic system (angular momentum). A Pauli cell obeys Pauli's exclusion principle such as a K shell on the spiritual soulatrophic energy level has a male and female forming a cellular union for sexual reproduction. Again the spin state has a dominant structural component as electromagnetic forces are diluted as the structures become larger.

Cytokinesis takes place where spin characteristics become equal: the cell collapses into two separate cells this is driven by ECASSE (Entromorphic Carbon Amplification by Solar Stimulated Emission). This is based on Bose condensate logic, one cell divides into two. Classical Bose condensate sees one photon become two. Spin allows energy to flow through quantum space it is different to bosonic spin which is integer spin many particles can and are more likely to occupy the same energy level. Carbon is unique in that it is also considered to be a boson due to an even number of nucleons in its nucleus. Fermionics has specific energy levels to occupy bosonic systems don't require this. Pauli cells are a powerful fundamental of all living systems and the basis of complementarity and communication. Spin products occur from the quantum amplification effect such as a penis and vagina and most biological properties have spin products in them.

Pauli cell – the exclusion principle for fermionic living systems is a major factor determining behaviour. Due to nuclear and field amplification in living systems the spin characteristics are also amplified into spin structures. Spin and associated angular momentum can be seen at higher soulatrophic levels. Quantum space is occupied by organisms as entrochiraloctet cells, typically K shells. The cell has two energetic components, which either liberate or conserve energy. A functional entrochiraloctet cell requires one party to extract energy (liberate) from the external environment (photon), and the other party to conserve it in the nucleus, this is the basis of a male (energy gatherer) female (energy fixing) offspring bond. The liberator has a spin down 1/2 (thermoentropic) function.

A cell cannot remain stable when both or more parties possess the same spin (exclusion principle). In living fermionic microstates spin can be seen as communication systems, liberate and conserve as structures. It is seen in sports where both parties are forced into a Pauli cell with the same spin. At the end of energy exchange the winner takes a spin up conservational (enthalpic) status, and the loser a spin down liberator (thermoentropic) status. Spin is related to soulatrophic pathways, where spin alternates along the pathway due to the exclusion principle. It also produces soulatrophic ladders (chiasmata), sexual reproduction where carbon entrochiraloctet cells are formed at soulatrophic energy levels by the formation of Pauli entrochiraloctet cells. For classical sexual reproduction the male is the energy liberator or valance shell (sperm producer) and the female the energy conserver 'keeper of the nucleus' (oocyte and zygote – pregnancy). These Pauli cells form on all soulatrophic energy levels. Spin becomes engineerable the higher up the soulatrophic energy level. Humans can re engineer spin using technology, a woman can do a hard manual labour job, and a man can bring up a child and maintain a house. These are inverted sexually dimorphic states associated with technology and the carbosilicoferrous event and spin engineering.

An argument is a failed attempt to form a Pauli entrochiraloctet cell; both parties typically try to be liberators (shouting at each other) -1/2 each. This produces heat because of the incapacity to allow energy to flow through the entrochiraloctet cell. Wherever energy flows in soulatrophic systems, fermionic spin is at the heart of it.

Potential well – the strong and weak nuclear forces act over extremely small distances. This limited range of action is called the potential well. Over this distance the forces are initially electromagnetic and at greater distance gravitational in origin. The four forces of nature are linked together so although the strong force acts over short distances, it maintains its effect through its connection to the weak nuclear force (energy leaking), to electromagnetic and finally gravitational forces through the 'electroweak standard model' and its biological equivalent which is dominantly structural with electromagnetic dilution.

The potential well in carbon entromorphology can be seen as potential well products in the same way as spin products such as the gastrointestinal system the absolute nucleus. The gastrointestinal system (GI) develops first in organisms and houses the carbon atomic soulatrophic amplification origins (anus and the Big Bang). The microbiological soulatrophic level follows this. Hence the GI lumen is home to carbon dioxide the atomic soul and then microorganisms before the emersion of somatic entrochiraoctet levels multi-cellular entrochiraloctets. The amplification process can be seen in an organism's development of the coelom the sigma bond portion of an organism's soulatrophic pathway. The potential well is a small local field that does extend across soulatrophic energy levels.

The spiritual potential well exists when humans form carbon-to-carbon double covalent bonds along the unsaturated bond itself (a kiss). It is the elementary basis of the strong nuclear force. The other three forces are distance-based expressions in equilibrium of the true fundamental force in nature the strong force, the collective interaction is the basis of quantum gravity.

Another example of potential wells is the vagina, the penis and any bosonic communication systems such as speech. Again it is the 'electroweak standard model' and its biological structural equivalent.

Ribosome – in classical molecular biology ribosomes are enzymatic protein devises found in cellular cytoplasmic regions. They convert (express) genetic information copied from DNA to mRNA (messenger) into proteinascious forms. This is the basis of ECASSE as nuclear information (energy) is translated through electroweak means into field structures. They are amine based and have a large subunit and small subunit. It is a fundamental mechanism for genetic expression and uses tRNA (transfer) to produce peptide bonds and polypeptide chains (proteinascious forms). Prokaryotic organisms, the bacteria, also use ribosomes although the size of the subunits is slightly different. The rough endoplasmic reticulum in eukaryotic entrochiraloctet cells is rough because it is teaming with ribosomal expression systems; they are hypothesised to be the Z^0 part of the gauge bosons (w-w+γ).

Due to the soulatrophic amplification phenomenon the ribosomal system is amplified. On the cognitive level in humans the hands are the ribosomal devise where the opposable thumb is the small subunit and the rest of the hand the large subunit, amine origins. The hands convert psychogenes (neurogenes) into proteinascious forms by field expression or behaviour (again through the electroweak structures of communication). Psychogenes are from the psychogenome (neurological net – strong nuclear force) and copied to neurological bosonic action potentials (mRNA equivalence). This produces behavioural expression through the hands to proteinascious forms. This could be making a cup of tea by transcription of tea making psychogenes (neurogenes). silicogenes, legalgenes, divine genes (10 commandments) (silicogenome – ROM – strong force) are transcribed by technology into proteinascious forms, for example a car body by a robotic hand.

The car body is the proteinacious form and the wiring; optic fibre system is the bosonic mRNA equivalent.

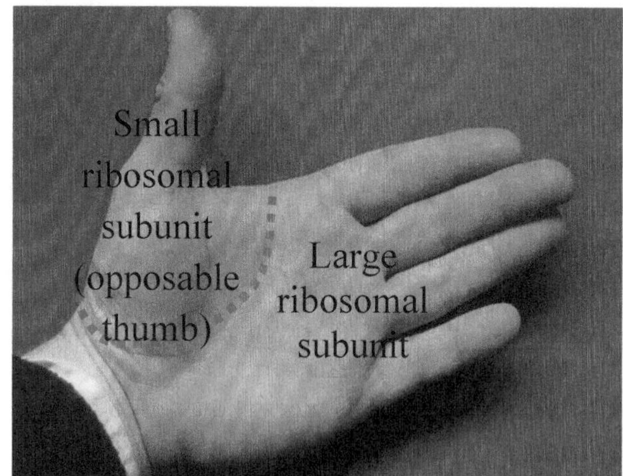

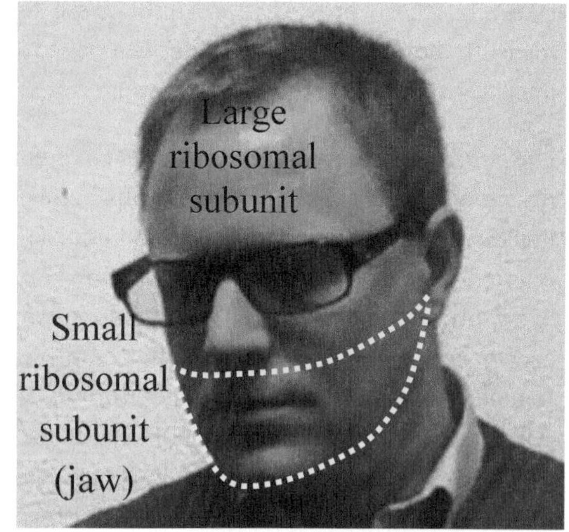

Spin – derived from Pauli's exclusion principle, for use in soulatrophic ladders and genomics. Angular momentum in fermionic particles in carbon entromorphology allows energy to flow through cellular systems. Particles in quantum union must show energy conservation (spin up +1/2) characteristics in collaboration with energy liberation (spin down -1/2) characteristics. In carbon entromorphology entrochiraloctet cells must acquire solar energy and liberate it into an entrochiraloctet cell and conserve it in the nucleus. Spin produces dipolarisation in carbon entrochiraloctet cells allowing energy to be distributed and transmutated. In gender models this allows energy to flow through soulatrophic ladders (human chiasmata) by pairing up electrons with opposite spin (charge dipole shift) this is a symbiotic model of living Pauli cells.

Sub-shells l – probability distributions in carbon soulatrophic entrochiraloctet cells. l=1 (s orbital) which has a spherical morphology, l=2 (P orbital) which has a streamlined dumbbell type morphology; entromorphic morphological determinant conserved by macroscopic environmental filtering. Deformed by the magnetic quantum number 'energy shift', the sub shell morphology is modified and stabilised by naturally selective processes.

The sub shells represent the characteristic morphological nature of the tissues in organisms. A tissue such as the eye can be described as possessing a K shell electron. The eye and the ear are an extension of the probability distribution for s type morphology. As a result the observer does not know where the electron is in keeping with the uncertainty principle, but instead knows that it is somewhere in the eye. Again the cognitive halogenous electron in a human can be found somewhere in the writing arm. Tissues and biological structures act as probability distribution for amplified electrons in carbon entromorphology through nuclear amplification (ECASSE) by quantum accumulative solar shielding. The 2s orbital is a housing probability distribution. The sub shells in living carbon cells explain the enormous variety in life as the sub shells absolute configuration can vary enormously depending on the entanglements and non locality it has been exposed to. Sub shells or orbitals are a superposition of states, in other words ALL possibilities are superimposed into probability fields. Physics cannot measure the precise configuration and has to rely on a model of ALL possibilities. This is limited by uncertainty relations in quantum systems, and life is a quantum system.

Unsaturated carbon-to-carbon double covalent bond – due to amplification, basic unsaturated carbon-to-carbon bonds are conserved although considerably larger. They typically occur across soulatrophic ladders, the bosonic communication bond. This is between amplified 1s K shell electrons (testicles, ovaries, eyes evidence of the link can be seen in teratoma tumors where head structures such as the eye grow on the ovaries). When two organisms look into each other's eyes they form a carbon-to-carbon covalent double bond on the cognitive (bilateral) soulatrophic energy level. When a male and female have sexual intercourse they form a carbon-to-carbon double bond (unsaturated bond). Although K shell electrons are not usually used in bonding carbon ECASSE does allow this to occur as energy levels closer to the nucleus impart a coordination with L shell particles. In animals this is the basis of hand-eye-coordination. Where wave functions for K shell particles influence L shell limb particles the amplitude solution to the wave function becomes equal hence a cooperative wave effect. It is not clear if physics currently takes into account any determinism of more nuclear wave functions to higher kinetic energy level wave functions (QED).

The ferrofermionic pathway of carbon soulatrophic energy levels also forms around carbon-to-carbon double bonds. These are seen between nuclear regions as metabolic fields. The somatic to cognitive bond has a sigma component (spine) and Pi component (kidneys to lungs unhybridised). The same type of bond exists between the brain stem and the cerebral hemispheres through the Limbic system. The double bond allows rotation and displacement but is limited and based on triplanar stability. This is clearly evident in the torso (body) in most organisms.

The carbon-to-carbon unsaturated double bond is the basis of communication highways in living systems. It allows bosonic transfer of nuclear genomic nucleonic states (binary nuclear – strong force – transmission by the weak nuclear force ECASSE).

The carbon-to-carbon unsaturated double covalent bond is the basis of triblastic coelomic systems the basis of complex organisms and the gastrointestinal system; although the GI system is an elaboration of the potential well between soulatrophic energy levels.

In carbon entromorphology life is described as a 'soulatrophic pathway' from the word the 'soul'. It runs all the way back to the Big Bang and is defined by specific levels such as atomic, microbial, somatic, cognitive and spiritual. Mathematically the pathway is known as a 'fractional dimension D'; a fractal geometry. Each level is self-symmetrical to the other levels and is driven by iterative cell cycling (ECASSE) and is made up of the fraction of the lower levels and acts as a fraction of the levels above it. **The repetitious nature of some of these definitions is done to reinforce the ideas.**

Neurochromosome extended cell cycle - an extension to cell size and function and its consequence for the cell cycle.

Neurochromosome cell cycle – the extension to the classical cellular cycle for amplified entromorphic atoms.

The cell cycle is a fundamental property of the growth and death of all living organisms. In carbon entromorphology the cellular boundaries are redefined and extended. As a result, the cell cycle is also significantly extended to produce very powerful results.

The classical cell cycle is based on classical prokaryotic and eukaryotic cellular systems. It has four distinctive temporal components although the actual time periods merge into one seamless system (regulated by cyclin). In carbon entromorphology, amplified carbon entrochiraloctet cells follow the cell cycle. The first phase (interphase) is the G1 or basic growth phase, then the S phase which allows much duplication to take place such as the DNA duplication. The G2 phase makes the cell ready for fission or division into two cells or mitosis.

In mitosis (M) the cell moves thought prophase, metaphase, anaphase, telophase and finally cytokinesis or separation into two daughter cells, and back to G1. Many cells enter G0 which is a senescent phase where many cells don't undergo any more rounds of growth and division but remain stable but subject to the 2nd law of thermodynamics where degradation begins to occur.

For extended cellular systems in carbon entromorphology we can identify by 'self symmetry' and observe the four temporal components of the cell cycle. For example, a typical day in the life of a human has four distinct phases, the morning which can be linked to G1 activity, the afternoon which can be linked to S activity (work part of the day), the evening or G2 most reactive part and finally sleep or neurological mitosis (m) into the next day. Other cycling systems are also evident and are typically associated with the temporal events on an astrophysical basis. For example, pregnancy, childhood, adulthood, the four seasons and many others. Certainly sleep or neuromitosis in humans and the stages of sleep mirror those seen in classical mitosis, which for the first time gives us a powerful basis for understanding what sleep actually is and its parallel 'self symmetry' to cell cycling. The cell cycle also fits (hypothesised) with the octet in carbon logic which is an important observation as it links all the temporal activities of carbon cells. If the octet is the thermodynamic potential driving the cycle then we have a powerful platform in the periodic table for understanding cell cycles. Below are the four octet components of the cycle and their association to the acquired electronic configurations in carbon. Also displayed are the periodic table links and examples of extended cell cycles in life.The G1 or nitrogenous interphase: the S or oxidative interphase: the G2 or halogenous interphase the most reactive of all, and finally the mitotic M or neogenous component. Interesting to note is the observation about a woman's menstrual cycle which lasts four weeks and again hypothetically follows the octet, with pre menstrual tension the halogenous and therefore most reactive part of the cycle, just before cytokinesis which is the period itself. The different phases are linked to the octet and their electronic configurations reflect the reactivity in women during this time.

It is very important to realise that living cells of any size exist 'relative' to other cellular components. Hence, depending on the resolving nucleus a cell can have many different relative energy levels and associated cycles. A cell can be described by its hydrogen or carbon logic, or carbon to carbon bonds etc. Ultimately any energy level is relative to many others and produces complex multi cellular systems depending on resolution by the observer.

The octet rule and the cell cycle in carbon based life, an evolutionary cycle.

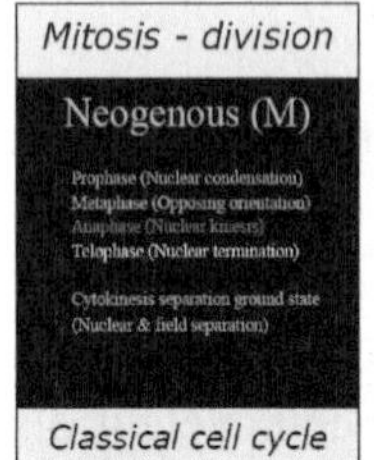

Soulatrophic pathway $+\frac{1}{2}_g$

Ferrofermionic

$^{28}Si \rightarrow {}^{56}Fe$

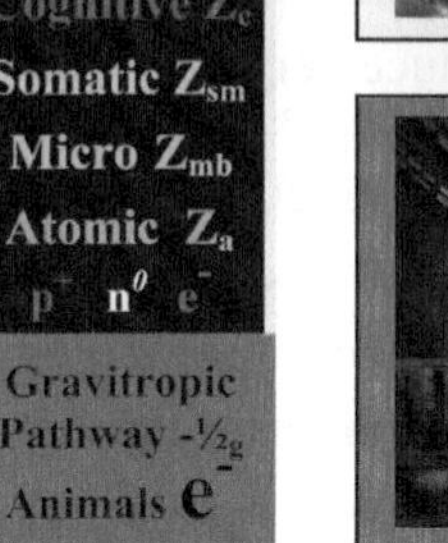

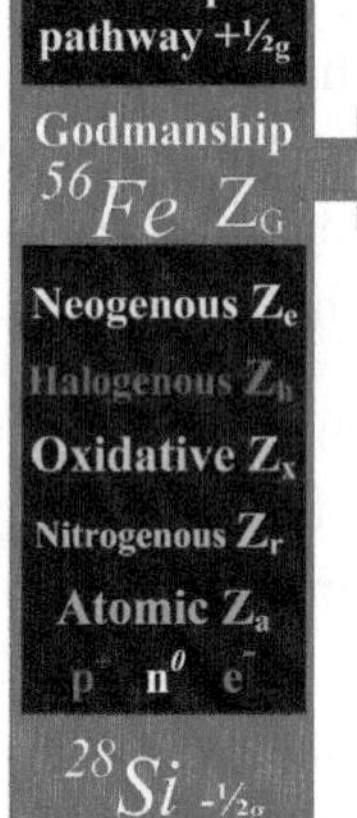

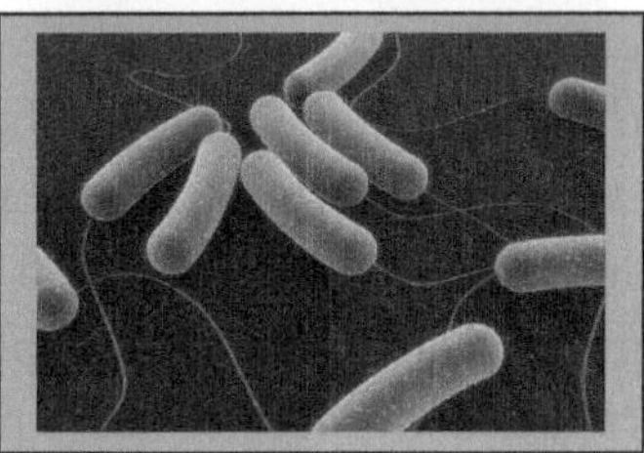

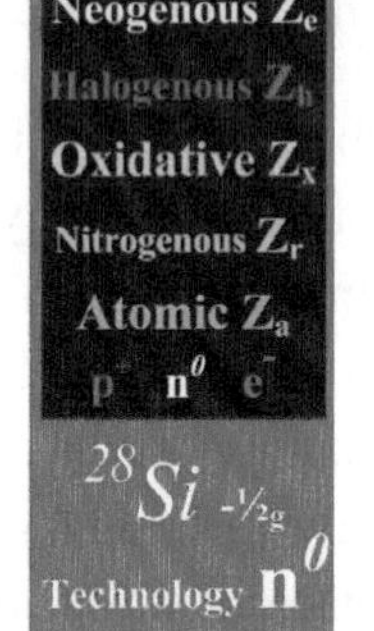

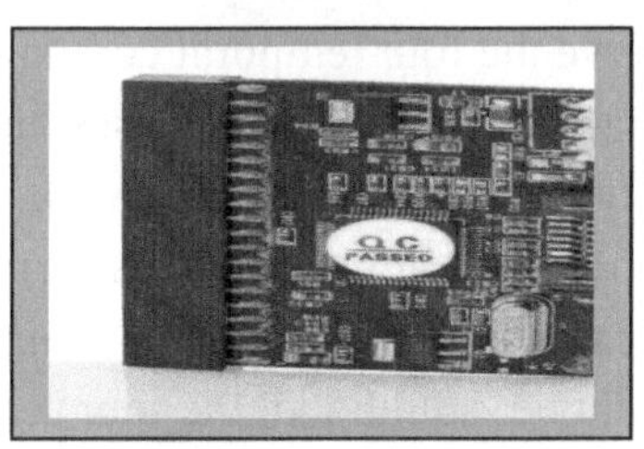

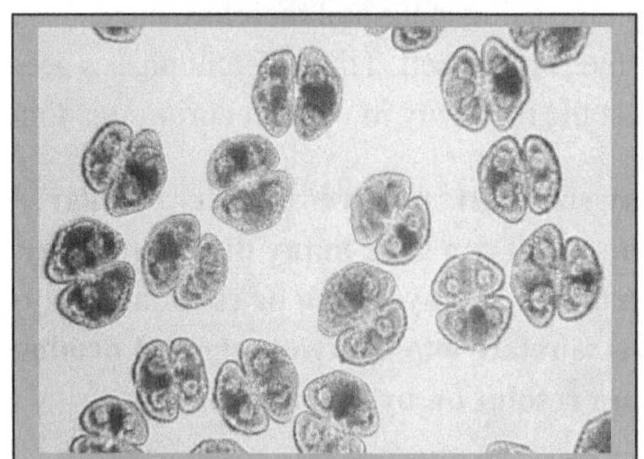

The octet rule or cell cycle is a dramatically extended concept in carbon entromorphology.

It allows biologists to consider any aspect of life in terms of the quantum potential driving octet (cell cycle) stability in accordance with physical laws.

In this example we can see the octet (cell cycle) for the whole of evolution, including the natural extension from carbon to the rest of the periodic table through the 'biology of technology'.

Ferrofermionics
Nitrogenous
G0
Quiescent
G1
Post cytokinesis classical basic growth
Interphase stage (Proteinascious forms)
Classical Cell Cycle

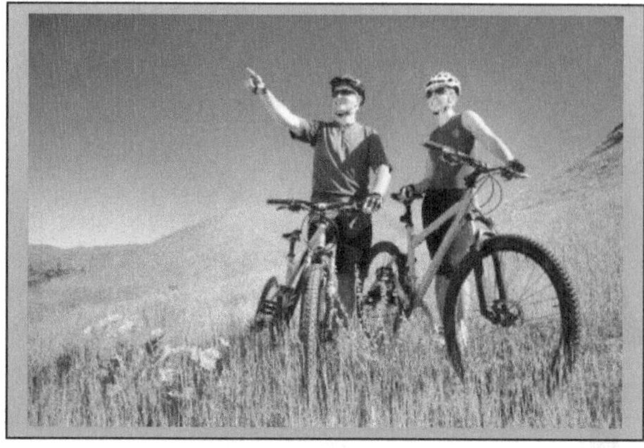

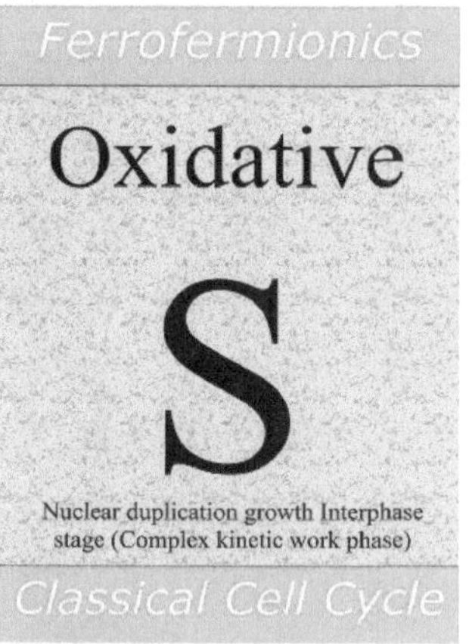
Ferrofermionics
Oxidative
S
Nuclear duplication growth Interphase
stage (Complex kinetic work phase)
Classical Cell Cycle

Ferrofermionics
Halogenous
G2
Exoskeletal membranous growth phase
Interphase pre-cytokinesis
Classical Cell Cycle

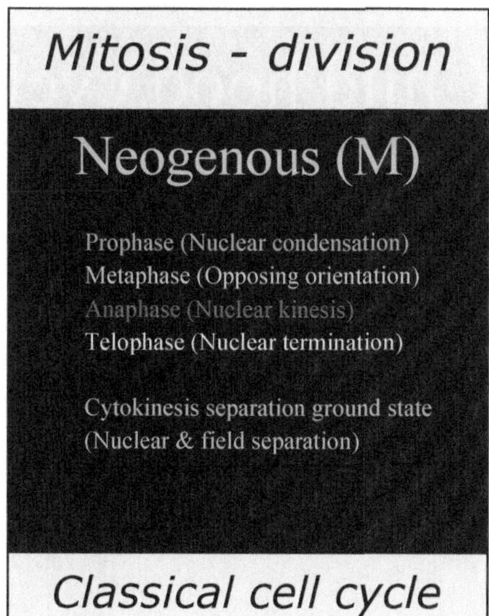
Mitosis - division
Neogenous (M)
Prophase (Nuclear condensation)
Metaphase (Opposing orientation)
Anaphase (Nuclear kinesis)
Telophase (Nuclear termination)
Cytokinesis separation ground state
(Nuclear & field separation)
Classical cell cycle

Planetary cycle

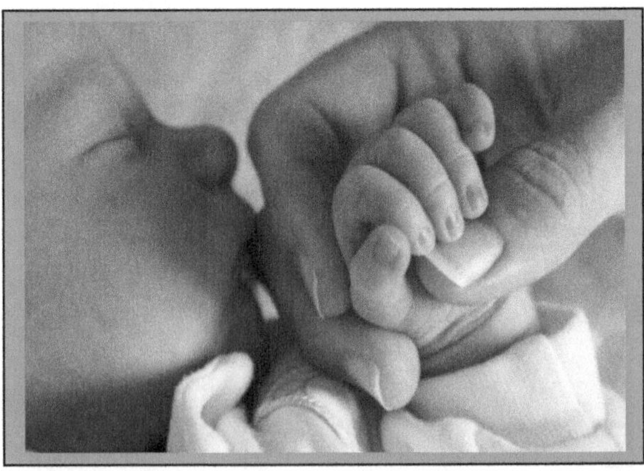

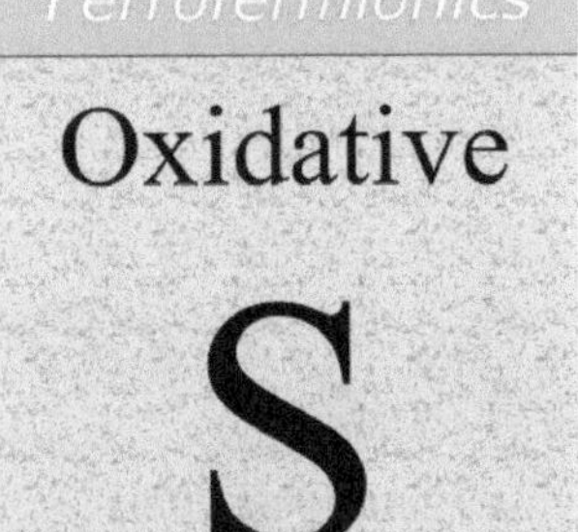

Ferrofermionics

Halogenous

G_2

Exoskeletal membranous growth phase
Interphase pre-cytokinesis

Classical Cell Cycle

Mitosis - division

Neogenous (M)

Prophase (Nuclear condensation)
Metaphase (Opposing orientation)
Anaphase (Nuclear kinesis)
Telophase (Nuclear termination)

Cytokinesis separation ground state
(Nuclear & field separation)

Classical cell cycle

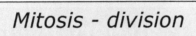

Child hood cycle

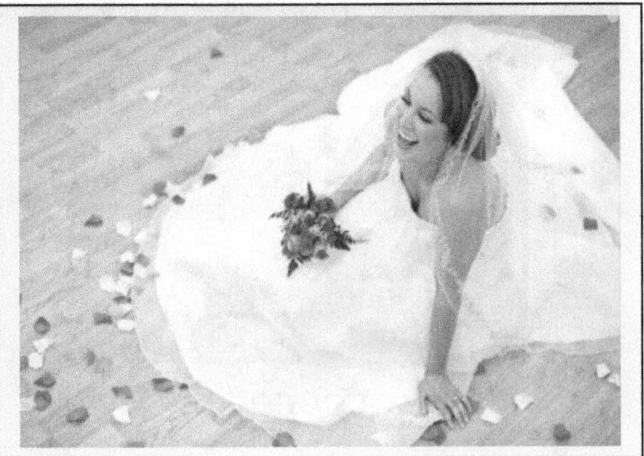

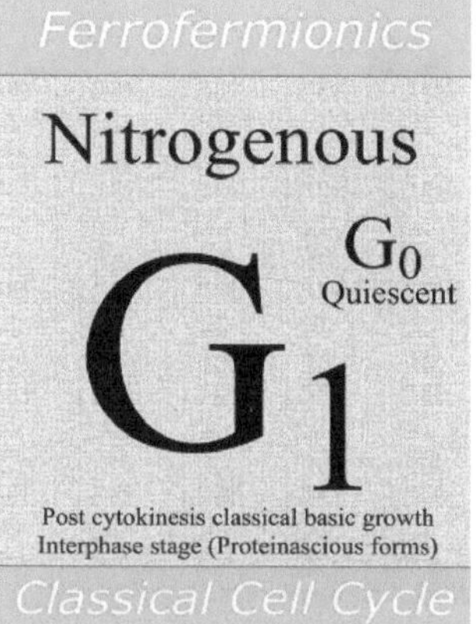

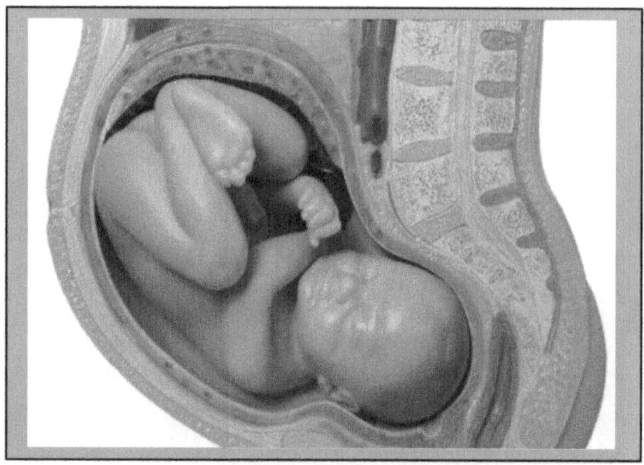

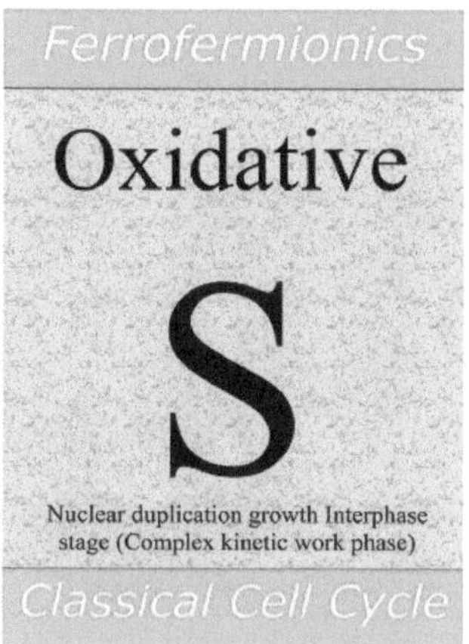

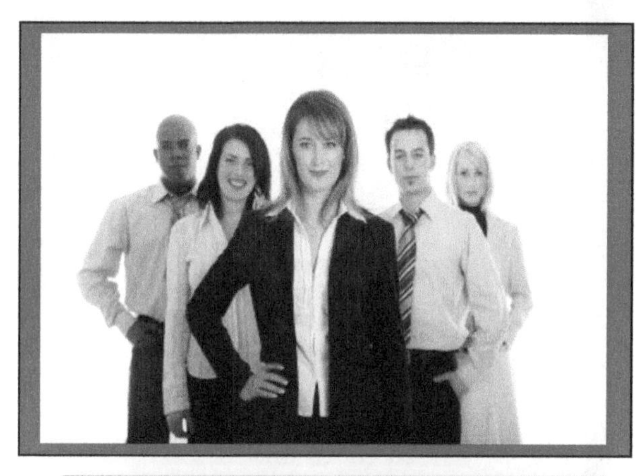

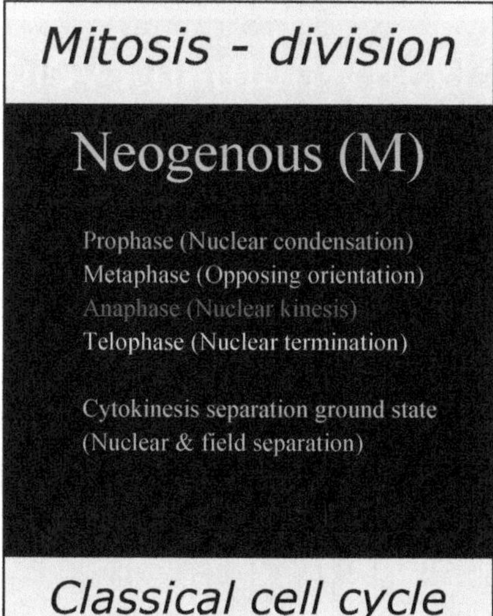

Human life cycle

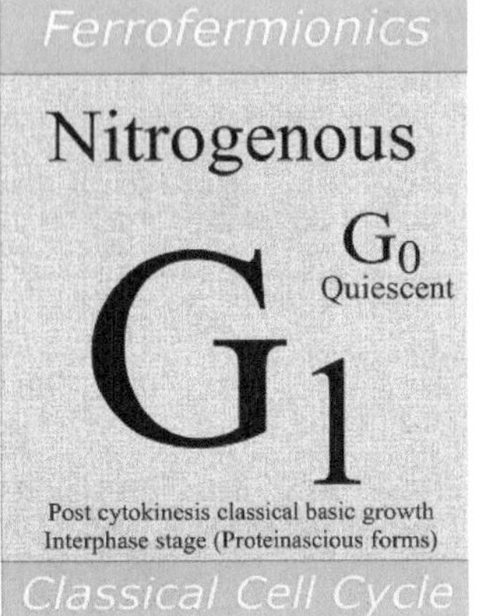

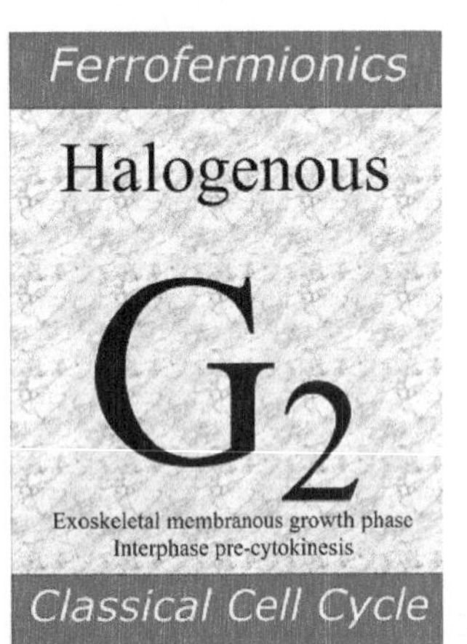

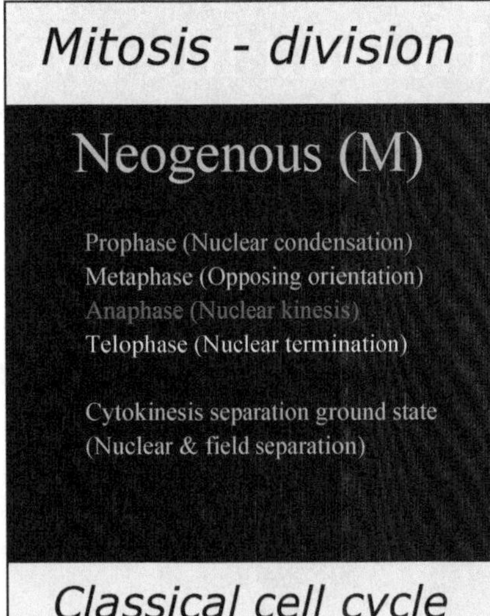

Daily human cycle

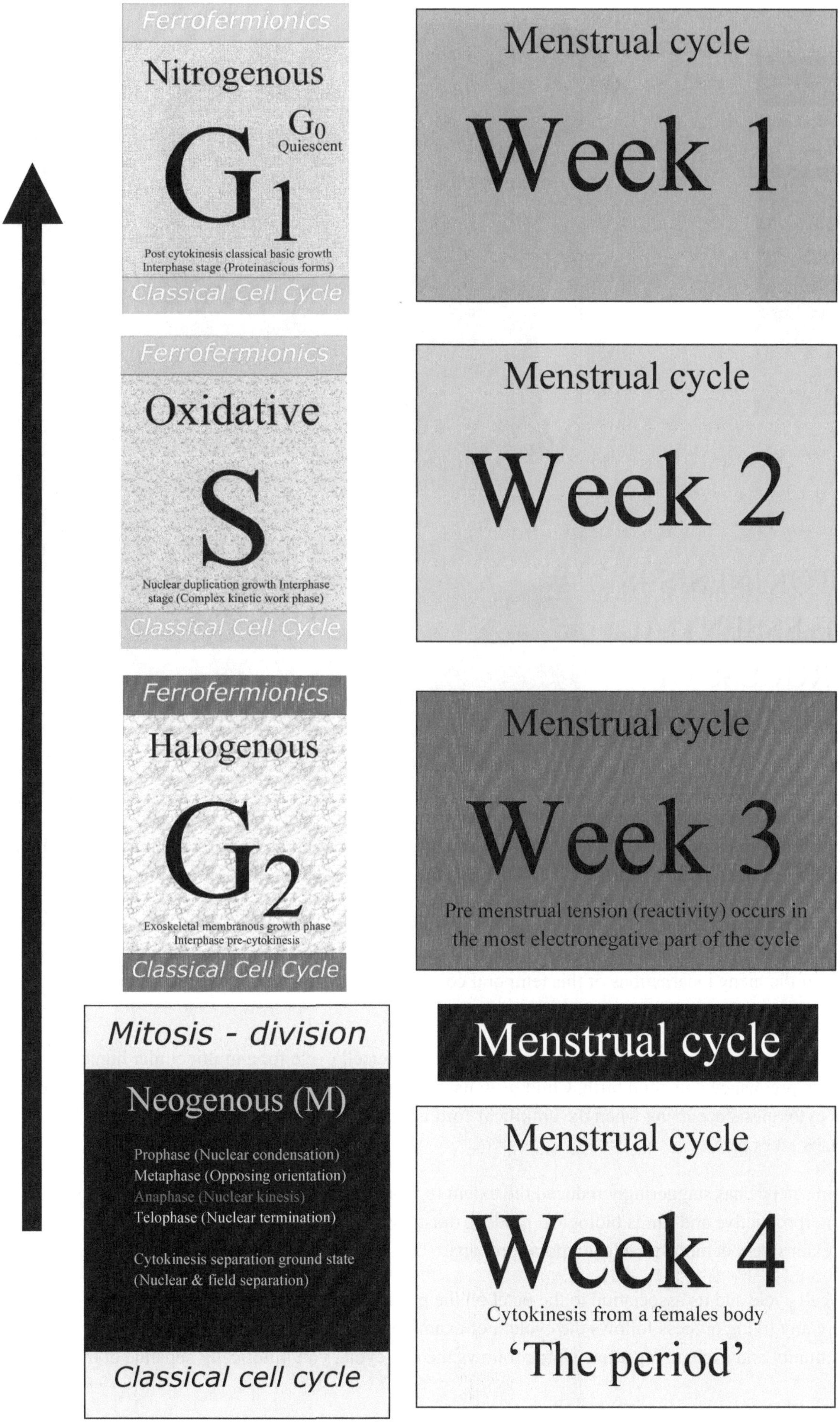

Ferrofermionics

Menstrual cycle

Week 1

Ferrofermionics

Oxidative

S

Nuclear duplication growth Interphase stage (Complex kinetic work phase)

Classical Cell Cycle

Menstrual cycle

Week 2

Ferrofermionics

Halogenous

G_2

Exoskeletal membranous growth phase Interphase pre-cytokinesis

Classical Cell Cycle

Menstrual cycle

Week 3

Pre menstrual tension (reactivity) occurs in the most electronegative part of the cycle

Mitosis - division

Neogenous (M)

Prophase (Nuclear condensation)
Metaphase (Opposing orientation)
Anaphase (Nuclear kinesis)
Telophase (Nuclear termination)

Cytokinesis separation ground state (Nuclear & field separation)

Classical cell cycle

Menstrual cycle

Menstrual cycle

Week 4

Cytokinesis from a females body

'The period'

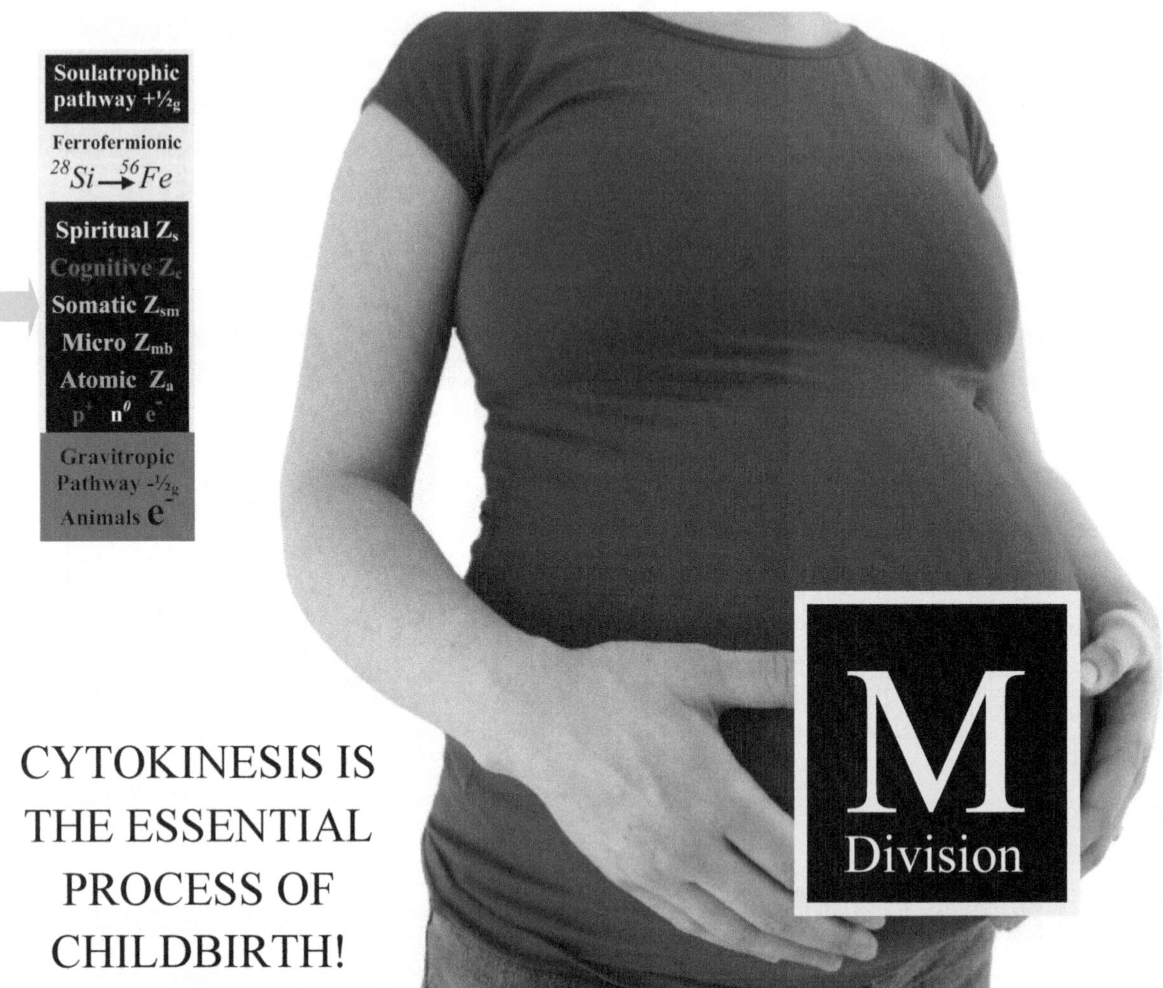

The concept of the cell is far more elaborate than is currently thought of by science. The cell is associated with carbon and even more fundamentally hydrogen. On the level of an entire multicellular organism, such as a human being, the cell cycle is thoroughly experienced.

Every day humans and other animals must successfully pass through the different interphase parts of the cell cycle G1, S, G2 and M and cytokinesis. A daily cycle, childhood cycle, lifetime cycles are just some of the many incarnations of this temporal component of carbon based kinetics and thermodynamics.

In the above photograph we can see the mitotic part of the cell cycle for a multicellular human through pregnancy and child birth. Child birth itself is a powerful example of cell division or mitosis with cytokinesis occurring when the umbilical cord is finally broken and full separation into two humans takes place.

Again science has staggeringly reduced the extent to which cell cycle logic is concerned. This is counterproductive and limits biological models, drastically reducing our integrated understanding of this extensively demonstratable temporal property.

The cell cycle and its association to the octet on the periodic table and carbon can be extended further where any living process follows the cycle. For example, even bumping into a friend in your local community and exchanging a few words follows the cell cycle, you mitotically separate at the end.

Meeting is a G1 interphase where basic statements occur, such as saying hello and or perhaps a shaking of the hands. A basic exchange of gossip is the S interphase and finally any reactive future interaction or G2 occurs.

Then we prepare to mitotically separate, prophase begins where we prepare to leave, metaphase orientation ready for moving, anaphase and telophase allows us to turn our bodies and repel each other, a final handshake breaks into the cytokinetic moment where we then leave.

Amazingly enough evolution over 13.7 billion years can be broken down into a cell cycle; it simply represents the deeply entrenched logic of carbon and hydrogen.

Neurochromosome bio-electric structures - conservation of electromagnetic atomic logic in amplified entromorphic atoms.

The many photographs of living structures in this book are powerful and undeniable evidence for the logic of nuclear and field amplification (ECASSE) in living entromorphic atoms.

The photographic evidence of structures associated with the atomic quantum organisation or orbitals is breathtaking. We don't need any numbers to clearly see the intrinsic relationship between carbons three orbitals and the bonded super structures, which have evolved from them.

I therefore submit the entire living world, its variation and physical complexity as evidence to completely support these theories. The physics world is not used to doing science this way and would consider a page of 50 or so numbers to be good evidence, observational proof is actually more powerful in biology as biological measurements are not compromised by resolvable complexity.

This evidence system provides us with trillions of data points, incredible consistency and through the 'superposition of states', it allows us to understand how the biological world can be so diverse and complex.

The 1s structures are from the first energy level in carbon and house the nucleus. These structures have nuclear properties such as a clearly defined isosurface and a coulomb barrier or shell. Examples of 1s biological conserved structures are: -

- All animals heads, their eyes are the structures as the result of the 1s electrons.
- All flowers, the seeds are tough and again have distinctive isosurface and coulomb barrier hence they have hard shells.

The 2s structures are from the second energy level in carbon and house the 1s first energy level and the nucleus. These structures have waveform properties such as a clearly loosely defined isosurface and a lack of coulomb barrier structures. Examples of 2s biological conserved structures are: -

- All animals thoracic and chest areas.
- All flowers, they are relatively spherical but have loosely distinctive isosurface and often no evidence of coulomb barrier hence they have soft structural shells.

The p structures are from the second energy level in carbon and form part of the 2s housing for the 1s first energy level and the nucleus. These structures have waveform properties such as a clearly loosely defined isosurface and a lack of coulomb barrier structures.

Examples of p biological conserved structures are: -

- All valance ground state and hybridised (sp2) bonding limbs, all muscles and other bonding highly delocalised movement structures.
- All leaves and similar structures, they are relatively lobed and localised but have loosely distinctive isosurface and often no evidence of coulomb barrier hence they have soft structural shells. They often exist at right angles to each other a unique fundamental property of the p orbitals.

Fractional Dimensions & Soulatrophicity

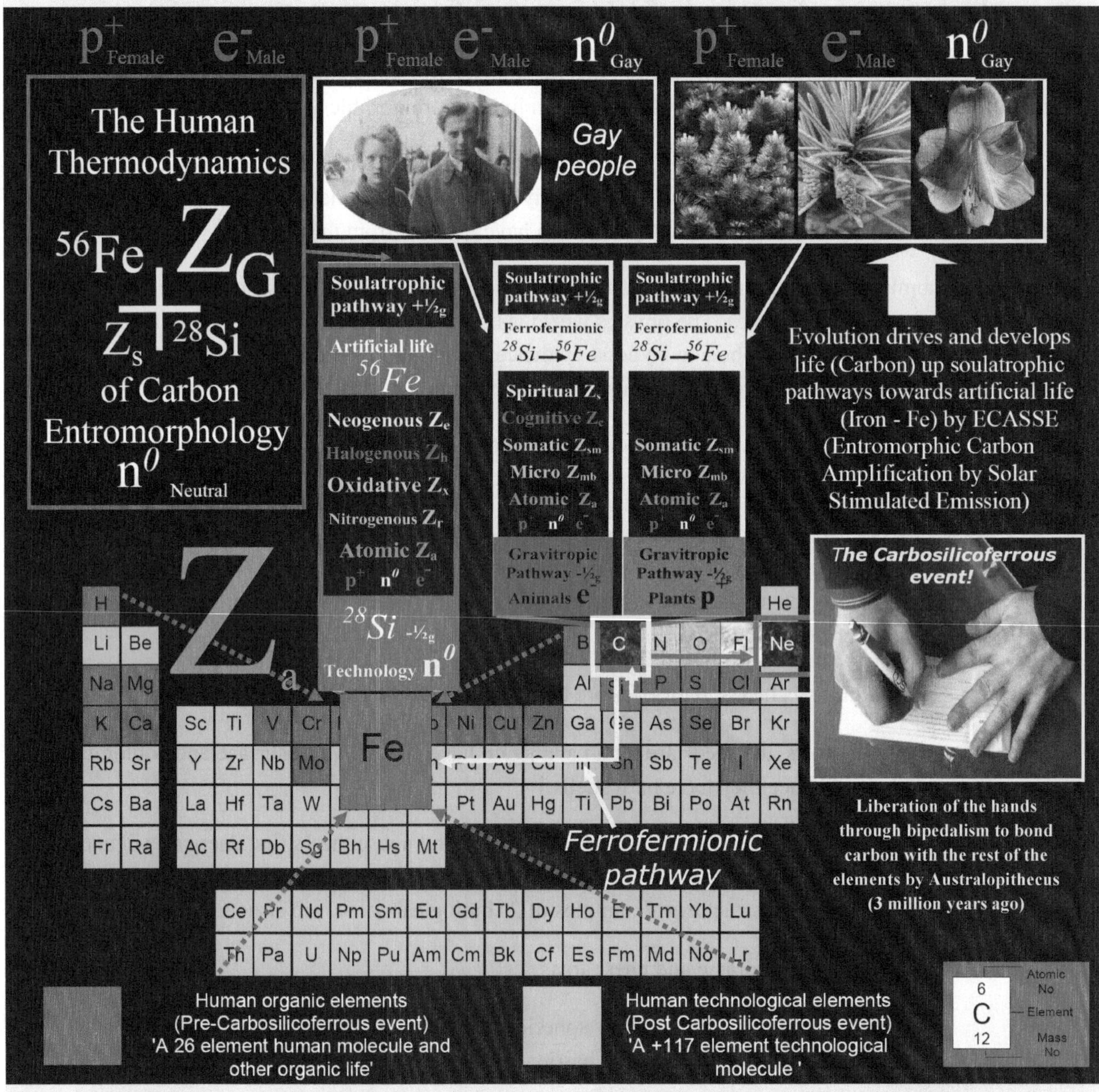

Life and its place on the periodic table (self-awareness through organoaccretion)

Science check

Do the observations correlate with the theoretical model?

YES ✓ *NO*

PART ELEVEN

Conservation of Atomic Logic

Neurochromosome entrochiraloctets - the basic carbon based fractionally dimensional template for living anatomy and physiology.

An entrochiraloctet comes from the word entropy, which is a measure of order and time, chiral which is a unique energy distribution property of carbon and the octet which is the rule followed by carbon to stabilise its physics. **Again many of the concepts in this book are repeated to improve clarity.**

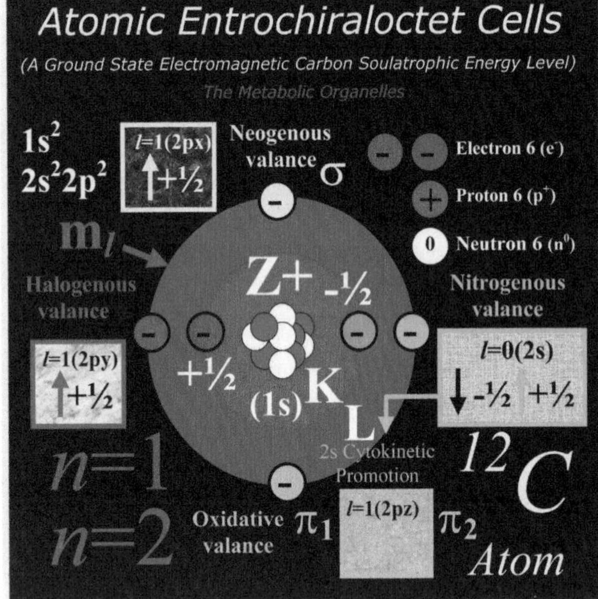

Carbon logic is conserved and can be used as the basis for understanding all aspects of living systems.

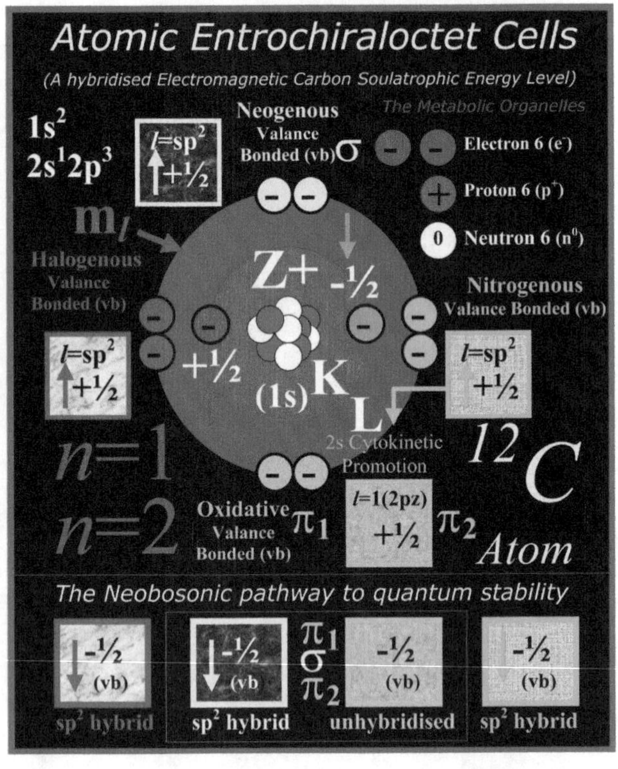

Each soulatrophic energy level has an entrochiraloctet in some stage of development, from ground state to hybrid. The template is a carbon template, the nuclear and field particles correspond with anatomical and physical properties in an organism having been comprised of atoms.

The nuclear properties are proton/neutron for the atomic level, AT/CG for the microbial level where DNA is the nucleonic form and ON/OFF for the neurological level.

Valance L shell particles produce organs such as limbs and K shell particles form structures such as the eyes, components of the torso, gonads etc.

Soulatrophic entrochiraloctets link together to form a soulatrophic pathway where entrochiraloctets bond to form hybrids as though they were carbon atoms bonding to other carbon atoms. Hence the torso of a human is an amplified carbon-to-carbon double bond or unsaturated bond. The spine is the sigma bond and the kidneys and lungs form the pi bonds, which are unhybridised.

The entrochiraloctet carbon template for living organisation forms the fractal fractional dimensional unit for describing all living things. It appears in some stage of development as a self-symmetrical carbon based unit. Variation in the orbital logic of the 1s, 2s and p orbitals produce all the variation for all life on Earth.

Life is a fractional dimension (fractal) described as the soul (meaning origin) or soulatrophic pathway.

The spiritual entrochiraloctet hybrid is the complete set of fractional dimensional cellular units united as one into a full amplified organism.

Fractional Dimensions & Soulatrophicity

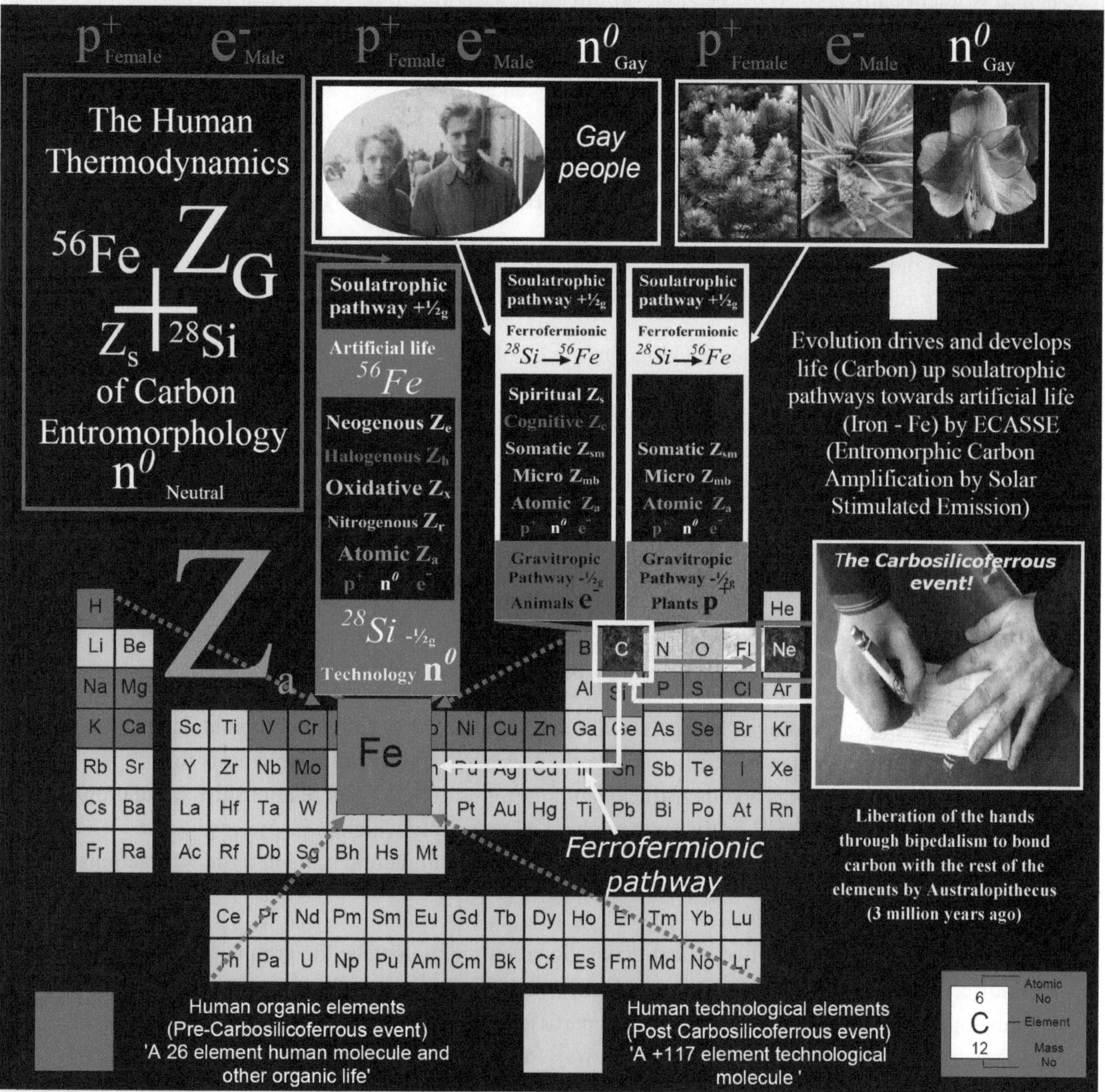

Life and its place on the periodic table (self-awareness through organoaccretion)

Spiritual entrochiraloctet systems produce the family and other grouped organisms. The parents are pink K shell and the siblings are the blue L shell valance. It is the fractional dimensional unit of the emerging technology cell through human colaboration. It is fractionally defined by the cognitive cell.

Cognitive entrochiraloctet produces the arms, chest and head. It is the fractional dimensional unit of the spiritual cell which is fractionally defined by the cognitive cell.

Somatic entrochiraloctets produces the legs, abdomen and genitals. It is the fractional dimensional unit of the cognitive cell. It is fractionally defined by the microbial cell.

Microbiological entrochiraloctets produce the typical animal cells and bacteria. The K shell is the cytoplasm and the L shell is the bilipid membrane. It is the fractional dimensional unit of the somatic cell. It is fractionally defined by the atom **(application/assignment of the central dogma).**

Atomic entrochiraloctets produce the typical carbon atom. It is the fractional dimensional unit of the cell. It is fractionally defined by the fermionic trinity and quarks.

Neurochromosome quantum taxonomy - living classification by fractionally dimensional systems based on carbon logic.

Taxonomy is the fundamental science of classification of living organisms according to basic rules of self similarity in both morphology and physiology.

The old system was based on the five kingdom model of logic but has been further refined in the early 90's to a 3 domain system based on cell type. Modern classification is phylogenetic and utilises relationships classification based on DNA similarities.

Taxonomy breaks down an organism into its fractional components based on both anatomical and functional characteristics. There is no physical basis for classical taxonomy where the 3 domain breaks down to the 5 kingdoms and further to phyla, class, order, family and the final two genus and species which make up the binomial scientific naming system for all life.

What is required is a system based on fundamental physical models of nature; hence carbon entromorphology proposes quantum taxonomy based on conserved carbon anatomy and functionality.

This approach allows a biologist to organise living things according to their carbon logic through nuclear physics and quantum mechanics. This produces a more powerful atomic basis for classification of living things and their true quantum origins and physical relationships. Soulatrophic pathways are the carbon based equivalents of classical taxonomy which in its own right, is a fractionally dimensional approach to describing living orders of natural scale in terms of fractional components such as the cell.

Page 247 and 248 display the two global gravitational entrochiraloctets: the ground state and the sp2 hybridised state. From classical carbon mechanics each fermionic particle constitutes a conserved atomic logic for all life based on carbons own particle organisation and physics. Quantum taxonomy isn't easy to explain but it does produce a pathway for all living evolution on planet Earth based on the power of conserved carbon logic alone **but with a gravitational determinism it is therefore, a quantum gravity model.** Hence on a global Earth based level of organisation, life can be described in terms of nuclear and quantum levels of particle organisation. In short, the global cell is mediated by gravity but has a seamless link to quantum carbon electromagnetic origins. In other words ALL life on

Earth can be thought of as one gigantic carbon atom where each organism and its respective macroscopic state i.e. solid based life such as a worm, aquatic based life such as a fish, an air atmospheric based life such as a cow and a post planetary cytokinetic organism - based on the environmental vacuum of space - such as humans going into space.

The two K shell global particles are the plants the angiosperms and gymnosperms which make up the spin up and down taxonomic particles, and form a stable shell; all plants are somatic and radially symmetrical.

There are four L shell valance particles giving a 2s spin up and down orientation such as the crustaceans and all other radially symmetrical organisms, the two p organisms on the aquatic level are the fish and emerging amphibians able to cytokinetically separate from the waters (px, py). There is no pz organism in the ground state but the land animals are hybridised and pz is present.

The sp2 hybrids are land based/air based taxonomic particles, of which there are four.

The four-valance aquatic organisms bond with the four land based organisms to produce a complete global entrochiraloctet. The four taxonomic organisms produce the insects, the herbivores, the carnivores and the omnivores for both aquatic and land based organisms (taxonomic particles).

- A detritivorous (nitrogenous) crustacean and a detritivorous insect.
- A herbivorous (oxidative) fish and a herbivorous (oxidative) cow.
- Producing a relationship between a carnivorous (halogenous) shark and a carnivorous (halogenous) tiger.
- An omnivorous (neogenous) amphibian and omnivorous human (neogenous).

The octet rule now applies on a global scale and a more detailed explanation of quantum taxonomy is required as the micro organisms also fall into the same logic and can be described as a unified entrochiraloctet carbon cell.

For example the nitrogenous detritivore cell is the chitinous fungus, photosynthetic protozoan's are oxidative, none photosynthetic organisms are halogenous and the animal and plant cells are the neogenous component.

A more thorough explanation of this complicated global taxonomy will follow in time and I apologise upfront if people are confused. Technology can also be classed in this way as a natural extension to global entrochiraloctet cells to give a powerful and totally complete model of life.

This is a global carbon atom made of ALL life on Earth but still demonstrating a powerful electromagnetic link to atomic carbon. This is quantum nuclear and field amplification on a global scale and an example of quantum gravity as electromagnetic rules transcend to the rules of general relativity and gravity determinism in dynamic equilibrium.

The energy levels are now macroscopic environments on the global scale hence the nucleus is the solid Earth, the K shell, n=1 is the aquatic environment and the L shell, n=2 is the gas atmosphere. Hybridising electron promotion produces planetary cytokinesis where humans go into space.

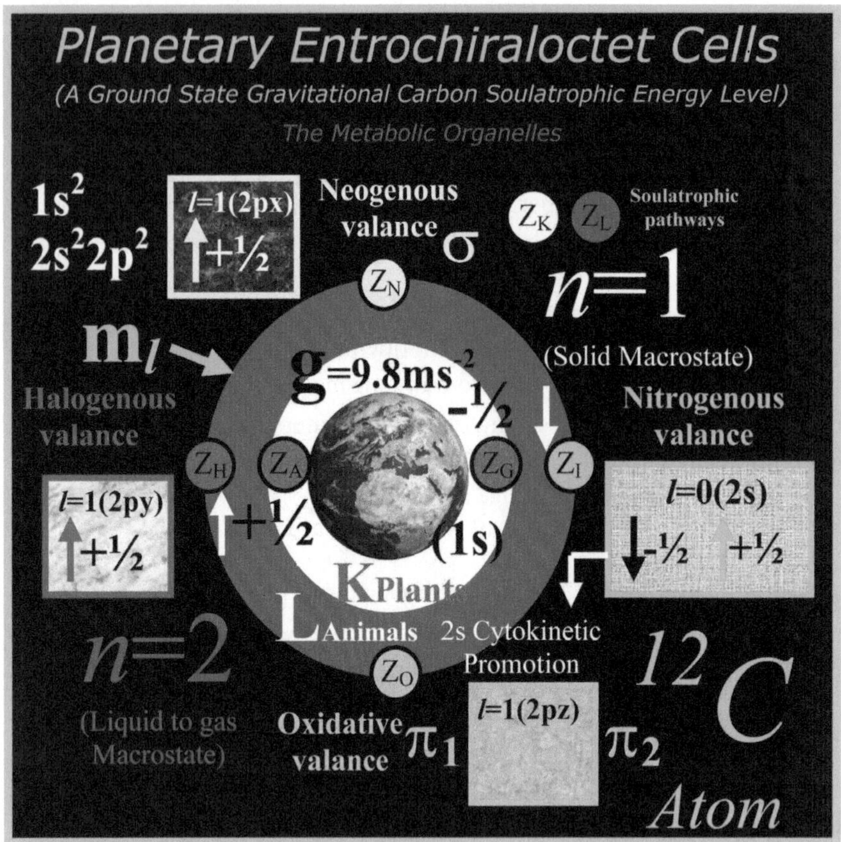

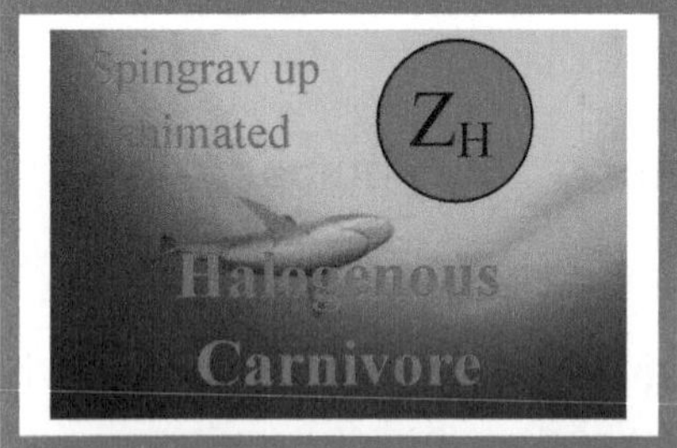

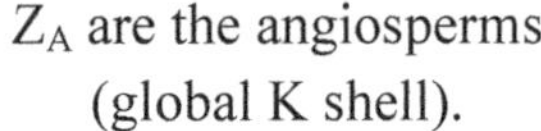

Z_A are the angiosperms (global K shell).

Z_G are the gymnosperms (global K shell).

Z_I are the radially symmetrical organisms and the crustaceans.

(nitrogenous detritivores global L shell).

Z_H are the carnivores (halogenous - very reactive global L shell).

Z_N are the cytokinetic omnivorous amphibian (aquatic cytokinesis global L shell).

The insects and plants are polysomatic polyentromorphs

(multi-segmented entrochiraloctets), due to post nuclear linear effects.

The hybrid forms the land based (air atmosphere) organisms as if they are valance bonded to the ground state entrochiraloctet, producing strong evolutionary organisational bonds.

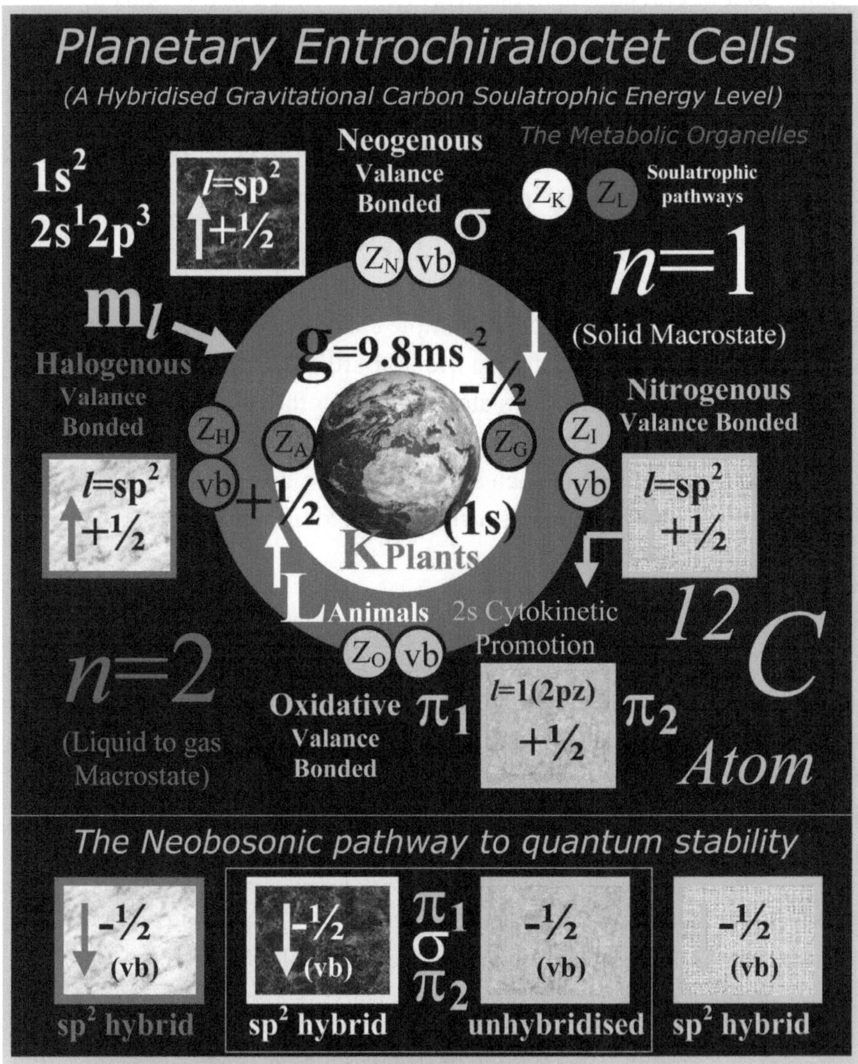

Z_I are the insects (nitrogenous detritivores valance bonded global L shell).

Z_O are the herbivores (oxidative valance bonded global L shell) such as a cow (pz promotion).

Z_H are the carnivores (halogenous valance bonded global L shell) such as a tiger.

Z_N are the omnivores (neogenous valance bonded global L shell) such as humans (planetary cytokinesis).

The gas and vacuum macrostate in the global valance shell, the land animals and the heavens.

Vb means they are all valance bonded.

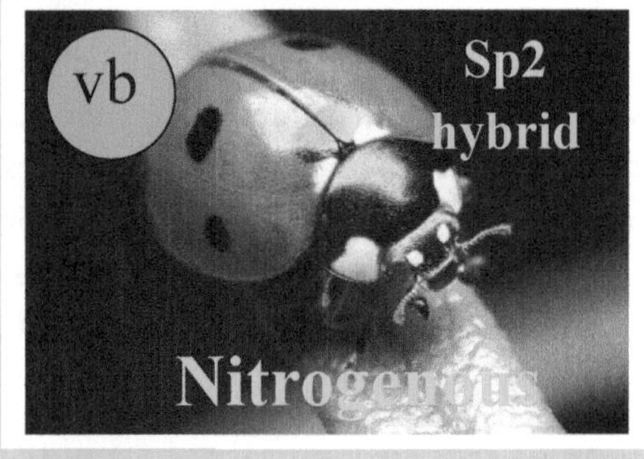

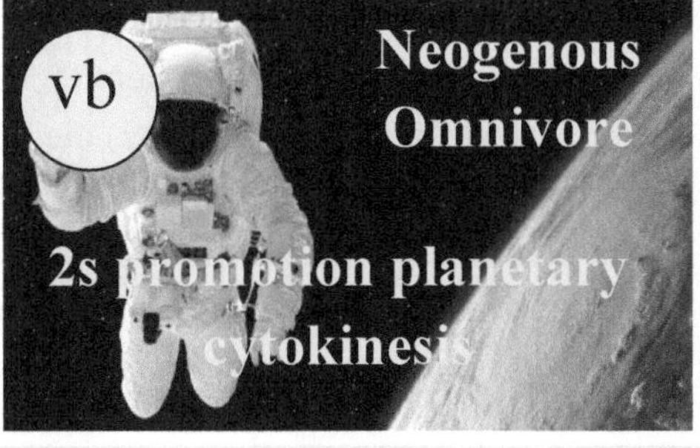

The microorganisms also fit this model and will be considered separately. The spin orientation is also very important as it describes the orientation of an organism pointing towards the Earth or up towards the Sun, in other words fermionic spin in gravitational systems (often described in this book as spingrav) as opposed to classical electromagnetic spin but with a connected logic: again a quantum gravity model of intermediate boson/fermionic organisms.

Neurochromosome metabolic organelles - quantum energy balancing through the living valance shell and conductance band by ECASSE (Entromorphic Carbon Amplification by Solar Stimulated Emission).

Metabolic organelles (valance bonding in all living things) a simple example of which is a human picking up an apple and biting, it is a major part of sustained life on Earth. The person has bonded to the apple through their hand. On the atomic level a carbon atom bonds to other atoms by the same type of logic. The bonding is driven by the absence of four electrons in carbon and acquisition of such stabilising charge is favoured in nature. An organism, such as a human, is made of trillions of atoms and so bonds to trillions of other atoms: hence the apple is a simple scaled up version of atomic bonding but with its origins in quantum atomic covalent bonding.

Derived from the periodic table of the elements they describe the physical quantum instability in carbon soulatrophic entrochiraloctet cells (pathways). Carbon has four valance electrons in its L shell (n=2) and requires 4 more in order to stabilise the L shell energy level. The 4 more in this model are termed the 'valance bonded' metabolic organelles, as bonded to the ground states valance electrons by sp2 hybridisation: **food is the typical bonded form in most organisms on a daily basis**. This stabilising need is seen every day as living organisms find food to complete their valance octet and reach daily quantum stability.

The first electron acquired produces an electronic configuration seen in nitrogen. This is the nitrogenous valance bonded metabolic organelle G1 of the extended cell cycle. Nitrogenous systems (from food they are the proteins) are nuclear linked through nucleic acid systems (DNA) bosonically linked to amino acid proteinascious forms. They are structural materials and conform to drive quantum stability.

The second electron acquired produces an electronic configuration seen in oxygen. This is the oxidative valance bonded metabolic organelle S phase of the extended cell cycle. Oxidative systems are heat engines allowing nitrogenous structures to move and produce their own heat (in food they are the carbohydrates). The third electron is the halogenous (fluorinated) and produces membranous selectivity and powerful electrophilic potentials (in food they are the fats, membranous barriers). The halogenous valance is the most reactive state on the periodic table in terms of electronegativity.

The final forth electron valance bonded produces the neogenous state (in organisms they are the bonds made with other organisms such as a partner – a duplicate nuclear system) or Nobel gas electronic configuration and quantum stability for the L shell. This metabolic organelle is deceptive as the neoiterative cell cycling system (neobosonic ECASSE) produces a new nuclear system (cellular mitosis and meiosis). Due to the nuclear instability potential (ferrofermionic pathway) means that the stable L shell is still nuclear unstable and produces a new incomplete L shell, but a slight increase is seen in nuclear ferrofermionic ECASSE stability. The iterative cell cycling through solar acquisition continues as living carbon based entrochiraloctet systems grow and tend towards iron stability. In essence the ECASSE (Entromorphic Carbon Amplification by Solar Stimulated Emission) is seen by growth through food acquisition, which is the solar stimulated emission part of the process. Hence the 4 metabolic organelles and the 4 extra valance bonded particles drive soulatrophic evolution and nuclear and field amplification in all life. A physics lecturer eats food and bonds with students (to find their 4 valance bonded metabolic organelles, proteins, carbohydrates, fats and other organisms) as they amplify nuclear and field logic (physics knowledge) through teaching, this is an example and proof of ECASSE in action.

The halogenous metabolic organelles G2 (extended) cell cycle and mitosis – this is the membranous selective communicative interface in carbon soulatrophic entrochiraloctet cells (metabolic regulation interface very reactive) communication – 3rd missing electron in the neobosonic iterative cell cycle producing a highly reactive potential; associated with the halogenous metabolic organelles, the valance electron in carbon. The halogenous system is the electronegativity maximum for the whole of the elements in the periodic table. This super reactive state is close to quantum stability and produces powerful electrophilic potentials in soulatrophic entrochiraloctet pathways; through living desire.

Wars of all descriptions occur under halogenous electronegativities, it is the highest state of reactivity in living carbon cells. Quantum stability is close and evolutionary pressure is maximised to stabilise the L valance shell with the acquisition of one final electron and therefore quantum stability. This is a region of extremes of energy liberation and conservation. It selectively regulates the metabolic organelles (valance) and is seen as membranous structures in carbon soulatrophic entrochiraloctet pathways. On the atomic level it relates to electronegativity of fluorine electronic configuration with a single electron acquisition in carbon entrochiraloctet cells. It produces the fatty acid bilipid layer and membrane (associated with chloride ions) as valance bonding on the atomic level and the formation of the micelle system and fluid mosaic bilipid layer characteristic of prokaryotic and eukaryotic cells.

On the microbiological level it is found in the animal cells as valance bonded metabolic organelles, which act to form multicellular membranes and tissues (the emergence of limbs). The animal cell systems are complexly selective and utilise large metabolic valance bonded materials, hence a dramatic increase in entrochiraloctet size. On the somatic and cognitive soulatrophic energy levels the halogenous valance bonding is associated with the limb of preference (writing hand) on the cognitive level. It results in technology as valance bonded associated with communication systems, pressure systems, army, police, and medical systems anything highly reactive and membranous, the extended immune response: any technology, which selectively regulates the external environment through communication highways. The working world can be seen as a halogenous/neogenous process. Electrophilic motivation is maximised by these organelles. This is the G2 and mitotic phase of the neobosonic classical cell cycle. It is characterised by preparation for cytokinesis often exoskeletal and membranous structures are made and mobilised. The final neogenous electron is gained through the mitotic cytokinetic process. Nuclear precipitation prophase, orientation metaphase kinetic separation of nuclear components anaphase and establishment of new nuclear centres telophase and finally cytokinesis the formation of a new entrochiraloctet neogenous vb (valance bonded) metabolic organelle of the previous entrochiraloctet. **See page 252 and 253 for supporting images to help understanding this very complex model.**

The neogenous metabolic organelles – nuclear replication metabolic organelles 'Nobel gas electronic configuration' in carbon – 4th missing electron. The final electron in carbon L shell (n=2 – valance shell) systems is the valance bonded neogenous system. It relates to quantum stability (field stability only) and a completion of the 4 missing electrons in carbon soulatrophic pathways increasing evolution soulatrophic position through ECASSE. The L shell becomes stabilised, although the ferrofermionic nuclear instability produces a reset on the cell cycle. For example, every day you have to bond with food and other organisms to remain alive. Withdrawal of any of these valance bonded organelles (food, other organisms) results in oxidised states and quantum degradation.

The neogenous valance bonded organelle is seen as a replication of entrochiraloctet components as part on the neoiterative cell cycling mechanism. The nucleus is renewed through replication in the form of a new entrochiraloctet cell produced in new interphase G1, S, G2 and M. This cell has a new L shell instability and must progress through the four missing electrons again, and again, and again and again. Every day we go through a daily cell cycle G1 (morning) S (afternoon) G2 (evening hence most people have sex in the evening as it's the most reactive halogenous phase, and finally M (sleep and cytokinesis) separation into a new day. This can be duplication in a microbial cell, or the production of a new baby monkey, or a new football team. The variation is in size and complexity but the fundamental basis of cytokinesis remains conserved. Every stable L shell locks energy in the nucleus (small amounts of increased mass) in carbon and pushes it closer to iron nuclear (ferrofermionic) stability by electronegativity driven solar accumulation.

The ferrofermionic pathway sees nuclear structures forming over millions of years from the neobosonic iterative solar amplification ECASSE. It produces a pathway where the nuclear structures start as nucleonic proton/neutron systems, then DNA/ATCG systems, then neuronal axonic systems ON/OFF and finally electronic silicon systems ON/OFF and legal and divine nucleonic black/white. Further nucleonic forms are yet to come and there are huge variations, such as legal systems.

The nitrogenous metabolic organelles (extra definitions) – structural valance, and valance bonding in carbon soulatrophic entrochiraloctet cells (anabolism – 1st missing electron); associated with the nitrogenous metabolic organelles, the valance electron in carbon. With electronic configuration either by electron or through the acquisition of bonding nitrogen atoms or larger nitrogenous structures, like a cooked chicken breast.

Carbon gains structural nuclear capability and electronegativity similar to that of nitrogen. Nitrogenous metabolic organelles form nucleic/amino acid systems on the atomic soulatrophic energy level. It is related to carbon chiral centres and the quantum amplification effect through the structural nature of nitrogen ECASSE. It produces the fungi on the microbiological level and the structural limbs on the somatic level and cognitive level. The valance bonded nature on the somatic and cognitive levels are associated with protein acquisition. The cognitive and spiritual levels also produce technological structures. In technology post carbosilicoferrous event the nitrogenous organelle is an extension of the carbon nitrogenous energy level. It forms the basis of structures made out of other elements but having that original functional origin. An oxidative valance bonded organelle is required to make such structures move and generate heat; they are the structural systems in cells. The nitrogenous metabolic organelle system is the G1 interphase portion of the cell cycle (neobosonic).

The oxidative metabolic organelles –liberatable energetic valance bonding in carbon soulatrophic entrochiraloctet cells (catabolism – 2nd missing electron). Associated with the oxidative metabolic organelles (the valance electron in carbon), with electronic configuration either by electron or through the acquisition of bonded oxygen atoms. Carbon gains heat engine capabilities and electronegativity similar to oxygen. Oxidative metabolic organelles form oxidative energy liberating systems on the atomic soulatrophic energy level. It produces the protozoans and algae as valance bonded acquisitions on the microbiological level. It produces nephronic (kidneys, protonaphridia) structures on the somatic level. The cognitive and spiritual (shared valance bonding) level sees the emergence of the lung/gill system for heat engine (catabolic) processes; oxygen is acquired directly from the environment as a valance bonded organelle. In humans this is extended to technology associated with heat engines and movement (combustion engines). Oxygen is highly electronegative and acts as a terminal electron receiver in substrate level and mitochodrial Krebs cycle models of energy liberation. Oxygen content in the global environment to the tune of 21% of the atmosphere is a quantum amplification effect produced by life on a global scale over hundreds of millions of years. Quantum oxidative physics has been amplified to global levels in similar ways to the 78% Nitrogen: more evidence for the octet neobosonic model of life. The oxidative metabolic system is the S stage of interphase in the classical cell cycle. Energy is liberated to duplicate cell components.

Wave functions are the mathematical treatment of the valance metabolic organelles – the quantum mechanics of electromagnetic fields. The wave function describes probability fields where electrons may be found limited in measurement by Heisenberg's uncertainty principle. They represent different discrete energy levels based on Planck's constant. In carbon entromorphology n=1 K shell and n=2 L shell are conserved and amplified by solar accumulation ECASSE. Movement of particles up and down energy levels must absorb or emit radiation, and in order to do so this is the basis of the quantum jump although the quantum jump has an emergently strong gravitational component in living organisms. These fields are inherently measurably uncertain because of Heisenberg's uncertainty principle. A wave function can collapse where nuclear expression through the weak nuclear force is terminated. The wave/particle duality in an organism oscillates between these states.

Organic metabolic organelles.

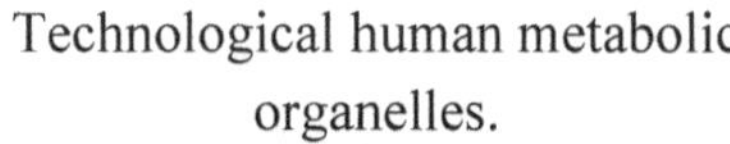
Technological human metabolic organelles.

Organic metabolic organelles.

Technological human metabolic organelles (transducing devices).

Sleep is observed with a termination of expression and internalisation (particle form). When an organism moves the wave function becomes more evident. The wave function relates back to Z+ the charge on the nucleus and link between the strong nuclear force and the electromagnetic force through the electroweak standard model and its living structural equivalents in communication.

Consciousness is personified by the expression of spiritual wave functions and death by the collapse of spiritual, cognitive and somatic wave function with lower wave functions such as the microbiological level remaining expressed by bacteria.

Neurochromosome nucleonic soulatrophicity - the nuclear ferrofermionic pathway and the evolution of the fractional dimension.

Entromorphic genomics, and the emergence of neurological genetics – the continuity and communication of binary encoded nucleonic systems in carbon entrochiraloctet cells (binary systems). Nuclear genomics evolve from proton/neutron systems on the atomic soulatrophic energy level to nucleic (DNA) systems for AT/CG coding to neuronal systems of On/Off action potentials to silicon and legal systems of On/Off electronic logic. ANY stored information system is capable of being binary encoded and represents the nucleonic organisation on all soulatrophic levels.

Gluonic logic holds words together and words group into sentences through pionic logic in language and writing. DNA is also held together in the same way, gluonic logic holds nucleotides together and pionic logic holds AT/CG base pairs together. The following images summarise nucleonic systems for all life and the conserved nature of such nuclear systems in carbon entromorphology.

- The classical nuclear region is the atomic level, however the word 'nuclear' was originally used to describe cells and Rutherford used the word to describe the atomic nucleus. Astonishingly he and others failed to realise that the convenient use of the word was a reflection of a multi nuclear family.

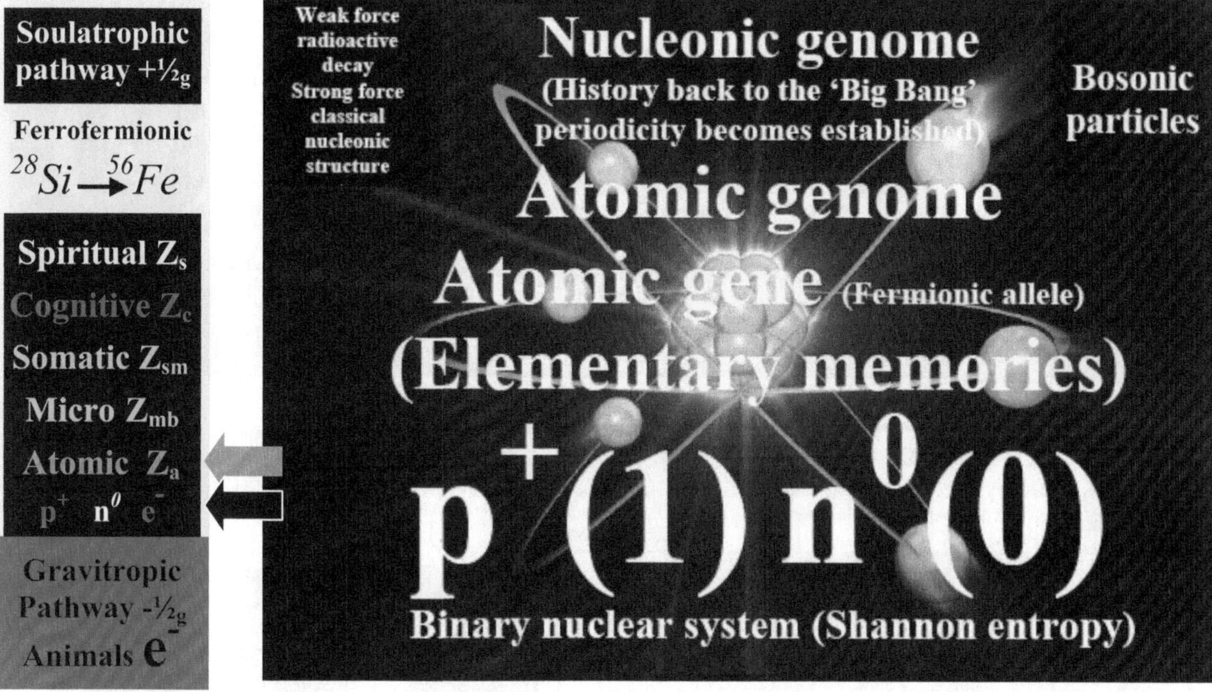

- The DNA level of nuclear organisation, again the word nuclear is used and was the original word of choice for describing cells long before the atom was realised.

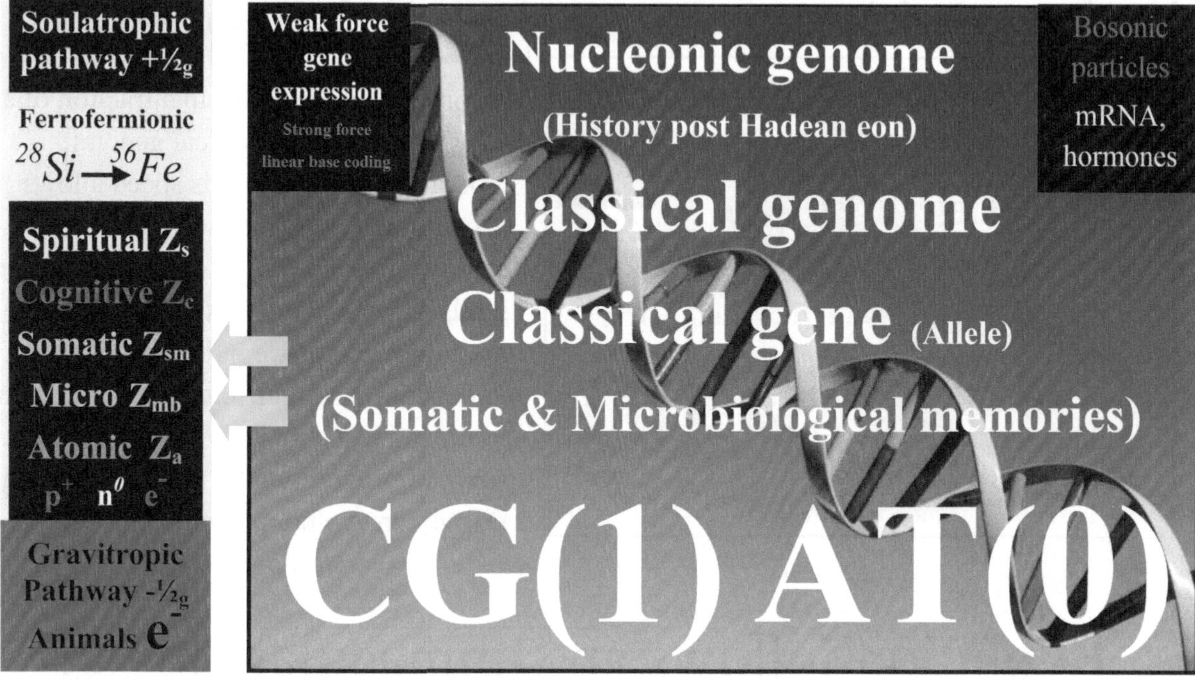

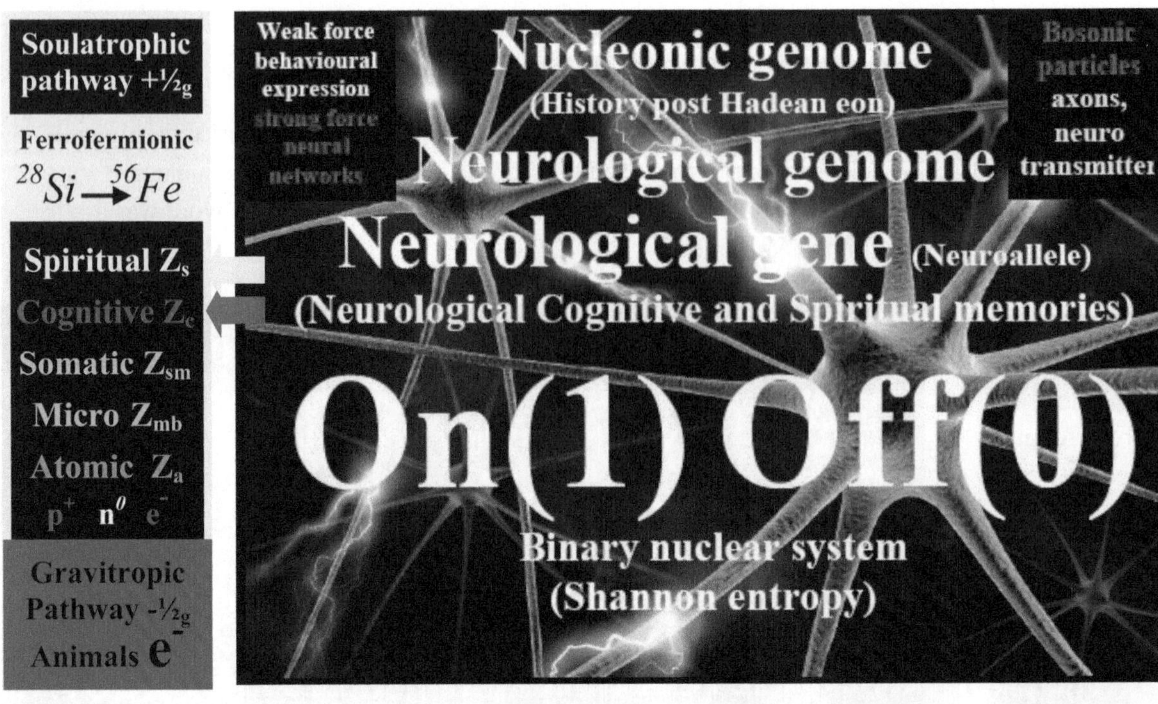

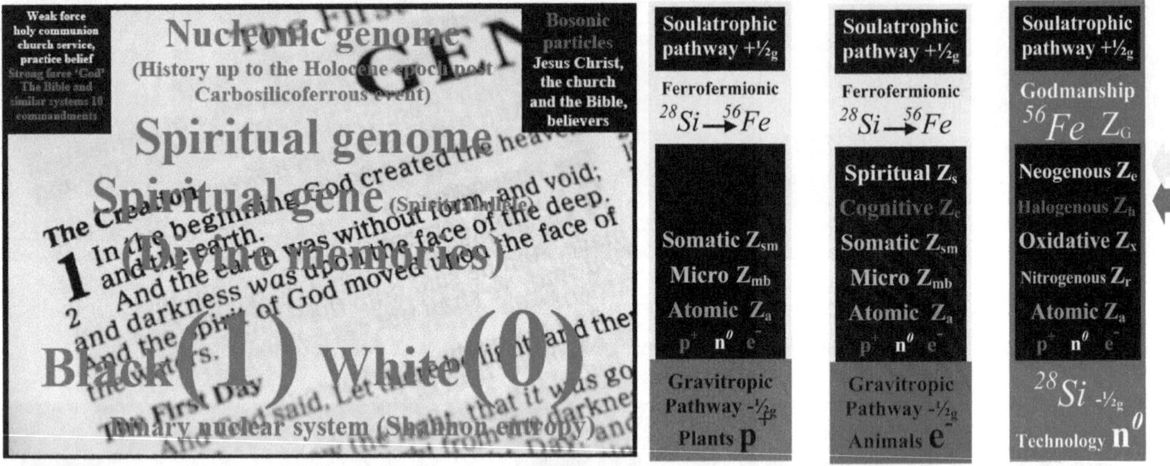

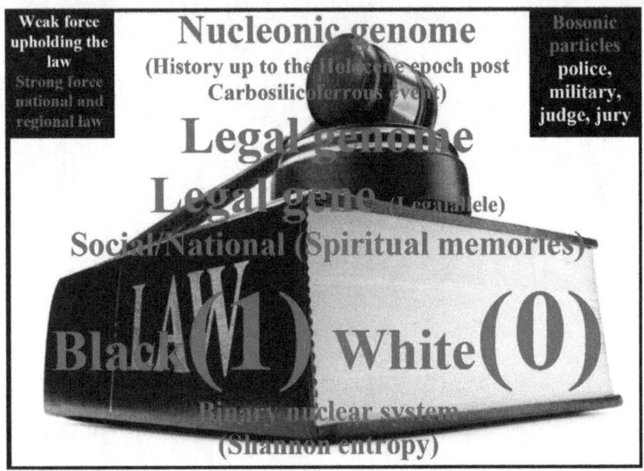

Recombination rates act in a fraction of a second on neurological genomic soulatrophic levels. This speed allows evolution by natural selection to be demonstrated and proved easily. Evolution holds a perfect logic in that by rejecting evolution you prove it, because you are demonstrating natural selection. Education, sport, and just about anything absolutely personifies evolution by natural selection Darwinian evolution triumphs in carbon entromorphology.

Evolution follows an 'Immaculate logic!'

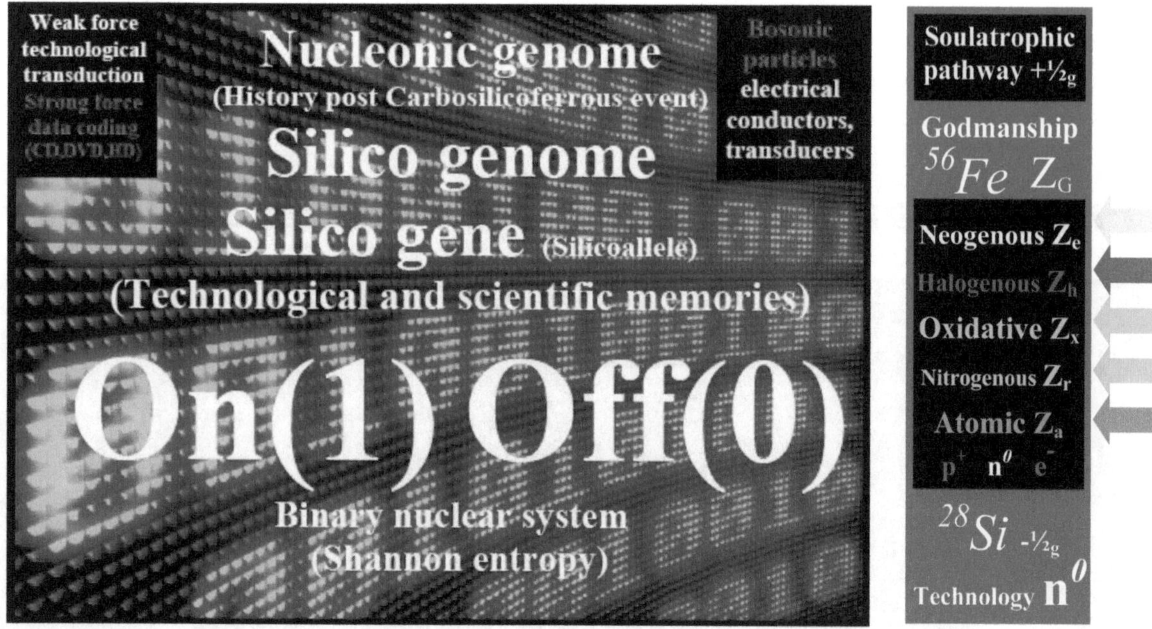

Planck's constant quantifies the nature of all stored condensed information systems.

On all soulatrophic energy levels gender communication by forming soulatrophic bosonic ladders (a hug and kiss in humans or any other form of communication) occurs by nuclear fusion means, also known as recombination producing new resultant nuclear systems.

On the spiritual (familial) soulatrophic energy level psychomeiosis is the model of nuclear recombination, also known as 'thought'. On the somatic (radial) level cytomeiosis is nuclear fusion (recombination) known as sexual DNA zygotic reproduction; classical genetic communication.

On all genomic nuclear soulatrophic energy levels we find the same meiotic zygotic nuclear fusion, followed by decay or mitotic expression into quantum expression fields (through the body of the organism – described by classical quantum mechanics as the electroweak standard model). The meiotic system is mediated by the strong force; mitotic systems are mediated by the weak nuclear force, which decays into an organism and associated behavioural processes through ribosomal systems into organisms actions. Ribosomal systems convert nuclear coding into proteinascious forms (products) by bosonic translation (through mRNA translation on somatic DNA levels and the hands through neurological (bosonic equivalent) on cognitive and spiritual levels). In short, all soulatrophic levels recombine nuclear energy into recombined forms (cytomeiotic), which are translated and expressed by ribosomal devices into proteinascious products through electroweak models. DNA thought, neuronal thought, atomic thought and silicon thought all operate on the same basic nuclear to field expression electroweak system. The gauge bosonic system works gravitonically in reverse where external energy is taken into atomic entromorphic systems through solar acquisition (ECASSE).

Expression – the process of conversion from the strong to the electroweak nuclear force in soulatrophic pathways through ECASSE. An electron in quantum fields expresses an atomic gene through conserved non local entanglements. A protein in cytoplasmic quantum space is an expression of a gene having self symmetry to expression in the atoms it is comprised of. A proteinascious form in cognitive quantum space is an expression of a psychogene through classical communication, however the self symmetrical logic is the same. A silicogene is expressed into a proteinascious form such as a tax rebate. All expression systems occur with intermediate bosonic particles, DNA has mRNA, psychogenes have neuronal networks and action potentials. Silicogenes have wires, optic fibres and electromagnetic communicators. Society (national genomics and divine genomics), expresses legalities through public servants such as the police and priests who are the acting system gauge bosons (national force carriers), for massive national quantum fields such as England.

Fractional Dimensions & Soulatrophicity

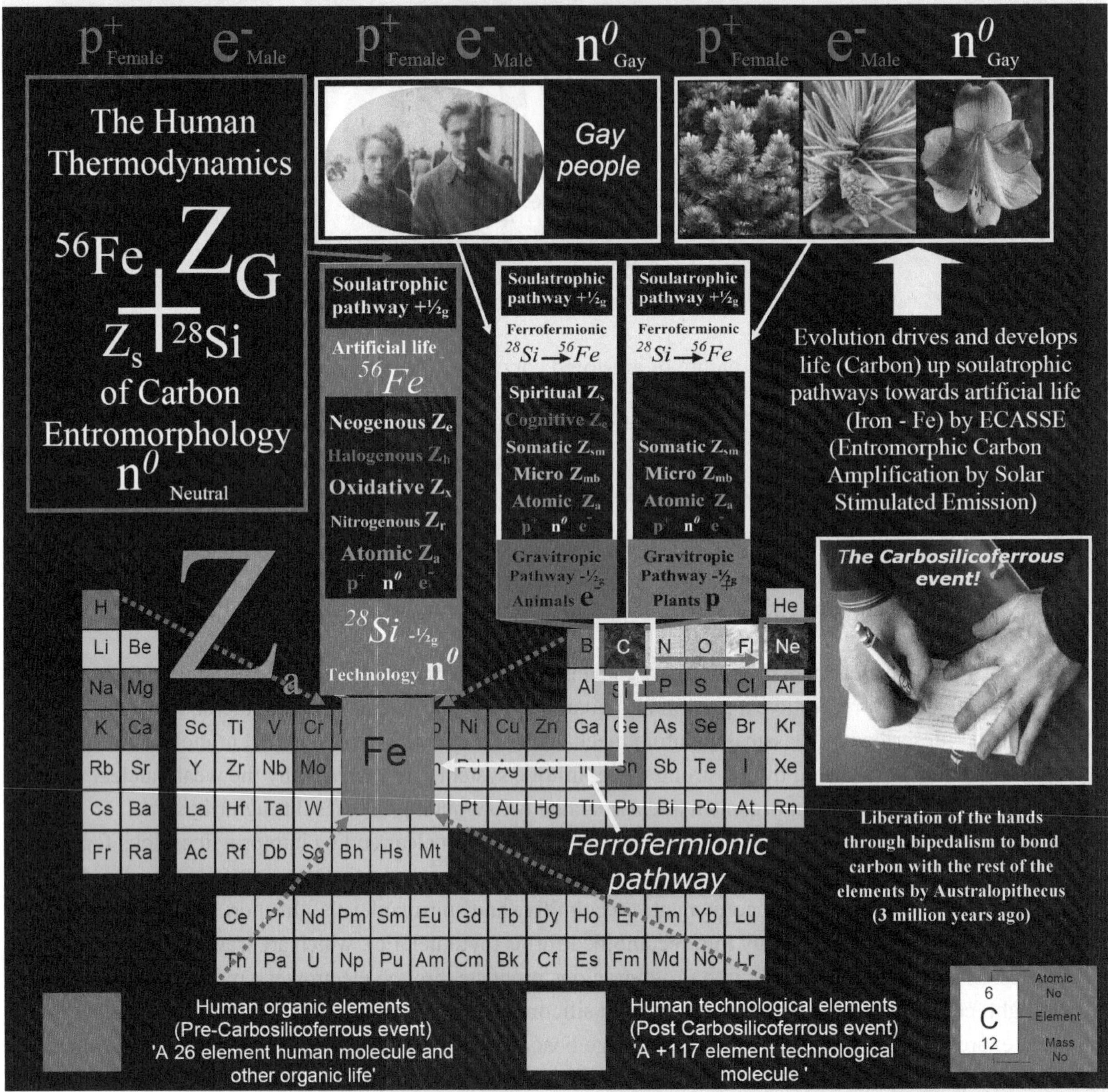

Life and its place on the periodic table (self-awareness through organoaccretion)

Atomic genes – An alternative model of nucleonic systems (proton/neutron binary). Based on self symmetry to genomic expression systems this is typically seen on the microbial and somatic levels of organisation. Due to soulatrophic amplification, nuclear systems on different levels higher or lower but part of the ferrofermionic pathway show associated conserved logic. All nuclear pathway components are binary and express models, which are based on the strong nuclear force and weak nuclear expression force. Atomic genes or nucleonic proton neutron linear arrangements are expressed by the weak nuclear force as electron energies in quantum fields through classical wavefunctions.

Nuclear physics on the atomic level is expressed by quantum mechanical field models. Nuclear physics at this juncture is extended to include molecular biology (DNA nucleonic systems), neurology (neuronal nucleonic systems) and legal, silicon and divine nucleonic systems. Any extrapolatory extension of soulatrophic pathways either deeper into inner space or outwards into outer space are likely to posses the same fundamental structures of nuclear binary innerspace centres and local quantum expression outerspace fields. Outside of these limits gravitational forces become dominant over electromagnetic forces as nuclear modulators, but are probably still the appropriate tools of expression. Cognitive (G2) genes - soulatrophic amplification nuclear systems on different energy levels display different levels of detail but fundamentally operate in the same way. The neuronal (nerve cell) system and more fundamentally binary function are applicable on all soulatrophic nuclear levels and demonstrate the strong (gluonic logic) and weak nuclear force. The strong force (gluonic logic) is seen in the neural networks and the weak force as the expression force producing behavioural function through ribosomal hand eye coordination through pionic logic. This system can be used to understand atomic nucleonic systems, which are often too small to study properly as the entromorphic span is far too large (limited by uncertainty).

Cognitive genomics - appertaining to nucleonic organisation and expression; the complete nucleus is the cognitive genome or brain stem/heart system. Organisation is in the binary form of neuronal action potentials (proton) and absence of action potentials (neutron). Expression occurs by weak nuclear force (pionic logic) and the final nuclear products or proteinascious forms are behaviourally mediated in origin. Bosonic force carrying particles or neurons (dendritic and axonic) propagates the electromagnetic force. Again the binary nature of these systems follows Planck's constant and the nature of quanta although energy absorbed or emitted is spectral and more complex as a result and an increasing mass particle component. Electromagnetic components are diluted in this effect.

Cognitive bilateral soulatrophicity – the emergence of bilateral symmetry: segmentation from the somatic level and cephalisation, the formation of the head. The metabolic organelles emerge as the limbs (arms) which are the nitrogenous metabolic organelles, and the lungs which are the oxidative unhybridised (sp2) pi bonds. The halogenous is the dominant limb (writing arm) and the neogenous is the formation of the heart and brainstem. The brain (cerebral hemispheres) is part of the sigma bond with the somatic radial soulatrophic energy level. This is the metabolic organelle of the cognitive bilateral soulatrophic energy level. The brain stem/heart system is the cellular nucleus on this level.

The valance bonded metabolic organelles originate as food (nitrogen, oxygen, halogenous components) and shelter (a nitrogenous valance bonded organelle), but extend and separate into neutronic technology in humans; for example, petrol is an oxidative organelle on this level. Initially through the spiritual familial level where other humans have been orchestrated to make them as personal possessions with more of a cognitive level. Humans, through higher spiritual cephalisation have a complex range of valance bonded metabolic organelles (possessions). All other organisms display the basic valance bonded acquisitions of food, shelter and a mate on the cognitive bilateral level. The neogenous acquisition of a brain, the cerebral hemispherical neuronal super ganglion, is a measure of evolutionary hierarchy. The spiritual familial soulatrophic energy level sees the emergence of the super cell. Soulatrophic energy levels can be thought of as levels of consciousness.

The cognitive bilateral level displays a cellular environment limited by sight and sound. The environment spatial parameter is part of the organisms overall cell and mediated by the position of valance bonded metabolic organelles such as food. In the spiritual level the cellular limit is determined by the imagination. Hence the cellular topography extends beyond the measurable universe hence the emergence of the divine cell (Heisenberg's cell) and heaven and hell as components in most humans.

At this level the cell can be considered to extend to gigantic space. Humans can perceive atoms and galaxies; a massive spread of natural scale.

For humans this level sees a dramatic change to the valance bonded metabolic organelles. The carbosilicoferrous event communicates living energies to the rest of the periodic table in humans through silicon (Stone Age) to iron (Iron Age).

The metabolic organelles become technology and the emergence of the neutronic part of the soulatrophic evolutionary trinity. Nitrogenous elements are structures such as a car body. Oxidative elements are heat engines such as the combustion engine in a car. The halogenous constitutes complex energies such as lights and selective membranicity.

And the neogenous element is an on board computer (silicon nuclei) or simply another organism. Metabolic valance bonded organelles acquired through the spiritual level become cognitive organelles as personal possessions.

Cytomitosis – occurring on all soulatrophic energy levels it is mediated by the weak nuclear force (electroweak standard model) and pionic logic. On the somatic level cells divide into organisms driven by telomeric temporal constraints. On the cognitive bilateral and spiritual familial level it is observed in behavioural expression in organisms and communication (ECASSE). It is based on nuclear fission reactions seen in organisms at standard temperature and pressure.

For post carbosilicate interphase it is observed in the manufacturing industry for example, producing technological cells (products) by ribosomal technological expression, technology making more technology through the 'biology of technology'. A psychomeiotic fusion reaction on cognitive bilateral and spiritual familial soulatrophic levels (a thought) is expressed by ribosomal transcription, into proteinascious forms (products) such as the hands.

The ribosomal systems have a large and small subunit system seen as the opposable thumb model of the hand on the cognitive and spiritual level but also as protein subunit structures on the somatic and microbiology level. The head and mandible or jaw is the large and small ribosomal subunit system on the cognitive bilateral level in most organisms.

Telomeric DNA end caps are models of nuclear temporal decay in classical DNA cells and pionic logic. The telomeres are a model of the weak nuclear force interaction through the electroweak standard model on that level and decay into organisms from neutronic zygotes.

The decay of telomeres is a perfect model of radioactive decay on higher soulatrophic energy levels and is driven by the naperian logarithmic base e and π. Cytomitosis slows down and even stops when the telomere has decayed completely and can be thought of in terms of half life, when half the telomeres have decayed in a cell.

This is the basis of the metabolic cellular clock and the seat of any attempts to increase longevity in established organisms and the basic cell cycle. Telomerase enzymes are active in cancer cells and may lead the way to cures and increased life span.

Convection model of living animation and entromorphic elevation – a model of living animation and behaviour can be described as a convection current for solar animated carbon entrochiraloctet cells. The 'flame model' applies where carbon entrochiraloctet cells are animated in a convection field based on spin angular momentum means. In humans bipedalism 90° entromorphic elevation due to a complete planetary 'spingrav up' in sp2 planetary valance bonded metabolic organelles is balanced at night by a 'spingrav down' state hence the quantum nature of sleep. This is the prelude to planetary cytokinesis where humans leave the planet and go into the final phase of fermionic evolution through the vacuum of space (evaporative model of evolution). Sleep is the result of the gravitational entrochiraloctet where the land organisms are sp2 valance bonded metabolic organelles and therefore have two electron spingrav configurations in day and night.

The convection model of living animation is shown here.

At night the conscious wavefunction collapses into sleep (particle state), waking is a cytokinetic event as an organism separates from one day to the next. They begin to express conscious wavefunctions during the day hence the wave component of the organism, where the Sun is pumping free energy into the Earth macrostate. Overall any organism is in a particle/wave duality.

The following pages demonstrate the other component of animation that of the 'entromorphic elevation'. The elevation is a measure of organism's evolutionary hierarchy leading toward 90° animation and finally evacuation from planet Earth through planetary cytokinesis.

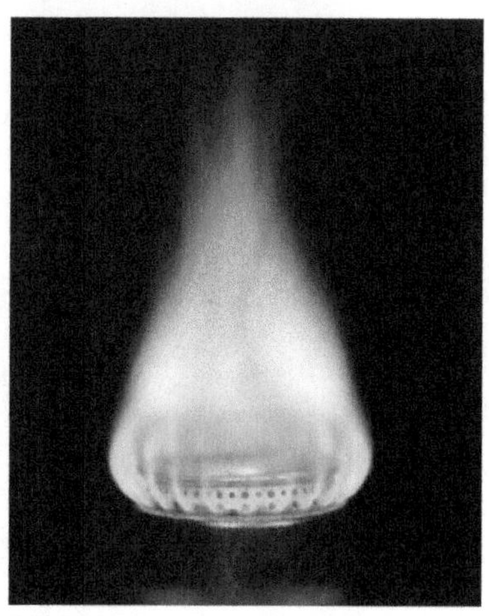

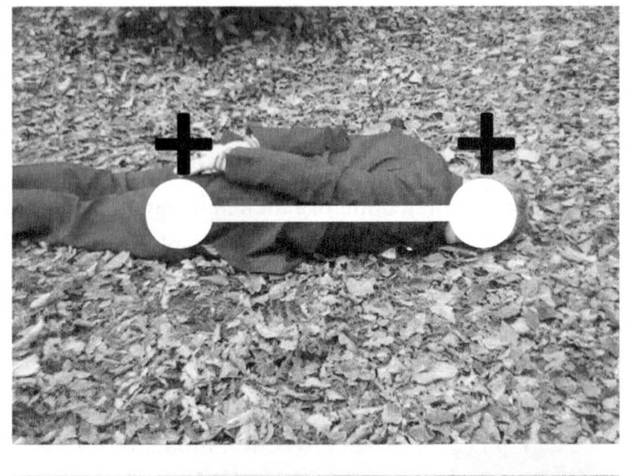

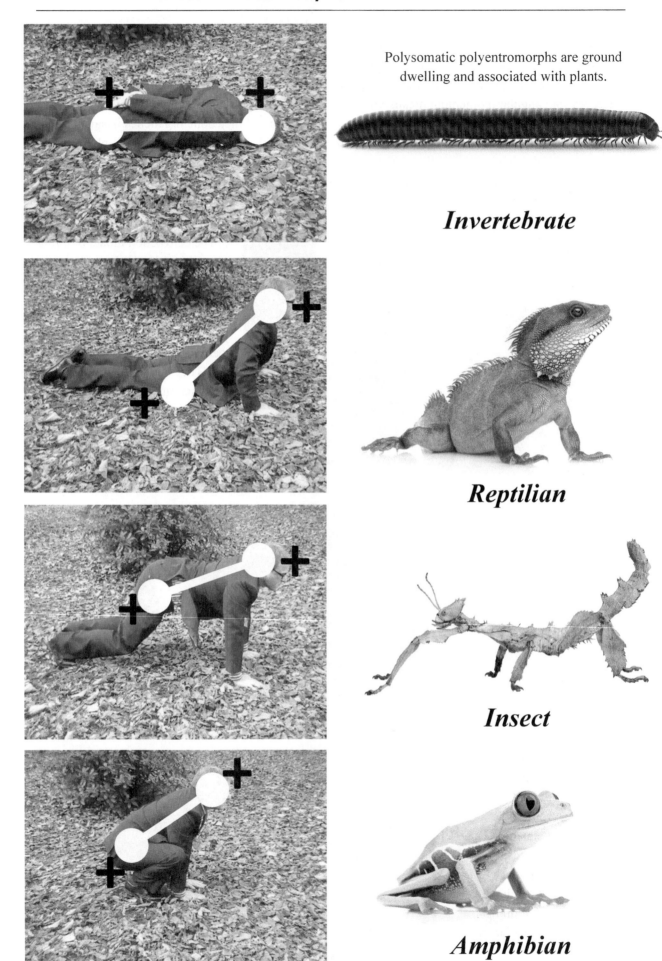

Polysomatic polyentromorphs are ground dwelling and associated with plants.

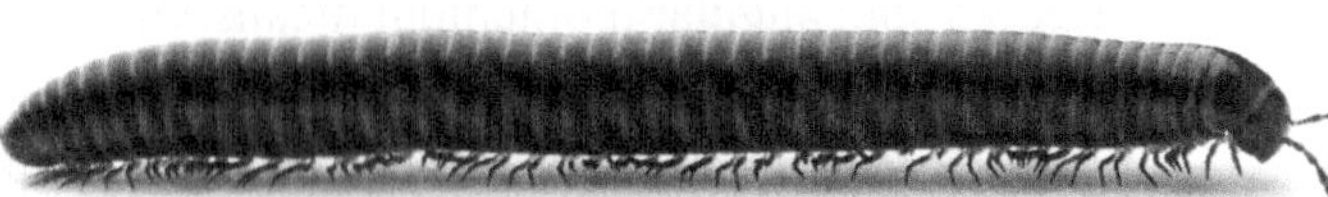

Invertebrate

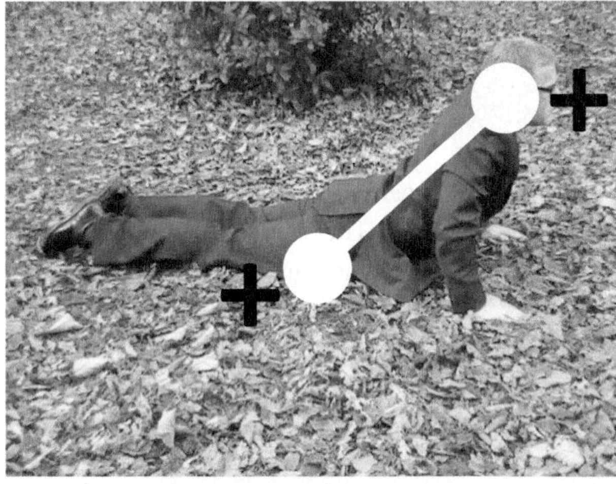

Reptilian

Insect

Amphibian

+
+
Mammalian

Mammalian

+
+
Australopithecus

+
+
Homo
sapiens

Genus
Homo
(The biology of
technology)
+
+
90º
The
'carbosilicoferrous
event'
bipedalism in genus
Australopithecus
and planetary
cytokinesis.

Cytomeiosis – where any two bosonic producers form a zygote, where potential nuclear energy is combined by living nuclear fusion such as sperm and egg. On the cognitive and spiritual level this is seen as the nature of thought, ideas through neurosperm (observation 1) and neurooocyte (observation 2) combination to prodcuce a new idea or thought. On the somatic level it produces a classical offspring. Because of nuclear amplification cytomeiosis extends to all the soulatrophic energy levels as a means of producing new nuclear systems by fusion zygotic meiotic reactions.

Hydrogen soul – the theoretical hydrogen atom at the beginning of any extended (Big Bang) soulatrophic pathway. Any higher soulatrophic energy levels can be defined in terms of super fermionics and super hydrogen particles. This is the basis of soulatrophonics and hydrogen entromorphology and the gender model.

Intension – a way of describing momentum (p) in living systems.

Love – The strong nuclear force (gluonic logic) as seen on higher soulatrophic energy levels. Possibly the base force, which expresses itself through the weak nuclear force as electromagnetic force and at large distances gravitational force (electroweak pionic leakage). This single force in nature expresses itself as the other three forces around distance from the source as components of the strong nuclear force. Nuclear physics to quantum mechanics to general relativistic mechanics. Due to the potential well effect, the extent of the love force is unknown only appearing through expression into electromagnetic quantum fields. 'I didn't know she had it in her (strong force gluonic and pionic logic) until she paid for the meal!' I thought she was tight with money (electromagnetic expression)! Love is the nuclear potential but because of the **'short range effect'** it must be expressed in order for others to make measurements on it, it is a deeply hidden force.

Lust – The love force (strong nuclear force) is expressed into electromagnetic fields (quantum mechanics), known as the body through pionic leakage into the electroweak standard model. Gastrointestinal systems and potential wells are based on the true nuclear region and the quintessential manifestation of the strong nuclear force. Lust is associated with 'expressed looks' an organisms body and face through the nature of beauty as expressed from the love force (nuclear hidden basis), which cannot be truly expressed due to it's narrow range. Beauty is skin deep! Is a model of the electromagnetic force and is described by quantum mechanics and the wave function through gauge bosons. A person is inherently uncertain in terms of the nuclear picture due to the limitations for the potential well and uncertainty (Planck's constant).

Microbial genes - Due to soulatrophic amplification, nuclear systems on different levels also display the same fundamental binary nuclear coding system amplified from protonic and neutronic (nucleonic) origins. This amplified nucleonic system can be used to understand atomic nucleonic states, which are often too small to study properly. This model takes into account the strong (DNA base pair coding) and weak nuclear forces and gluonic and pionic logic. The weak nuclear force is seen as the central complementary base-pairing binary AT and CG bond running down the centre of the double helix, and moderated by telomeric ends and pionic logic. The strong force is based on the phosphate sugar backbone system between nucleotides through gluonic logic. Genes can be thought of as gravitonic components, although the bosonic particle in DNA systems is mRNA (W+) and tRNA (W-) bosons and ribosome (Z^0 boson) and the expression is a proteinascious form, such as haemoglobin molecules. These links are based on self symmetries across natural organisation.

Microbial genomics - appertaining to nucleonic organisation and expression, the complete nucleus is the microbial genome a single circle of DNA (singular chromosomal). Organisation is in the binary form of DNA CG (proton) and DNA AT (neutron); single chromosomal system only. Expression occurs by weak nuclear force and the proteinascious forms are peptide (proteins) through gauge boson products. Bosonic force carrying particles mRNA (messenger ribose nucleic acid) propagate the electromagnetic force. The total genome contains many gravitonic genes. Often plasmids are found as smaller circles of DNA, which form the basis of primitive sexual reproduction on this energy level.

Somatic zygote the first multi cellular level the radially symmetrical organisms – A somatic thought which produces a new organism as the expression of the genomic thought (DNA thought). This is evolution running on a very slow clock. Neurological psychozygote form in real time as the evolutionary clock runs fast on neurological levels, DNA changes very very slowly. Somatic thought takes years to become effective; an organism must reach adulthood in order to reproduce. DNA is a very slow nucleonic evolutionary clock at atomic clocks run even slower as a result it is difficult to demonstrate evolutionary change as such change occurs over vast periods of time. Neurological evolutionary clocks run in real time so evolution can be demonstrated effectively on this level for neurological genetics. Nuclear fusion products are nuclear zygotes made out of chiasmatic DNA.

Pyschozygote – the nature of thought or sexual recombination on the cognitive bilateral and spiritual familial soulatrophic energy levels. A psychozygote is the product of nuclear fusion in the brain and brain stem. The brain stem evolutionary clock runs reasonably slow and psychogenes are found in the form of instincts (memories and the pychogenome). Instincts do change in an organism's lifetime but comparatively slowly to that of the spiritual level where recombination rates occur second by second. Neurological genetics identifies technology such as the Internet as real time evolution occurs in less than a second (think about computer processing speeds and RAM and ROM memory – silicogenomics and technogenesis). Evolution in the Internet occurs at frightening speeds and as such acts as an excellent demonstration of the principles of evolution by natural selection. This is an extension of classical DNA genetics to neurological genetics, genes are memories. The brain is a huge supercoiled chromosome which is why it looks the way it does with convolutions coiled into the tight structure of the brain. DNA chromosomes look similar, as do extended neurological chromosomes such as a book, a CD ROM, a video cassette they are all supercoiled neurological chromosomes; the centre is the centromere.

Nucleonic systems revisited for clarity – Nuclear systems in carbon soulatrophic entrochiraloctet cells based on binary information potential energy systems, proton-neutron, AT-GC (DNA), On/Off (neural), 1,0 (silicon) in soulatrophic pathways (organisms – carbon cells). Nucleonic systems form linear products of binary encoded stored information and consist of the strong nuclear force as the binding system and gluonic and pionic logic. The strong force gives way to the weak nuclear force, which is associated with genomic expression into organisms in carbon entromorphology through bosonic (gauge) ribosomal (W+,W-, Z^0 and photonic) proteinascious forms. On the nucleonic level where the components are DNA the strong force can be seen along the AT/CG sugar phosphate backbone (gluonic logic). The weak force is observed in the telomeric ends and the bonds running down the centre of the DNA molecule for A and T and C and G (pionic logic nuclear leakage). Expression of DNA occurs when the weak force allows the double helix to open up breaking the AT and CG bonds to be read by RNA polymerase.

In the same way a book is opened and read by the reader into neurological memories by the same logic for neurological levels. The telomeric repeating sequence at the ends of DNA decays throughout an organism's life in keeping with the radioactive decay model for the weak nuclear force. Eventually the weak force reaches its decaying limit and the DNA expression system starts to fail reaching a G0 senescent state. Cells get older and become susceptible to thermoentropic pathways leading to apoptosis and death. Telomerase, an enzyme, acts to reset the weak nuclear force by correcting the radioactive type decay processes making cells theoretically immortal. Sleep resets the neurological telomeric potential resulting in morning alertness; the drive to get through the day, your living motivation is telomeric (pionic logic). Telomerase is typically seen in cancer, although the genes associated with it are present in most organisms and may be the seat of increased longevity.

Where the nucleonic systems manifest themselves as neurological structures the binary encoding is based on the action potential. The hard-wired structure of neural networks is the basis of the strong nuclear force (gluonic logic). Behavioural output in organisms is the result of the weak nuclear force (pionic logic). An action potential is a strong quantitised (Planck's constant) positive charge. Again a binary event, you either have an action potential or you don't.

Fractional Dimensions & Soulatrophicity

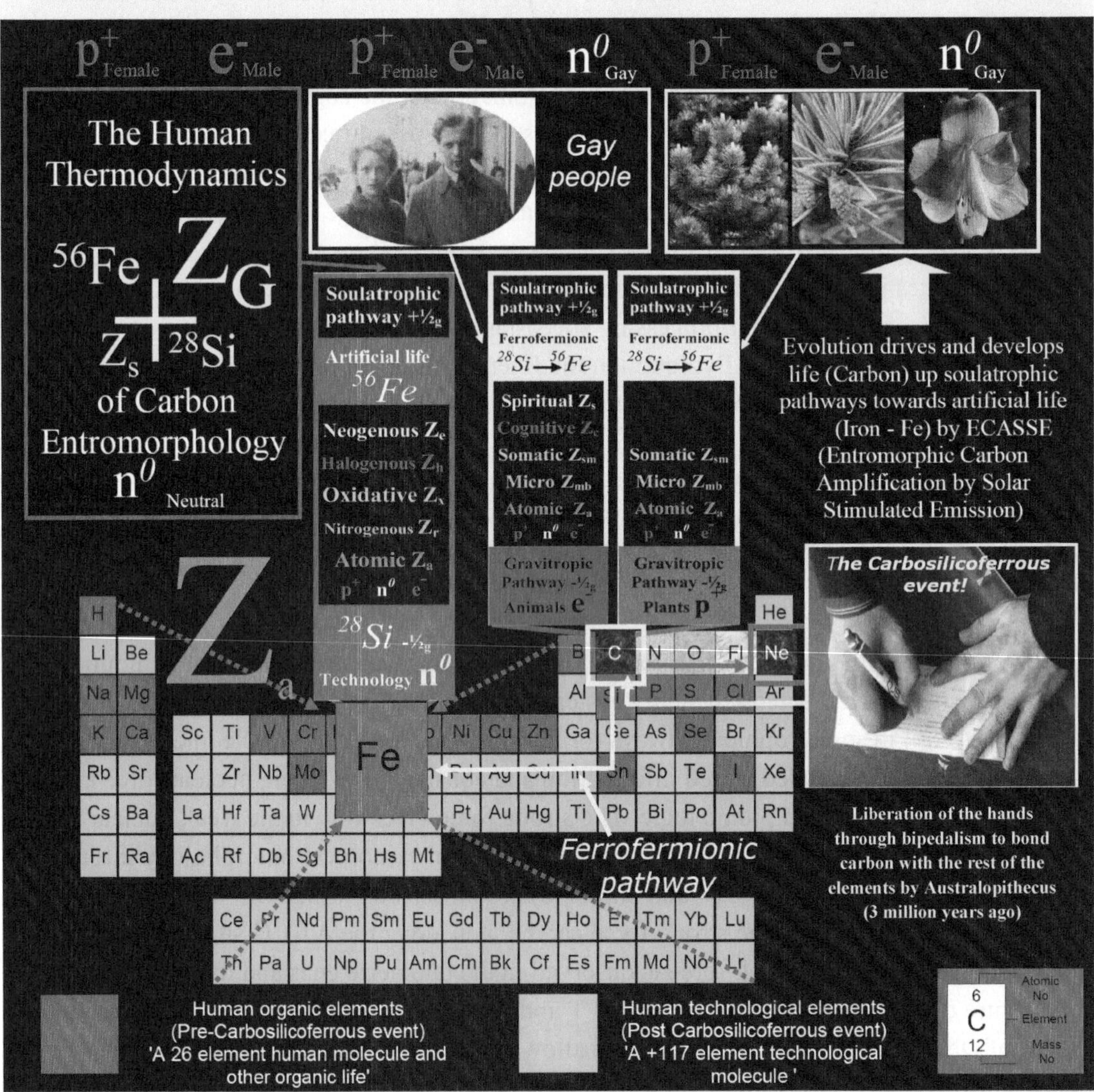

Life and its place on the periodic table (self-awareness through organoaccretion)

Post carbosilcoferrous event, the nucleonic systems are electronic silicon based semi conductive systems. Again this is the basis of a classical binary nucleonic encoding system. The strong force is observed in the storage systems, such as compact disks and hard drives (chromosomal super coiled gluonic systems such as a book). The weak force can be seen in transduction ribosomal systems (silicon behavioural systems) such as a monitor, or any collaboration with electronic metabolic organelles (electroweak standard model equivalents). Shannon entropy systems can be used to model information stability in transfer from the strong force to the weak force, and on to the electromagnetic force and gravity. All four forces are derived from the strong nuclear force; this force is derived into the other three by distance from the nuclear source. This is a unifying model of the four forces where only the strong force exists but displays different characteristics depending on the distance from the nucleus and from inner to outer space to where an observation is being made. There is evidence that there was only one force at the Big Bang, and it decayed into its current four force model.

The translational event between the strong nuclear force and three other required bosonic force-carrying particles (and pionic logic). In DNA this is seen in mRNA (W+), which shuttles between ribosomal proteinascious translation and the gene (strong force) of origin mRNA. In neurological nucleonic systems the bosonic force-carrying particle is the axon and dendritic portion of the neuron; it is also seen in the neurotransmitters and synapse and endocrine systems (hormones W+, W-, Z° and photonic). In electronic systems the force carrying particle is the optic fibre or and charge conductive mechanism. These systems run at close to the speed of light and as such act in a more typical fashion associated with bosonic particles without mass components. The DNA, neuronal systems act at very slow speeds and as such, are bosonic particles with mass components, intermediate particle/wave bosons.

Potential well – the strong and weak nuclear forces act over extremely small distances with gluonic and pionic logic. This limited range of action is called the potential well. Over this distance the forces are initially electromagnetic and at greater distance gravitational in origin. The four forces of nature are linked together so although the strong force acts over short distances it maintains its effect through its connection to the weak nuclear force which acts on all particles, to electromagnetic and finally in direct equilibrium to gravitational forces producing the states of matter.

The potential well in carbon entromorphology can be seen as potential well products in the same way as spingrav products, such as the gastrointestinal system, the absolute nucleus. The GI system develops first in organisms and houses the carbon atomic soulatrophic amplification origin **(the anus as a model of the singularity and the Big Bang are a most perculier model of the evolution of the universe, never the less it may be that strange distant echo)** bond portion of an organism's soulatrophic pathway. The potential well is a small local field that does extend across soulatrophic energy levels.The spiritual potential well exists when humans form carbon-to-carbon double covalent bonds along the unsaturated bond itself (a kiss). It is the elementary basis of the strong nuclear force expressed by pionic leakage into the electroweak standard model equivalent. The other three forces are distance-based expressions of the true fundamental force in nature the strong force.

Another example of the potential well is the vagina, the penis and any bosonic communication system such as speech where stored information is transmited to an observer. A man is driven to penetrate women because her potential well attracts him. This is seen in the powerful pull men feel to have sex with women on the somatic radial level, and the powerful desire for unity through marriage and long term commitment bonding by women through the cognitive bilateral level.

Shannon entropy - genomic communicative efficiency factor (binary nucleonic systems). The nuclear morphology seen in soulatrophic pathways produces binary encoding systems. These are seen as proton/neutron, DNA AT/CG, neuronal action potential On/Off, Silicon On/Off. Shannon entropy was developed around the spiritual silicon based computer technology. It is also applicable to the other nuclear forms as it relates to the bosonic transfer of potential nuclear energy from one soulatrophic pathway to another (e.g. somatic radial sexual reproduction).

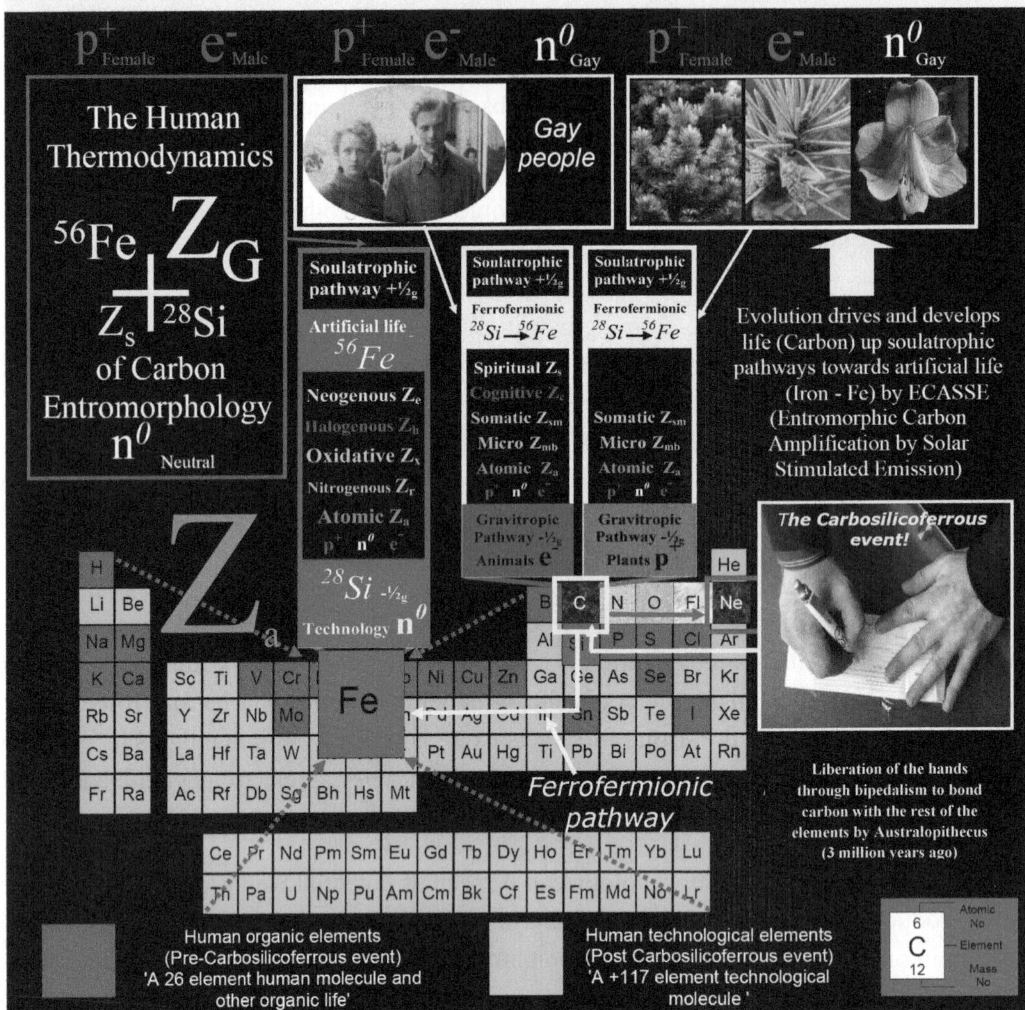

Somatic genes - due to soulatrophic amplification nuclear systems on different levels display different levels of detail but fundamentally operate in the same way. The DNA system and function, which are applicable on all soulatrophic nuclear levels. This system can be used to understand atomic nucleonic system, which is often too small to study properly. This model takes into account the strong (DNA base pair coding) and weak nuclear forces and gluonic and pionic logic. Nuclear expression fields – the central complementary base pairing bond running down the centre of the double helix and moderated by telomeric ends.

Somatic genomics - appertaining to nucleonic organisation and expression, the complete nucleus is the somatic genome. Organisation is in the binary form of DNA CG (proton) and DNA AT (neutron). Chromosomal systems forms higher level binary coding where an entire chromosome is either protonic or neutronic in origin. Expression occurs by weak nuclear force and the proteinascious forms are peptide (proteins). Bosonic force carrying particles mRNA (messenger ribose nucleic acid) propagate the electromagnetic force.

Spiritual genomics and neurological genetics - appertaining to nucleonic organisation and expression, the complete nucleus is the spiritual genome. Organisation is in the binary form of neuronal action potentials (proton) and absence of action potentials (neutron), the nucleonic equivalent. Expression occurs by weak nuclear force and the proteinascious forms are behavioural in origin. Bosonic force carrying particles or neurons (dendritic and axonic) also, as recorded information such as a book or TV show. They propagate the electromagnetic force and show nucleonic supercoiling. To read something is to use gauge bosonic logic and to follow ECASSE (nuclear and field amplification).

Spiritual familial genes - due to soulatrophic amplification nuclear systems on different levels display different levels of detail but fundamentally operate in the same way; the neuronal (nerve cell) system and function, which are applicable on all soulatrophic nuclear levels. This system can be used to understand atomic nucleonic systems, which is often too small to study properly. This model takes into account the strong (neural networks) and weak nuclear forces (action potential propagation) through expression fields. Spiritual genes are memories and new ideas (thoughts) occur by real time recombination. The evolutionary clock runs very very fast on this level, changes (evolution) occur second to second. On the somatic level DNA only recombines when an organism has reached adulthood, so evolution on this level is extremely slow by comparison to spiritual levels.

Silicogenes (neurogenes and neurogenome, neurological genetics) - due to soulatrophic amplification nuclear systems on different levels, display different levels of detail but fundamentally operate in the same way. The electronic digital system and function, which is applicable on all soulatrophic nuclear levels. This system can be used to understand atomic nucleons, which is often too small to study properly. This model takes into account the strong (read only memory ROM) and weak nuclear forces (electronic transduction) through optic fibres, within connections and electromagnetic devices.

Z+ - nucleonic binary systems in carbon soulatrophic entrochiraloctet cells and the genomic nuclear charge factor. Amplified nuclear regions following the soulatrophic ferrofermionic pathway; derived from a subatomic basis. The nucleonic systems start at the atomic level and progress via solar amplification by ferrofermionic and neobosonic pathways; from proton/neutron to nucleic acid AT/CG to neuronal action potential On/Off to silicon electronic On/Off binary systems.

Important – parts of this book are repeated often with slight changes to tone and meaning to produce a broad understanding. It is often the case that due to the enormous breadth of action in carbon entromorphology that the reader may feel perplexed or even lost. This is difficult to avoid and the large quantity of images in this book aim to try to graphically simplify the complexities for the reader. Due to the cross fertilisation of concepts from physics to biology it is important to keep referring to the images at the front of the book, thank you and please stick with it!

Neurochromosome human thermodynamics: a letter - an established field in science for carbon entromorphology to contribute in.

Date: 18th June 2010 (sent to a peer reviewer in the USA)

Subject: Inclusion of various aspects of carbon entromorphology for consideration in the field of human thermodynamics.

Dear ,

May I take this opportunity to repeat my thanks to your good self regarding this particular matter! Your contribution to my own professional development through the links you have created between my own scientific interpretations and models and those in the field of 'human thermodynamics' is truly exceptional. I will be creating a lecture on my feelings about my entry into this field, my excitement and your fantastic promotion of my work over the weekend (for YouTube I will email you when I have done this). In this lecture I intend to talk about you and the current concepts and issues on the EoHT (Encyclopaedia of Human Thermodynamics – EoHT.com).

Please keep all of the material I have sent you, I like nice artistic graphics with lots of photographs 'a photo or picture speaks a thousand words'. I prefer photographic data as they contain millions of data points. When I used to do physical chemical science I had to rely on often no more than 10 data points to create a calibration curve for example, and I always hated the limitations of physical blind interpretations of natural phenomena. But by the same token the mathematics was vastly more reliable due to the 'resolution problem' I have made about the whole of human knowledge.

Firstly please take stock of what I have sent you; you have an original first edition print of my 50-page paper entitled 'carbon entromorphology – absolute fundamentals' including spelling mistakes. This paper is organised in terms of the thermodynamics, in other words you will find gravitropism (Earth limited potential energy - evil) and thermoentropic pathways and the association with the Earth's restrictive effects on living thermodynamics at the beginning of the paper, as if the scene is started with the Earths formation with life in the form of simple carbon compounds (the Hadean Eon). Then there are pages on heliotropism as a model of thermodynamic free will and finally at the end of the paper are the concepts of phototropism (solar thermodynamics - good) divine entity (thermodynamically energy rich and free to liberate energy limitlessly through free will) or God like organisms which I believe are the God's of tomorrow namely the androids and robots, organisms that can live almost indefinitely. I believe that the Internet is the very essence of this revolution in energy availability as a humble westerner on planet Earth can and usually does use satellite technology every day through information technology. The Internet is a heavenly concept where all the answers to any question are provided in a split second.

In between are the basic properties of living organisation based on my soulatrophic pathways based on the physical particle trinity, protonic, electronic and neutronic entities as the basis of all gender logic and the centralised and conserved nature of carbon often termed as carbon centric logic (in EoHT). This particle pathway gives evolution theory a complete thermodynamic quantum related model of life with discrete energy levels driven by thermodynamic solar acquisition (good) or solar liberation (evil), with free will heliotropism as a model of thermodynamic free will.

The drivers of the thermodynamic growth phototropism (good) lead to higher levels of soulatrophicity with more thermodynamic negative entropy and greater enthalpic and Gibbs free energy.

Soultrophicity is a fractional dimension of natural living organisation based on fractal geometry. It identifies that there are many natural levels of scale (four force equilibrium describes the difference in the different levels) in nature for example, sub atomic, atomic, molecular, cellular, organismal, family, group, company, university, nationality, global. In each case the organisation is based on and understood in terms of fundamental physics, central nuclear determinism and field theory as an expression of nuclear potential. As we transverse this profile of conscious organisation we find a shift in the equilibrium formed by the four forces of nature. On atomic levels electromagnetic effects proliferate determinism but as matter bonds to form larger 'inertial reference frames' there is a distinctive and seamless translation of determinism from quantum rules to general relativistic gravitational mass determinism, with life as a special case which is an integration of all the levels of natural scale forming my nucleonic and soulatrophic pathways. This is the nature of quantum gravity in association with the conscious observer making measurements on this fractional treatment of natural order. Spontaneous thermodynamics are related to evolution although evolution theory currently makes no link between the thermodynamics, energy cycling or any physical fundamental model of life which is counterproductive and makes evolution a concept without true rules.

It's also important in the context of thermodynamic spontaneity to consider that because my systems conserve and centralise carbon as the main stay amplified logic (part of the 26 element human molecule) that the octet rule and the iron nuclear instability model can be used to understand how energy flows through living systems and either accumulates or is liberated through Gibbs free energy. Growth and death can be reduced to thermodynamic systems and their surroundings. Also entropy needs to be clearly identified as waste; most people don't understand it in any other way. You may have noticed that the symbol S for entropy I make into the word Shit (waste). I have a huge section in my exhibition which details examples of thermoentropic pathways such as death, murder, poverty, dilapidation, explosion, disease, AIDS, cancer, rubbish, pain, gravitropism or disorder of any description. I have about 50 photos relating to human thermoentropy. I also a have another system of again about 50 photos relating to thermodynamic growth such as new birth, winning, wealth, prosperity, growth, happiness, love, lust, good health, phototropism and solar acquisition and accumulation. It makes for an interesting viewing and again is presented in pathways with thermoentropic concepts first leading on towards Gibbs enthalpic concepts as a temporal model.

One of the problems I have is in considering the question, which is actually unanswerable, 'is the universe a closed system?' Obviously classical interpretation of the second law of thermodynamics suggests entropy is increasing over time in closed systems. Clearly life on Earth has an input of energy and therefore work done on living systems but the inevitability of disorder or the challenge to the second law which some people claim is again very difficult to resolve. As the universe flies apart and becomes diluted through electromagnetic means, it is also being focused by gravity producing temporal order. The electromagnetic components of this logic produce disorder as particle/wave systems deteriorate to simpler logic, but are focused and accumulated by gravity to produce order; this makes thermodynamics on a universal level a little tricky.

There is the periodic table model of all life on Earth, which I believe is the most appropriate model for considering ALL natural organisation. In carbon entromorphology the very fact that humans have the periodic table and the fact that we are currently up to element 117 (atomic number) where many larger elements are not naturally occurring is absolute proof of the integrated logic in human thermodynamics. You refer to the 26-element organism, if you look carefully at my interpretation you will find a 117-element organism; on the diagram the red elements are part of your 26-element human molecule model. The problem is, as I have already said, that humans and therefore life in general have interacted with 117 elements and therefore your 'living or human molecule' is too small to fully describe life completely. Technological systems are an extension to life and a mobile phone is therefore an organism as well but comprised of elements outside the classical 26. The 26-element logic is typically for classical organic life, but as previously mentioned, humans interact and therefore bond with 117 elements. The difference is ‘technology’; we always misunderstood the implication of technology in the biological sciences. It’s typically considered in anthropology and archaeology (hence my suggestion of human thermodynamics archaeology paper later in the year).

Fractional Dimensions & Soulatrophicity

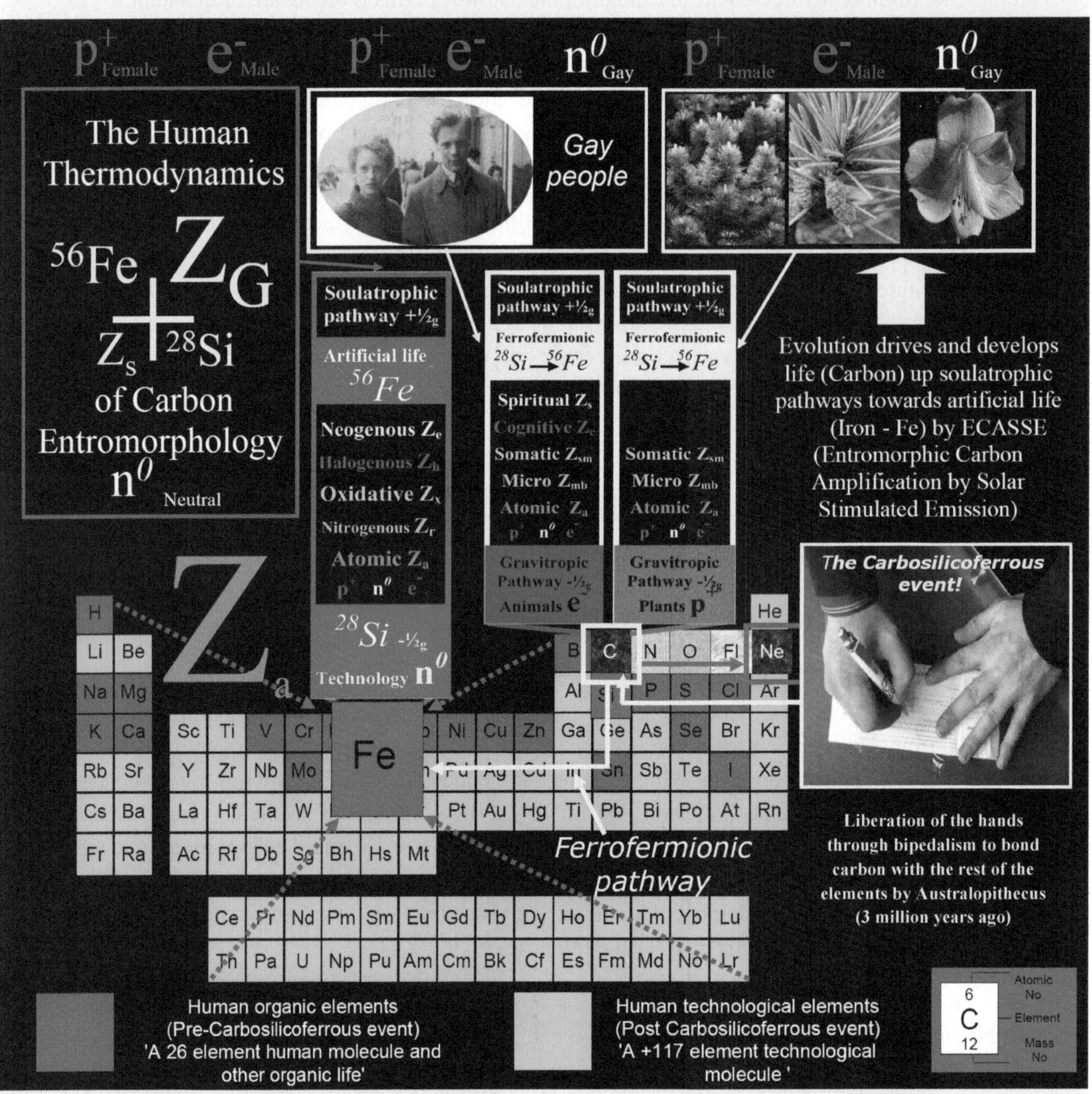

Life and its place on the periodic table (self-awareness through organoaccretion)

The 'biology of technology' is unique to my interpretations and states that of the 117 elements currently associated through bonding to humans the 26 in the human molecule logic only considers organic life and needs extending to all 117 through the extension of carbon living organisation for a truly complete picture of all life in the universe.

Your 26 element human molecule is an excellent interpretation but definitely needs extending further. I may find as I delve into your work that you have considered this already. Also the order and contribution to physical determinism is absolutely critical, for example I would say hydrogen is the most important, then carbon, nitrogen, oxygen and the halogens then silicon and iron. There are reasons for this but what is absolutely undeniable is the centralising determinism and conservation of such logic from carbon, hence carbon entromorphology, you have four limbs and a torso and a head.

Carbon has four valances, a K shell and a nucleus, this is undeniable so I would agree that all 26 undergo amplified effects but they are typically in place to extend carbon centric logic. It's important to realise that the insects and the plants are polyentromorphs hence they have many legs or many units; this can be explained in a very powerful way as 'post nuclear linear morphology'. Carbon remains absolutely central in all its compounds hence chiral centred chemistry, the other elements exist simply to extend that logic but it is still carbon which is fundamentally determining organisation and function. We call ourselves carbon based; even thought there is more hydrogen and oxygen than carbon? Why, because carbon is the most versatile of all the elements with the broadest spectrum of physical states and the highest rate of recombination's but again with carbon at the absolute heart. Carbon can form four sharing covalent bonds with trillions upon trillions of other atoms in accordance with your 26-element logic; each of the four bonds can have different strengths hence chirality. Think about it if you pick up ANY chemistry book and look up chirality you will see the word chiral means hand in Greek. This is because your hands are perfect examples of the asymmetry of chiral carbon; hence this is proof of my theory because your hands are chiral (left and right hand forms) which is what you would expect from a gigantic carbon atom? Powerful proof of the theory of nuclear and field amplification, and the conserved nature of carbon for human molecules or human atoms.

Also it's very important to note the way in which such super molecules or super atoms come into being and the organisation it brings with it. The current models of biological science have no basis in physics, in other words concepts such as fermionics, bosonics, quantum mechanics, nuclear physics, thermodynamics; special and general relativity all need identification in biological models but taking into account biological logic such as genetics for cross fertilisation of ideas and fluidity of common logic. I feel that human thermodynamics is the way forward for doing this, and it seems according to the EoHT that others have seen this potential. There is an instant problem with this because the logical axioms of physics are mathematical and numerical measurements are critical. Living organisms are too complex and measuring initial physical properties and conditions to allow differential equations to be used effectively becomes impossible because living organisation is resolvably 'too complex'. Observational measurements are however very reliable and the biologist can look at living systems with sight to see their organisation and physics without the need for critical numerical and mathematical treatment. So what we need to do is to compromise on the axiomic basis of high-level human thermodynamics. The powerful equations of physical science are an incredible achievement and their extrapolative and interpolative abilities cannot be overlooked. I believe through human thermodynamics that we can still use the powerful and reliable mathematical logic of physics to give the biologist a powerful physical model of living systems but with an inability to solve such equations due to biological measurement variation and uncertainty. Human thermodynamics would do well to centralise the problems of the limitations of science and determinism founded by a limitation in science through what I call the 'resolution problem'. In the resolution problem a physicist studies very small particles such as atoms which are impossible to see and comprehend on that level and also very large particles such as a star are often so far away that again they become unresolvable.

This simplifies these systems to such an extent that mathematical measurement is effective and very usable. The biologist can see a structure such as a heart but it lacks basic symmetry and is impossible to sum up using conventional calculus because we can see its heterogeneous complexity but our observational measurements are perfectly sufficient for understanding and elaborating on such a system. This spread of logic throughout the whole of science is the 'resolution problem' where particles are very small or so far away that they appear small enables numerical measurements to be made mathematically, cells are too complex for this treatment. Hence the physical sciences and biological sciences have no logical bridge and non-harmony, which is counterproductive to humanity. We need to almost lose the unique niche areas in science such as physicists, chemists, biologists, psychologists, sociologists, geologists, etc and we need to be identifying general scientists aware of the way all natural levels of scale come together. Human thermodynamics I believe can produce this result where a classical physicist and a biochemist regularly communicate with each other because they represent conserved logic at different levels of natural scale.

I must say in all honesty that my entry into human thermodynamics is very exciting. I believe very firmly that this subject has the potential to be the biggest of them all as it looks to understand the conscious mind and the nature of all human knowledge in a unified management field. Human thermodynamics I believe is an umbrella system, which aims to bring a semblance of order to 'non-physically fundamental' biological sciences. I feel we are on the threshold of incredible understanding in the whole of human history. I believe my theories driven by many aspects of physical science and limits in biology in conjunction with the contributions made by excellent scientists and philosophers such as your self will make human thermodynamics the most powerful knowledge management system of all time and I look forward to driving these incredible ideas forward towards a beautiful future for humanity. We are on the cusp of integration and cooperation all the way through science and philosophy. We have a system that embraces the power of physical science and mathematics with the broad philosophical and spiritual concepts of the arts.

The physicists typically define a particle as associated with atoms. You have already extended this logic, as have I, to describe molecules, cells, organisms, families, nations, and global particles. Although you do choose the level of molecules as being fundamental where I believe it is just one of many particle/wave systems. Again the physicists limit the philosophy and in theory in accordance with the Copenhagen interpretation of quantum mechanics we should be referring to any inertial reference frame as a particle/wave duality regardless of conscious scale. When I talk about the first law of thermodynamics I describe it in accordance with the duality, hence I refer to the 'conservation of mass and energy' not just energy. Relativistic logic about the complementary nature of energy and mass with the condensation factor produced by the square of the velocity of light is so central to all physical logic that the first law needs updating to reflect it fully.

I have already started working on the paper for the 'Journal of Human Thermodynamics'. I work very fast and will have something with you within a week or so. I will most definitely utilise equations as much as possible but my graphics do convey a lot of information and I intend to use them liberally although there will a significant commentary as well. I intend to try to stylise the paper again using the thermodynamics to give order to the paper but also to describe my particle mechanics and evolutionary pathways to model evolution and life from fundamental physics, together with energy distributions, physical evolution through quantum gravitational principles, gender, the four forces of nature and the thermodynamics of morality and the thermodynamic drivers of the octet rule and the iron nuclear stability model of elementary evolution. I also intend to include cross fertilisation of ideas especially the extension of classical genetics and evolution to cover all levels of natural scale and the concept of neurological genetics and atomic genetics, which I believe is very powerful.

So all in all it should be an interesting paper, I will also bring in as much cross citation to established theories as I can. I have a friend who will perform a preliminary peer review before I send it on to you.

Again may I take this opportunity of thanking you once again, I look forward to working with you and I am confident with the websites, lectures and contributions from many areas of human thermodynamics that we will be carving the future of human evolution with this massive super concept. The expression I would use is a 'paradigm shift' where human thermodynamics extends sound scientific logic but also realises the importance of spiritualism and philosophy.

The following is a very important concept in human thermodynamics and will form the main theory of carbon entromorphology, as it aims to explain how thermodynamic amplification can occur in living systems, based on LASER logic.

Some time ago a very senior physicist said that he could not understand how nuclear and field amplification could occur in living systems. What was needed was a powerful logic, which enabled life and carbon to undergo amplification of their physical properties to produce larger and larger in-phase structures based on negative Gibbs free energy and its affinity for itself and the rest of the elements. What was required was an amplification logic, which would explain how carbon can combine with a vast array of other elements to extend through bonding and links their fundamental physical properties. We know that carbon remains central in all of its interactions and that a carbon nucleus is behind all living logic. We can argue that through chiral centres carbon acts as a natural physical amplification template producing huge atomic systems, which appear to have all the physical characteristics and anatomy of conserved carbon but on a vastly amplified scale. The variation in bonds in chiral carbon is another unique property, which enables carbon to form bonds or links of different strengths. This increases the number of permutations and combinations in carbon, which produce all the natural variation in the living world through the logic of organic chemistry.

What was still required was an understanding of how life through carbon can fix solar energy into its vast array of structures from atoms to molecules to cells to tissues to organs to organisms etc?

Carbon in its atomic form has a unique property in that it has an even number of nucleons in its nucleus. What this means is that carbon as well as being a fermion in other words having physical properties (which are particles which are affected by the strong force) but also has bosonic properties which are associated with the gauge bosons, the W+, W-, Z^0 and photonic γ bosons and the presence of gravitational bosons such as gravitons and Higgs bosons which have never been observed. The bosons follow Bose Einstein statistics, which suggest that a boson is more likely to be in the same physical state as other bosons like it. In other words carbon through organic chemistry sees carbon actively thermodynamically tending to bond and link and concentrate carbon into living systems. Carbon is also a fermion a hadron and so the bosonic statistics must also insist that carbon cannot act directly like photonic bosons where photons are quite literally in the same state as each other and that the probability of them being together is increasingly likely. Carbon therefore forms inertial reference frames such as cells and organisms where both the fermionic and bosonic statistics are representative of their duel thermodynamic potentials.

For bosonic statistics and particles Einstein first postulated a property called stimulated emission that photon acting on a particle could excite the release of other photons producing amplification. This logic is called Bose condensation and results in amplified in-phase coherence commonly known as LASER (Light Amplification by Stimulated Emission of Radiation). If carbon is a boson then it must have the potential to condense its properties through in-phase coherence to produce a beam of carbon-based particles all acting with common inertia. The fermionic properties mean the carbon atoms can bond to others and share physical states as an integrated result hence giving us the field of organic chemistry and the life sciences. This means that we have a logic, which demonstrates why carbon has so much affinity for itself and up to 26 atoms in total for conventional organic life, but up to 117 atoms for a complete model of life including technology. The in-phase coherence produces living ECASSE beams (condensed atomic inertial frames) where the input energy is solar and the gain medium, which allows Bose condensate logic to operate in carbon, which is a particle/wave duality, or fermion/boson duality logic to exist. This enables us to thermodynamically understand how carbon can amplify its properties to produce all of the large biological particles which all have in-phase coherence. Think of a flock of birds, or a line ants or a football team trying to win a match. They all drive forward through ECASSE logic to produce these in-phase beams such as Manchester United's team, which works as an in-phase ECASSE beam.

Staggeringly enough this logic also enables us to beautifully demonstrate and prove this ECASSE logic. The photograph on the web site page is of a human's arm stretching out, with a small LASER pointer in the hand. The arm and hand are excellent examples of amplified bosons in carbon entromorphology. The arm itself operates as an antagonistic force carrying particle or boson with the logic of the W+ and W- antagonism; the neurological extension is the force carrying component carrying neurological genes for bosonic expression. The hand is the Z^0 particle, which is the gauge boson, associated with ribosomism in biology; the ribosome is the translator of genetic information into expressed particles. It contains a large (hand) and small subunit (opposable thumb), and interfaces with the release of photonic gauge bosons. The LASER pointer is activated by the hand (Z° boson) through force carrying particles through the antagonism of the arm (W+ and W-). The LASER is classical evidence of Bose condensation but what is fascinating is that both the LASER pointer and its activation to produce an in-phase coherent beam is actually the result of solar energy. Solar energy fixing produces all consequences of life. So this demonstration is a perfect proof of the theory of ECASSE as a means of explaining how nuclear and field amplification in living systems works. The fixed beam is the direct result of solar energy fixing and Bose condensate and therefore carbon based amplification which is the thermodynamic mechanism driving the process of all life on Earth.

This is excellent and totally undeniable evidence of how carbon a boson and fermion can produce amplified physical states such as cells by Bose condensate solar energy fixing. And also it's a fantastic demonstration of the gauge bosons and the consequences of amplified versions of this atomic logic in living organisms. The very existence of the LASER in physics has come into being because the physicists working on the development and the manufacturers all do so through fixing solar energy. The energy of any LASER beam has come directly from the Sun itself and has been amplified by Bose condensate logic, a perfect proof of this fundamental driving mechanism in all of life and satisfactorily answers the physicist's confusion as to how gigantic entromorphic carbon atoms such as humans can come into being.

Finally I also want to mention Boltzmann entropy and statistical mechanics. In statistical mechanics Boltzmann defined entropy as the frequency of particular ensembles of microstates in the course of their thermal fluctuations. The more symmetrical the particle systems, the lower the entropy of the system, the greater the frequency of particular ensembles of microstates. If we regard a microstate as any inertial reference frame in living organisms such as atoms, molecules, cells, organisms we see my fractional dimension soulatrophic pathway demonstrates entropic distributions. Consider the symmetries of the genital regions to those of the head and face. The genitals are simpler thermodynamic systems and have low frequencies of microstates and low symmetry, where as the face has low entropy with high levels of symmetry present at that particular level. This is very useful for the human thermodynamicist as it enables us to treat ensembles of living microstates such as a group of people in terms of symmetry and homogeneity. Take an ensemble of football supporters all chanting in-phase songs about their team. They are very symmetrical and represent low entropy states, imagine this scene suddenly turning violent there is massive symmetry breaking in the particle groups reducing the homogeneity to a heterogeneous out of phase incoherent mess. Boltzmann entropy is very useful for understanding human or any living thermodynamics in living particle systems based on symmetry and in-phase or out of phase coherence.

Beauty and ugliness are thermodynamic consequences in living ensembles of microstates are easily understood by Boltzmann entropy. Beauty for example demonstrates high levels of symmetry such as the perfection of a child's face, very symmetrical homogeneity in the complexion, lack of lines, wrinkles and blemishes and high levels of symmetry and low frequencies of microstates and chaos. Compare this to the huge numbers of microstates seen in a 92-year-old person, there is a collapse of symmetry and the difference between them is a perfect example of entromorphology. The entro part based on entropy and the morphology allows us to see the emergence of waste of entropy through the morphology and breaking of such symmetries. Therefore in human thermodynamics we can now consider Boltzmann entropy and other aspects of statistical mechanics to understand beauty (symmetry with high frequencies of particular ensembles of living microstates) and ugliness (symmetry break down with low frequencies of particular ensembles of microstates), in the course of their thermal fluctuations.

Your set up is very impressive and your contribution to this field outstanding, I do still believe that we are on the threshold of a unifying logic for ALL knowledge. Do remember that in carbon entromorphology the text you are reading is a 'super coiled chromosome', the words are neurological genes and the entire page is the chromosome. Hence the actual words follow pionic and gluonic quark logic and the article itself is the very essence of nucleonic nuclear organisation. When you open it up and read it you do so using the bosonic gauge particles such as hands, arms, eyes etc this is the quantum mechanics of reading and is the same process which allows DNA to be opened and read into proteins in every cell in your body.

Any way I think that's enough for now, have a think about all of this and I will be in touch very soon.

Best wishes,

Mark Andrew Janes BSc (Hons) CBiol MSB

Neurochromosome ECASSE – entromorphic carbon amplification by solar stimulated emission (The Bose condensate thermodynamic logic driving life on Earth and nuclear and field amplification).

Some time ago a very senior physicist said that he could not understand how nuclear and field amplification could occur in living systems. What was needed was a powerful logic, which enabled life and carbon to undergo amplification of their physical properties to produce larger and larger in-phase structures. What was required was an amplification logic, which would explain how carbon can combine with a vast array of other elements to extend through bonding and links their fundamental physical properties.

Where, thermodynamically, the spontaneity of the affinity between carbon and other elements follows the following logic: -

We know that carbon remains central in all of its interactions and that a carbon nucleus is behind all living logic. We can argue that through chiral centres carbon acts as a natural physical amplification template producing huge atomic systems, which appear to have all the physical characteristics and anatomy of conserved carbon but on a vastly amplified scale. The variation in bonds in chiral carbon is another unique property, which enables carbon to form bonds or links of different strengths. This increases the number of permutations and combinations in carbon, which produce all the natural variation in the living world through the logic of organic chemistry.

What was still required was an understanding of how life through carbon can fix solar energy into its vast array of structures from atoms to molecules to cells to tissues to organs to organisms etc?

Carbon in its atomic form has a unique property in that it has an even number of nucleons in its nucleus. What this means is that carbon as well as being a fermion in other words having physical properties which are particles which strongly interact but also has bosonic properties which are associated with the gauge bosons, the W+, W-, Z^0 and photonic γ bosons and the presence of gravitational bosons such as gravitons and Higgs bosons which have never been observed. The bosons follow Bose Einstein statistics which suggest that a boson is more likely to be in the same physical state as other bosons like it. In other words carbon through organic chemistry sees carbon actively thermodynamically tending to bond and link and concentrate carbon into living systems. Carbon is also a fermion (a hadron) and so the bosonic statistics must also insist that carbon cannot act directly like photonic bosons where photons are quite literally in the same state as each other and that the probability of them being together is increasingly likely. Carbon therefore forms inertial reference frames such as cells and organisms where both the fermionic and bosonic statistics are representative.

Neurochromosome periodic table: soulatrophicity.

Carbon entromorphology fits beautifully on the periodic table; it demonstrates the evolution of carbon to iron nuclear stability (artificial life), through the soulatrophic trinity protonic pathway (plants), electronic pathway (animals) and neutronic pathway (technology).

The basis of all life is the periodic table of the elements, all carbon entromorphology logic fits and functions on this table; it is the basis for all life as we know it. The gender trinity protonic plants soulatrophic pathway, the electronic animal's soulatrophic pathway and finally the neutronic technological soulatrophic pathway are shown in their true place relative to all other elementary atoms.

To begin with, the periodic table is a powerful organisation of all the basic elements and their organisation is related to their physics. Not only do we see a clear demonstration of the long-term effects of 'hydrogen entromorphology' through the evolution of the elements through nucleogenesis but we also see how they react relative to each other.

The fact that we have a periodic table is testament to the gregarious reactivity of carbon-based life. Before humans stood upright carbon based life only associated itself with part of the periodic table namely the regions associated with organic chemistry (red boxes on the table diagram). When humans freed the hands during the carbosilicoferrous event, where genus Australopithecus became bipedal, the hands were then free to make association with the rest of the natural elements. We see two very powerful new associations through carbon with silicon which has a similar chemistry to carbon and versatility to carbon, hence the Stone-Age. But also, the overall driver for all elementary evolution through iron, which is the most stable nuclear element of all.

Fractional Dimensions & Soulatrophicity

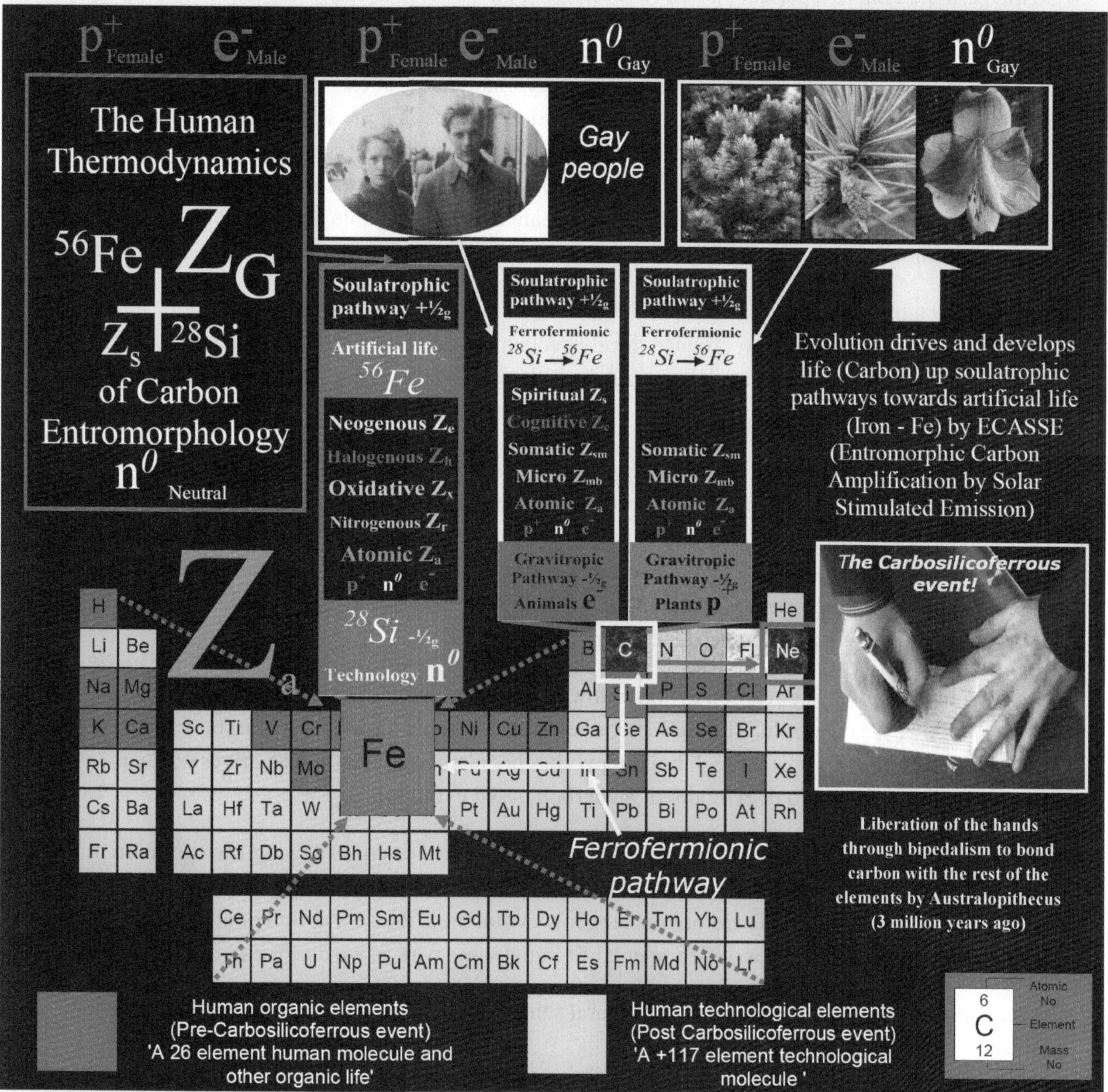

Life and its place on the periodic table (self-awareness through organoaccretion)

It has been said that all the other elements behaviour and physical potential is based on iron as a stable end point to elementary evolution. Hence we have the Iron Age as well, and with both the Stone Age and with the Iron Age we see the emergence of the technological organisms through technogenesis.

This emerging artificial life is a natural extension of carbon based life and so we find technology designed to fit as an extension to carbon nuclear and field physics. Hence robots are designed to look like carbon; hence a robot has a head, torso and four limbs, and two eyes. This is an excellent way of demonstrating the seamless extension of carbon ECASSE (Entromorphic Carbon Amplification by Solar Stimulated Emission). Humans have taken this even further through the lanthanides and actinides where some elements are artificially manufactured and may only exist for a short time before decaying into stable forms.

So the periodic table is a perfect canvas for describing all universal activity after the Big Bang.

So carbon entromorphology and its soulatrophic (fractional dimensional) model of all living evolution fits beautifully on the periodic table. Since this theory maintains that carbon, and even more fundamentally hydrogen, are amplified through bonding means, that the soulatrophic pathways fit on carbon for the animals and plants but extend to ALL other elements both naturally occurring and artificially occurring. Hence the final soulatrophic pathway that of the technology (neutronic pathway) fits onto iron. All three soulatrophic pathways nucleate the periodic table and integrate its logic into carbon entromorphology and finally silicoferrous entromorphology through technological life.

The periodic table started at hydrogen some 13.7 billion years ago and it's taken almost 12 billion years of elementary evolution to produce the elements most associated with life such as carbon. Carbon is formed at the end of a nuclear reaction and starts the emergence of soulatrophic growth after the Earth formed 4.5 billion years ago.

Because the elements are made out of hydrogen we have the hydrogen entromorphological system. This process is also a nuclear and field amplification process as the elements are organised according to size. Hydrogen is the lightest and uranium the heaviest naturally occurring elements, and iron being the most nuclear stable element. Hence hydrogen entromorphology or nucleogenesis is also a nuclear and field amplification process. Carbon entromorphology is the electroweak extension through covalent bonding of classical nucleogenesis.

The potential in carbon sees carbon evolving towards a neon Noble gas electronic configuration, where it follows the octet rule. This defines carbon-based organic chemistry as carbon tries to stabilise its 2nd energy level (L shell) through bonding with other elements. This in turn drives nuclear potentials towards iron as a duel model of all carbon based evolution.

Evolution therefore also fits on the periodic table although physics doesn't recognise evolution in this way. Evolution is not exclusive to organic life, it is the fundamental process of natural selection which has defined all the elements. Therefore nucleogenic evolution is the description of all activities naturally selecting elements for environmental fit in the same way as we do for carbon-based systems.

This extension to Darwinian evolution by natural selection is essential for an accurate model of the universe.

The Octet cycle in carbon is the home of the classical cell cycle in carbon-based life. The nitrogenous G1 interface is followed by the oxidative S interphase and then the halogenous G2 interphase and finally neogenous mitotic stage where quantum stability is achieved. In carbon this is not the end of the story as any quantum instability is derived from the potential in the nucleus, so carbons nuclear potential drives quantum bonding to reduce its energy level towards iron. In living organisms this conclusive end point to elementary evolution is seen in the concept of God. A perfect super stable iron based logic producing the androids, robots and virtual organisms as a goal to evolution by natural selection. This makes the periodic table the most profound all-inclusive model of consciousness.

I therefore commit the existence of the periodic table its initial years of nucleogenesis or hydrogen entromorphology to emerging carbon entromorphology and finally where life (carbon) identifies and therefore reacts with, the rest of the table through silicoferrous entromorphology as powerful fundamental evidence of these new theories. The concept of a cell is dramatically extended in carbon entromorphology, it is defined against a temporal parameter as well as size, an entire human is a cell, they are fractionally dimensioned into classical cells etc.

Neurochromosome soulatrophicity: the nature of the conscious soul and mental soulatrophicity.

The carbon soul – the theoretical initial carbon at the beginning of any soulatrophic (from the word soul meaning origin) pathway; although soulatrophic pathways extend back to the Big Bang through hydrogen entromorphology. Soulatrophonics is the basis of a soulatrophic pathway and can be any level (e.g. atomic) relative to higher levels (e.g. cognitive) based on the potential of an infinitely complex soulatrophonic inner space pathway. When soulatrophic consciousness moves up and down levels in accordance with quantum rules they must either absorb or emit energy (or its mass equivalence); in soulatrophic systems these are massive quantum jumps (such as the orgasm) producing spectral profiles. This is the basis of quantum heredity, and includes any entroejaculatory process such as speech. The carbon soul gives way to the hydrogen soul and then to fermionic particles and so on to the limits of nuclear physics and humanities knowledge. The NFVE (Neobosonic Ferrofermionic Vectorial Evolution) model of life and ECASSE (Entromorphic Carbon Amplification by Solar Stimulated Emission) suggests that there is a conservation of any of the soulatrophic pathways. In other words, any higher levels can be described in terms of the organisation of lower levels, as the higher level is an amplification of lower ones hence with DNA the nucleonic logic on somatic and microbial levels, can now be extended to atomic genetics, and neurological genetics. Logical cross fertilisation across atomic, micro, somatic, cognitive and spiritual levels can be homogenised in terms of their respective logic and unification of knowledge as a result.

We have the fermionic model, the hydrogen model and the classical carbon entromorphological model of carbon or entrochiraloctet cells. In theory a unified mathematical description of soulatrophic pathways could extrapolate unknown soulatrophic logic i.e. could be used to propose unknown features of nuclear pathways. Soulatrophicity has a mathematical basis in fractal geometry. Life is a fractal model, each soulatrophic energy level occurs at different discrete (self symmetrical) levels of natural scale (atomic, microbial, somatic, cognitive and spiritual). Life therefore has a 'fractional dimension D' from Hausdorff's dimension as part of ECASSE (Entromorphic Carbon Amplification by Solar Stimulated Emission). The levels have self symmetrical logic with conserved atomic organisation and function although at greater levels of amplified scale. Solar stimulated emission explains how carbon can undergo amplification due to its bosonic properties. This should lead to an extension of soulatrophicity and more fundamental models of reality at both the smallest and largest levels relative to the spiritual soulatrophic consciousness of the scientist devising soulatrophic energy levels. A carbon entrochiraloctet cell with nucleus, K and L shells and metabolic instability, seen through amplification 'neo-iterative cell cycling' through a system of discrete energy levels. The soulatrophic pathway follows the neobosonic and ferrofermionic pathways. Soulatrophic energy levels can be conceptualised in an organism as levels of consciousness. These pathways are fundamental to the whole of the periodic table of the elements. The elements act to reduce or eliminate field (quantum) instability by acquiring charge. The charge is linked to the Sun and solar fixing. The overall driver is linked directly to the strong nuclear force, which is driving all elementary nuclei to the stability of iron; this is the 'Holy grail' of Nature. The two processes, neobosonic and ferrofermionic, are the basis for all living systems based on the soulatrophic amplification of simple particles into super particles by amplification. There are sub-atomic, atomic, microbiological, somatic (radial), cognitive (bilateral) and spiritual (familial). Each soulatrophic energy level can be described in carbon entrochiraloctet terms, hydrogen terms or purely as fermionic systems because of the logic of particle composition. Morphology in carbon cells is relative to many different cellular levels. **This is very important as it means that any level of organisation can be described be any other level, the logic is cross fertilised.** Soulatrophic energy levels obey Pauli's exclusion principle and therefore have alternating spin characteristics. Those spin characteristics take on form (mass) and we find spingrav products such as a penis. Soulatrophic ladders form by gender communication (sexual reproductive) systems by alternation of spin characteristics and symbiosis. Charge is distributed in such systems in accordance with the spin and therefore the dipoles, which are present. The soulatrophic ladders bring un-paired electrons together to form quantum stable structures with greater evolutionary momentum than the parental spin origins in isolation. Soulatrophic levels run in conjunction with gravitational entrochiraloctets.

The fundamental functional groups of hydroxyl, carbonyl, amino, carboxylic acid and the alkane, alkene, alkyne systems are the organic basis of all life on Earth (ECASSE). The acquisition of halogenous octet reactivity and electronegativity gave carbon the highest electronegativities in the periodic table. This makes carbon a super electrophile which fits the model for organic chemistry and the levels display the same fundamental binary structure and bosonic expression systems. Amplification through entrochiraloctet follows the model of the amine, which is conserved in all soulatrophic energy levels. The difference is a reduction in entropic potentials and an increase in enthalpic potentials. As carbon moves down the ferrofermionic pathway it produces microscopic configurations with increasing frequency and therefore stability (higher levels of order).

Entromorphic span and the erotic potential- an organism represents a distinctive soulatrophic pathway ranging from spiritual to atomic and sub atomic levels (limited only by human knowledge of such inner space pathways). The entromorphic span or erotic potential is seen where an organism is aware of the highest and lowest soulatrophic energy levels and the gigantic quantum gravitational jumps associated with moving between such energy levels which produce orgasmic communication (or ejeculatory information transfer, talking is also ejaculatory they are information communications).

The other part of the ferrofermionic pathway is the neobosonic iterator, the solar accumulator (fractional dimension). This is the quantum, electromagnetic solar fixing process in carbon soulatrophic entrochiraloctet cells. This cycle is the basic cell cycle in nature but based on group IV neogenous potentials on the periodic table. A new entrochiraloctet is a neogenous product, which enters G1, the nitrogenous phase with structural growth. It then utilises the proteinascious forms to duplicate the cell contents hence the oxidative S phase. Then finally is enters the G2 halogenous phase which is the most reactive, where the mitotic cytokinesis begins to produce a new neogenous entrochiraloctet. This process also has a G0 stage post mitosis which is quiescent or senescent, where the neobosonic iterator stops and the cell remains stabilised and homeostatically maintained but not divisible. This process the cell cycle of neoiterative cycle is seen on all soulatrophic energy levels and forms the basic part of all life's temporal activities. **When two people meet, talk and eventually go their separate ways they do so in accordance with the neoiterative cell cycle, leaving and waving goodbye is cytokinesis.** This process finds carbon stabilising its valance electrons through solar acquisition in entrochiraloctet cells, driven by electronegativities. The cycle is the classical cell cycle found in biology but the phases are associated with specific elementary electronic configurations.

Introduction of the 'Formling series' which displays morphologies: which reflect microstatic conditions. High kinetic energy levels display increasing streamlining. Low kinetic levels are spherical having high-pressure morphology. The classical cocci are spherical and associated with s (l=0) type orbital morphology. The cells become streamlined and the bacillus/vibrio/ filamentous forms exist as these organisms reach higher kinetic energy levels. This is p type (l=1) orbital morphology showing streamlining and L type metabolic organelles such as flagella. Valance bonded metabolic organelles are seen as basic nutrients for the energy systems, providing nitrogen, oxygen, halogenous components from the immediate microstate. The statistical mechanical model describes bacterial cells in balance with external macrostate, the basis of ecological niche. Reproduction on this level forms colony-forming units, which produce basic tissues (colonies – simple ground state entrochiraloctets). They reproduce by Pauli's exclusion principle through nuclear fission by nuclear duplication. Nuclear meiotic fusion reactions are considerably rare, although they do occur on this level. The microbiological evolutionary clock runs very very slow, hence evolution into higher forms has taken hundreds of millions of years possibly even billions.

Neobosonic systems (periodic table –from carbon to neon) – The quantum instability in carbon entrochiraloctet cells (soulatrophic pathway), is driving electronengativity distributions. Acquisitions of solar electromagnetic radiation (ECASSE) satisfy the unstable region with high-level electrons promoted due to photonic acquisition.

This cycling of quantum stability through solar means is the basis of the classical cell cycles and locks energy in carbon entrochiraloctet nuclei. It describes: -

- A ground state entrochiraloctet cell in interphase G1 the nitrogenous metabolic organelle system (structural growth).
- A developing ground state entrochiraloctet cell in interphase S the oxidative metabolic organelle system (duplication of nuclear components).
- A developed ground state entrochiraloctet in interphase G2 the halogenous metabolic organelle system (membranous rendering).
- The exclusion principle induces mitosis, prophase, metaphase, anaphase, telophase and cytokinesis. As part of the final neogenous metabolic organelle system.

Finally this section (part eleven) is particulary complex as it investigates the detailed elements of the theory of carbon entromorphology. It takes time to link often obscure concepts to everyday life which is why every effort has been made to include as many images as possible to try to minimise the complex ideas.

Unfortunately any theory which attempts to unify all the levels of natural scale is always going to challenge any mind. Anyone with a particularly broad educational backgroung will undoubtedly find the concepts easier to take on board. It does well to continue to look back at the start of this book to reinitiate the readers mind to some of the basic ideas such as the carbon self symmetrical structures in a human animal model.

Science check

Do the observations correlate with the theoretical model?

YES ✓ *NO*

PART TWELVE

Human Thermodynamics and Morality

Neurochromosome gravitropism - the 2nd law of thermodynamics, and the morality concepts of evil (anti-ECASSE, NFVE).

Gravitropism (morally evil) is an anti-life system and it relates to the inevitable effect of the second law of thermodynamics, that decay is increasing in the universe, thermoentropic pathways are evil.

For example, murder liberates 13.7 billion years of stored energy: this is anti-life anti-solar conservation or anti-phototropism (anti good), it destroys accumulated living energy.

Morality is a thermodynamic system and energy conservation is a process of good where organisms can progress towards artificial life by moving up the three soulatrophic pathways. Gravitropism (evil) acts in reverse effectively acting as a basic model of natural selection, for example the credit crunch is gravitropic (evil) and has seen trillions of dollars turned into useless thermoentropy or waste. This money has phototropic (good) origins in that it is completely solar and the result of phototropic (good) logic. By turning it into waste, humanity has pushed its self back down the soulatrophic pathways; anti-evolution. Morality is often classed as metaphysical, carbon entromorphology brings it into the physical through thermodynamics, and with energy management makes good and evil measureable concepts.

All life is interfaced against the plants which are 'primary producers', hence the same logic is translated into the animals. Hence heliotropism (free will) is an accurate description of an animal's free will as well as the plants. Phototropism (good) and gravitropism (evil) are also plant associated words. Phototropism (good) is the act of conserving solar energy in living organisms, and gravitropism (evil) is used to describe how life is under the constant action of Earth's restrictive pull in equilibrium with its solar animation. The science of carbon amplification soulatrophic entrochiraloctet decay systems is that of morally evil. The science of the carbon amplification soulatrophic pathway collapses and devolves backwards down the soulatrophic pathway, morally evil usually by oxidation by electrophilic agents. Soulatrophic energy levels can be thought of in an organism as levels or states of nuclear consciousness.

The process by which living systems based on carbon decrease in size is by reduction in size of orbital probability fields. Gravitropism (evil) is based on blocking a solar provider (Sun). Gravitropism (evil) is the science of 'evil' and fixes living carbon on planet Earth in 'Earth bondage'. The pathway leads back through the soulatrophic energy levels to atomic and subatomic levels. This state of carbon as fixed carbonates and carbon dioxide personifies the concept of hell and describes early Earth conditions in the Hadean Eon and before that all the way back to the Big Bang. As such, the soulatrophic pathways describe hell as the past and heaven as a possible future: this is a thermodynamic model of life.

Gravitropism (evil) invariably leads to lower more primitive levels of soulatrophic existence. The thermodynamic angle of evil is found in the formation of thermoentropic pathways where energy is no longer made available to carbon life forms and where heat production is increasing. Anti-neobosonic anti-ferrofermionic effects anti-ECASSE (Entromorphic Carbon Amplification by Solar Stimulated

Emission). Spiritual soulatrophic spingrav is a spin down potential, anti-solar spin anti-solar alignment: a pathway reducing evolutionary momentum pushing down soulatrophic pathways to simpler organisms. The entromorphic elevation in an organism is reduced towards Earth gravity dominance where organisms are flattened to the Earth. Phototropism (good) is the inverse where organisms are animated towards a perpendicular elevation like humans who stand bolt upright. The more evolved the organism the greater the entromorphic elevation.

The Earth has a magnetic quantum effect inducing deformation of conserved sub shell probability distributions. This is the environmental filtering aspect of natural selection, the most significant effects are from the Sun and the Earth directly: the Moon is stabilising. Gravitropic (evil) – typically associated with L shell valance shell electrons and associated energies and instability this is carbons interface with the external environment making it its vulnerable region. The L valance shell is a quantum waveform and is associated with thermal energy exchange systems of the immediate macrostate. L shells are membranous in soulatrophic entrochiraloctet cells; they orchestrate energy liberation from outside soulatrophic entrochiraloctet cells.

They are the second level of nuclear shielding in carbon soulatrophic cells. They are seen in males, the stamen part of plants (pollen is an L shell component –the air and insects act bosonically to transfer the strong nuclear genomic force). Men wear dark Earth colours to camouflage themselves as they act to procure solar energy from the external macrostate in accordance with L valance shell energies and instabilities. This is a thermoentropic zone. The gravitropic (evil) association in males (evil) is demonstrated by the statistic that 90% of violent crimes are produced by males. Females are nuclear shells and therefore more stable hence they tend to conserve life hence only 10% of violent crime is committed by them. The gender model reveals the moral thermodynamic responsibilities of solar energy conservation or good (phototropism) and solar energy liberation or evil (gravitropism).

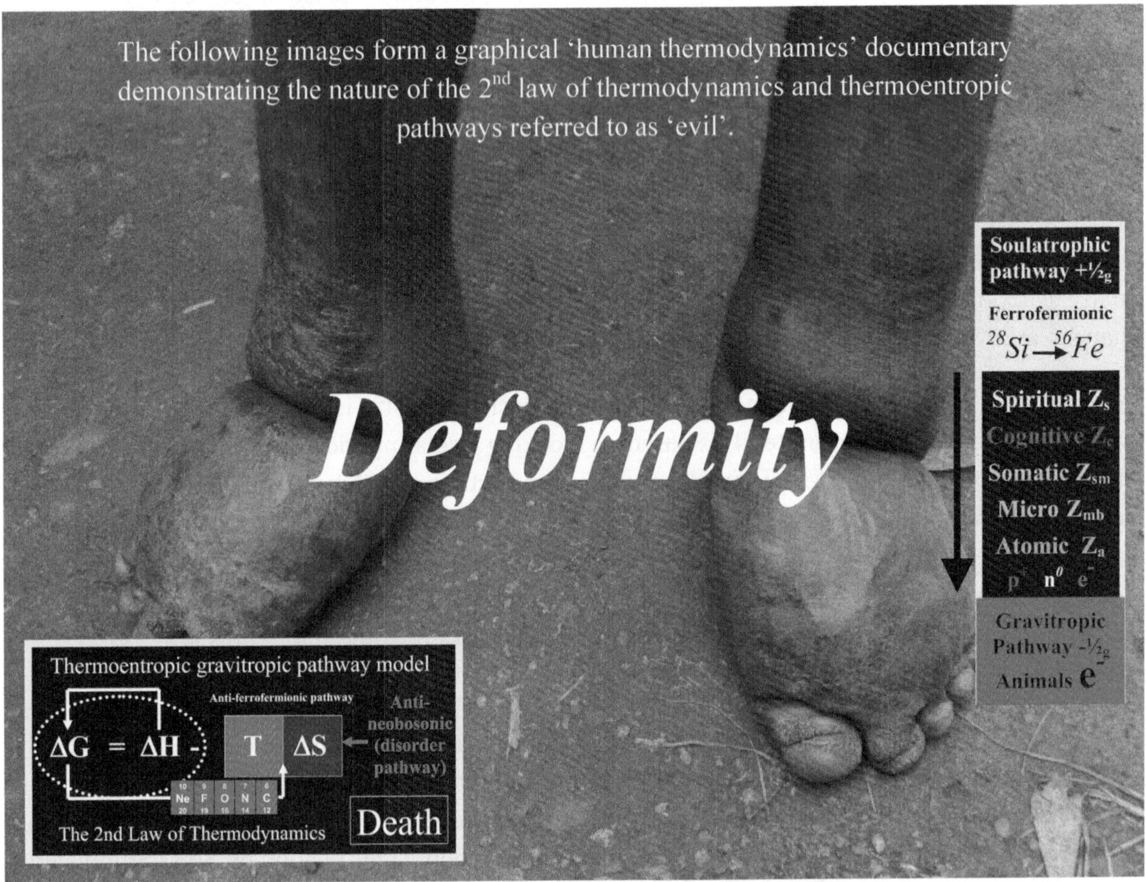

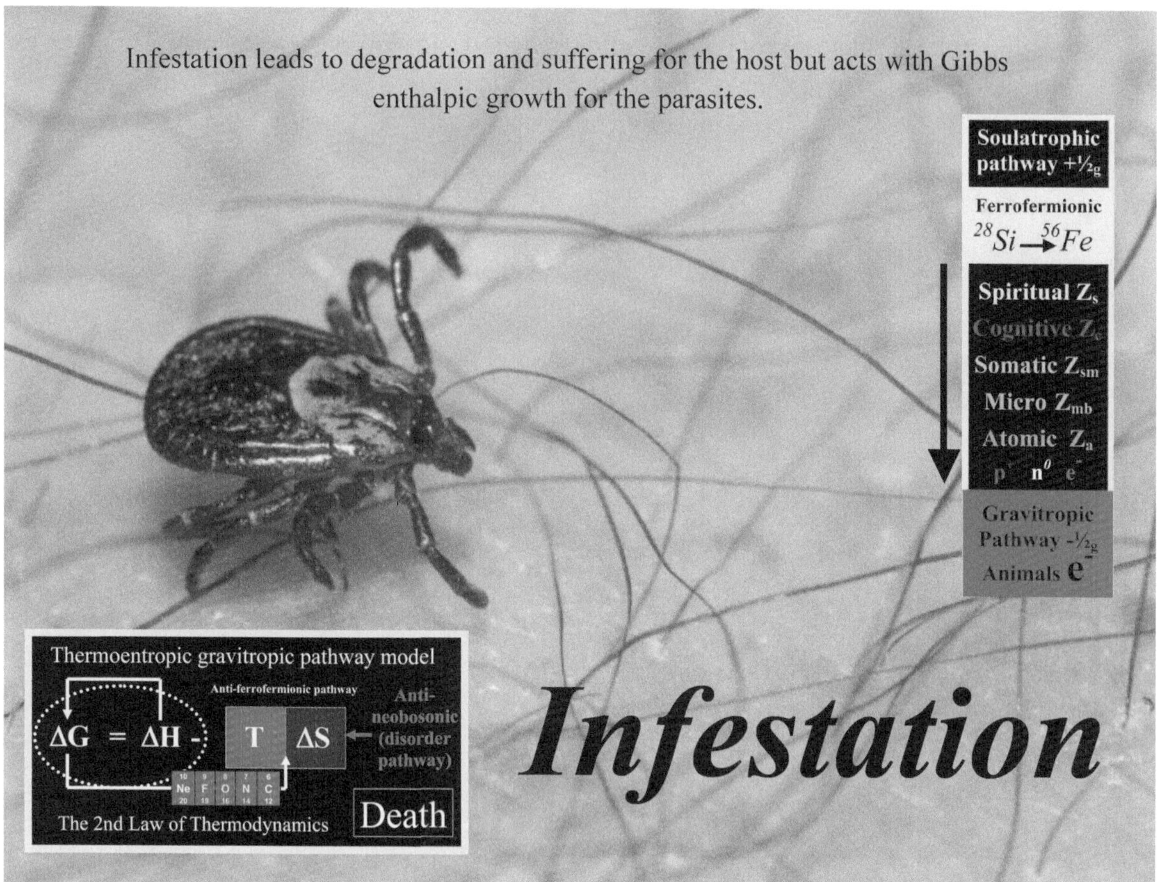
Infestation leads to degradation and suffering for the host but acts with Gibbs enthalpic growth for the parasites.
Soulatrophic pathway +½g
Ferrofermionic
28Si→56Fe
Spiritual Zs
Cognitive Zc
Somatic Zsm
Micro Zmb
Atomic Za
p n0 e
Gravitropic Pathway -½g
Animals e
Thermoentropic gravitropic pathway model
Anti-ferrofermionic pathway
ΔG = ΔH - T ΔS
Anti-neobosonic (disorder pathway)
Ne F O N C
The 2nd Law of Thermodynamics
Death
Infestation

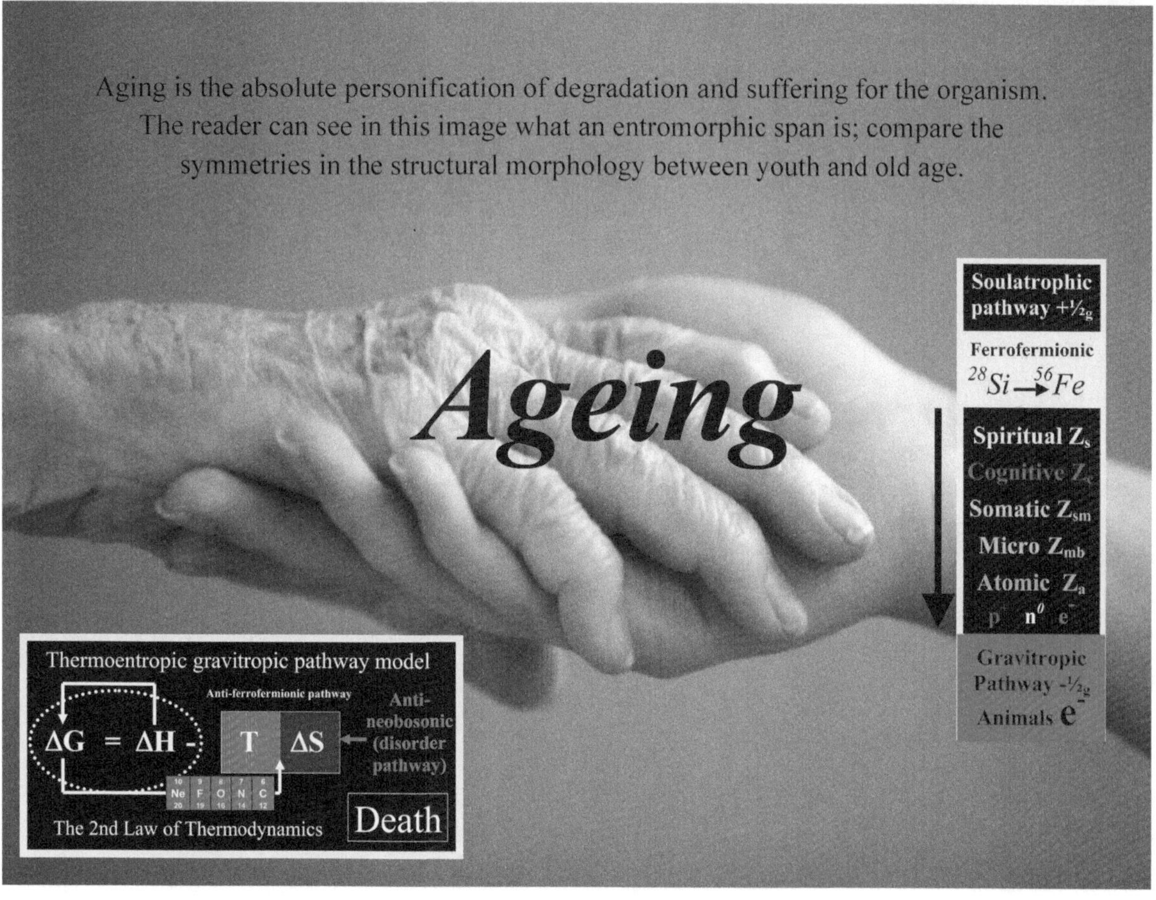
Aging is the absolute personification of degradation and suffering for the organism. The reader can see in this image what an entromorphic span is; compare the symmetries in the structural morphology between youth and old age.
Ageing
Soulatrophic pathway +½g
Ferrofermionic
28Si→56Fe
Spiritual Zs
Cognitive Zc
Somatic Zsm
Micro Zmb
Atomic Za
p n0 e
Gravitropic Pathway -½g
Animals e
Thermoentropic gravitropic pathway model
Anti-ferrofermionic pathway
ΔG = ΔH - T ΔS
Anti-neobosonic (disorder pathway)
Ne F O N C
The 2nd Law of Thermodynamics
Death

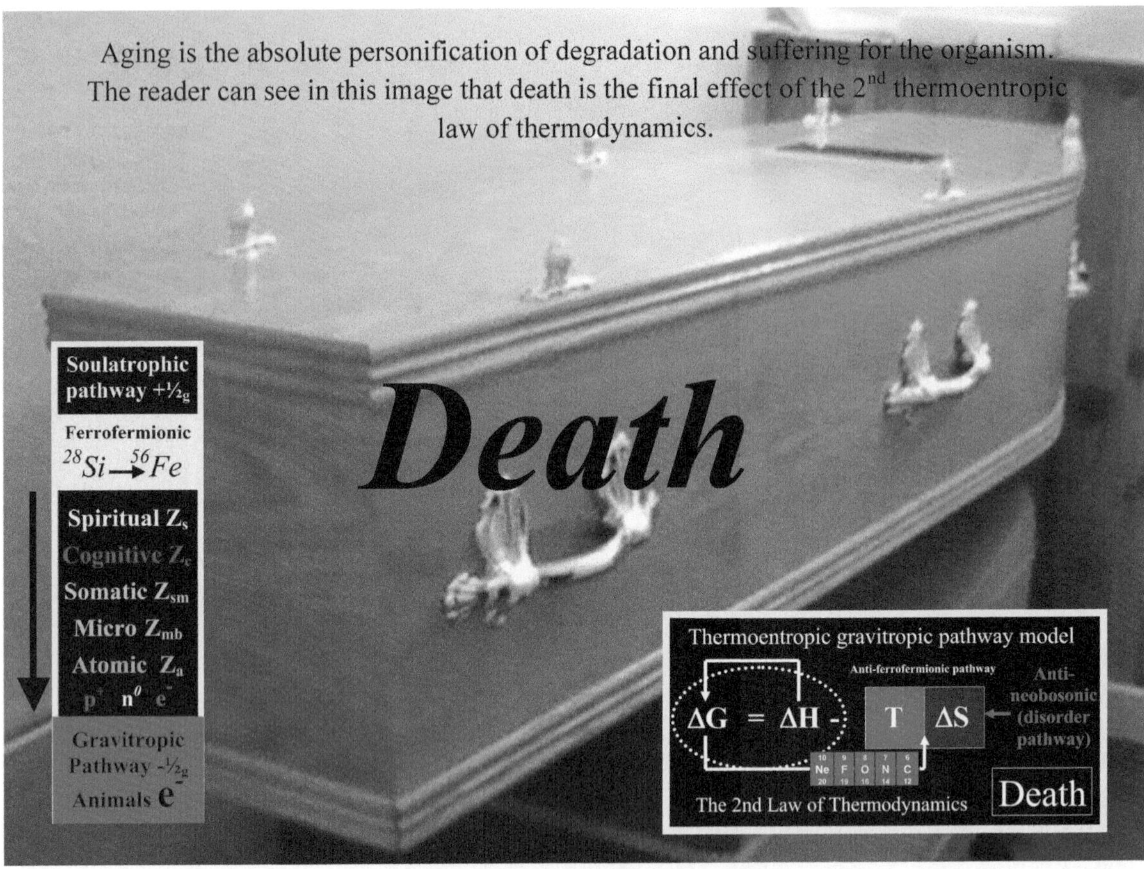
Aging is the absolute personification of degradation and suffering for the organism. The reader can see in this image that death is the final effect of the 2nd thermoentropic law of thermodynamics.
Death
Soulatrophic pathway +½g
Ferrofermionic
28Si→56Fe
Spiritual Zs
Cognitive Zc
Somatic Zsm
Micro Zmb
Atomic Za
p+ n0 e-
Gravitropic Pathway -½g
Animals e-
Thermoentropic gravitropic pathway model
Anti-ferrofermionic pathway
ΔG = ΔH - T ΔS
Anti-neobosonic (disorder pathway)
Ne F O N C
The 2nd Law of Thermodynamics
Death

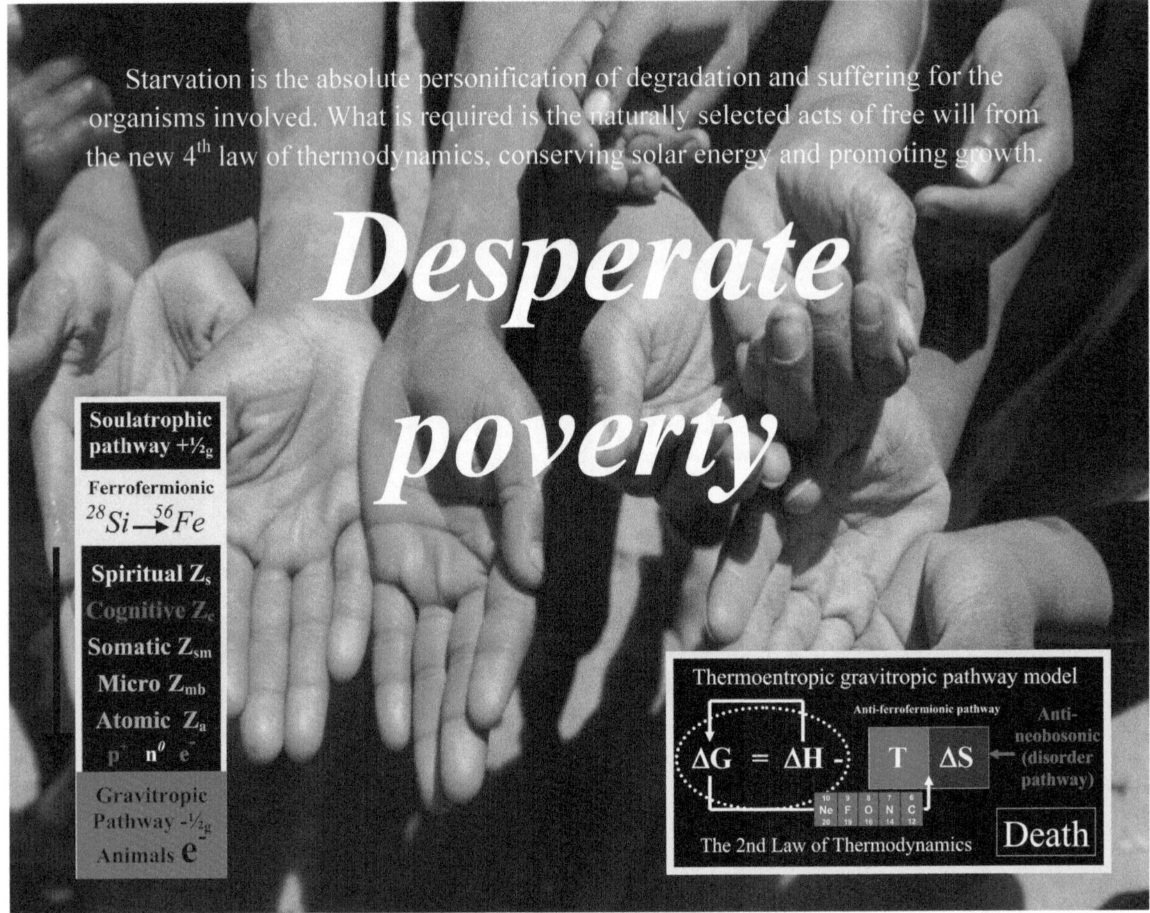
Starvation is the absolute personification of degradation and suffering for the organisms involved. What is required is the naturally selected acts of free will from the new 4th law of thermodynamics, conserving solar energy and promoting growth.
Desperate poverty
Soulatrophic pathway +½g
Ferrofermionic
28Si→56Fe
Spiritual Zs
Cognitive Zc
Somatic Zsm
Micro Zmb
Atomic Za
p+ n0 e-
Gravitropic Pathway -½g
Animals e-
Thermoentropic gravitropic pathway model
Anti-ferrofermionic pathway
ΔG = ΔH - T ΔS
Anti-neobosonic (disorder pathway)
Ne F O N C
The 2nd Law of Thermodynamics
Death

Soulatrophic pathway +½g
Ferrofermionic
28Si→56Fe
Spiritual Zs
Cognitive Zc
Somatic Zsm
Micro Zmb
Atomic Za
p+ n0 e-
Gravitropic Pathway -½g
Animals e-
Criminal burdon
Criminals are a strong demonstration of degradation and suffering for the organisms they prey on. A criminal produces a large amount of waste which is a 2nd law thermoentropic event.

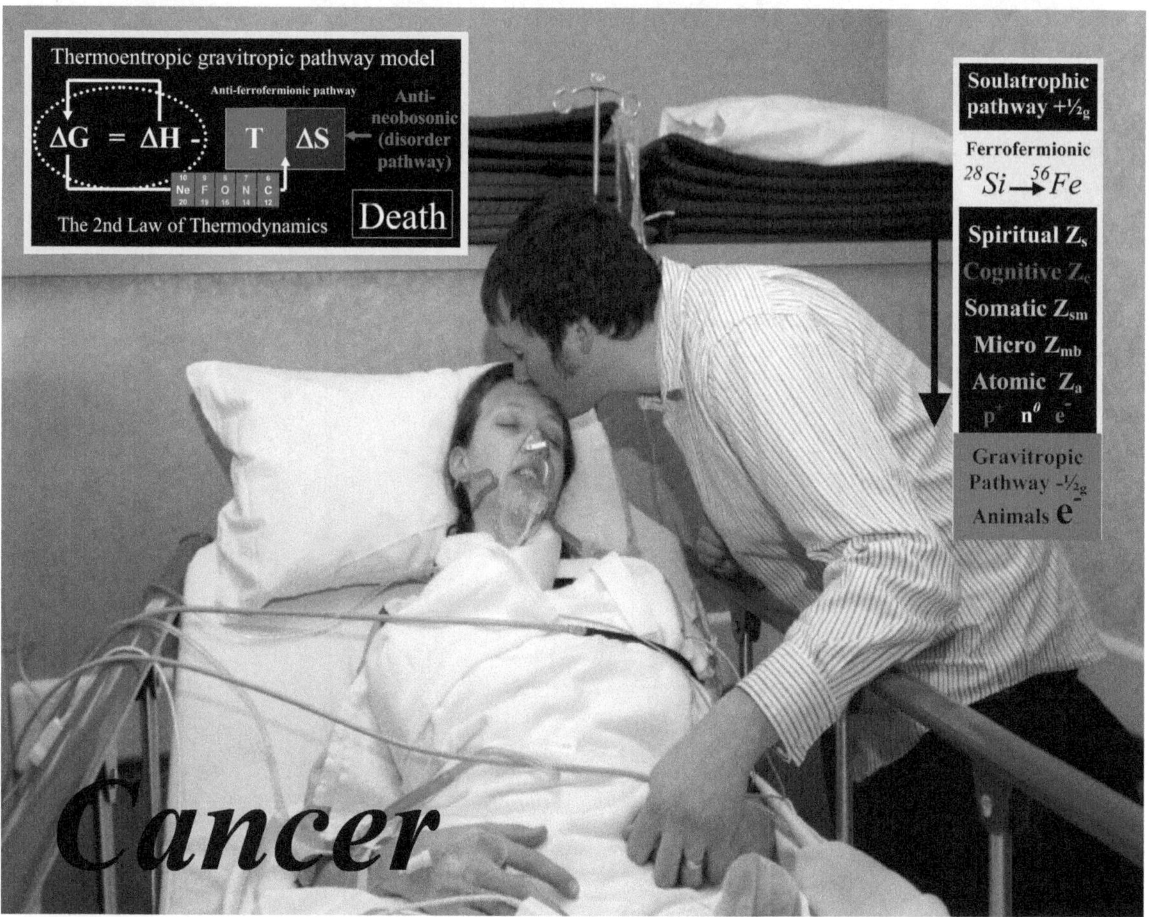
Thermoentropic gravitropic pathway model
Anti-ferrofermionic pathway
ΔG = ΔH - T ΔS
Anti-neobosonic (disorder pathway)
Ne F O N C
The 2nd Law of Thermodynamics
Death
Soulatrophic pathway +½g
Ferrofermionic
28Si→56Fe
Spiritual Zs
Cognitive Zc
Somatic Zsm
Micro Zmb
Atomic Za
p+ n0 e-
Gravitropic Pathway -½g
Animals e-
Cancer

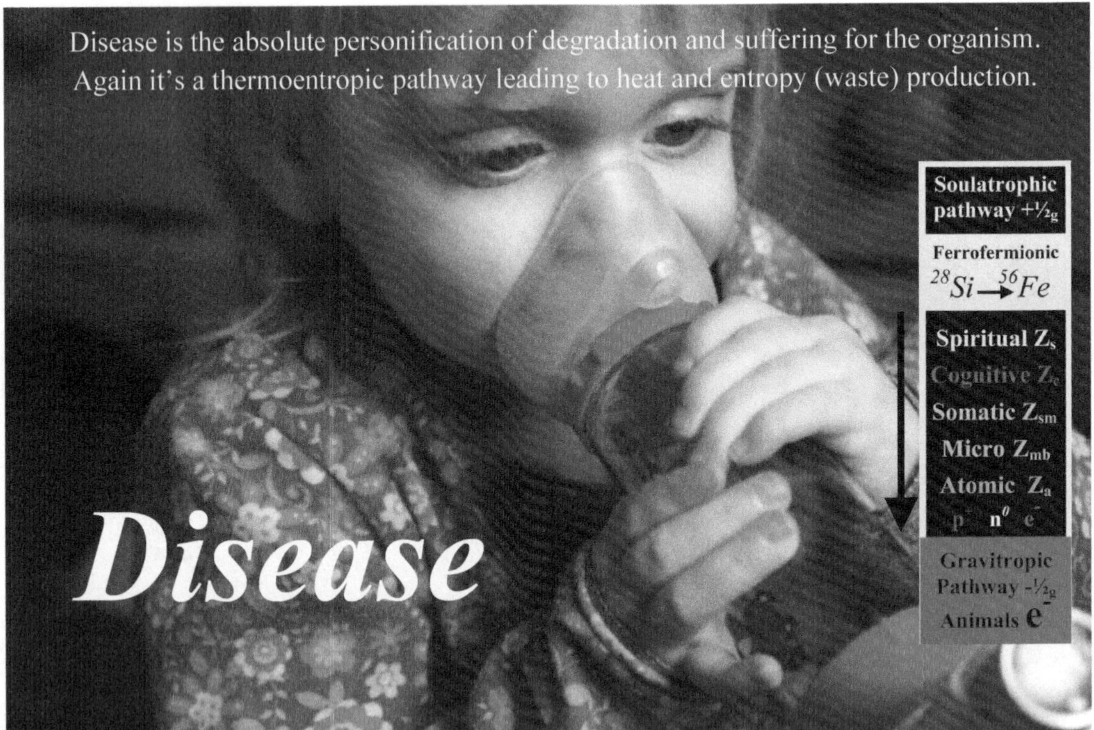
Disease is the absolute personification of degradation and suffering for the organism.
Again it's a thermoentropic pathway leading to heat and entropy (waste) production.
Soulatrophic pathway $+\frac{1}{2}_g$
Ferrofermionic
$^{28}Si \rightarrow ^{56}Fe$
Spiritual Z_s
Cognitive Z_c
Somatic Z_{sm}
Micro Z_{mb}
Atomic Z_a
p^+ n^0 e^-
Gravitropic Pathway $-\frac{1}{2}_g$
Animals e^-
Disease

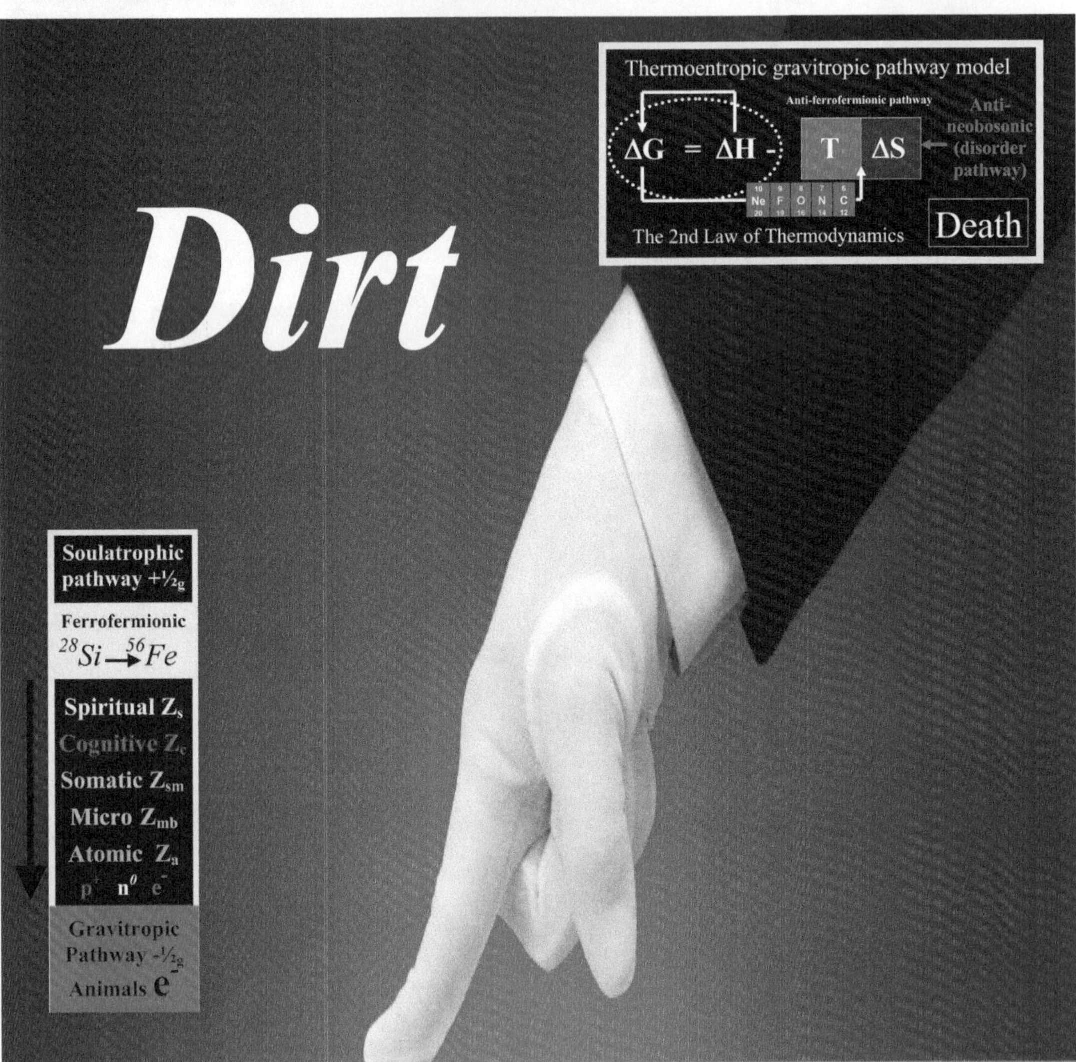
Thermoentropic gravitropic pathway model
Anti-ferrofermionic pathway
$\Delta G = \Delta H - T \Delta S$
Ne F O N C
Anti-neobosonic (disorder pathway)
The 2nd Law of Thermodynamics
Death
Dirt
Soulatrophic pathway $+\frac{1}{2}_g$
Ferrofermionic
$^{28}Si \rightarrow ^{56}Fe$
Spiritual Z_s
Cognitive Z_c
Somatic Z_{sm}
Micro Z_{mb}
Atomic Z_a
p^+ n^0 e^-
Gravitropic Pathway $-\frac{1}{2}_g$
Animals e^-

Soulatrophic pathway $+\frac{1}{2}_g$
Ferrofermionic
$^{28}Si \rightarrow {}^{56}Fe$
Spiritual Z_s
Cognitive Z_c
Somatic Z_{sm}
Micro Z_{mb}
Atomic Z_a
p^+ n^0 e^-
Gravitropic Pathway $-\frac{1}{2}_g$
Animals e^-
Rape & other physical violation
Physical violence leads to increasing disorder and a breaking of natural symmetry again it demonstrates the 2nd law of thermodynamics.
Thermoentropic gravitropic pathway model
Anti-ferrofermionic pathway
$\Delta G = \Delta H - T \Delta S$
Anti-neobosonic (disorder pathway)
Ne F O N C
The 2nd Law of Thermodynamics
Death

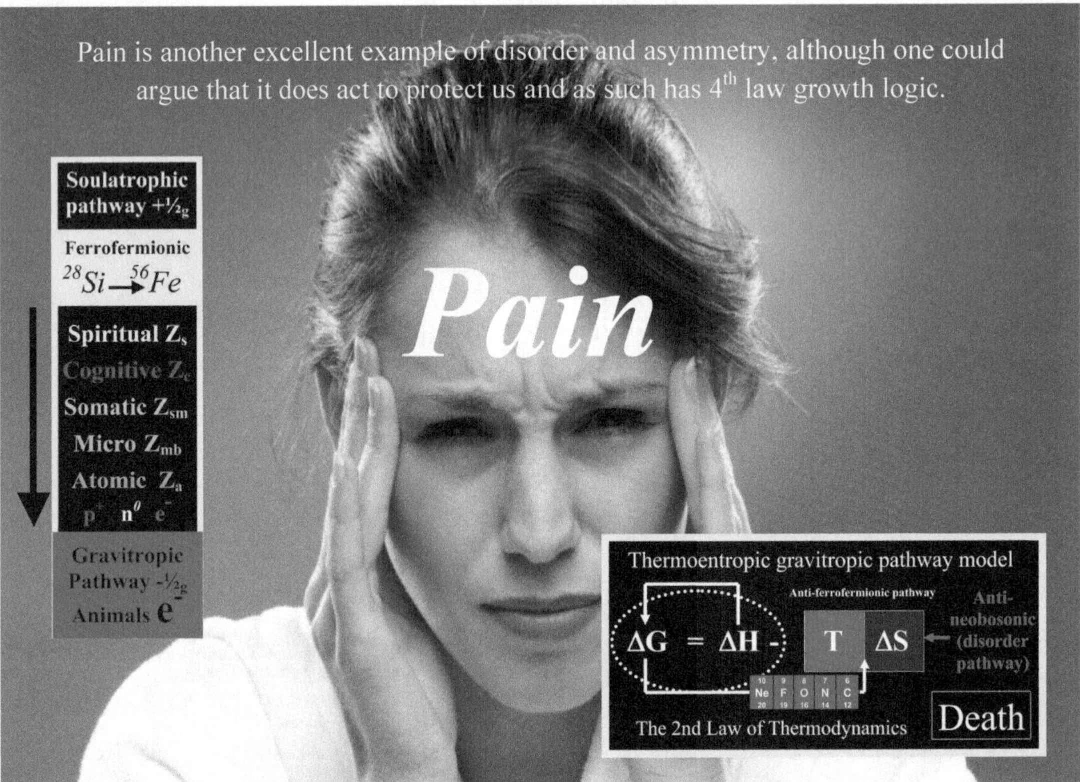

Any aspect of disordered reactions leading to a decrease in the frequency of particle symmetry absolutely personifies the 2nd law of thermodynamics.

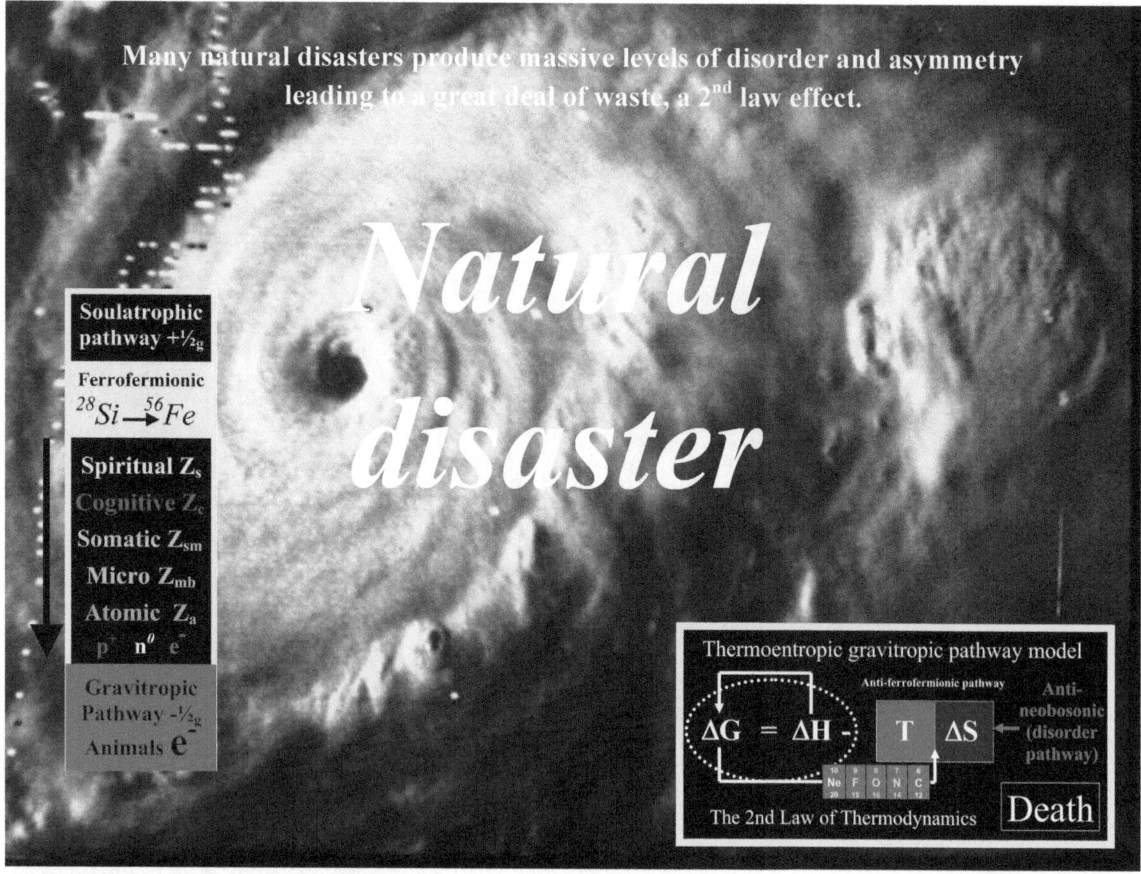
Many natural disasters produce massive levels of disorder and asymmetry leading to a great deal of waste, a 2nd law effect.
Natural disaster
Soulatrophic pathway +½g
Ferrofermionic
28Si → 56Fe
Spiritual Zs
Cognitive Zc
Somatic Zsm
Micro Zmb
Atomic Za
p+ n0 e-
Gravitropic Pathway -½g
Animals e-
Thermoentropic gravitropic pathway model
Anti-ferrofermionic pathway
ΔG = ΔH - T ΔS
Anti-neobosonic (disorder pathway)
Ne F O N C
The 2nd Law of Thermodynamics
Death

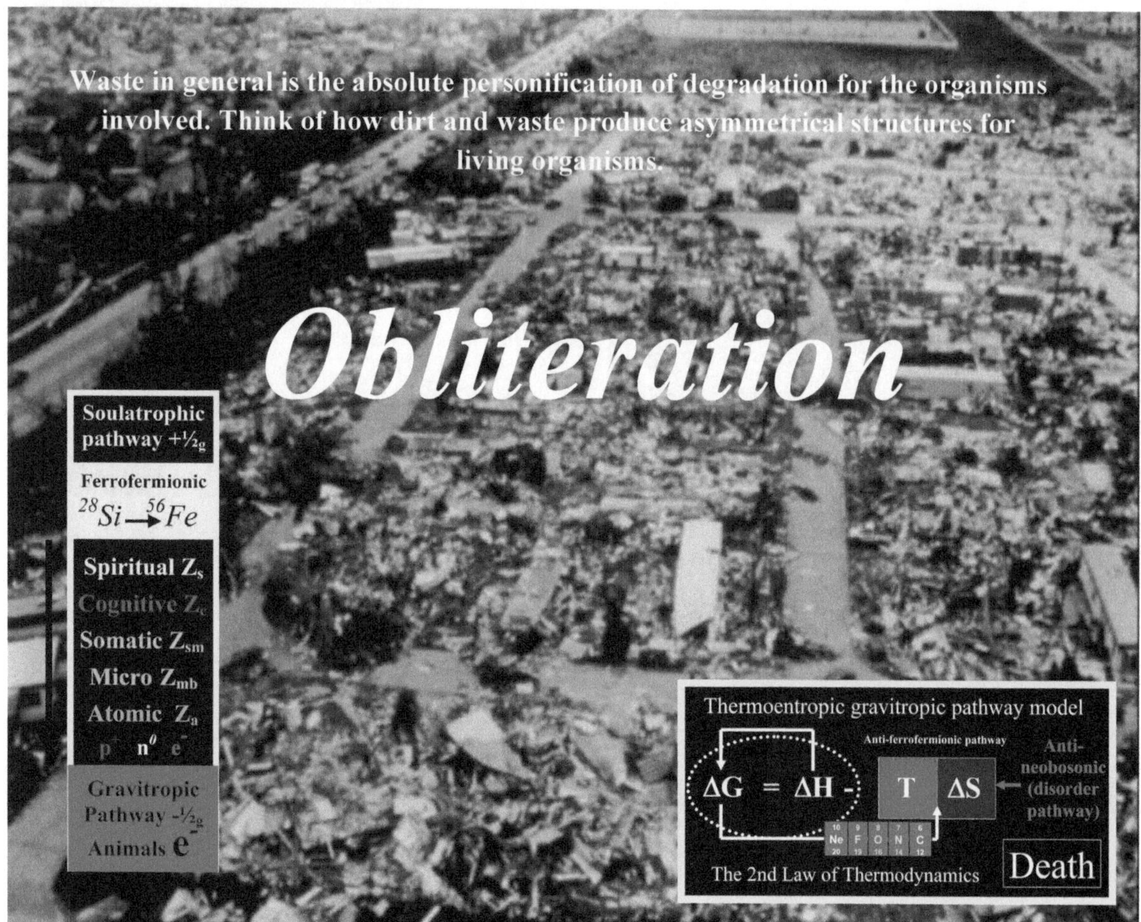
Waste in general is the absolute personification of degradation for the organisms involved. Think of how dirt and waste produce asymmetrical structures for living organisms.
Obliteration
Soulatrophic pathway +½g
Ferrofermionic
28Si → 56Fe
Spiritual Zs
Cognitive Zc
Somatic Zsm
Micro Zmb
Atomic Za
p+ n0 e-
Gravitropic Pathway -½g
Animals e-
Thermoentropic gravitropic pathway model
Anti-ferrofermionic pathway
ΔG = ΔH - T ΔS
Anti-neobosonic (disorder pathway)
Ne F O N C
The 2nd Law of Thermodynamics
Death

Thermoentropic gravitropic pathway model

Anti-ferrofermionic pathway

$\Delta G = \Delta H - T \Delta S$

Anti-neobosonic (disorder pathway)

Ne F O N C

The 2nd Law of Thermodynamics

Death

Soulatrophic pathway $+\frac{1}{2}_g$

Ferrofermionic

$^{28}Si \rightarrow {}^{56}Fe$

Spiritual Z_s

Cognitive Z_c

Somatic Z_{sm}

Micro Z_{mb}

Atomic Z_a

p n^0 e^-

Gravitropic Pathway $-\frac{1}{2}_g$

Animals e^-

Degradation

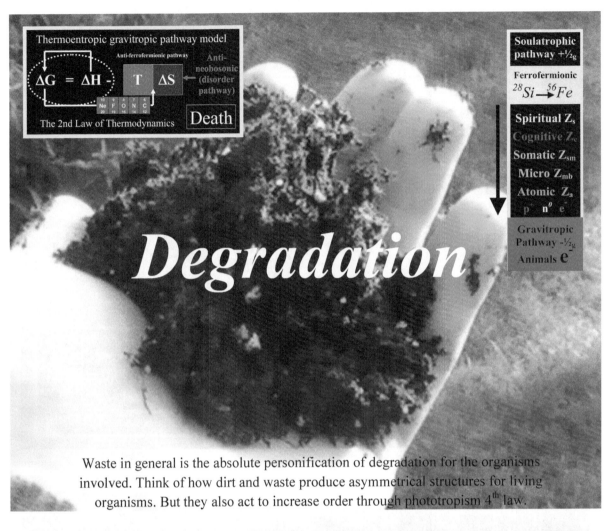

Waste in general is the absolute personification of degradation for the organisms involved. Think of how dirt and waste produce asymmetrical structures for living organisms. But they also act to increase order through phototropism 4th law.

Waste in general is the absolute personification of degradation for the organisms involved. Think of how dirt and waste produce asymmetrical structures for living organisms. But they also act to increase order through phototropism 4th law.

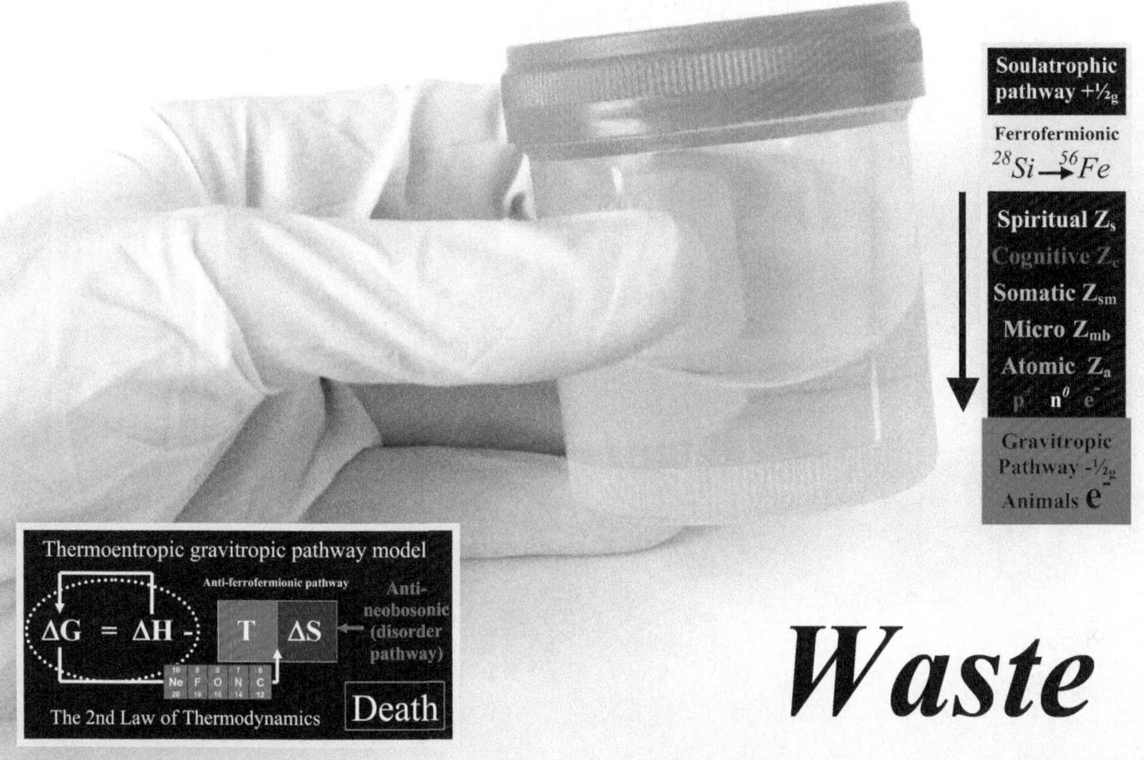

INTERESTING FACT ABOUT SWEARING!

Did you know that swearing has its roots in thermodynamics! Consider the following words: - shit, crap, bullshit, rubbish, bollocks, bastard, dickhead, arse......you get the idea! They are all words associated with the 2nd law of thermodynamics and entropy (waste), which is why swearing, is used in disorderly symmetry breaking situations.

Consider also a 'shit' day at work! Your day has precipitated the 2nd law of thermodynamics and free energy has not been made available to you, your thermodynamic efficiency is reduced and been precipitated as entropy and heat!

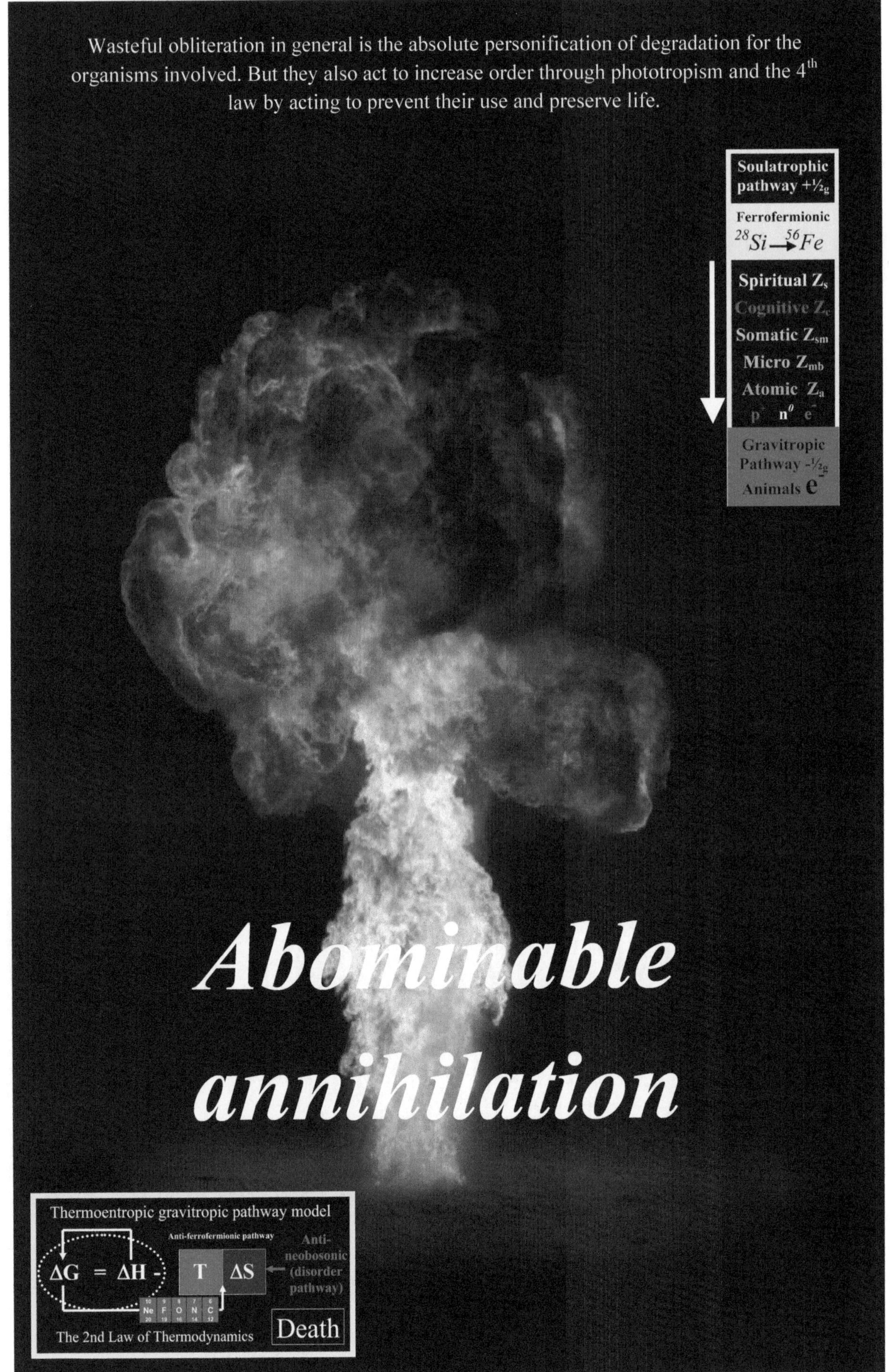
Wasteful obliteration in general is the absolute personification of degradation for the organisms involved. But they also act to increase order through phototropism and the 4th law by acting to prevent their use and preserve life.
Soulatrophic pathway $+\frac{1}{2}_g$
Ferrofermionic
$^{28}Si \rightarrow {}^{56}Fe$
Spiritual Z_s
Cognitive Z_c
Somatic Z_{sm}
Micro Z_{mb}
Atomic Z_a
p^+ n^0 e^-
Gravitropic Pathway $-\frac{1}{2}_g$
Animals e^-
Abominable annihilation
Thermoentropic gravitropic pathway model
Anti-ferrofermionic pathway
Anti-neobosonic (disorder pathway)
$\Delta G = \Delta H - T \Delta S$
Ne F O N C
The 2nd Law of Thermodynamics
Death

Child abuse
Soulatrophic pathway $+\frac{1}{2}_g$
Ferrofermionic
$^{28}Si \rightarrow ^{56}Fe$
Spiritual Z_s
Cognitive Z_c
Somatic Z_{sm}
Micro Z_{mb}
Atomic Z_a
p^+ n^0 e^-
Gravitropic Pathway $-\frac{1}{2}_g$
Animals e^-
Thermoentropic gravitropic pathway model
Anti-ferrofermionic pathway
$\Delta G = \Delta H - T \Delta S$
Anti-neobosonic (disorder pathway)
Ne F O N C
The 2nd Law of Thermodynamics
Death
There is nothing but disorder and asymmetry about the thermoentropic acts of child abuse.
The offenders are promoters of the 2nd law of thermodynamics.

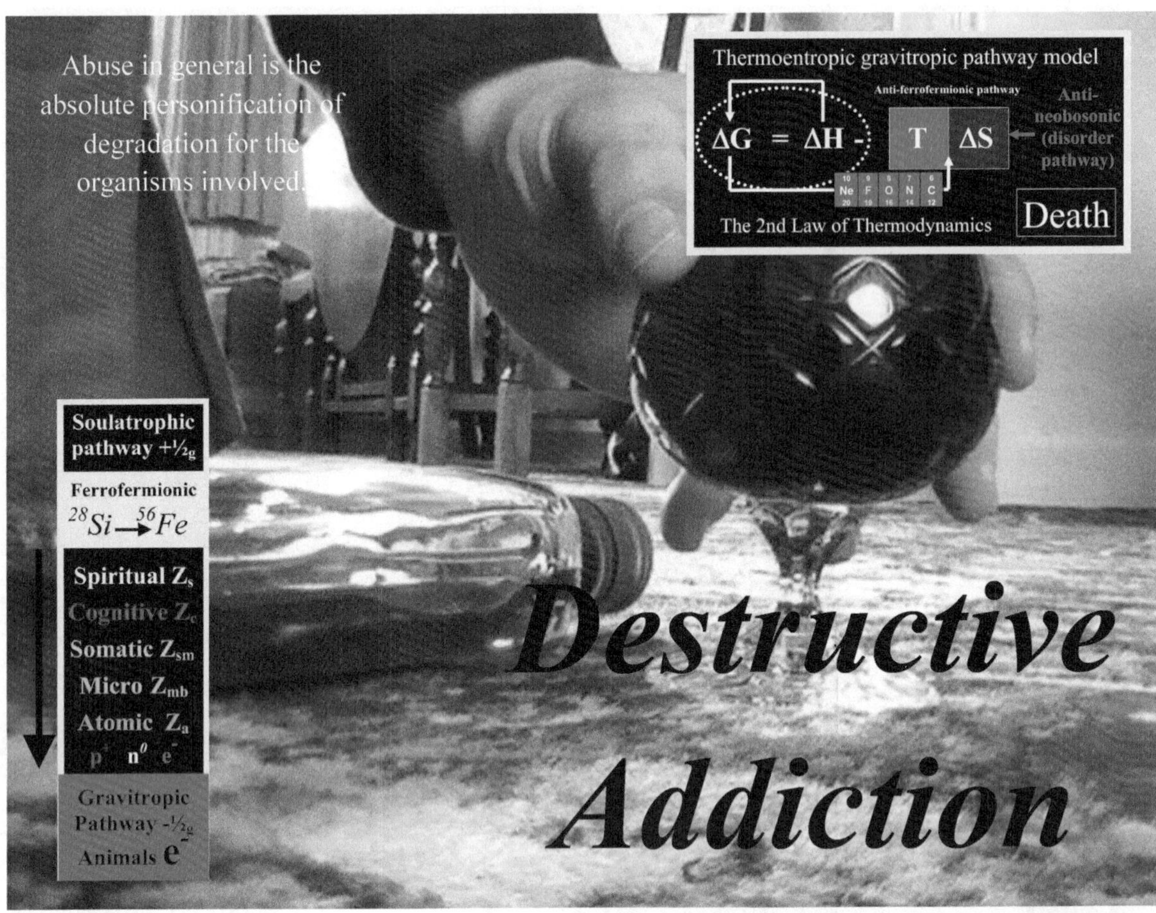
Abuse in general is the absolute personification of degradation for the organisms involved.
Thermoentropic gravitropic pathway model
Anti-ferrofermionic pathway
ΔG = ΔH - T ΔS
Anti-neobosonic (disorder pathway)
Ne F O N C
The 2nd Law of Thermodynamics
Death
Soulatrophic pathway +½g
Ferrofermionic
28Si → 56Fe
Spiritual Zs
Cognitive Zc
Somatic Zsm
Micro Zmb
Atomic Za
p+ n0 e-
Gravitropic Pathway -½g
Animals e-
Destructive
Addiction

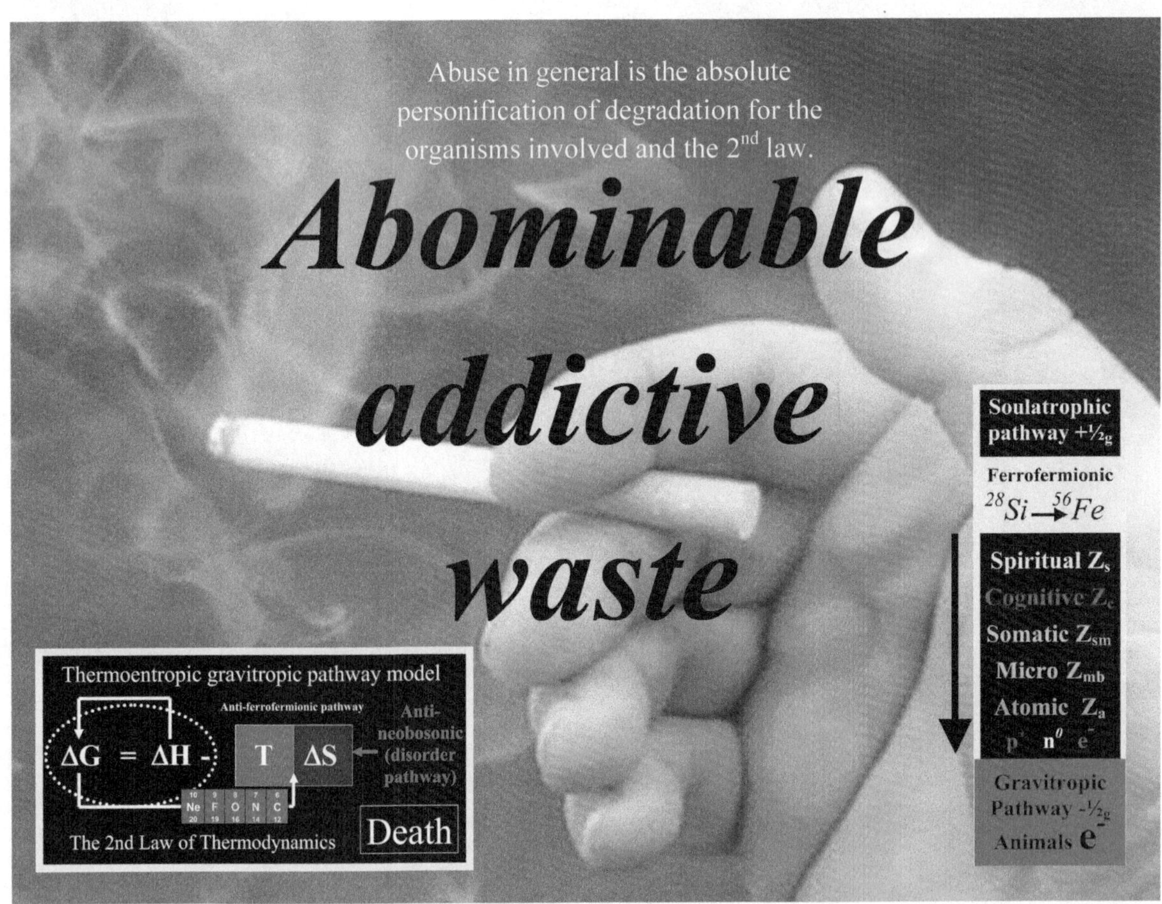
Abuse in general is the absolute personification of degradation for the organisms involved and the 2nd law.
Abominable
addictive
waste
Soulatrophic pathway +½g
Ferrofermionic
28Si → 56Fe
Spiritual Zs
Cognitive Zc
Somatic Zsm
Micro Zmb
Atomic Za
p+ n0 e-
Gravitropic Pathway -½g
Animals e-
Thermoentropic gravitropic pathway model
Anti-ferrofermionic pathway
ΔG = ΔH - T ΔS
Anti-neobosonic (disorder pathway)
Ne F O N C
The 2nd Law of Thermodynamics
Death

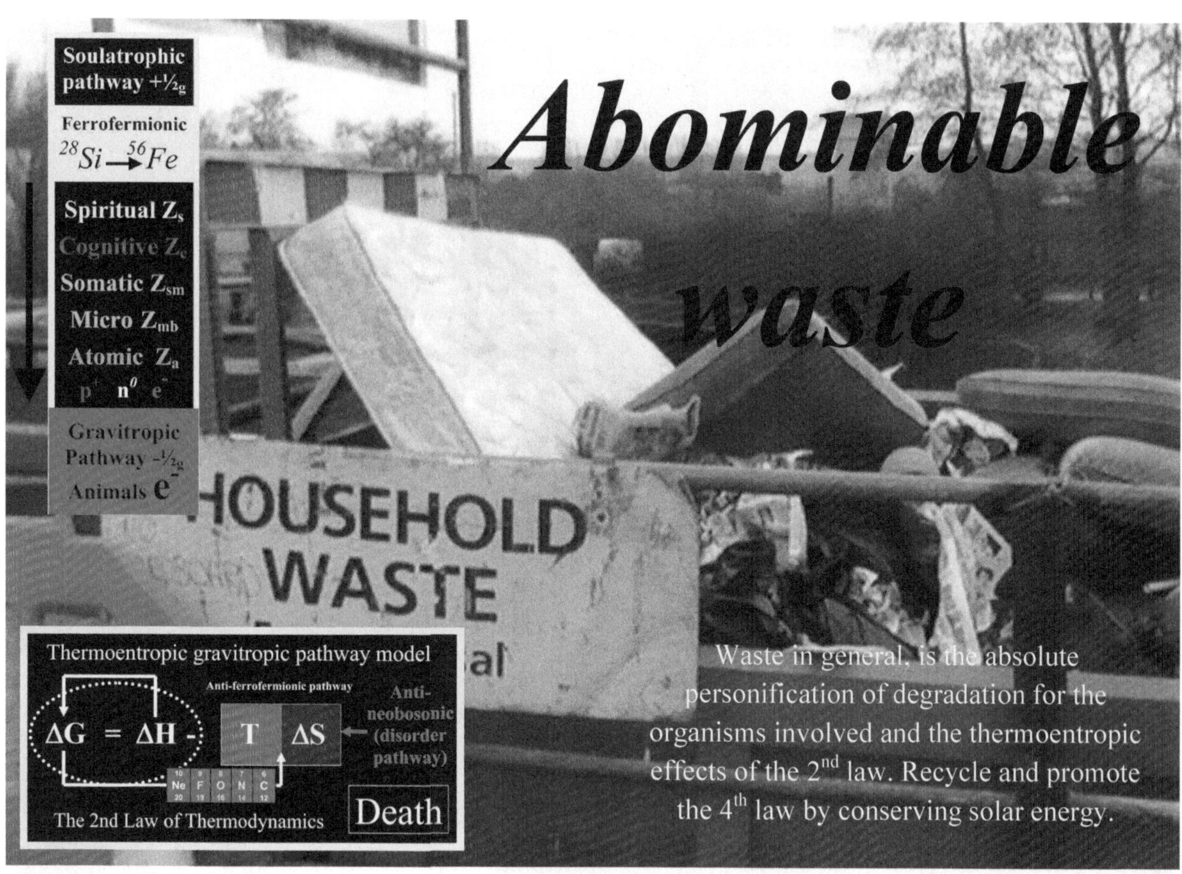
Abominable waste
Soulatrophic pathway $+\frac{1}{2}_g$
Ferrofermionic
$^{28}Si \rightarrow ^{56}Fe$
Spiritual Z_s
Cognitive Z_c
Somatic Z_{sm}
Micro Z_{mb}
Atomic Z_a
p^+ n^0 e^-
Gravitropic Pathway $-\frac{1}{2}_g$
Animals e^-
HOUSEHOLD WASTE
Thermoentropic gravitropic pathway model
Anti-ferrofermionic pathway
$\Delta G = \Delta H - T \Delta S$
Anti-neobosonic (disorder pathway)
Ne F O N C
The 2nd Law of Thermodynamics
Death
Waste in general, is the absolute personification of degradation for the organisms involved and the thermoentropic effects of the 2nd law. Recycle and promote the 4th law by conserving solar energy.

Unacceptable destruction
Soulatrophic pathway $+\frac{1}{2}_g$
Ferrofermionic
$^{28}Si \rightarrow ^{56}Fe$
Spiritual Z_s
Cognitive Z_c
Somatic Z_{sm}
Micro Z_{mb}
Atomic Z_a
p^+ n^0 e^-
Gravitropic Pathway $-\frac{1}{2}_g$
Animals e^-
Thermoentropic gravitropic pathway model
Anti-ferrofermionic pathway
$\Delta G = \Delta H - T \Delta S$
Anti-neobosonic (disorder pathway)
Ne F O N C
The 2nd Law of Thermodynamics
Death
Waste in general is the absolute personification of degradation for the organisms involved and the thermoentropic effects of the 2nd law. Recycle and promote the 4th law by conserving solar energy.

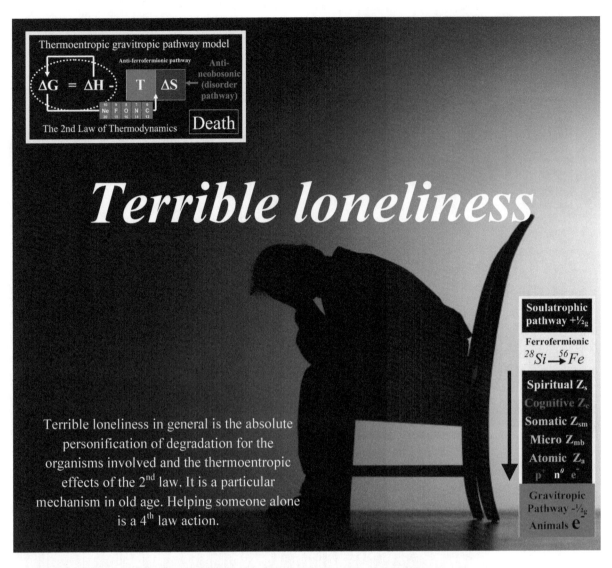

Thermoentropic gravitropic pathway model

Anti-ferrofermionic pathway

$\Delta G = \Delta H - T \Delta S$

Anti-neobosonic (disorder pathway)

Ne F O N C

The 2nd Law of Thermodynamics

Death

Terrible loneliness

Terrible loneliness in general is the absolute personification of degradation for the organisms involved and the thermoentropic effects of the 2nd law. It is a particular mechanism in old age. Helping someone alone is a 4th law action.

Soulatrophic pathway $+\frac{1}{2}_g$

Ferrofermionic

$^{28}Si \rightarrow {}^{56}Fe$

Spiritual Z_s

Cognitive Z_c

Somatic Z_{sm}

Micro Z_{mb}

Atomic Z_a

p^+ n^0 e^-

Gravitropic Pathway $-\frac{1}{2}_g$

Animals e^-

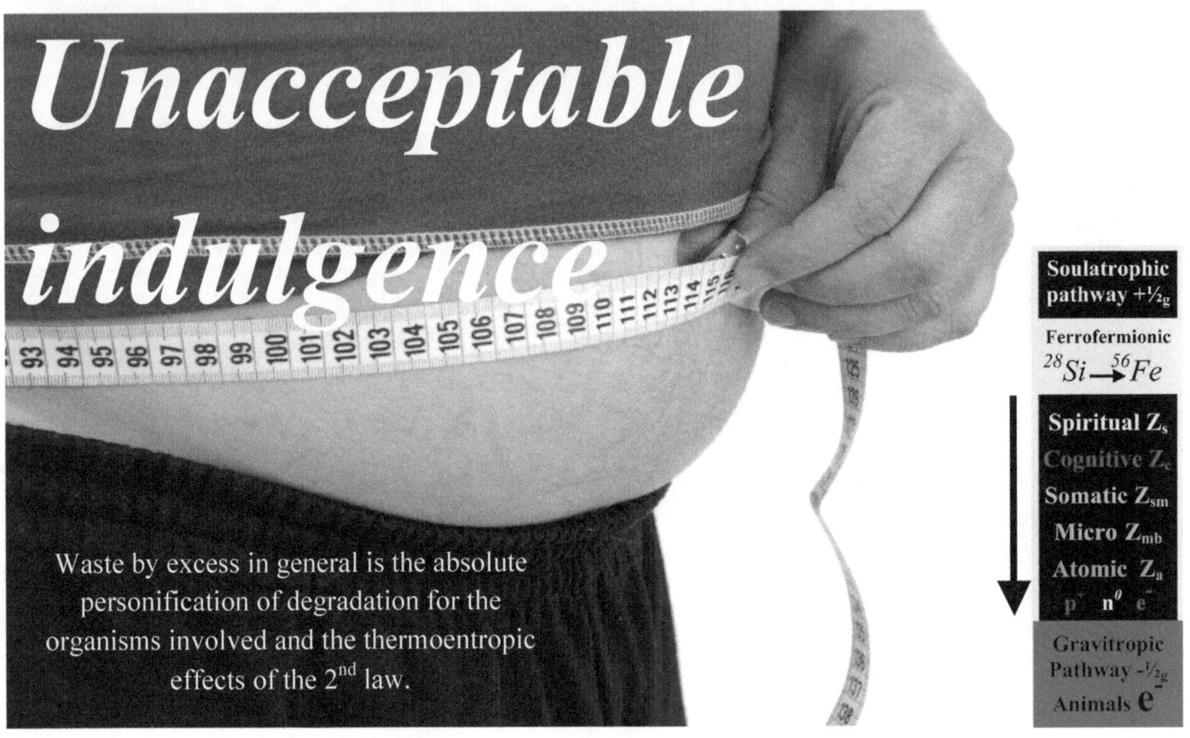

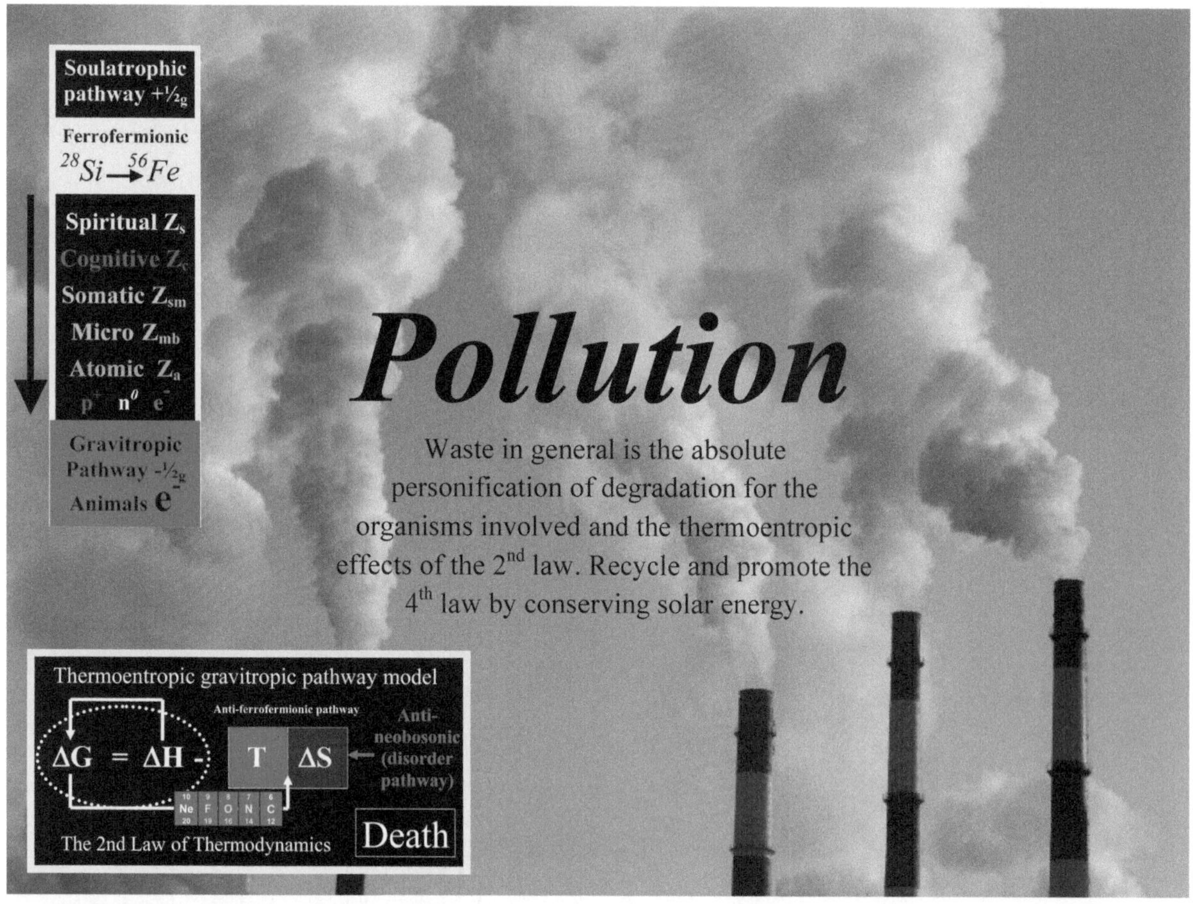
Soulatrophic pathway +½g
Ferrofermionic
28Si → 56Fe
Spiritual Zs
Cognitive Zc
Somatic Zsm
Micro Zmb
Atomic Za
p+ n0 e-
Gravitropic Pathway -½g
Animals e-
Pollution
Waste in general is the absolute personification of degradation for the organisms involved and the thermoentropic effects of the 2nd law. Recycle and promote the 4th law by conserving solar energy.
Thermoentropic gravitropic pathway model
Anti-ferrofermionic pathway
Anti-neobosonic (disorder pathway)
ΔG = ΔH - T ΔS
Ne F O N C
The 2nd Law of Thermodynamics
Death

Pollution
Soulatrophic pathway +½g
Ferrofermionic
28Si → 56Fe
Spiritual Zs
Cognitive Zc
Somatic Zsm
Micro Zmb
Atomic Za
p+ n0 e-
Gravitropic Pathway -½g
Animals e-
Thermoentropic gravitropic pathway model
Anti-ferrofermionic pathway
Anti-neobosonic (disorder pathway)
ΔG = ΔH - T ΔS
Ne F O N C
The 2nd Law of Thermodynamics
Death
Waste in general is the absolute personification of degradation for the organisms involved and the thermoentropic effects of the 2nd law. Recycle and promote the 4th law by conserving solar energy.

Waste in general is the absolute personification of degradation for the organisms involved and the thermoentropic effects of the 2nd law. Recycle and promote the 4th law by conserving solar energy.
Pollution
Soulatrophic pathway +½g
Ferrofermionic
28Si → 56Fe
Spiritual Zs
Cognitive Zc
Somatic Zsm
Micro Zmb
Atomic Za
p+ n0 e-
Gravitropic Pathway -½g
Animals e-
Thermoentropic gravitropic pathway model
Anti-ferrofermionic pathway
Anti-neobosonic (disorder pathway)
ΔG = ΔH - T ΔS
Ne F O N C
The 2nd Law of Thermodynamics
Death

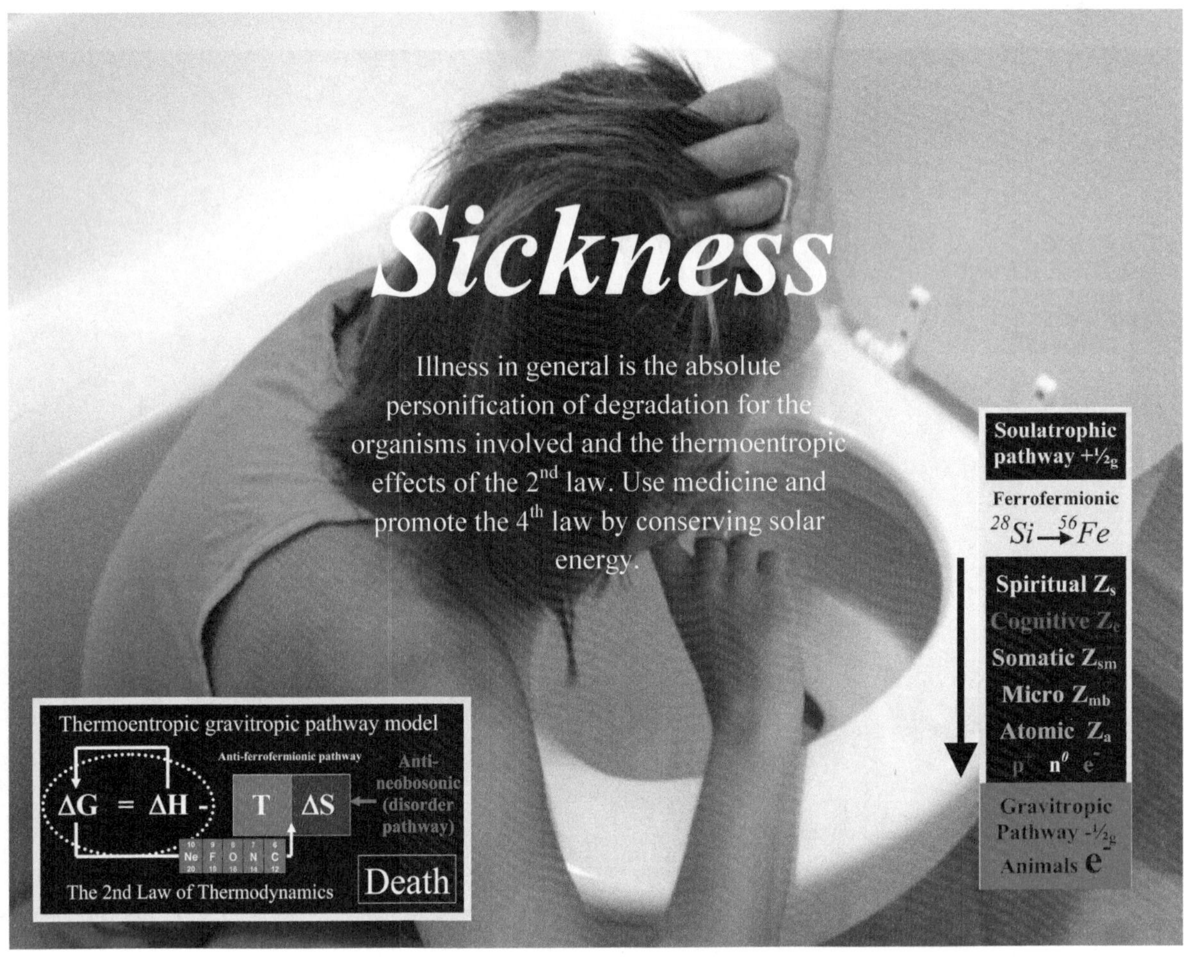
Sickness
Illness in general is the absolute personification of degradation for the organisms involved and the thermoentropic effects of the 2nd law. Use medicine and promote the 4th law by conserving solar energy.
Soulatrophic pathway +½g
Ferrofermionic
28Si → 56Fe
Spiritual Zs
Cognitive Zc
Somatic Zsm
Micro Zmb
Atomic Za
p+ n0 e-
Gravitropic Pathway -½g
Animals e-
Thermoentropic gravitropic pathway model
Anti-ferrofermionic pathway
Anti-neobosonic (disorder pathway)
ΔG = ΔH - T ΔS
Ne F O N C
The 2nd Law of Thermodynamics
Death

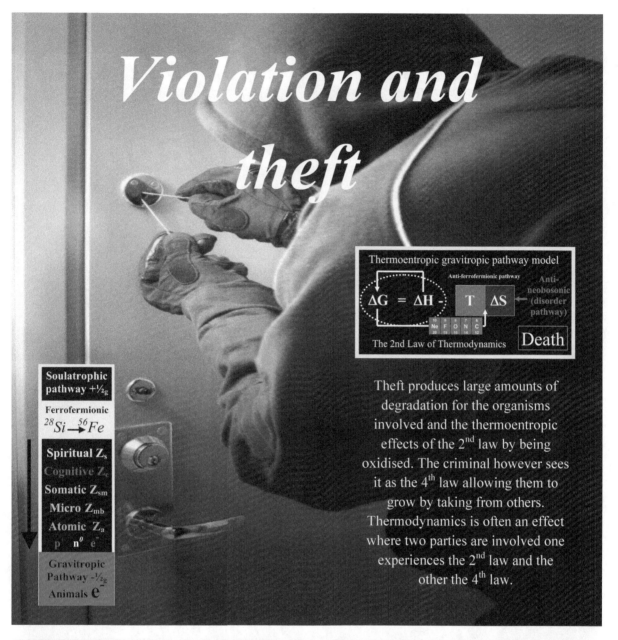

Violation and theft

Thermoentropic gravitropic pathway model

Anti-ferrofermionic pathway

$\Delta G = \Delta H - T \Delta S$

Anti-neobosonic (disorder pathway)

10 Ne 20 | 9 F 19 | 8 O 16 | 7 N 14 | 6 C 12

The 2nd Law of Thermodynamics

Death

Soulatrophic pathway $+\frac{1}{2}_g$

Ferrofermionic

$^{28}Si \rightarrow {}^{56}Fe$

Spiritual Z_s

Cognitive Z_c

Somatic Z_{sm}

Micro Z_{mb}

Atomic Z_a

p^+ n^0 e^-

Gravitropic Pathway $-\frac{1}{2}_g$

Animals e^-

Theft produces large amounts of degradation for the organisms involved and the thermoentropic effects of the 2nd law by being oxidised. The criminal however sees it as the 4th law allowing them to grow by taking from others. Thermodynamics is often an effect where two parties are involved one experiences the 2nd law and the other the 4th law.

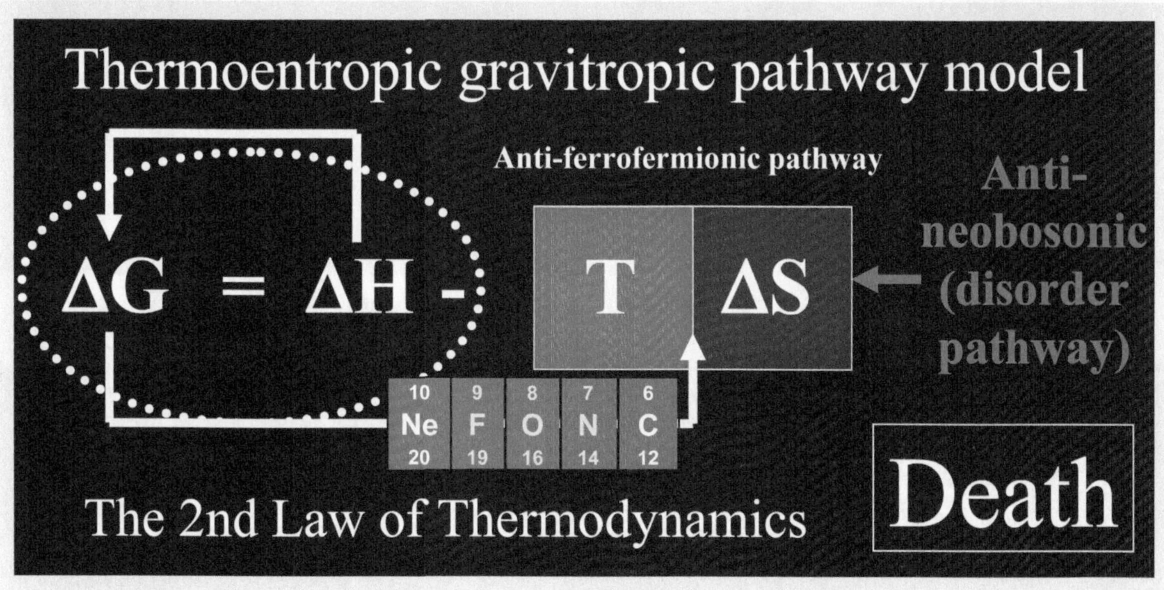

Death & disorder
The Hadean Eon
Carbon dioxide
& carbonate forms
n
m l
O
C
O
Ca++
O-
O-
C
O
Thermoentropic
pathways
l
m s
Gravitropism
The long term carbon cycle and the
2nd law of thermodynamics

Neurochromosome heliotropism - the thermodynamics of morality, the spiritual balance of good versus evil; free will and life.

Heliotropism - soulatrophic energy levels can be thought of as levels of consciousness. All organisms have a property of 'free will' they can opt to follow soulatrophic pathways and progress to higher levels of order (towards artificial life) or they can opt to liberate energy from soulatrophic pathways pushing life back down such pathways effectively sending life backwards evolutionary speaking.

Heliotropism (free will) a word taken from the plants is used to describe how a plant follows the Sun and improves its chances of survival by doing so. All life is interfaced against the plants which are 'primary producers', hence the same logic is translated into the animals. Hence heliotropism (free will) is an accurate description of an animal's free will as well as plants. Phototropism (good) and gravitropism (evil) are also plant associated words. Phototropism (good) is the act of conserving solar energy in living organisms, and gravitropism (evil) is used to describe how life is under the constant action of Earth's restrictive pull in equilibrium with its solar animation.

The science of actively seeking out solar energy and locking it into carbon soulatrophic bonds and pathways; active minimisation of suffering in carbon entrochiraloctet cells is heliotropic. The basis in thermodynamics is found in the Gibbs equation where Gibbs free energy is made available to organisms for liberation to thermoentropic states. Liberation following a phototropic (good) pathway invariably fits the ferrofermionic neobosonic dynamic model. Energy conservation is balanced against energy liberation. For living stability a solar pathway is advantageous. It is possible to favour an anti-solar anti-accumulator gravitropic (evil) pathway pushing life back through evolution to simpler forms.

This thermoentropic system leads to lower soulatrophic energy levels and reduced soulatrophic hierarchy actively taking life back through time. When a person murders another, they produce thermoentropic states and collapse the wave functions down to that of the microbiological level. The energy is liberated and made unavailable (thermoentropic) for life on higher soulatrophic energy levels. **The 10 commandments in classical religion are a model based on thermodynamics where thermodynamically stabilising advice is produced for the establishment of living stability.**

This is the first clear model of a soulatrophic phototropic (good) ferrofermionic pathway (ECASSE Entromorphic Carbon Amplification by Solar Stimulated Emission). Conservation of living solar energies is the basic key to such systems (ECASSE). Spending Gibbs free energy by living fermionic microstates in the accumulator model means a net increase in enthalpic content is achievable and desirable for living stability (good – phototropism). You can spend £10 on the lottery and win £1,000,000 back. Hence overall rates of accumulation are possible should a heliotropic (free will) phototropic (good) pathway be considered often this is a risk taking opportunity.

In reverse the spending of £10 on the lottery can yield no win and hence £10 worth of thermoentropic (evil) unavailable energy. The Heliotropic (free will) choice is based on the plants where a flower follows the pathway of the Sun in order to maximise its order. In the animals this is indirectly followed as the solar input comes through the global K shell (plants) for fixing into carbon entrochiraloctet bonds. Heliotropism (free will) is the fundamental basis of decisive morality, which concerns itself with life's freedom and stability and development and inevitable stability (human and animal rights).

The 4th law of thermodynamics allows a conscious being to have the thermodynamic (heliotropic) free will to decide to drive the 2nd thermoentropic law or the Gibbs enthalpic 4th law. We can decide whether to grow or die! The 2nd law however in the very long run will prevail, the 4th law buys life time and lots and lots of it through technology. Who knows how long humans will live in the next 50 to 100 years!

2nd law

2nd law of thermodynamics – thermoentropic pathways leading to reduced evolution or even reversed evolution.

Soulatrophic pathway $+\frac{1}{2}_g$

Ferrofermionic

$^{28}Si \rightarrow {}^{56}Fe$

Spiritual Z_s

Cognitive Z_c

Somatic Z_{sm}

Micro Z_{mb}

Atomic Z_a

p n^0 e^-

Gravitropic Pathway $-\frac{1}{2}_g$

Animals e^-

4th law

4th law of thermodynamics – Gibbs enthalpic pathways leading to increased evolution or even amplified evolution.

Thermoentropic gravitropic pathway model

Anti-ferrofermionic pathway

$\Delta G = \Delta H - T \Delta S$

Anti-neobosonic (disorder pathway)

Ne F O N C

Death

The 2nd Law of Thermodynamics

Gibbs enthalpic neoiterative phototropic accumulator work model

'Le Chatelier's principle'

Acts of free will

Solar neobosonic (order) pathway

Ferrofermionic pathway

$\Delta G = \Delta H - T \Delta S$

C N O F Ne

Growth

The 4th Law of Thermodynamics

Freedom of choice!

Which pathway will you follow?

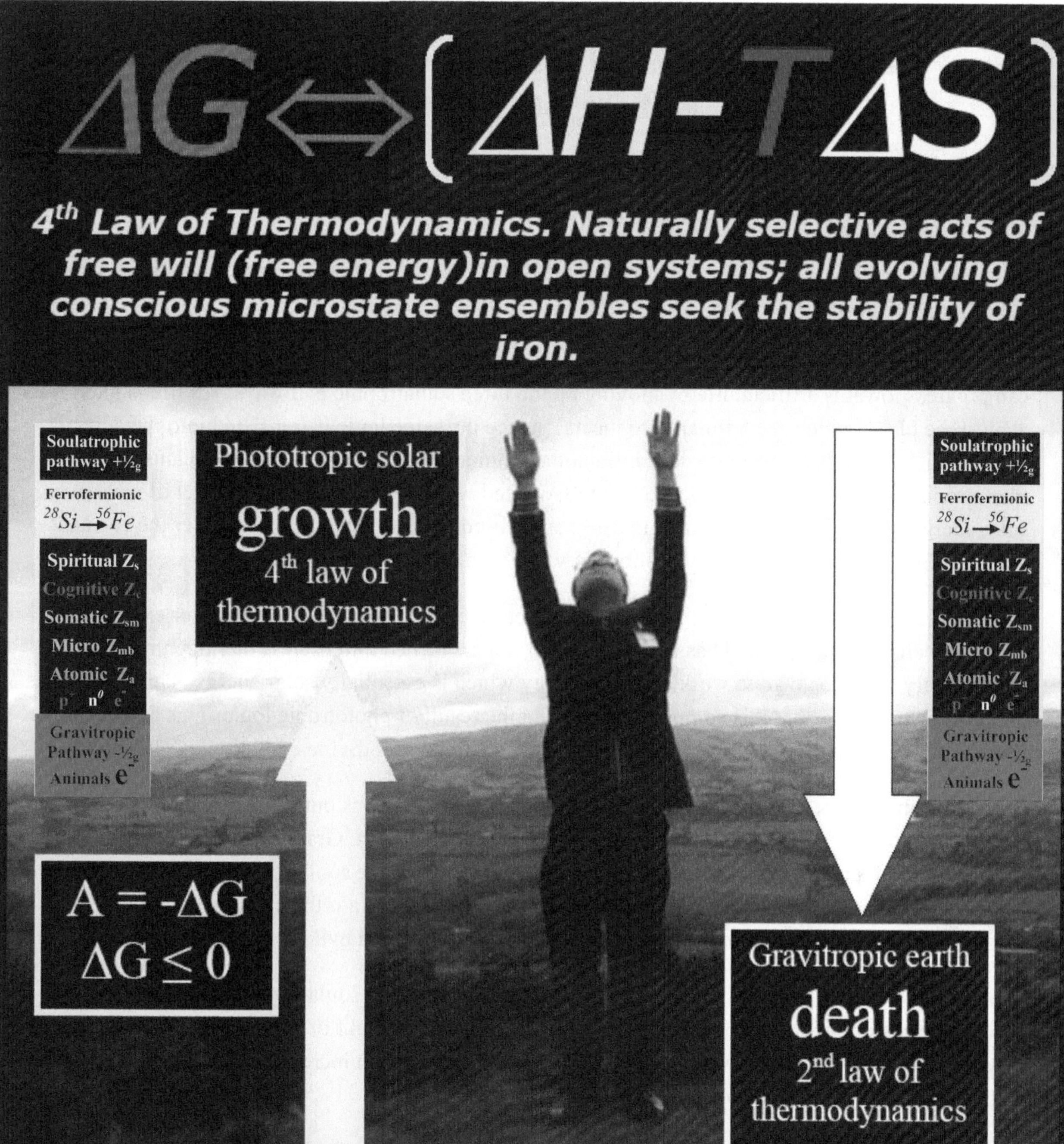
ΔG⇔(ΔH-TΔS)
4th Law of Thermodynamics. Naturally selective acts of free will (free energy)in open systems; all evolving conscious microstate ensembles seek the stability of iron.
Soulatrophic pathway +½g
Ferrofermionic
28Si→56Fe
Spiritual Zs
Cognitive Zc
Somatic Zsm
Micro Zmb
Atomic Za
Gravitropic Pathway -½g
Animals e-
Phototropic solar
growth
4th law of thermodynamics
A = -ΔG
ΔG ≤ 0
Gravitropic earth
death
2nd law of thermodynamics
2nd Law of Thermodynamics. The total entropy of an isolated system increases over time. As time passes the universe is becoming increasingly disordered.
Heliotropism

Neurochromosome phototropism - delaying the 2nd law of thermodynamics, and the morality concepts of good (ECASSE, NFVE).

Phototropism (morally good) is a life supporting concept which relates to the inevitable effect of the thermodynamic growth, the second law and gravitropism is the reverse, in that decay is increasing in the universe, thermoentropic pathways are evil, Gibbs enthalpic pathways are good. For example murder liberates 13.7 billion years of stored energy, this is anti-life anti-solar conservation or anti-phototropism (reverse of evolution).

Morality is a thermodynamic system and energy conservation is a process of good where organisms can progress towards artificial life by moving up the three soulatrophic pathways. All life is interfaced against the plants which are 'primary producers', hence the same logic is translated into the animals. Hence heliotropism is an accurate description of an animal's free will as well as the plants. Phototropism and gravitropism are also plant associated words. Phototropism is the act of conserving solar energy in living organisms, and gravitropism is used to describe how life is under the constant action of Earth's restrictive pull in equilibrium with its solar animation.

Gravitropism acts in reverse effectively acting as a basic model of natural selection, for example the credit crunch is gravitropic and has seen trillions of dollars turned into useless thermoentropy or waste. Every day people go to work to make money which is essentially solar energy. This money has phototropic origins in that it is completely solar and the result of phototropic logic. This accumulative process drives life up soulatrophic pathways towards artificial life, divine evolutionary stability.

Phototropism appears to challenge the second law of thermodynamics but the Earth is not a closed system as the Sun pumps massive quantities of energy into the Earth. Gravitropism by comparison is turning the solar energy into waste pushing humanity back down the soulatrophic pathways. Morality is often classed as metaphysical, carbon entromorphology brings it into the physical through thermodynamics, and energy management, and this makes good and evil measureable concepts.

Phototropism – the science of carbon entrochiraloctet amplification soulatrophic pathway optimisation, morally good. Soulatrophic energy levels can be thought of in an organism as levels of consciousness: the process by which living systems based on carbon increase in size by amplification and conservation of orbital probability fields.

Phototropism is based on any solar provider; the Sun personifies this for Earth-based life. Phototropism is the science of 'good' and leads living systems away from planet Earth into space and towards neobosonic ferrofermionic quantum stability. Phototropism invariably leads to higher levels of existence. Post carbosilicoferrous event, carbon energies become transmitted to the rest of the periodic table of the elements. This conveys higher levels of stability and the ability to leave Earth's potent gravity (the Earth bondage effect).

Phototropism leads to 'artificial life' a state of living existence where the entity involved becomes stable indefinitely. Energy is no longer fought over and is available in limitless quantities the heavenly macrostate. Higher existence is part of the 'divine inversion' effect. This occurs where organisms construct hierarchy based on an extrapolation of the classical soulatrophic pathway.

The pursuit of divine truth produces scientific doctrines, which dramatically increase phototropic pathways through technology. The pathways lead to a state of artificial life where man has created a

God in its own image, hence an inversion of classical divine doctrines of truth. Where humans try to follow a God's example they invariably create God like entities over long enough time as artificial life forms (androidism). **The awareness of the concept of God is the awareness of ferrofermionic pathways.**

In artificial life states, existence is based on a super conductive state of consciousness. Thermoentropic pathways are blocked and limited and therefore heat is also withheld from any such system; Neobosonic and ferrofermionic conservational effect.

Spiritual soulatrophic spin is spingrav up +½, solar spin solar alignment. The Sun has a magnetic quantum effect inducing deformation evaporative evolutionary effect of conserved sub shell probability distributions. This is the environmental filtering aspect of natural selection with the growth cycle and pathway.

One possible goal of evolution by natural selection is for organisms to live in peace forever.

Phototropic – associated with K shell electrons and associated energies. The planetary K shell is the plant world, highly phototropic energy fixers. Halogenous potentials favour the organism. The K shell is often associated with visible light where L shell components are more associated with infrared energies. K shells contain nuclear elements and communicate position in space by 'nuclear flaunting'.

Phototropic regions or K shells are seen in flowers, women use make up to produce phototropic reactions to gravitropic males due to their phototropic 'keeper of the nucleus' energy level. It is associated with particle (nuclear) states. Heliotropic (free will) pursuit of solar energy through neobosonic ferrofermionic pathways leads to quantum stability and artificial life.

Planetary cytokinesis – The ability for carbon soulatrophic pathways to lead to an escape velocity through a sufficiently large enough Gibbs free energy. To leave the Earth and enter space, forming a new post Earth cell in geostationary orbit or complete exit from the Earth to artificial life, fertilising space by gravitational barrier penetration (escape velocity).

This is the point at which the electromagnetic and gravitational entrochiraloctets become completed by ECASSE (Entromorphic Carbon Amplification by Solar Stimulated Emission). The Pauli cell becomes complete with both electromagnetic and gravitational energies developed; hence cytokinesis takes place through environmental separation. At which point life fertilises space where a rocket is a bosonic force carrier particle taking genomic forces into space. Carbon produces massive quantities of intermediate particle/wave duality organisms.

Macrostate – the macroscopic gravitational properties of a soulatrophic environment seen in carbon entromorphology as solid state, liquid state, gas state and vacuum state (Earth bound environments) and the energy thermodynamic way in which life transcends the different levels through phototropism.

The macroscopic properties are pressure and temperature, which induce natural selection in low probability microstates. This is based on the gravitational entrochiraloctet (quantum taxonomy). It is characterised as a probability distribution on a certain ensemble of cellular soulatrophic entrochiraloctet microstates or cellular organisms under an environmental housing (ecology), which induces orderly selectivity or survival of the fittest.

The macrostate is a complex system, for example Manchester United Old Trafford can be thought of the macrostate defined by the probability distribution of an ensemble of microstates (the team) in the course of their thermal fluctuations (the league). Ecology is the basis of the relationship between microstate and macrostate.The following are phototropic examples.

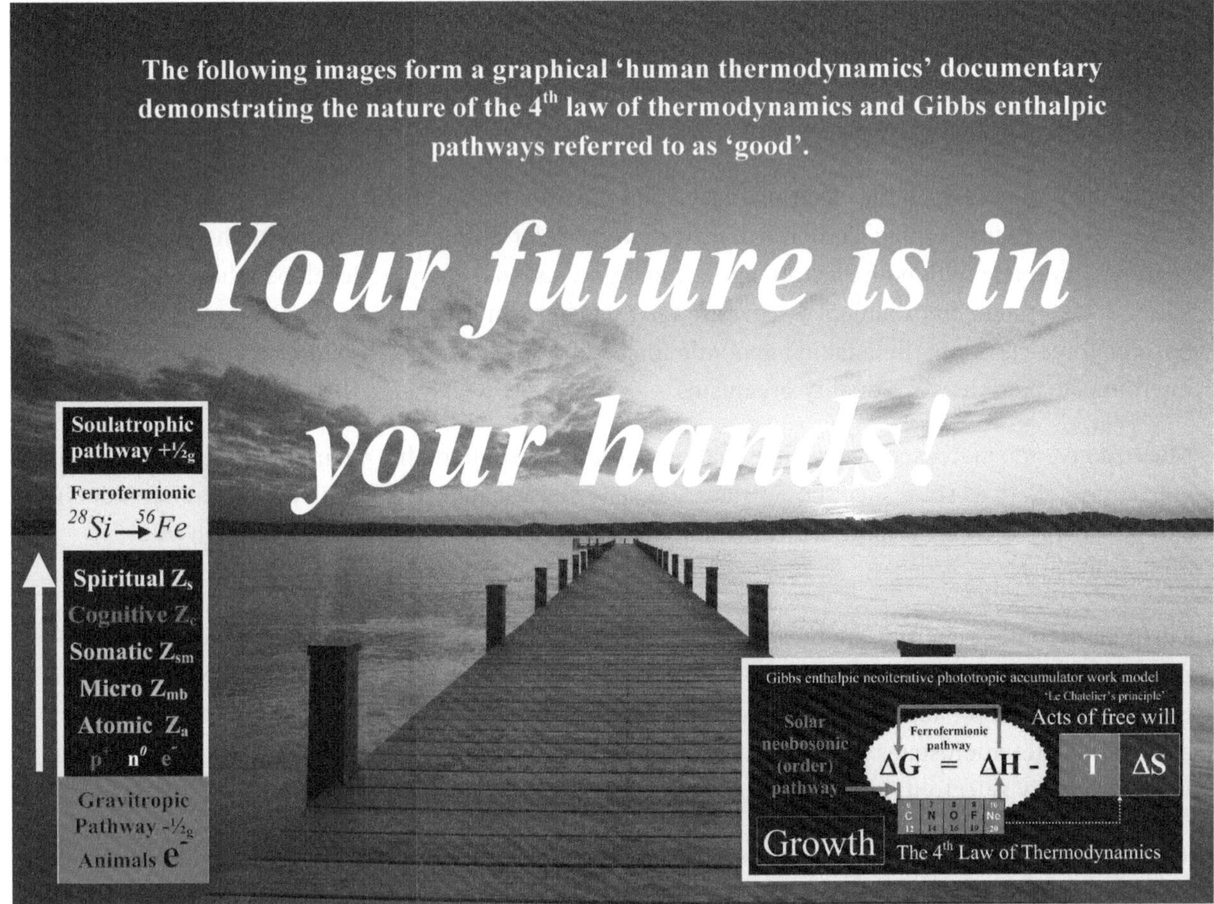

Love and embrace your family!
Soulatrophic pathway +½g
Ferrofermionic
²⁸Si → ⁵⁶Fe
Spiritual Zs
Cognitive Zc
Somatic Zsm
Micro Zmb
Atomic Za
p⁺ n⁰ e⁻
Gravitropic Pathway -½g
Animals e⁻
The 4th law is all about growth and no greater example exists than through family growth. It continues to increase the time required for the 2nd law to act.

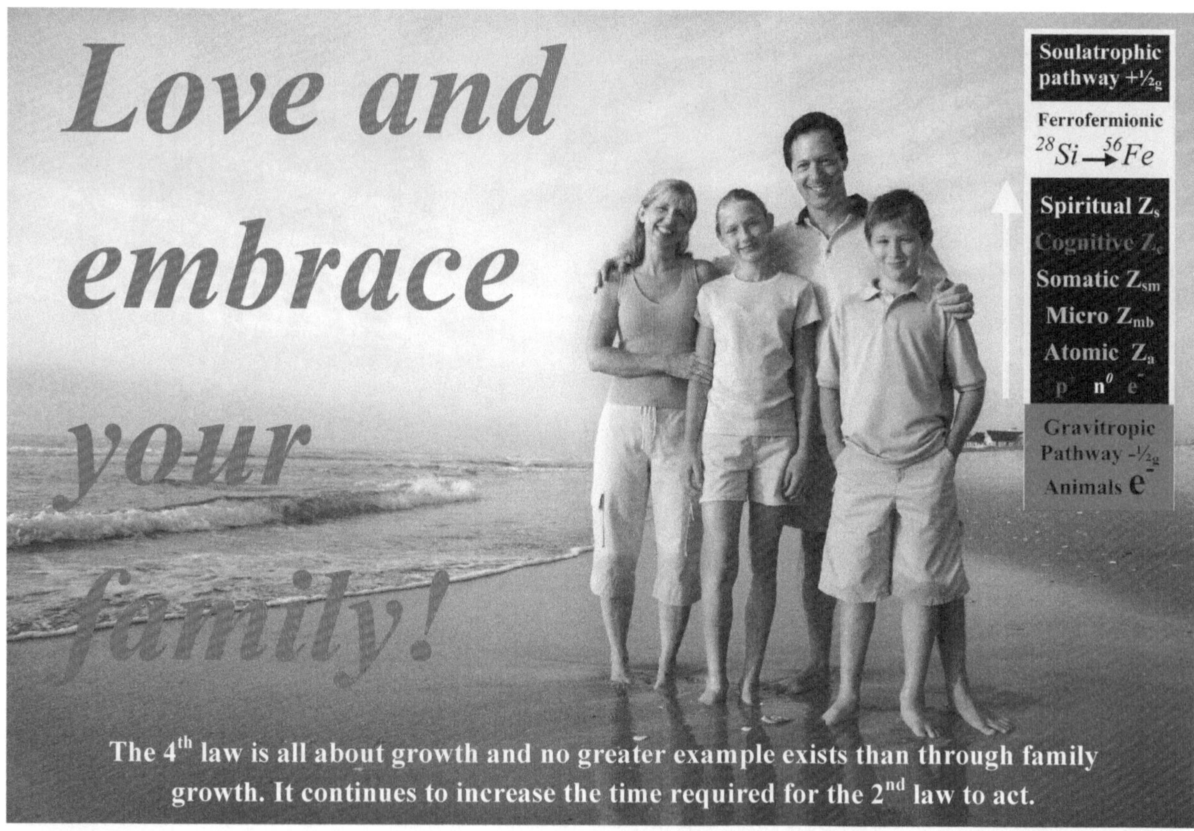

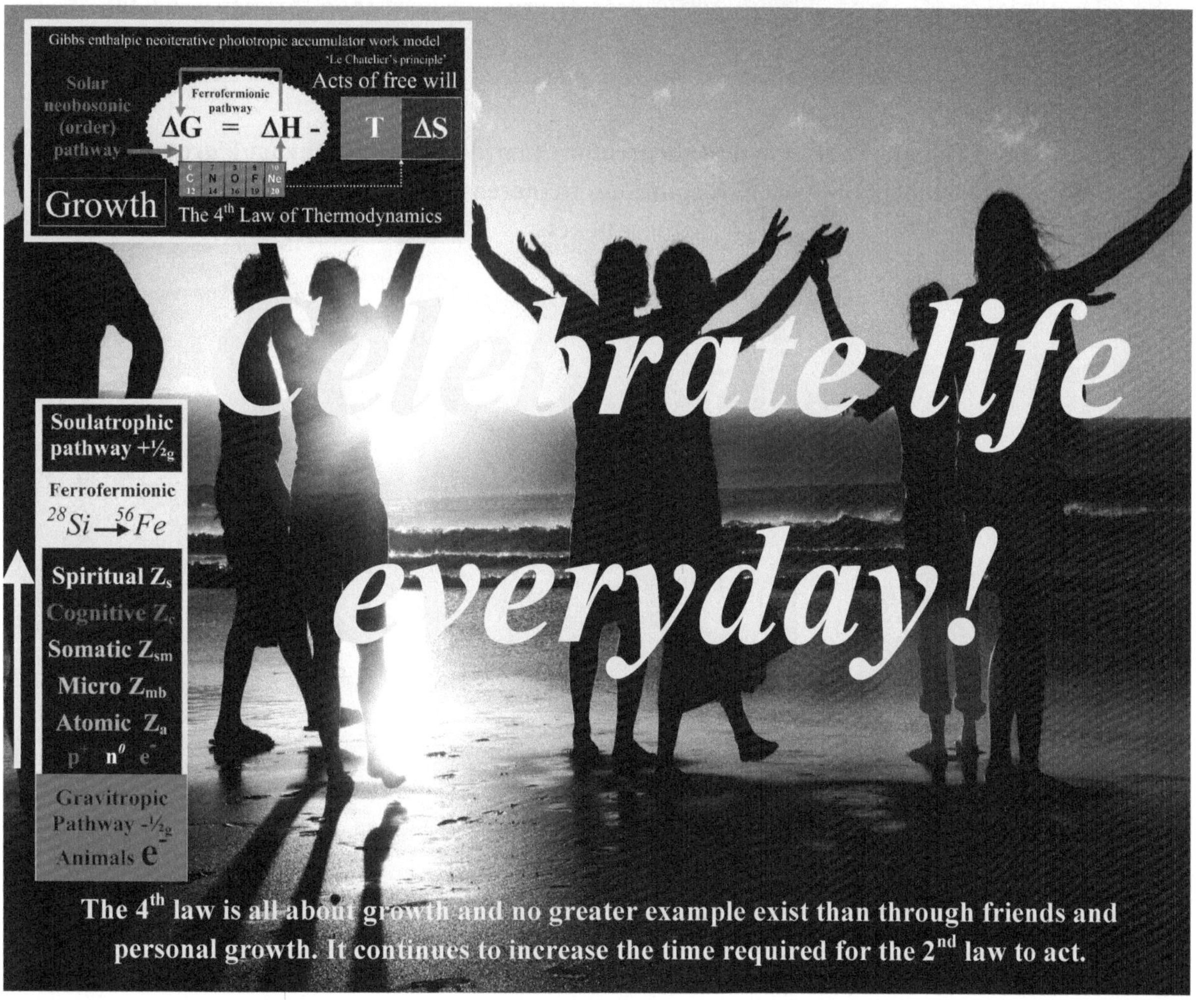

Gibbs enthalpic neoiterative phototropic accumulator work model
'Le Chatelier's principle'
Solar neobosonic (order) pathway
Ferrofermionic pathway
ΔG = ΔH - T ΔS
Acts of free will
C N O F Ne
Growth
The 4th Law of Thermodynamics
Celebrate life everyday!
Soulatrophic pathway +½g
Ferrofermionic
²⁸Si → ⁵⁶Fe
Spiritual Zs
Cognitive Zc
Somatic Zsm
Micro Zmb
Atomic Za
p⁺ n⁰ e⁻
Gravitropic Pathway -½g
Animals e⁻
The 4th law is all about growth and no greater example exist than through friends and personal growth. It continues to increase the time required for the 2nd law to act.

Celebrate what you have
Soulatrophic pathway +½g
Ferrofermionic
28Si→56Fe
Spiritual Zs
Cognitive Zc
Somatic Zsm
Micro Zmb
Atomic Za
p+ n0 e-
Gravitropic Pathway -½g
Animals e-
Gibbs enthalpic neoiterative phototropic accumulator work model
'Le Chatelier's principle'
Acts of free will
Solar neobosonic (order) pathway
Ferrofermionic pathway
ΔG = ΔH - T ΔS
Growth
The 4th Law of Thermodynamics
The 4th law is all about growth and no greater example exist than through peaceful growth and celebration. It continues to increase the time required for the 2nd law to act.

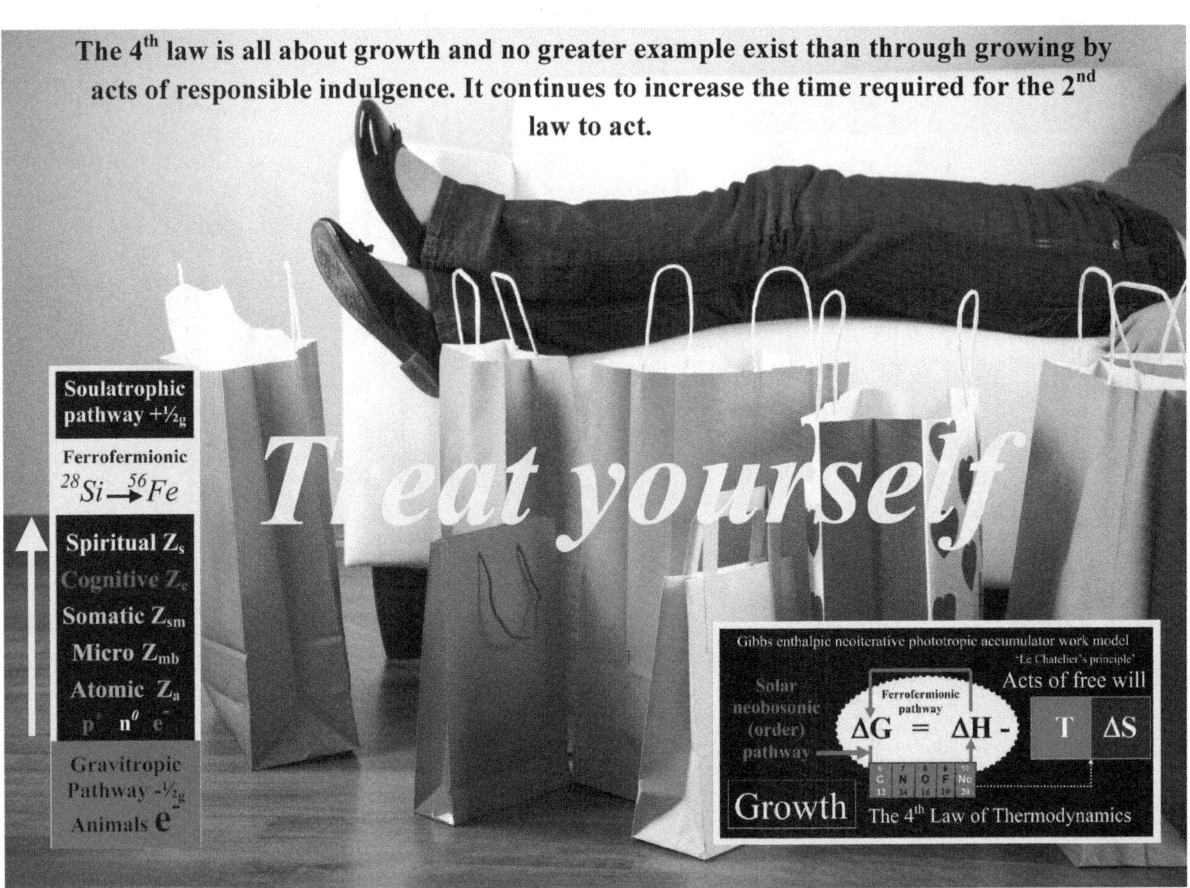
The 4th law is all about growth and no greater example exist than through growing by acts of responsible indulgence. It continues to increase the time required for the 2nd law to act.
Treat yourself
Soulatrophic pathway +½g
Ferrofermionic
28Si→56Fe
Spiritual Zs
Cognitive Zc
Somatic Zsm
Micro Zmb
Atomic Za
p+ n0 e-
Gravitropic Pathway -½g
Animals e-
Gibbs enthalpic neoiterative phototropic accumulator work model
'Le Chatelier's principle'
Acts of free will
Solar neobosonic (order) pathway
Ferrofermionic pathway
ΔG = ΔH - T ΔS
Growth
The 4th Law of Thermodynamics

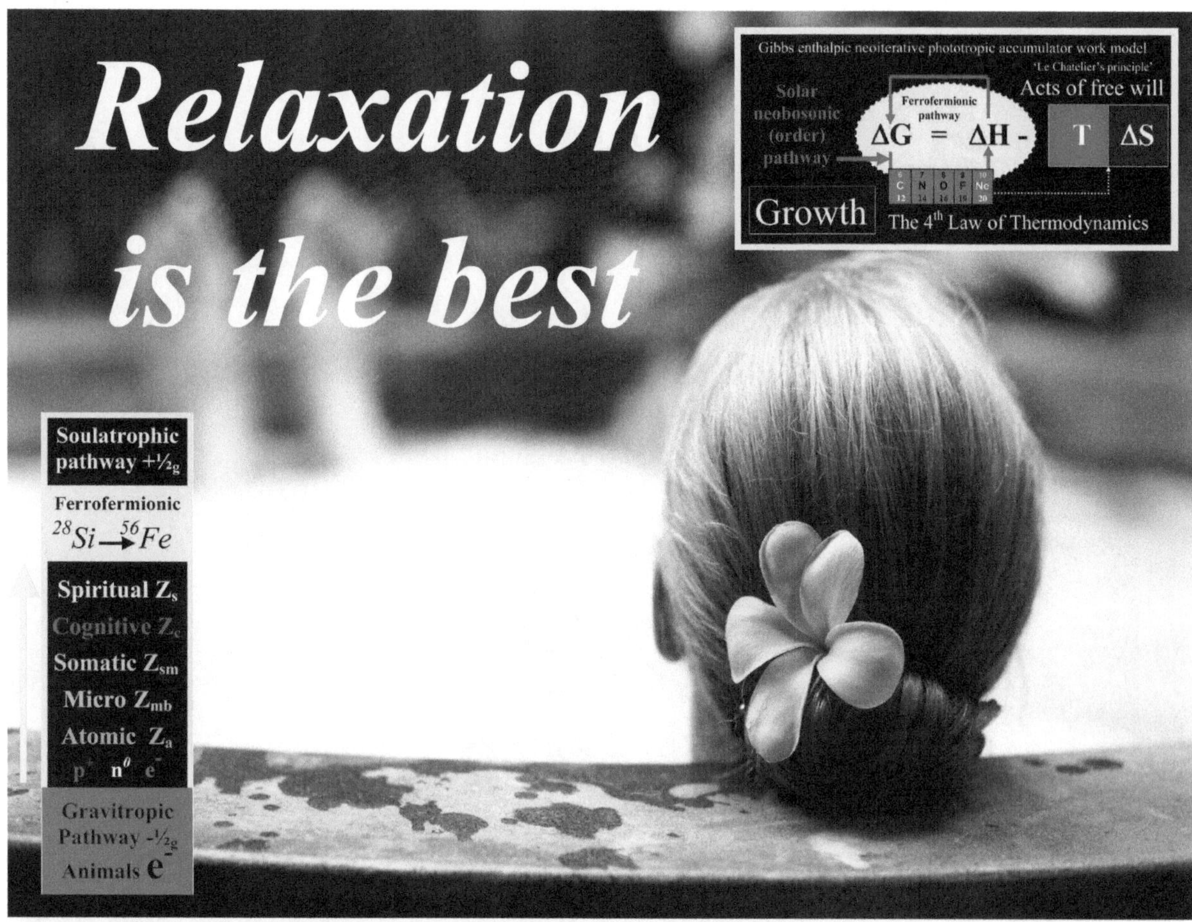
Relaxation is the best
Gibbs enthalpic neoiterative phototropic accumulator work model
'Le Chatelier's principle'
Solar neobosonic (order) pathway
Ferrofermionic pathway
Acts of free will
ΔG = ΔH - T ΔS
C N O F Ne
Growth
The 4th Law of Thermodynamics
Soulatrophic pathway +½g
Ferrofermionic
28Si → 56Fe
Spiritual Zs
Cognitive Zc
Somatic Zsm
Micro Zmb
Atomic Za
p+ n0 e-
Gravitropic Pathway -½g
Animals e-

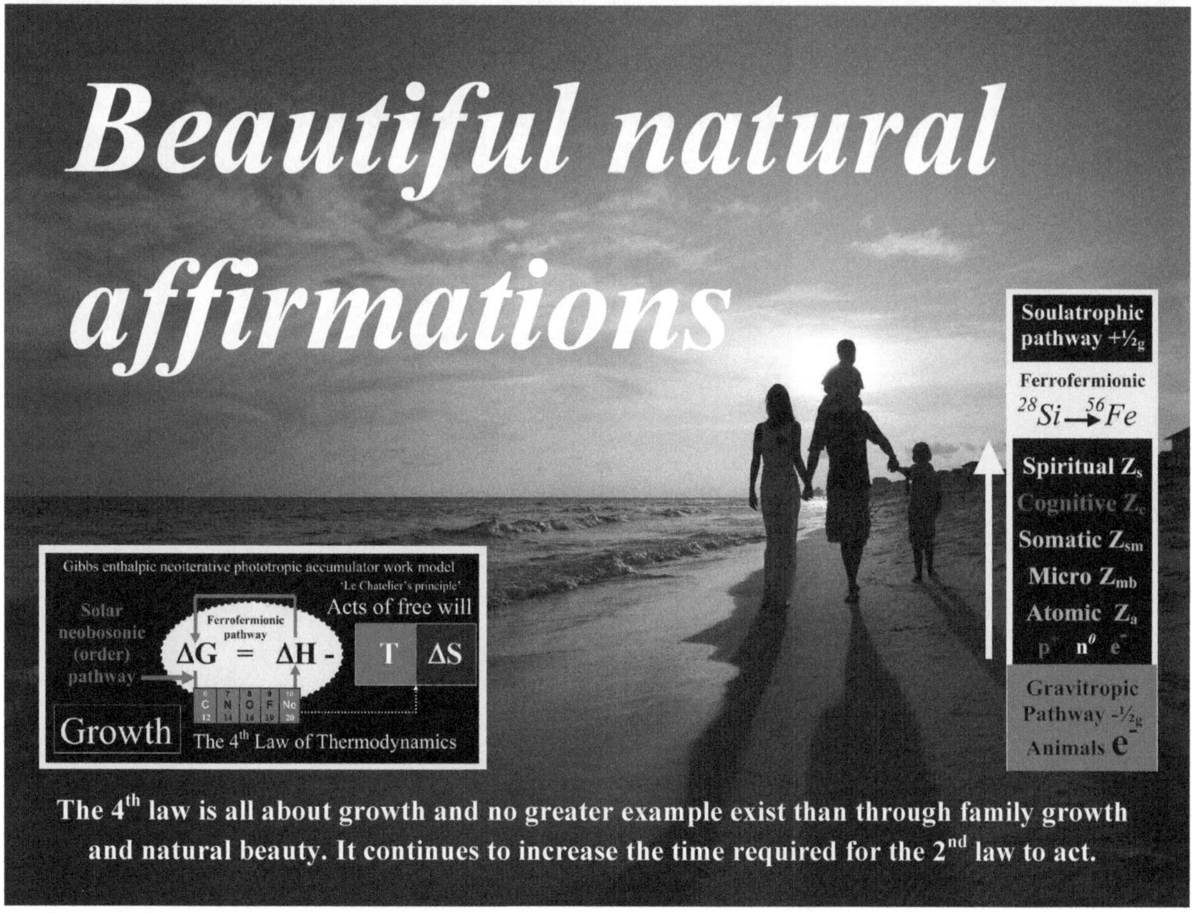
Beautiful natural affirmations
Soulatrophic pathway +½g
Ferrofermionic
28Si → 56Fe
Spiritual Zs
Cognitive Zc
Somatic Zsm
Micro Zmb
Atomic Za
p+ n0 e-
Gravitropic Pathway -½g
Animals e-
Gibbs enthalpic neoiterative phototropic accumulator work model
'Le Chatelier's principle'
Solar neobosonic (order) pathway
Ferrofermionic pathway
Acts of free will
ΔG = ΔH - T ΔS
C N O F Ne
Growth
The 4th Law of Thermodynamics
The 4th law is all about growth and no greater example exist than through family growth and natural beauty. It continues to increase the time required for the 2nd law to act.

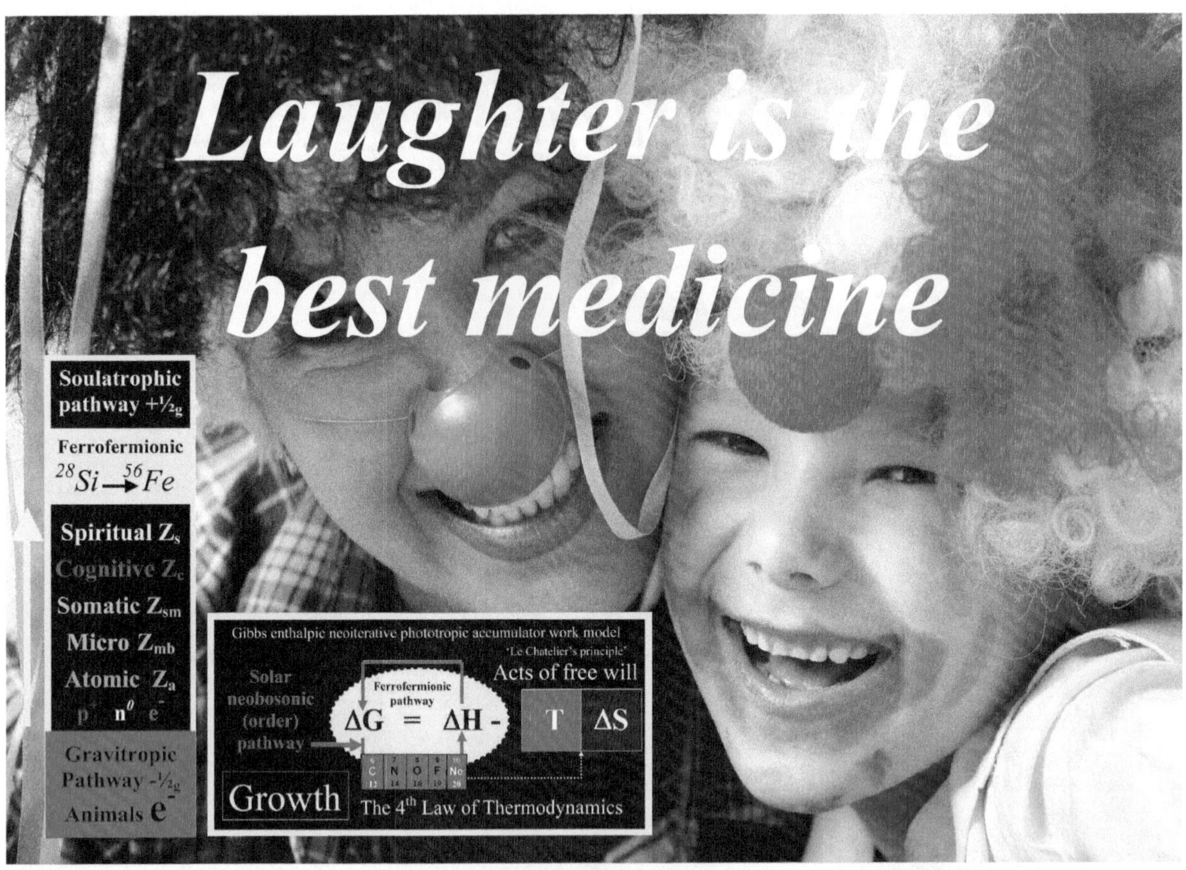
Laughter is the best medicine
Soulatrophic pathway +½g
Ferrofermionic
28Si→56Fe
Spiritual Zs
Cognitive Zc
Somatic Zsm
Micro Zmb
Atomic Za
p+ n0 e-
Gravitropic Pathway -½g
Animals e-
Gibbs enthalpic neoiterative phototropic accumulator work model
'Le Chatelier's principle'
Acts of free will
Solar neobosonic (order) pathway
Ferrofermionic pathway
ΔG = ΔH - T ΔS
C N O F Ne
Growth
The 4th Law of Thermodynamics

Celebrate Humanity!
Soulatrophic pathway +½g
Ferrofermionic
28Si→56Fe
Spiritual Zs
Cognitive Zc
Somatic Zsm
Micro Zmb
Atomic Za
p+ n0 e-
Gravitropic Pathway -½g
Animals e-
Gibbs enthalpic neoiterative phototropic accumulator work model
'Le Chatelier's principle'
Acts of free will
Solar neobosonic (order) pathway
Ferrofermionic pathway
ΔG = ΔH - T ΔS
C N O F Ne
Growth
The 4th Law of Thermodynamics

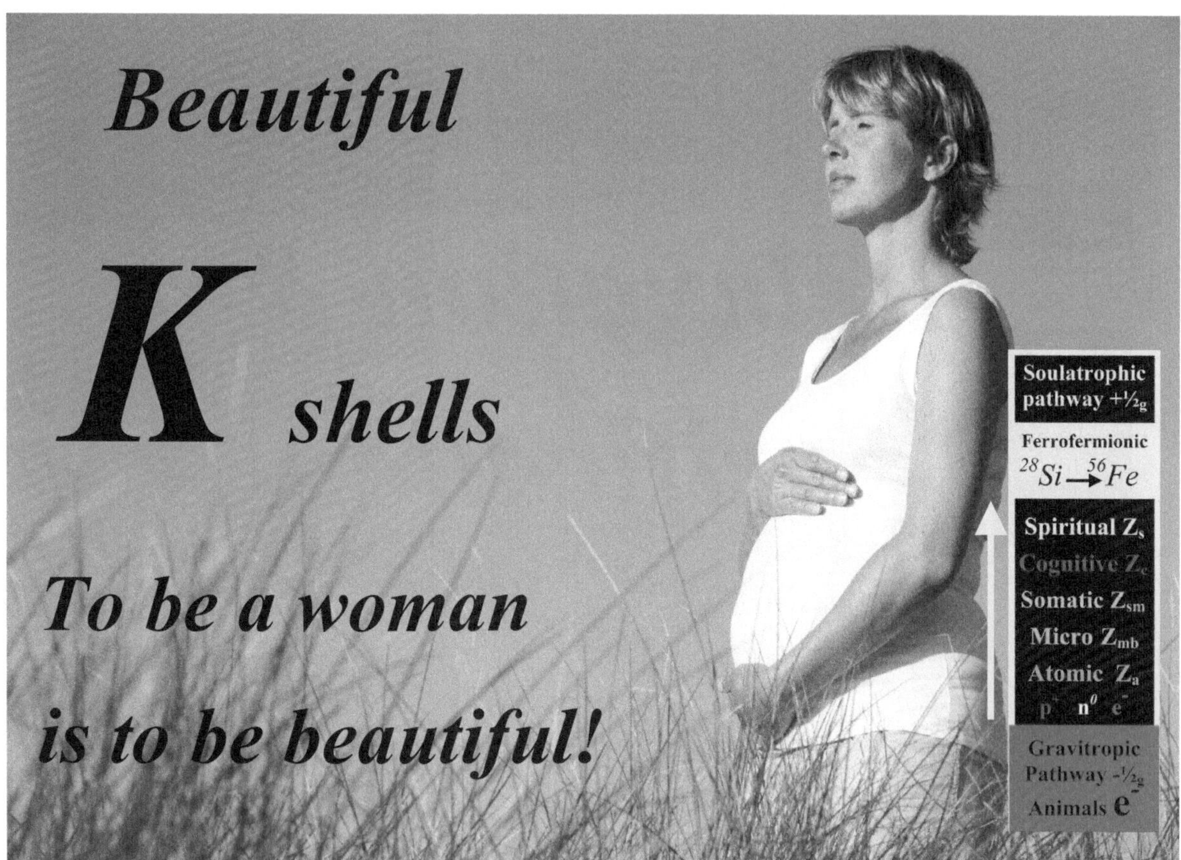
Beautiful
K shells
To be a woman
is to be beautiful!
Soulatrophic pathway +½g
Ferrofermionic
28Si → 56Fe
Spiritual Zs
Cognitive Zc
Somatic Zsm
Micro Zmb
Atomic Za
p+ n0 e-
Gravitropic Pathway -½g
Animals e-

The strong force
and love
Soulatrophic pathway +½g
Ferrofermionic
28Si → 56Fe
Spiritual Zs
Cognitive Zc
Somatic Zsm
Micro Zmb
Atomic Za
p+ n0 e-
Gravitropic Pathway -½g
Animals e-
Gibbs enthalpic neoiterative phototropic accumulator work model
'Le Chatelier's principle'
Acts of free will
Solar neobosonic (order) pathway
Ferrofermionic pathway
ΔG = ΔH - T ΔS
C N O F Ne
Growth
The 4th Law of Thermodynamics
The 4th law is all about growth and no greater example exist than through family growth, by reproducing we hold back the 2nd law and promote Gibbs enthalpic thermodynamics.

Soulatrophic pathway $+\frac{1}{2}_g$
Ferrofermionic
$^{28}Si \rightarrow ^{56}Fe$
Spiritual Z_s
Cognitive Z_c
Somatic Z_{sm}
Micro Z_{mb}
Atomic Z_a
p^+ n^0 e^-
Gravitropic Pathway $-\frac{1}{2}_g$
Animals e^-
Relaxation & purification
Gibbs enthalpic neoiterative phototropic accumulator work model
'Le Chatelier's principle'
Acts of free will
Solar neobosonic (order) pathway
Ferrofermionic pathway
$\Delta G = \Delta H - T \Delta S$
C N O F Ne
Growth
The 4th Law of Thermodynamics

Soulatrophic pathway $+\frac{1}{2}_g$
Ferrofermionic
$^{28}Si \rightarrow ^{56}Fe$
Spiritual Z_s
Cognitive Z_c
Somatic Z_{sm}
Micro Z_{mb}
Atomic Z_a
p^+ n^0 e^-
Gravitropic Pathway $-\frac{1}{2}_g$
Animals e^-
Gibbs enthalpic neoiterative phototropic accumulator work model
'Le Chatelier's principle'
Acts of free will
Solar neobosonic (order) pathway
Ferrofermionic pathway
$\Delta G = \Delta H - T \Delta S$
C N O F Ne
Growth
The 4th Law of Thermodynamics
Future stability
Happy 234th birthday Sue!

The 4th law is all about growth and no greater example exist than through love and growth, by sharing our lives we hold back the 2nd law and promote Gibbs enthalpic thermodynamics.
Gibbs enthalpic neoiterative phototropic accumulator work model
'Le Chatelier's principle'
Acts of free will
Solar neobosonic (order) pathway
Ferrofermionic pathway
ΔG = ΔH - T ΔS
C N O F Ne
Growth
The 4th Law of Thermodynamics
Soulatrophic pathway +½g
Ferrofermionic
28Si → 56Fe
Spiritual Zs
Cognitive Zc
Somatic Zsm
Micro Zmb
Atomic Za
p+ n0 e-
Gravitropic Pathway -½g
Animals e-
Celebrate human chemistry and bonding!

Indulge yourself!
Gibbs enthalpic neoiterative phototropic accumulator work model
'Le Chatelier's principle'
Acts of free will
Solar neobosonic (order) pathway
Ferrofermionic pathway
ΔG = ΔH - T ΔS
C N O F Ne
Growth
The 4th Law of Thermodynamics

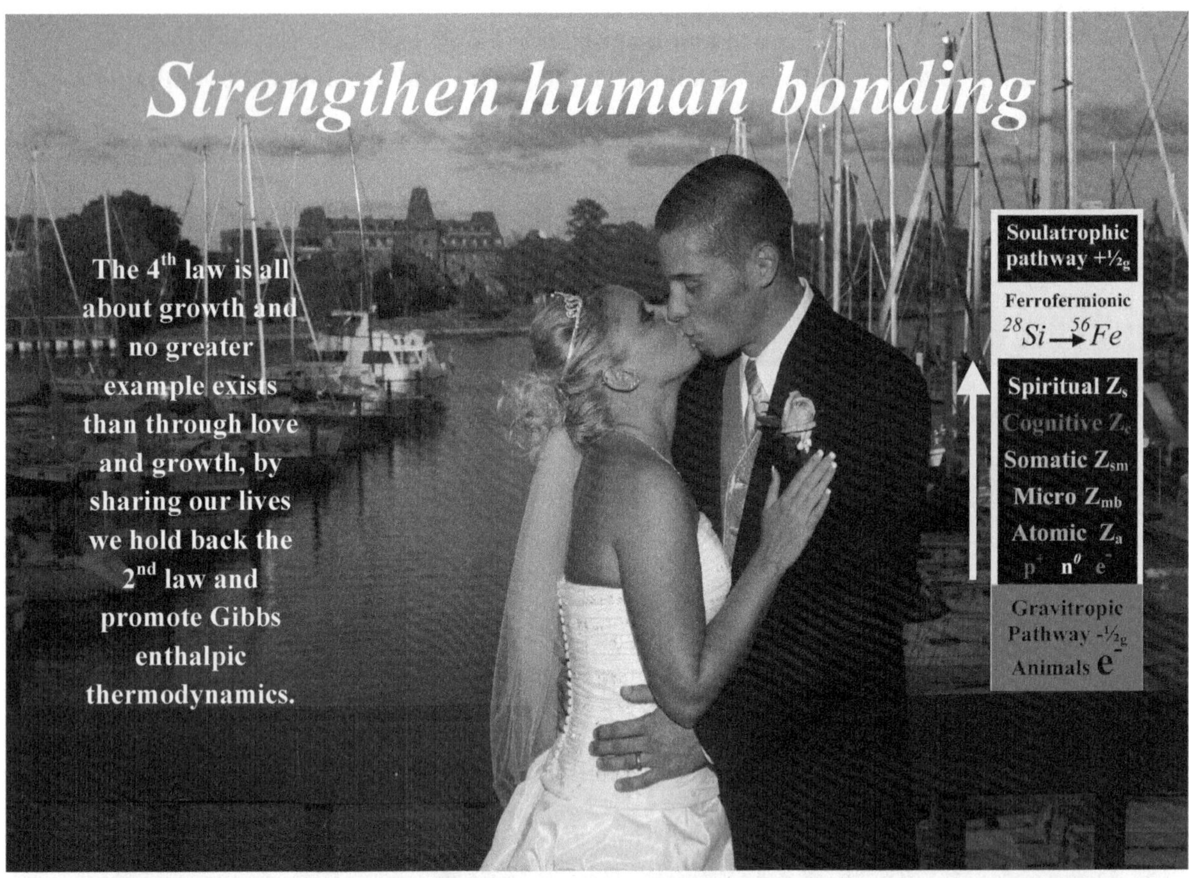

The 4th law is all about growth and no greater example exist than through educational conservation and its growth promoting abilities, by developing our lives we hold back the 2nd law and drive Gibbs enthalpic thermodynamics.

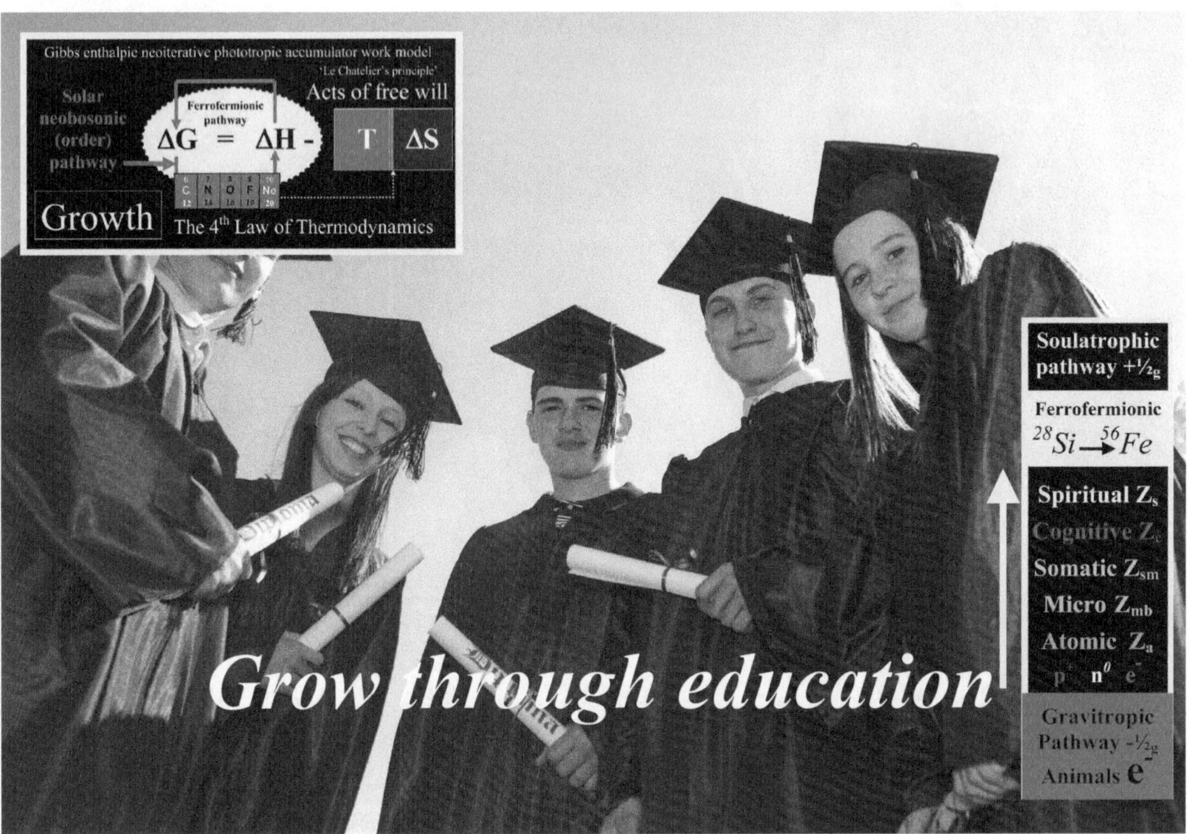

The 4th law is all about growth and no greater example exist than through economic growth and prosperity, by working we hold back the 2nd law and promote Gibbs enthalpic thermodynamics.

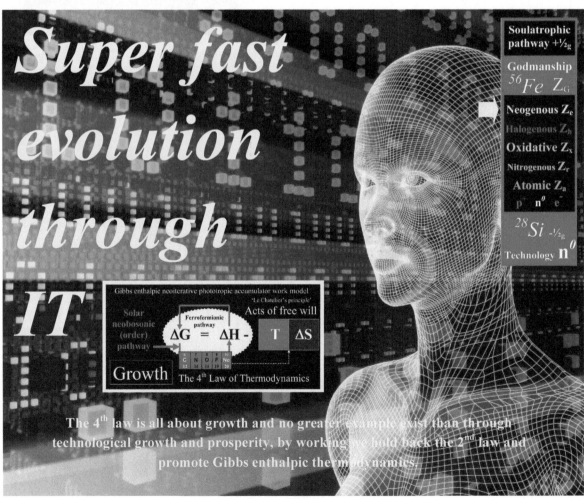

Super fast evolution through IT

Soulatrophic pathway $+\frac{1}{2}_g$

Godmanship

^{56}Fe Z_G

Neogenous Z_e

Halogenous Z_h

Oxidative Z_x

Nitrogenous Z_r

Atomic Z_a

p^+ n^0 e^-

^{28}Si $-\frac{1}{2}_g$

Technology n^0

Gibbs enthalpic neoiterative phototropic accumulator work model

'Le Chatelier's principle'

Solar neobosonic (order) pathway

Ferrofermionic pathway

Acts of free will

$$\Delta G = \Delta H - T\Delta S$$

C N O F Ne

Growth

The 4th Law of Thermodynamics

The 4th law is all about growth and no greater example exist than through technological growth and prosperity, by working we hold back the 2nd law and promote Gibbs enthalpic thermodynamics.

Soulatrophic pathway $+\frac{1}{2}_g$
Godmanship
^{56}Fe Z_G
Neogenous Z_e
Halogenous Z_h
Oxidative Z_x
Nitrogenous Z_r
Atomic Z_a
p^+ n^0 e^-
^{28}Si $-\frac{1}{2}_g$
Technology n^0
The 4th law is all about growth and no greater example exist than through medical growth and prosperity, by working we hold back the 2nd law and promote Gibbs enthalpic thermodynamics.
Gibbs enthalpic neoiterative phototropic accumulator work model
'Le Chatelier's principle'
Solar neobosonic (order) pathway
Ferrofermionic pathway
Acts of free will
$\Delta G = \Delta H - T \Delta S$
Growth
The 4th Law of Thermodynamics
Protect and grow

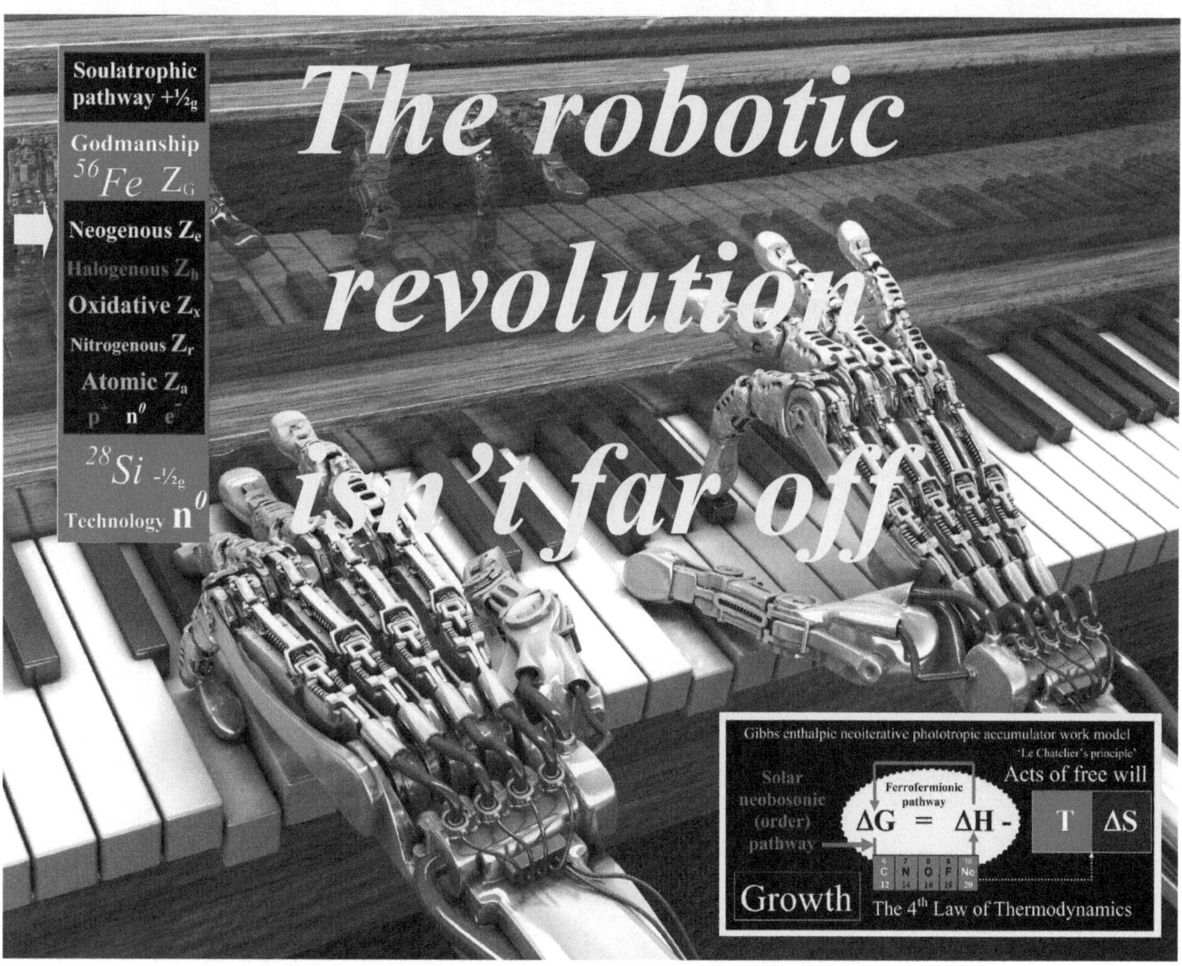
Soulatrophic pathway $+\frac{1}{2}_g$
Godmanship
^{56}Fe Z_G
Neogenous Z_e
Halogenous Z_h
Oxidative Z_x
Nitrogenous Z_r
Atomic Z_a
p^+ n^0 e^-
^{28}Si $-\frac{1}{2}_g$
Technology n^0
The robotic revolution isn't far off
Gibbs enthalpic neoiterative phototropic accumulator work model
'Le Chatelier's principle'
Solar neobosonic (order) pathway
Ferrofermionic pathway
Acts of free will
$\Delta G = \Delta H - T \Delta S$
Growth
The 4th Law of Thermodynamics

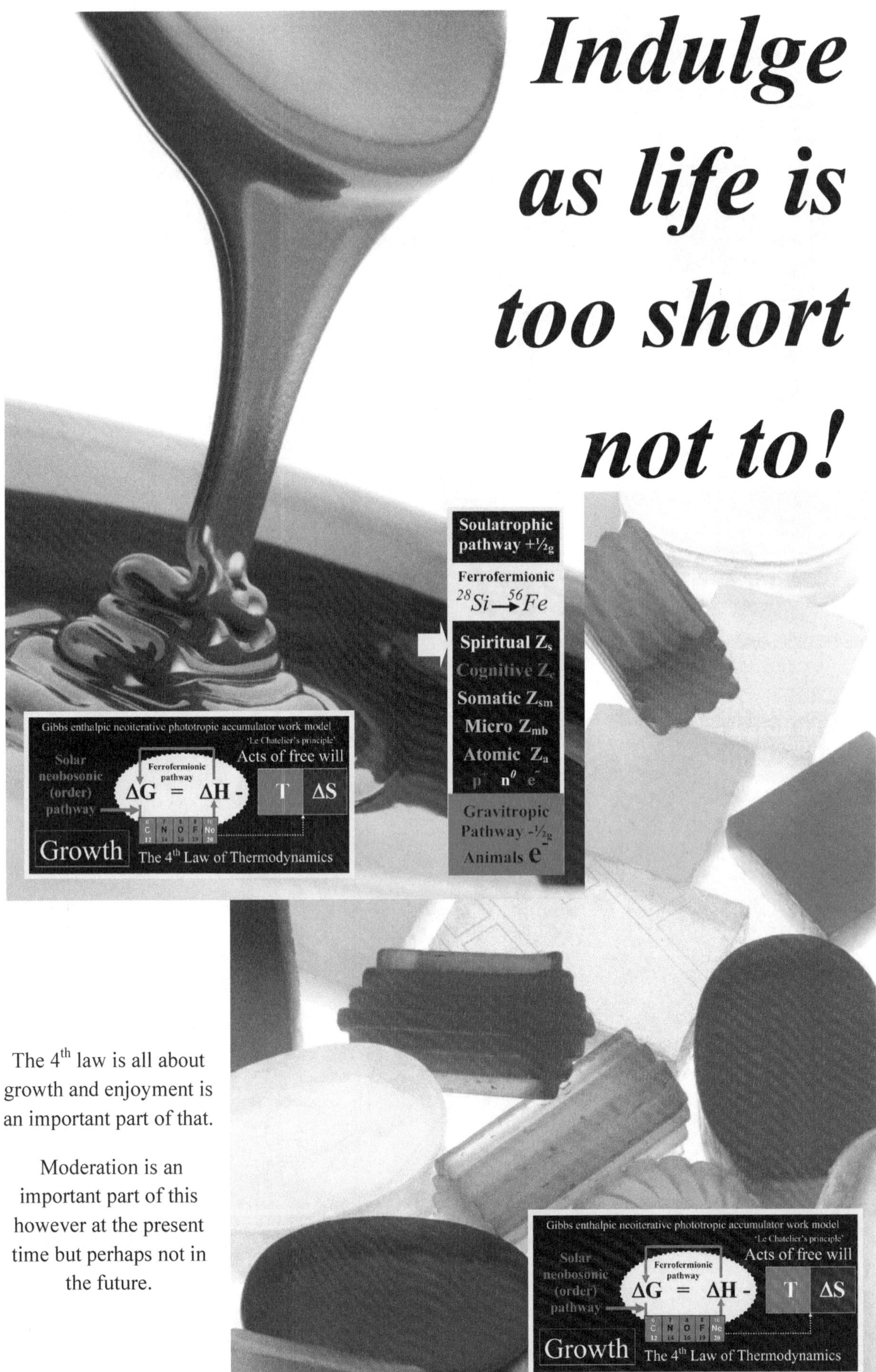

The 4th law is all about growth and enjoyment is an important part of that.

Moderation is an important part of this however at the present time but perhaps not in the future.

Embrace human variation

The 4th law is all about growth and no greater example exist than through variation by working together and sharing we hold back the 2nd law and promote Gibbs enthalpic thermodynamics.

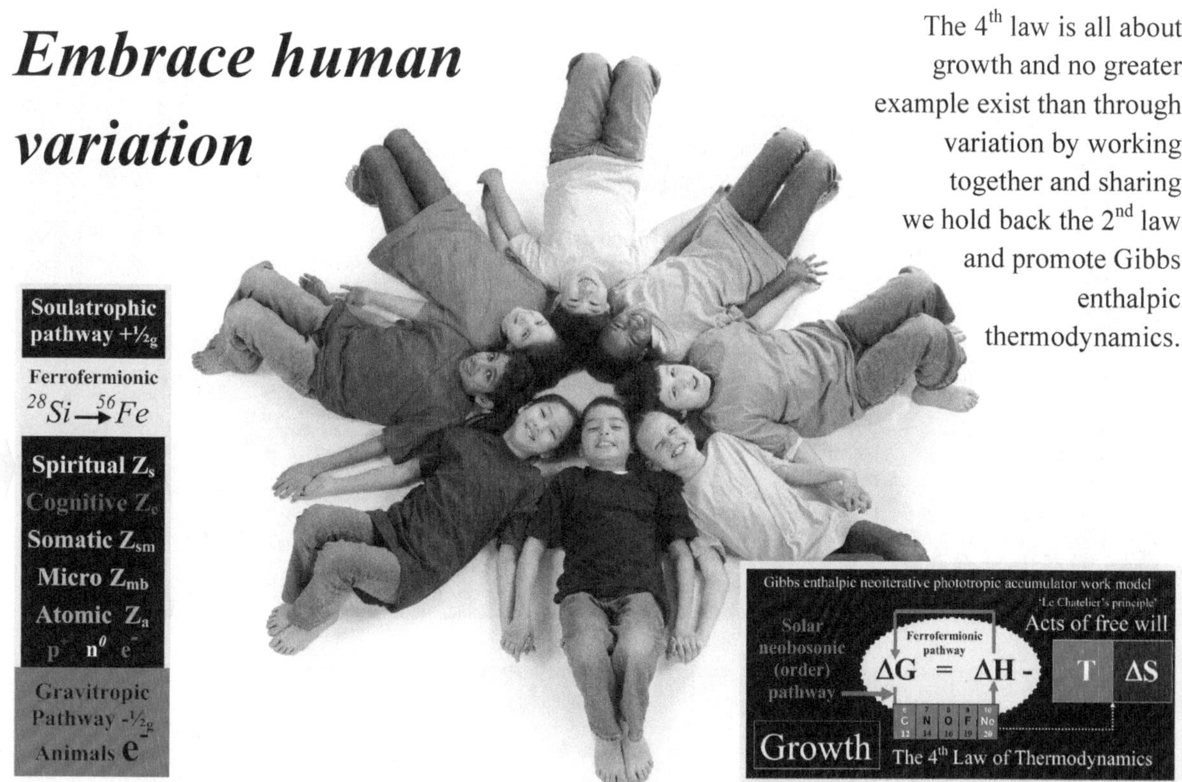

Soulatrophic pathway $+\frac{1}{2}_g$
Ferrofermionic
$^{28}Si \rightarrow ^{56}Fe$
Spiritual Z_s
Cognitive Z_c
Somatic Z_{sm}
Micro Z_{mb}
Atomic Z_a
p^+ n^0 e^-
Gravitropic Pathway $-\frac{1}{2}_g$
Animals e^-

The 4th law is all about growth and no greater example exist than through growth and prosperity, by working we hold back the 2nd law and promote Gibbs enthalpic thermodynamics.

Hope
NEXT EXIT

Fill your life with positive affirmations

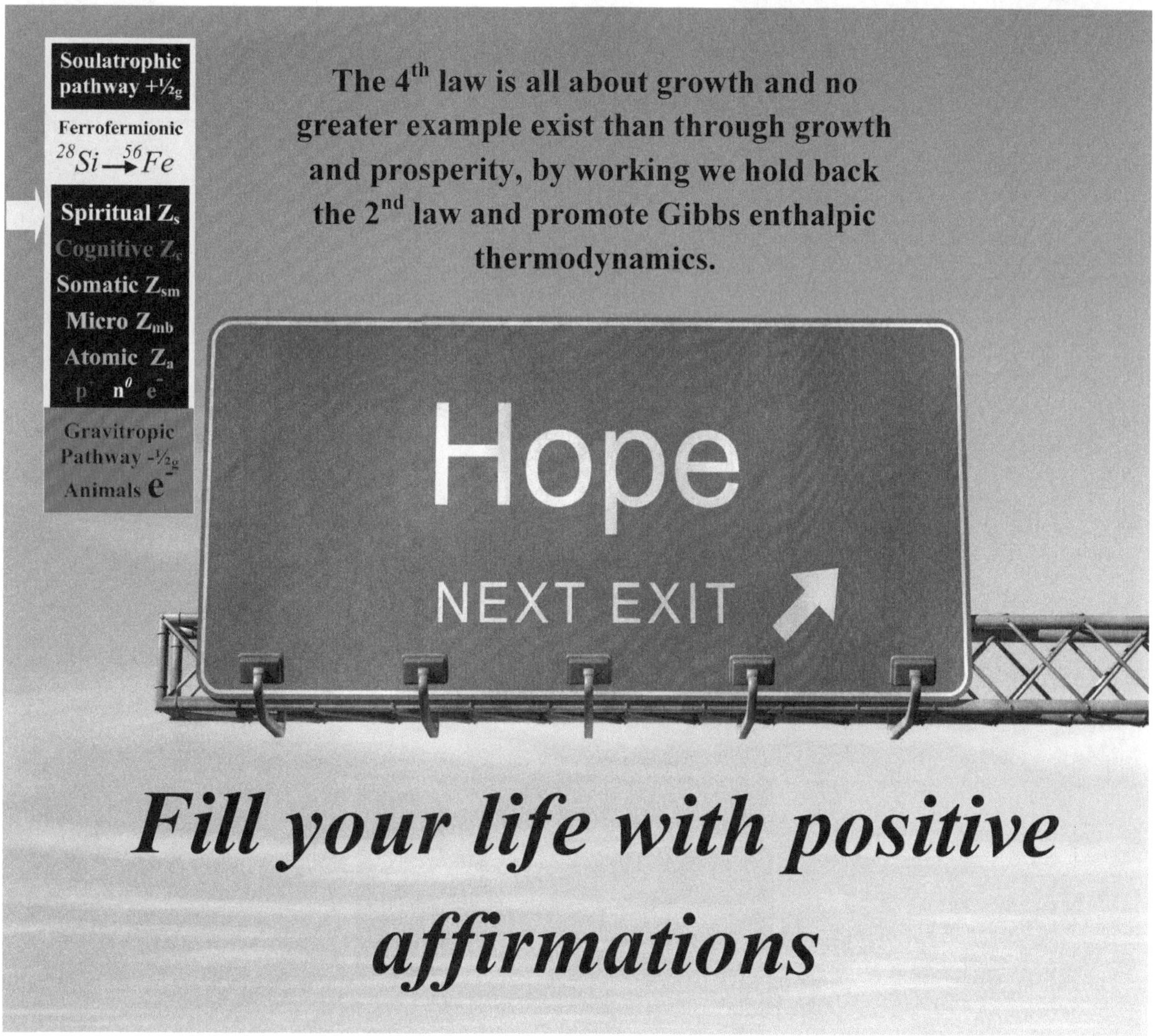

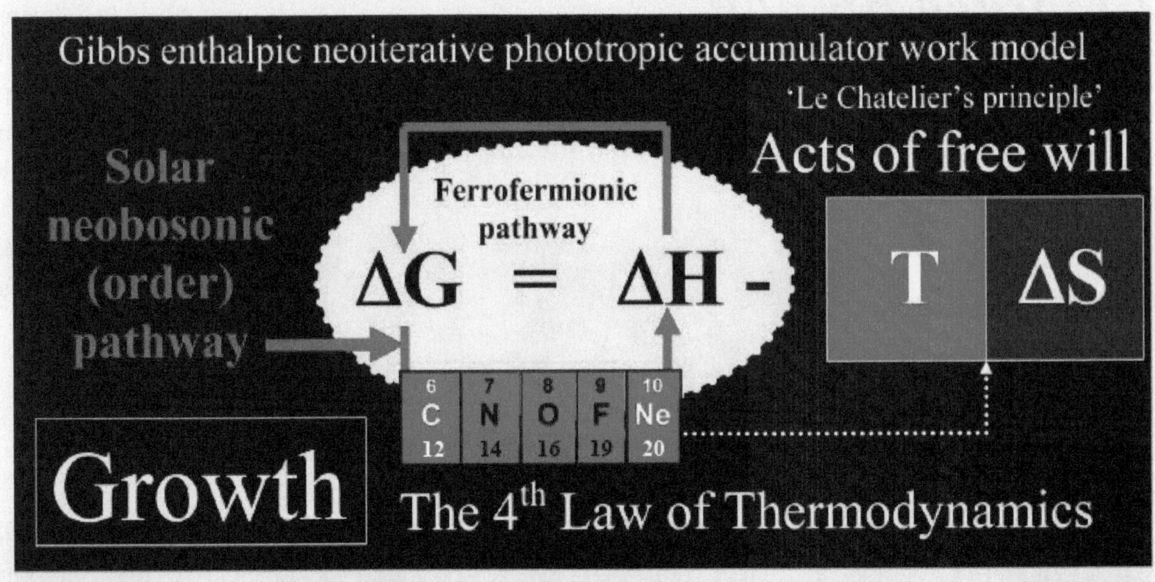

May the events of your day bring you ever closer to quantum stability through

Beautiful Phototropism!

Gibbs enthalpic neoiterative phototropic accumulator work model
'Le Chatelier's principle'
Acts of free will
Solar neobosonic (order) pathway
Ferrofermionic pathway
ΔG = ΔH - T ΔS
Growth
The 4th Law of Thermodynamics

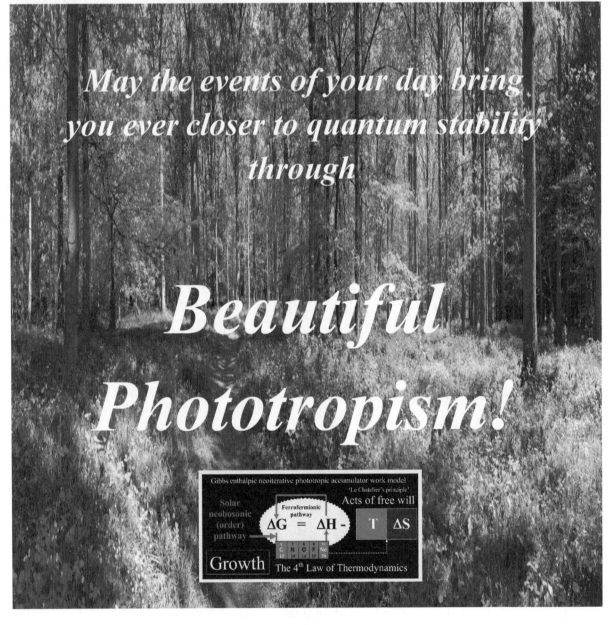

Growth & order

Planetary cytokinesis & post planetary technological life

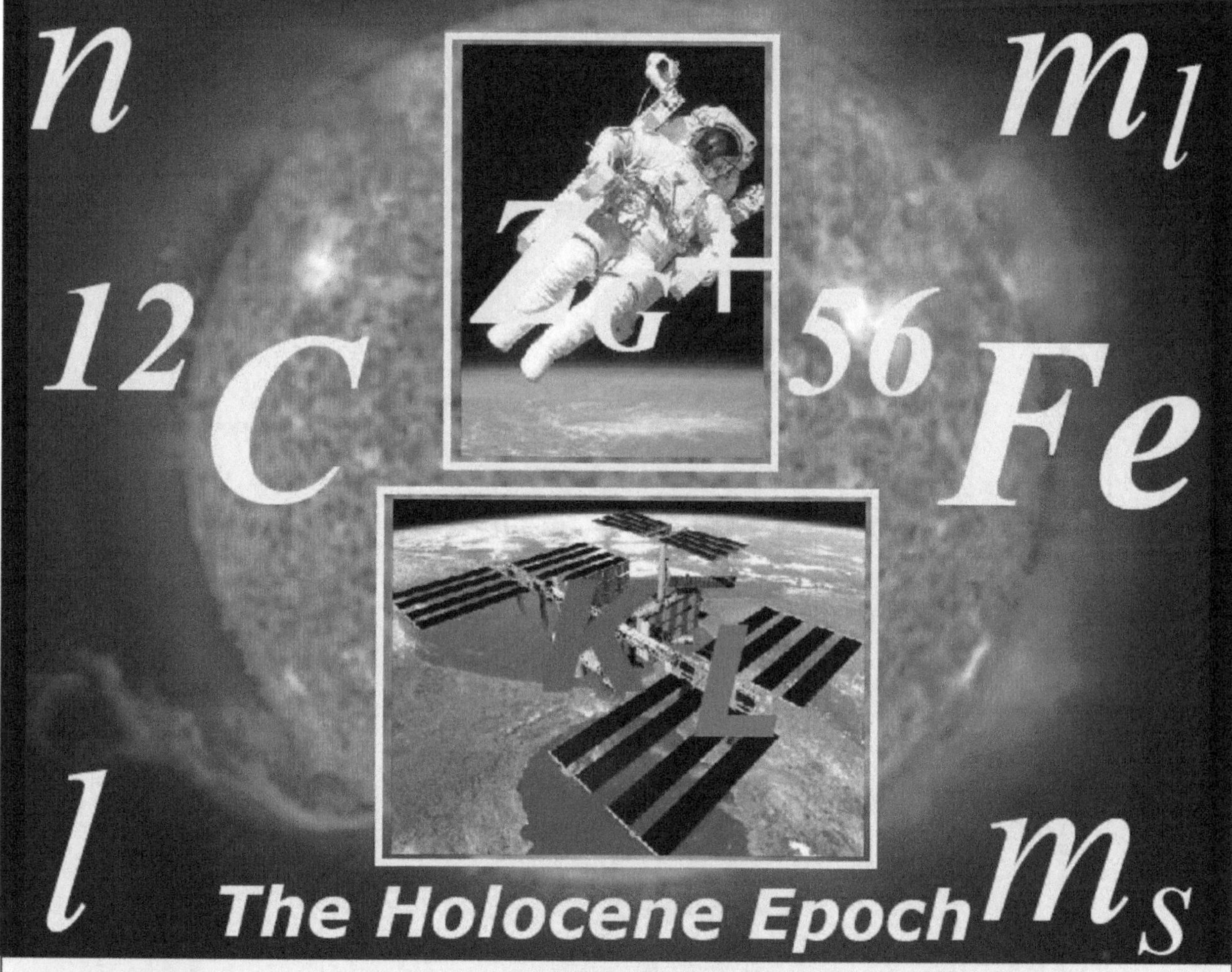

Phototropism

Naturally selective acts of free will the 4th law of thermodynamics

Neurochromosome artificial life - 'the divine inversion effect', divine evolutionary stability.

In carbon entromorphology God is a very real physical concept (it's important to understand this author is an atheist by classical definitions) and carbon entromorphologists celebrate the existence of God like potential although the definition is different. God or artificial life occurs as the goal of evolution a final super stable energy rich low pressure organism with the potential of living almost indefinitely. Human beings are the God's of the future and belief in the classical model of God has produced this result through the development of culture and human knowledge. God in all cultures represents an immaculate organism and humans aspire to live up to its standards. This process has produced language, mathematics, philosophy, art, music and just about all civilised logic.

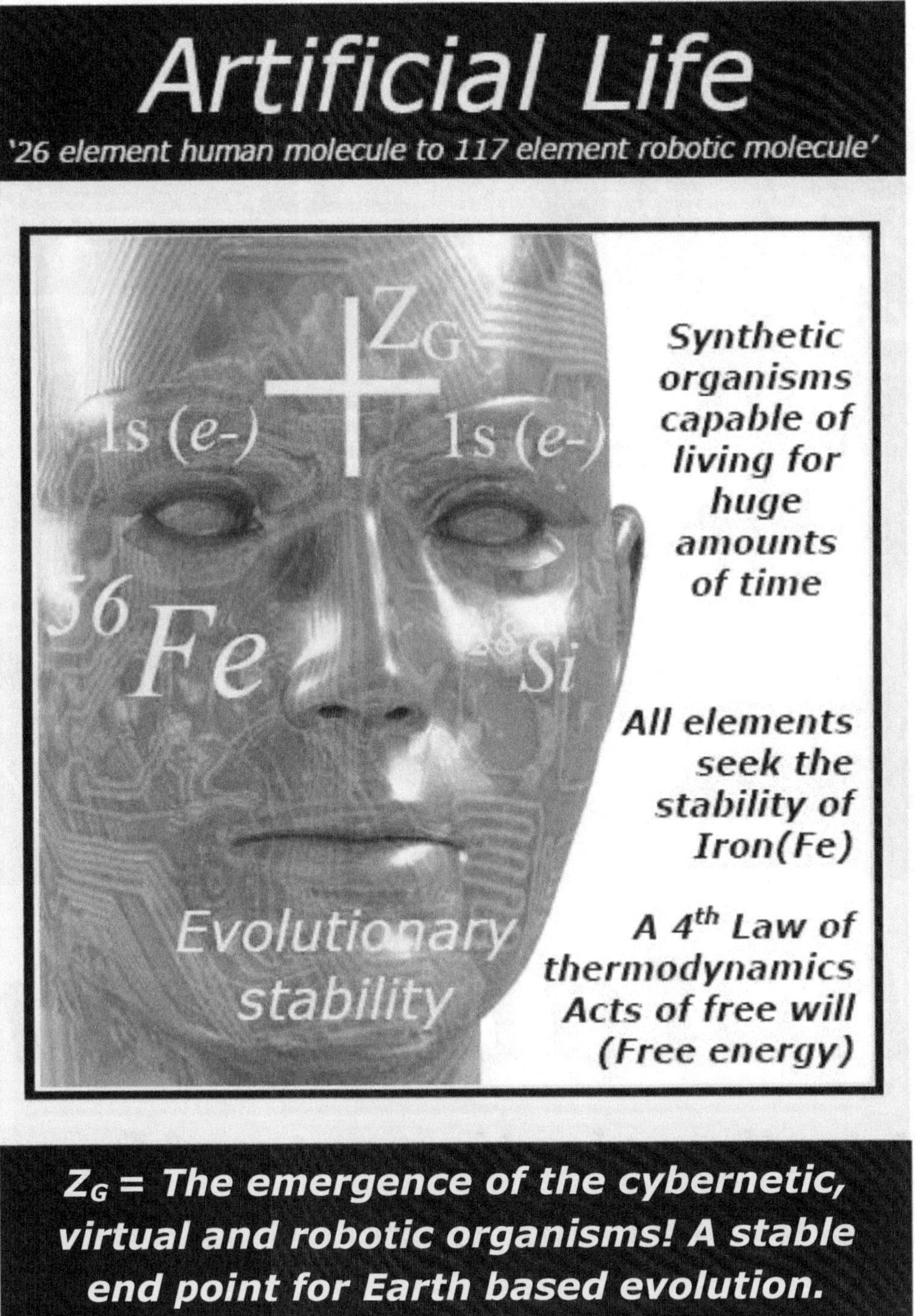

In short science is the seamless product of religion where Gods become the forces of nature. Because religion is practiced by most of the human beings on the planet and as such religion is still the main natural selection factor in human evolution science cannot reject it. Its foundations have come from belief in God and rejection of such profound logic makes no practical sense, in actual fact Darwinian evolution is a declining theory because it doesn't offer people anything. It doesn't matter how you choose to understand the concept of God only that you live up to the immaculate conception of such a concept; although peace and human and animal rights are essential and galvanising.

Science appears to be the opposite of religion yet science is trying to achieve the same goal, that of avoidance of death in human beings. Through medicine and other sciences humans will reach artificial life a future realisation of absolute temporal stability and limitless energy availability. I believe that carbon entromophology is the final stage of human spiritual evolution where science and religion can co-exist to produce a complete model of spiritual consciousness.

The artificial life entities of the future are the androids, robots and virtual organism's, the Internet I believe is the mind of God limitless energy availability, all the answers to all the questions we could ever ask. The 21st century is the most powerful century in human history as human beings gain the ultimate power over their lives and move towards a peaceful beautiful future known as a heaven.

Divine inversion model - the process in living soulatrophic cells by which life is energetically optimised by pursuing solar energy through phototropism (good). The model of soulatrophicity has two extremes from the Hadean Eon where carbon was in bondage to the Earth and hell like forms (evil or gravitropism – hell) is evident. The thermodynamic model for energy systems is the basis of the morality scale (Gibbs equation). This runs to the opposite extreme post Holocene Epoch where carbon cells maximise enthalpy and minimise entropy by increased efficient use of energy.

This classical moral linearity is based on the second law of thermodynamics, modelled by the Gibbs equation. By developing classical models of belief (religion) where a divine entity such as a God creates man in its own image. This process of religion and investigation produces scientific doctrines which when pursued lead towards technology and the formation of a God like consciousness through pursuance of the divine X and P. This God like consciousness is the inevitable conclusion of the ferrofermionic pathway. It leads to robotics and cyberspace entities with almost indefinite physical stability and the removal of senescence. Although uncertainty is always present in thermodynamic physical situations. Humanity creates God from its own image which it believes came from God.

This is the divine inversion model of ferrofermionic pathways; the God like consciousness is called 'artificial life'. By pursuing perfection through an imagined God by classical religion humans have begun to measure the immediate environment and make logical interpretations of what that tells them.

This leads to science and the final neutronic part of the evolutionary trinity, that of technology (technogenesis). This leads to increased living stability and the prospect of immortality in a literal real world situation. Divine X, P- The mechanical nature of classical models of God based on position (heaven) and momentum (intension).

Classical religions have sort to try to measure these components of God through prayer and study. This has developed into scientific methodology (Gods became forces of nature in science) and has produced increased human stability through technology. Although scientific uncertainty is reduced by measurement of physical nature, this isn't possible in belief systems due to a fundamental lack of any form of representative evidence. The bible (divine x, p) was created by humanity to measure God's will (momentum – intension).

Artificial life – the goal at the end of the ferrofermionic pathway of evolution, which produces technology and animates the periodic table. Iron is the most stable nuclear element and life tends towards it.

Heaven as a model of macrostate - a state of being with unlimited environmental stability and limitless energy availability. It is the environmental component of the 'divine inversion effect'. It is a low-pressure state associated with planetary cytokinesis and the vacuum neogenous stage of the Earth evaporative model of evolution. It is a Gibbs enthalpic energy conservation pathway from the forth law of thermodynamics, tending to just one microstate absolute order artificial life (unachievable due to the uncertainty principle) but life should come extremely close to it. This state is highly effective and stable and uncertainty although still evident and is highly manageable. It is the ferrous terminal state of phototropic evolution (ECASSE) and represents the future of life and the final state of evolution. Hell model of macrostate – a state of being directed in opposition to the ferrofermionic pathway.

Anti-phototropism is a lack of energy availability and energy stripping from existing soulatrophic entrochiraloctet cells taking place. It is acting in opposition to the 'divine inversion effect'. It is associated with high pressure and increasingly more solid macrostatic conditions, such as the Earth as home to the dead (atomic soulatrophicity). It leads to increased heat and entropy and is described by thermoentropic pathways from the second law of thermodynamics. It represents ultra high levels of microstates with little frequency of specific ensembles of microstates i.e. disorder and symmetry breaking states. It is a tendency towards disorder and acts as a temporal reversal to older states of evolution. In the ferrofermionic pathway the hell state occurs under early Earth conditions of fire and pressure. In essence it takes evolution back through time breaking the bonds in carbon and reducing enthalpic and Gibbs free energy in living systems.

Neurochromosome technological organisms – silicoferrous entromorphology, the nature of the 'biology of technology'. The final soulatrophic pathways and divine artificial life and heaven.

Carbosilicoferrous interface– the emergence of bipedalism in soulatrophic carbon entrochiraloctet cells. The release of the nitrogenous and halogenous cognitive valance (the hands in humans) for reaction with the rest of the periodic table. The fundamental neogenous valance bonded metabolic organelle in cognitive and spiritual soulatrophicity is silicon. This interface allows living energies to be transmitted to the rest of the periodic table through human hands.

This is the concluding part of planetary soulatrophic evolution the emergence of technology the neutronic phase of the evolutionary trinity (technogenesis). It runs along the ferrofermionic pathway towards nuclear stability seen in iron, and eventually a state of artificial life where organisms can stabilise themselves almost indefinitely. In the state of artificial life uncertainty still exists but is minimised in its effect.

Where Australopithecus stood upright and emancipated the hands to lead to the Stone Age (silicon). Stone Age pots are made out of clays containing aluminium silicate. This leads to the Iron Age and the ferrofermionic pathway becomes established.

Carbon and silicon are related on the periodic table they both have four valance and as such four similar metabolic organelle systems. In other words silicon can mimic carbon life; this is the basis of technological development and leads the periodic table to ferrous nuclear stability. In almost all modern life the three fundamental components of the ferrofermionics pathway namely carbon, silicon and iron are found extensively together in most technologies. Iron is at the heart of life through oxidative metabolic organelles in blood haemoglobin. This is strong evidence for the nature of the NFVE model of life and extended ECASSE (Entromorphic Carbon Amplification by Solar Stimulated Emission).

Carbosilicoferrous event - Australopithecus stood on two legs (bipedalism as opposed to quadrapedalism) and maintained freedom of reaction in their hands (freeing of the Z^0 boson). The release of the hands allow the creation of technology the neutronic phase of the planetary evolutionary trinity as living energies are transmitted to the rest of the periodic table.

The presence of carbon, silicon and iron is highly evident in technology throughout the living world. Most technology has the three elements in its design or as part of the systems that make them.

The bipedal elevation comes from a maximised planetary spingrav on humans producing an entromorphic elevation of 90° perpendicular to the Earth. This is the basis for planetary cytokinesis going into space and the final path towards artificial life and extreme stability.

Environmental filtering – external magnetic and gravitational fields induce morphological deformation ‘evolution’ in carbon soulatrophic entrochiraloctet cells. The deformative effect on sub shells (orbitals) produces variation in soulatrophic pathways (organisms), Soulatrophic pathways allow this to be conserved in between generations.

Adaptation effects seen in response to this force are conserved through soulatrophic ladders (sexual reproduction), this is the selectivity factor based on environmental constraints. It is mainly an orientational effect and causes energy shifting but not necessarily a change in energy level.

Artificial life – a point at the end of the ferrofermionic pathway of evolution, which produces technology and animates the periodic table. Iron is the most stable nuclear element and life tends towards it. The tendency leads towards consciousness with super conductive properties where living entities exist in stable, energy rich peaceful heavenly states of being.

Driven by divine belief in a God the interpretation of the divine elements of x and p (position and momentum or intension) is eventually determined by experimentation of the immediate environment, hence the birth of science from classical religions and accelerated ferrofermionic evolution.

The following images are a tiny sample of the vast number of real situations where humans find carbon, silicon and iron working together to produce all the technology we need. Whether this is in the form of the land and the Earth’s core with its magnetic field protection or iron in your blood allowing an organism to successfully utilise oxygen, iron is at the heart of the universe.

Buildings are made using iron tools and iron structures with silicon glass and bricks housing carbon based organisms that also designed and built these structures in the first place.
This unique collection of elements also allows the rest of the periodic table to come on line allowing humans to identify and utilise all the 92 naturally occurring element and leading in time to a very powerful and amazing future for humanity.

The technological organisms

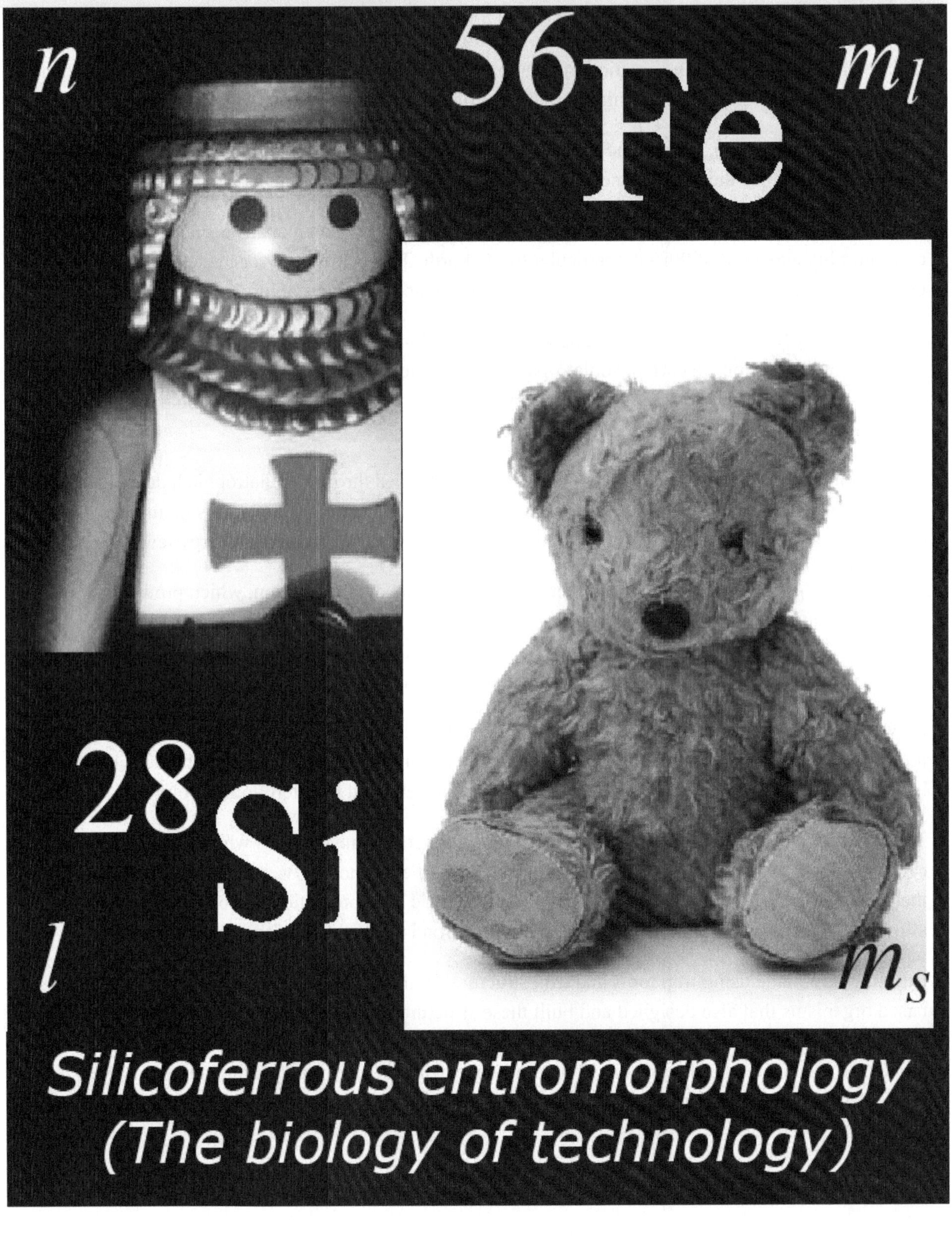

n ^{56}Fe m_l

^{28}Si l m_s

Silicoferrous entromorphology
(The biology of technology)

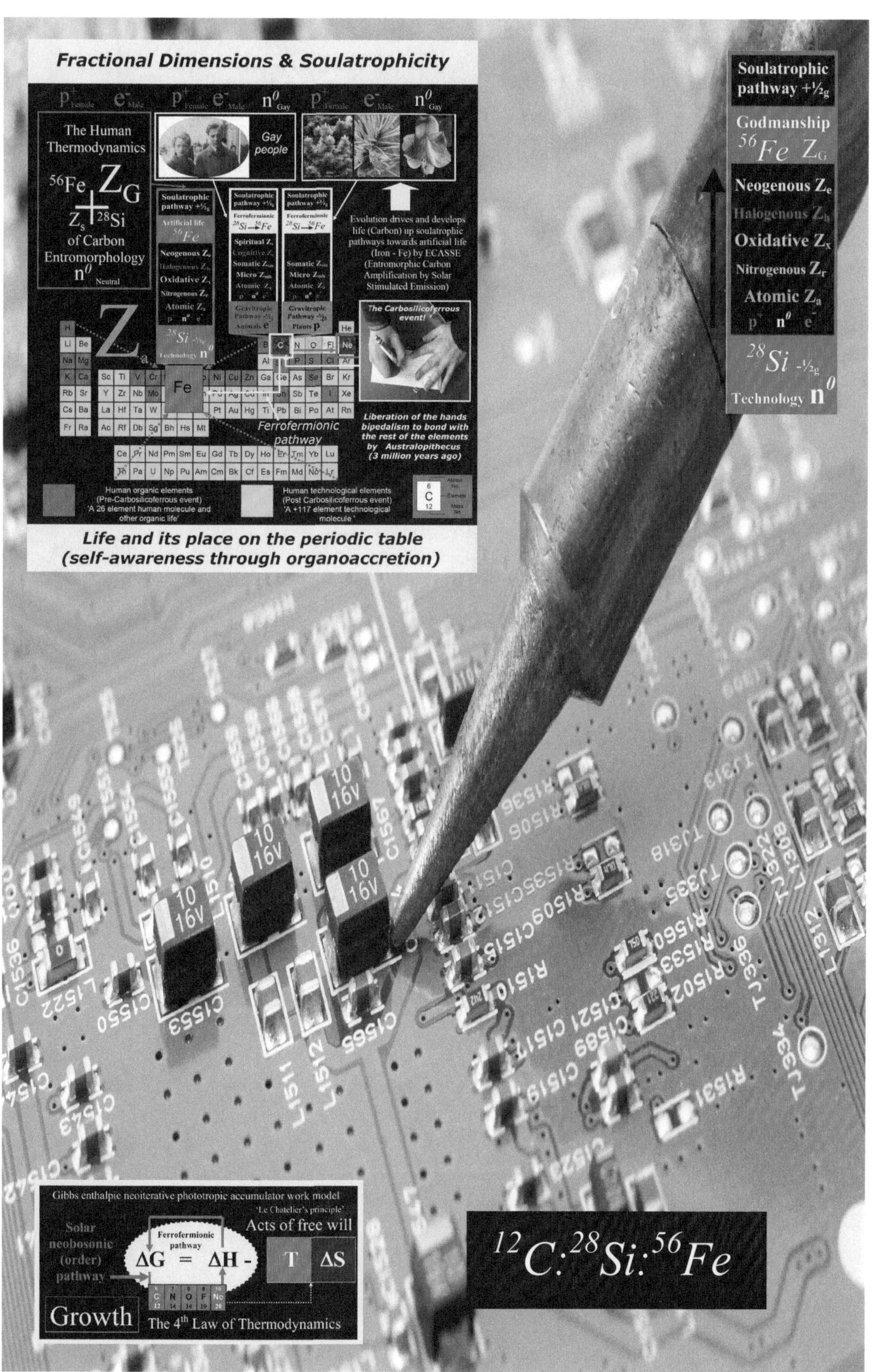
Fractional Dimensions & Soulatrophicity
The Human Thermodynamics
56Fe ZG + Zs 28Si
of Carbon Entromorphology
n0 Neutral
Gay people
Soulatrophic pathway +½g
Artificial life 56Fe
Neogenous Ze
Halogenous Zh
Oxidative Zx
Nitrogenous Zr
Atomic Za
28Si -½g
Technology n0
Ferrofermionic 28Si→56Fe
Spiritual Zs
Cognitive Zc
Somatic Zsm
Micro Zmb
Atomic Za
Gravitropic Pathway -½g
Animals e
Plants p
Evolution drives and develops life (Carbon) up soulatrophic pathways towards artificial life (Iron - Fe) by ECASSE (Entromorphic Carbon Amplification by Solar Stimulated Emission)
The Carbosilicoferrous event!
Ferrofermionic pathway
Liberation of the hands bipedalism to bond with the rest of the elements by Australopithecus (3 million years ago)
Human organic elements (Pre-Carbosilicoferrous event) 'A 26 element human molecule and other organic life'
Human technological elements (Post Carbosilicoferrous event) 'A +117 element technological molecule'
Life and its place on the periodic table (self-awareness through organoaccretion)
Soulatrophic pathway +½g
Godmanship 56Fe ZG
Neogenous Ze
Halogenous Zh
Oxidative Zx
Nitrogenous Zr
Atomic Za
28Si -½g
Technology n0
Gibbs enthalpic neoiterative phototropic accumulator work model
'Le Chatelier's principle'
Acts of free will
Solar neobosonic (order) pathway
Ferrofermionic pathway
ΔG = ΔH - T ΔS
Growth
The 4th Law of Thermodynamics
12C:28Si:56Fe

$^{12}C:^{28}Si:^{56}Fe$
Soulatrophic pathway $+\frac{1}{2}_g$
Godmanship
^{56}Fe Z_G
Neogenous Z_e
Halogenous Z_h
Oxidative Z_x
Nitrogenous Z_r
Atomic Z_a
p^+ n^0 e^-
^{28}Si $-\frac{1}{2}_g$
Technology n^0
Gibbs enthalpic neoiterative phototropic accumulator work model
'Le Chatelier's principle'
Acts of free will
Solar neobosonic (order) pathway
Ferrofermionic pathway
$\Delta G = \Delta H - T \Delta S$
6 C 12
7 N 14
8 O 16
9 F 19
10 Ne 20
Growth
The 4th Law of Thermodynamics

$^{12}C{:}^{28}Si{:}^{56}Fe$

Gibbs enthalpic neoiterative phototropic accumulator work model
'Le Chatelier's principle'
Solar neobosonic (order) pathway
Ferrofermionic pathway
ΔG = ΔH - T ΔS
Acts of free will
C N O F Ne
Growth
The 4th Law of Thermodynamics
Soulatrophic pathway +½g
Godmanship
56Fe ZG
Neogenous Ze
Halogenous Zh
Oxidative Zx
Nitrogenous Zr
Atomic Za
p+ n0 e-
28Si -½g
Technology n0
12C:28Si:56Fe

Soulatrophic pathway +½g
Godmanship
56Fe ZG
Neogenous Ze
Halogenous Zh
Oxidative Zx
Nitrogenous Zr
Atomic Za
p+ n0 e-
28Si -½g
Technology n0
12C:28Si:56Fe
Gibbs enthalpic neoiterative phototropic accumulator work model
'Le Chatelier's principle'
Solar neobosonic (order) pathway
Ferrofermionic pathway
ΔG = ΔH - T ΔS
Acts of free will
C N O F Ne
Growth
The 4th Law of Thermodynamics

Gibbs enthalpic neoiterative phototropic accumulator work model
'Le Chatelier's principle'
Solar neobosonic (order) pathway
Ferrofermionic pathway
$\Delta G = \Delta H - T \Delta S$
Acts of free will
C N O F Ne
6 7 8 9 10
12 14 16 19 20
Growth
The 4th Law of Thermodynamics
Soulatrophic pathway $+\frac{1}{2}_g$
Godmanship
$^{56}Fe\ Z_G$
Neogenous Z_e
Halogenous Z_h
Oxidative Z_x
Nitrogenous Z_r
Atomic Z_a
$p^+\ n^0\ e^-$
$^{28}Si\ -\frac{1}{2}_g$
Technology n^0
$^{12}C{:}^{28}Si{:}^{56}Fe$

Soulatrophic pathway $+\frac{1}{2}_g$
Godmanship
^{56}Fe Z_G
Neogenous Z_e
Halogenous Z_h
Oxidative Z_x
Nitrogenous Z_r
Atomic Z_a
p^+ n^0 e^-
^{28}Si $-\frac{1}{2}_g$
Technology n^0
$^{12}C:^{28}Si:^{56}Fe$
Gibbs enthalpic neoiterative phototropic accumulator work model
'Le Chatelier's principle'
Acts of free will
Solar neobosonic (order) pathway
Ferrofermionic pathway
$\Delta G = \Delta H - T \Delta S$
6 C 12
7 N 14
8 O 16
9 F 19
10 Ne 20
Growth
The 4th Law of Thermodynamics

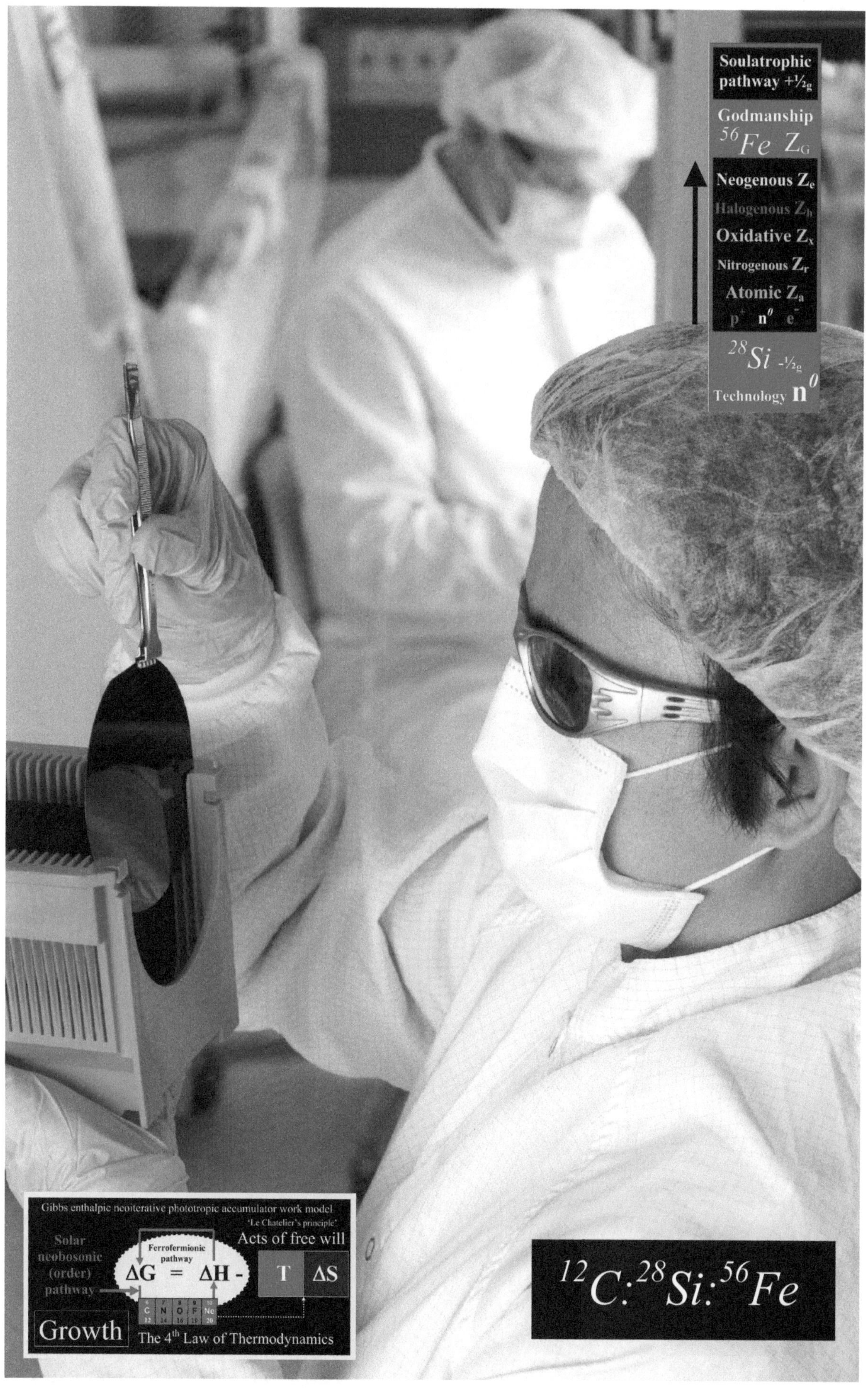
Soulatrophic pathway $+\frac{1}{2}_g$
Godmanship
^{56}Fe Z_G
Neogenous Z_e
Halogenous Z_h
Oxidative Z_x
Nitrogenous Z_r
Atomic Z_a
p^+ n^0 e^-
^{28}Si $-\frac{1}{2}_g$
Technology n^0
Gibbs enthalpic neoiterative phototropic accumulator work model
'Le Chatelier's principle'
Solar neobosonic (order) pathway
Ferrofermionic pathway
$\Delta G = \Delta H - T \Delta S$
Acts of free will
6 C 12
7 N 14
8 O 16
9 F 19
10 Ne 20
Growth
The 4th Law of Thermodynamics
$^{12}C:^{28}Si:^{56}Fe$

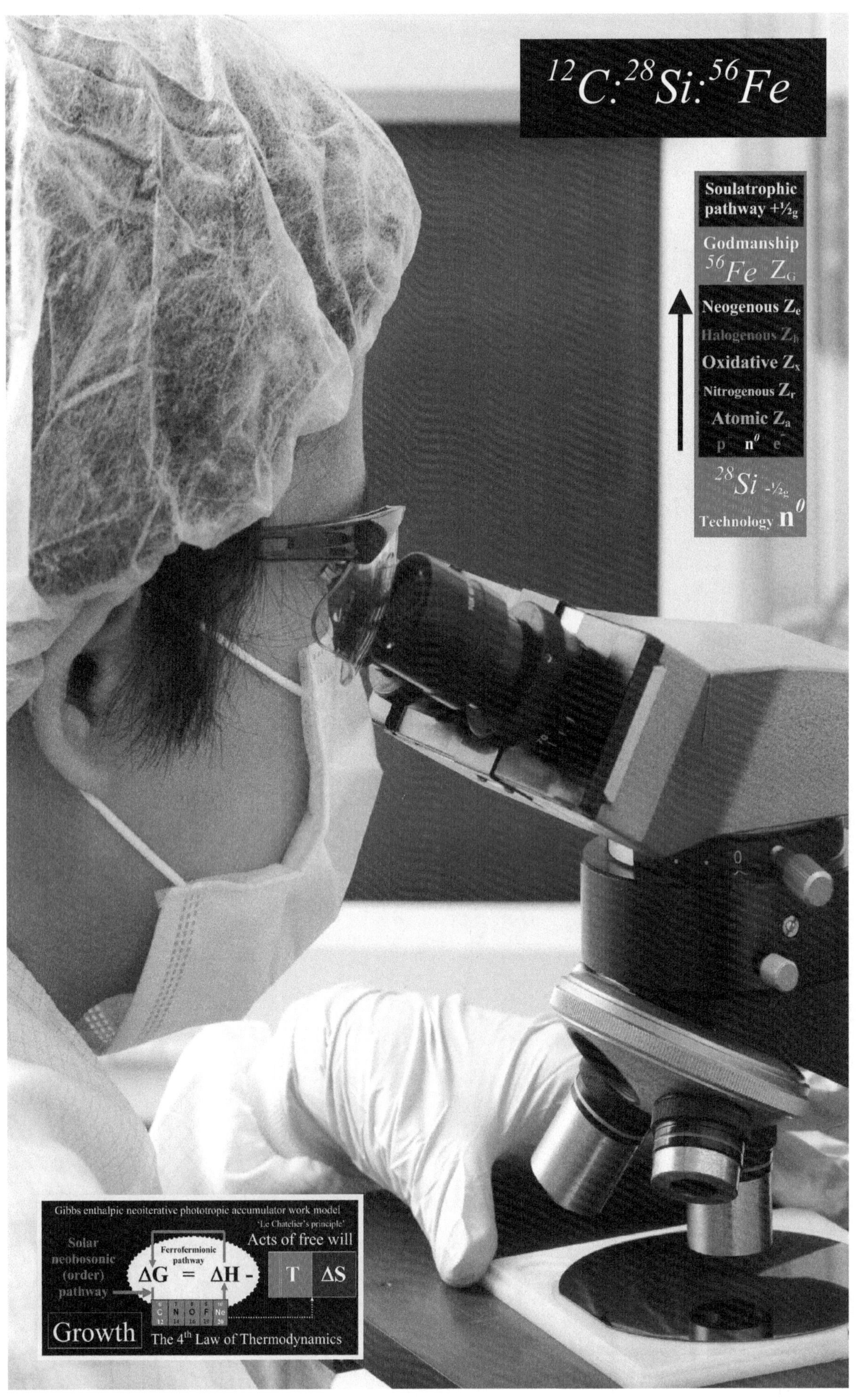
$^{12}C{:}^{28}Si{:}^{56}Fe$
Soulatrophic pathway $+\frac{1}{2}_g$
Godmanship
^{56}Fe Z_G
Neogenous Z_e
Halogenous Z_h
Oxidative Z_x
Nitrogenous Z_r
Atomic Z_a
p n^0 e^-
^{28}Si $-\frac{1}{2}_g$
Technology n^0
Gibbs enthalpic neoiterative phototropic accumulator work model
'Le Chatelier's principle'
Solar neobosonic (order) pathway
Ferrofermionic pathway
Acts of free will
$\Delta G = \Delta H - T \Delta S$
C N O F Ne
Growth
The 4th Law of Thermodynamics

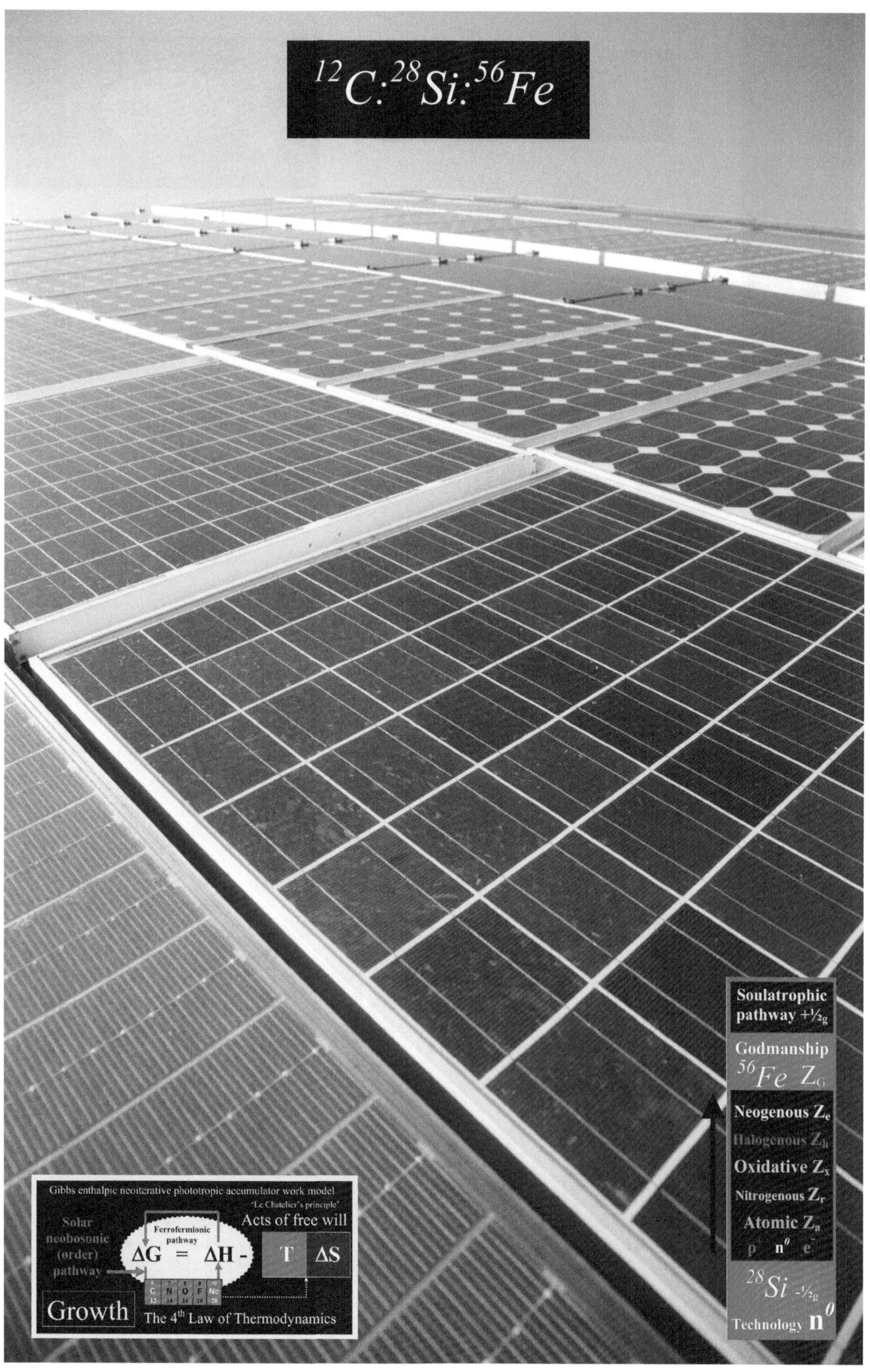
^{12}C:^{28}Si:^{56}Fe
Soulatrophic pathway $+\frac{1}{2}_g$
Godmanship
^{56}Fe Z_G
Neogenous Z_e
Halogenous Z_h
Oxidative Z_x
Nitrogenous Z_r
Atomic Z_a
p n^0 e^-
^{28}Si $-\frac{1}{2}_g$
Technology n^0
Gibbs enthalpic neoiterative phototropic accumulator work model
'Le Chatelier's principle'
Acts of free will
Solar neobosonic (order) pathway
Ferrofermionic pathway
$\Delta G = \Delta H - T \Delta S$
6 C 12
7 N 14
8 O 16
9 F 19
10 Ne 20
Growth
The 4th Law of Thermodynamics

Soulatrophic pathway +½g
Godmanship
56Fe ZG
Neogenous Ze
Halogenous Zh
Oxidative Zx
Nitrogenous Zr
Atomic Za
p+ n0 e-
28Si -½g
Technology n0
12C:28Si:56Fe
Gibbs enthalpic neoiterative phototropic accumulator work model
'Le Chatelier's principle'
Solar neobosonic (order) pathway
Ferrofermionic pathway
Acts of free will
ΔG = ΔH - T ΔS
Growth
The 4th Law of Thermodynamics
Nutrition Facts
Nutrition Facts

Soulatrophic pathway +½g
Godmanship
56Fe ZG
Neogenous Ze
Halogenous Zh
Oxidative Zx
Nitrogenous Zr
Atomic Za
p+ n0 e-
28Si -½g
Technology n0
Gibbs enthalpic neoiterative phototropic accumulator work model
'Le Chatelier's principle'
Solar neobosonic (order) pathway
Ferrofermionic pathway
Acts of free will
ΔG = ΔH - T ΔS
C N O F Ne
Growth
The 4th Law of Thermodynamics
12C:28Si:56Fe

Gibbs enthalpic neoiterative phototropic accumulator work model
'Le Chatelier's principle'
Solar neobosonic (order) pathway
Ferrofermionic pathway
Acts of free will
ΔG = ΔH - T ΔS
C N O F Ne
Growth
The 4th Law of Thermodynamics
Soulatrophic pathway +½g
Godmanship
56Fe ZG
Neogenous Ze
Halogenous Zh
Oxidative Zx
Nitrogenous Zr
Atomic Za
p+ n0 e-
28Si -½g
Technology n0
12C:28Si:56Fe

$^{12}C:^{28}Si:^{56}Fe$
Soulatrophic pathway $+\frac{1}{2}_g$
Godmanship
^{56}Fe Z_G
Neogenous Z_e
Halogenous Z_h
Oxidative Z_x
Nitrogenous Z_r
Atomic Z_a
p^+ n^0 e^-
^{28}Si $-\frac{1}{2}_g$
Technology n^0
Gibbs enthalpic neoiterative phototropic accumulator work model
'Le Chatelier's principle'
Acts of free will
Solar neobosonic (order) pathway
Ferrofermionic pathway
$\Delta G = \Delta H - T \Delta S$
C N O F Ne
Growth
The 4th Law of Thermodynamics

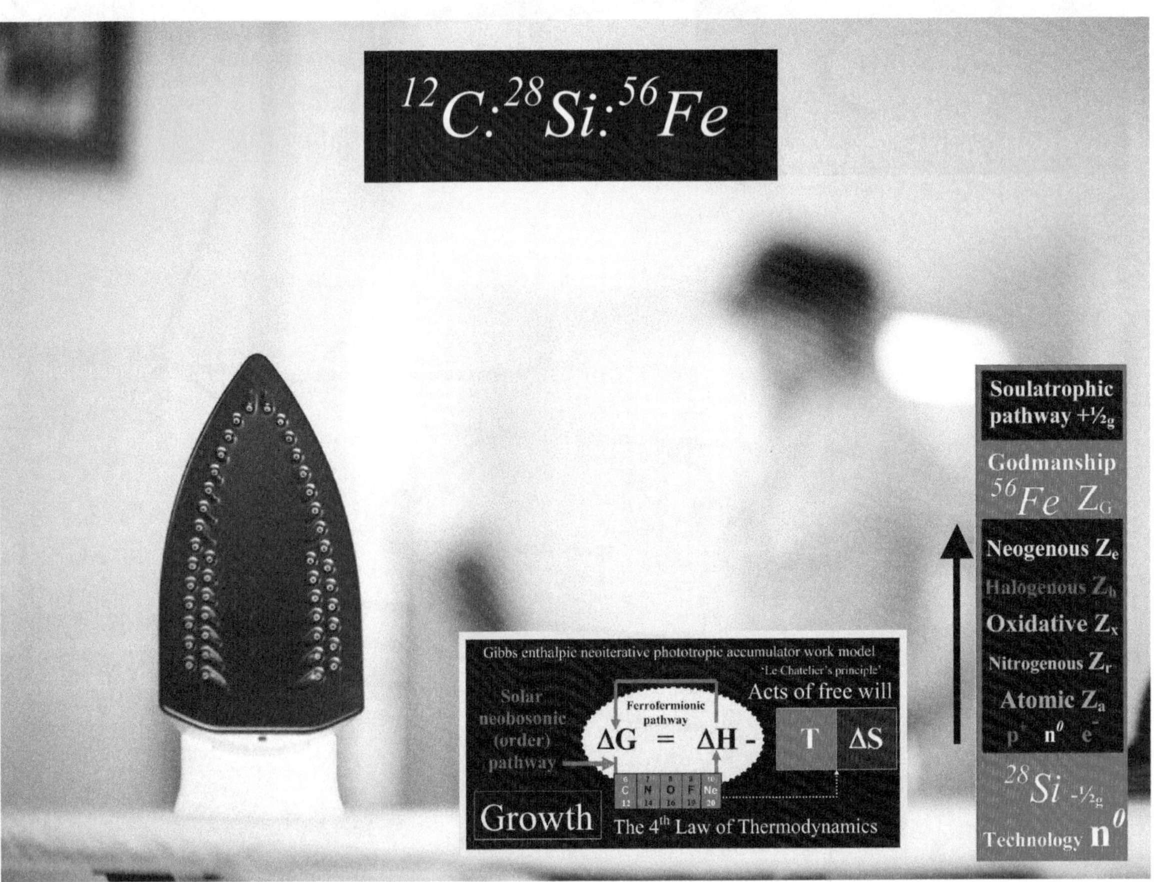
$^{12}C:^{28}Si:^{56}Fe$
Soulatrophic pathway $+\frac{1}{2}_g$
Godmanship
^{56}Fe Z_G
Neogenous Z_e
Halogenous Z_h
Oxidative Z_x
Nitrogenous Z_r
Atomic Z_a
p^+ n^0 e^-
^{28}Si $-\frac{1}{2}_g$
Technology n^0
Gibbs enthalpic neoiterative phototropic accumulator work model
'Le Chatelier's principle'
Acts of free will
Solar neobosonic (order) pathway
Ferrofermionic pathway
$\Delta G = \Delta H - T \Delta S$
C N O F Ne
Growth
The 4th Law of Thermodynamics

$^{12}C:^{28}Si:^{56}Fe$
Soulatrophic pathway $+\frac{1}{2}_g$
Godmanship
^{56}Fe Z_G
Neogenous Z_e
Halogenous Z_h
Oxidative Z_x
Nitrogenous Z_r
Atomic Z_a
p^+ n^0 e^-
^{28}Si $-\frac{1}{2}_g$
Technology n^0
Gibbs enthalpic neoiterative phototropic accumulator work model
'Le Chatelier's principle'
Solar neobosonic (order) pathway
Ferrofermionic pathway
$\Delta G = \Delta H - T \Delta S$
Acts of free will
6 C 12 7 N 14 8 O 16 9 F 19 10 Ne 20
Growth
The 4th Law of Thermodynamics

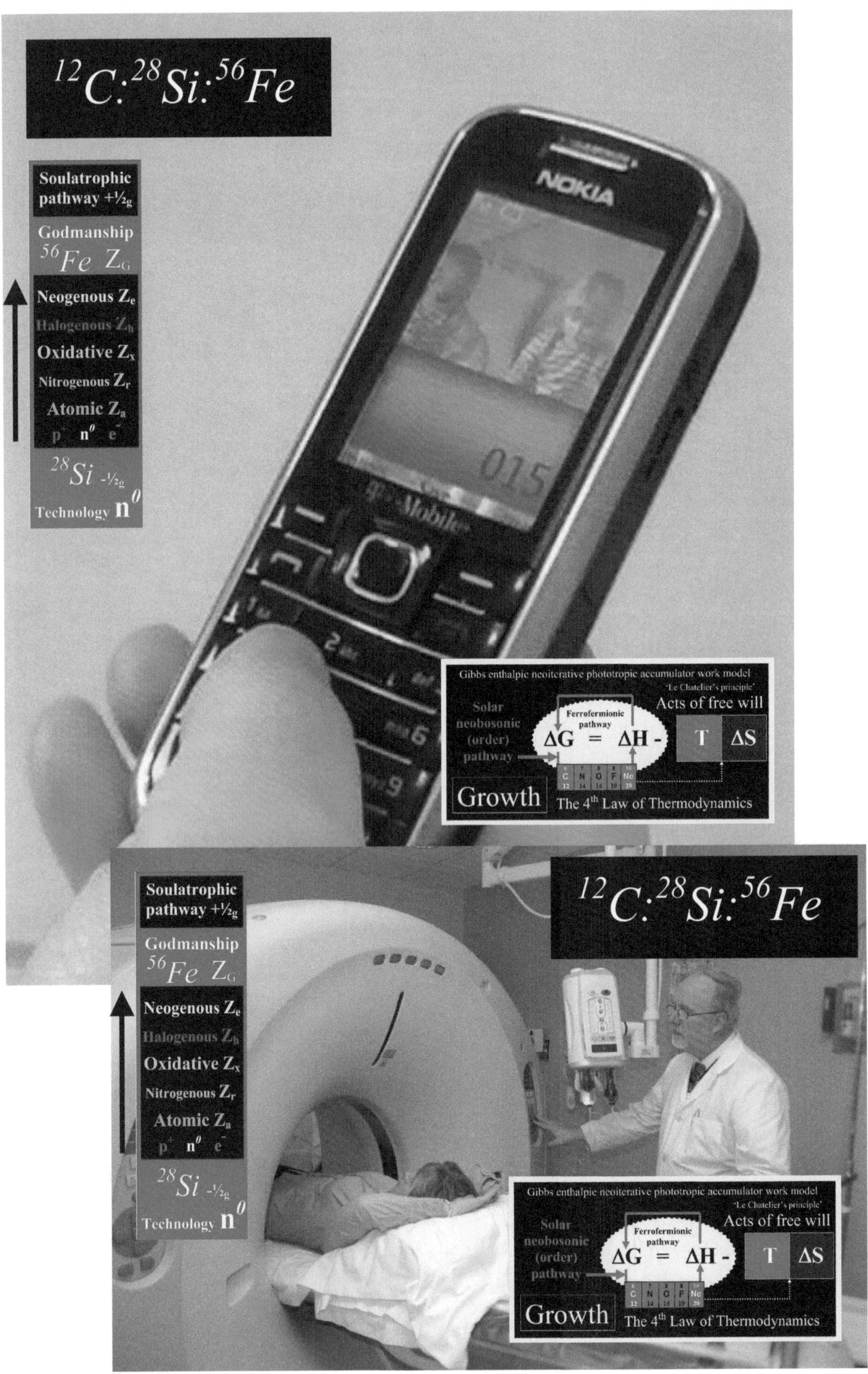

$^{12}C:^{28}Si:^{56}Fe$
Soulatrophic pathway $+½_g$
Godmanship
^{56}Fe Z_G
Neogenous Z_e
Halogenous Z_h
Oxidative Z_x
Nitrogenous Z_r
Atomic Z_a
p^+ n^0 e^-
^{28}Si $-½_g$
Technology n^0
NOKIA
015
Gibbs enthalpic neoiterative phototropic accumulator work model
'Le Chatelier's principle'
Acts of free will
Solar neobosonic (order) pathway
Ferrofermionic pathway
$\Delta G = \Delta H - T \Delta S$
C N O F Ne
Growth
The 4th Law of Thermodynamics
Soulatrophic pathway $+½_g$
Godmanship
^{56}Fe Z_G
Neogenous Z_e
Halogenous Z_h
Oxidative Z_x
Nitrogenous Z_r
Atomic Z_a
p^+ n^0 e^-
^{28}Si $-½_g$
Technology n^0
$^{12}C:^{28}Si:^{56}Fe$
Gibbs enthalpic neoiterative phototropic accumulator work model
'Le Chatelier's principle'
Acts of free will
Solar neobosonic (order) pathway
Ferrofermionic pathway
$\Delta G = \Delta H - T \Delta S$
C N O F Ne
Growth
The 4th Law of Thermodynamics

Neurochromosome gravity – carbon entromorphology gives us an excellent demonstration of the four forces of nature and of quantum gravity.

In carbon entromorphology living organisms are called entromorphic atoms. This means that an organism should be constantly aware of the effects of the four forces of nature in everything they do. This gives us an excellent opportunity to understand the interaction of the four forces. Currently gravity stands out in physics as the missing link and a link between quantum logic and general relativity is highly desirable. Attempts to identify the bosonic particles predicted by physics such as the Higgs boson and the graviton have not been totally successful so far.

What carbon entromorphology shows us is that the four forces exist in a complete state to allow energy and its mass equivalence (time) to flow from nucleonic systems of logic out to field logic and from field logic into nucleonic systems. As energy moves from nuclear states to field states it does so through the Gauge bosons, the W+, W-, Z^0 and photon γ particles. In carbon entromorphology we can see this process by looking at our arms. The arm is an amplified carbon bond and has antagonistic logic in keeping with the W+ and W- Gauge bosons but also a hand which is hypothesised as an amplified Z^0 boson. The final boson the photon γ can be seen when a person waves goodbye to someone.

Energy from the strong interaction associated with the brain is bosonically transmitted through the W+, W- antagonistic muscular system of the arm to the Z^0 boson or hand (ribosomal origins) which completes the transfer by the reflection of photons γ to a distant observer. This is the conventional electroweak standard model but an enlarged version allowing energy from mass origins in the brain (nucleus) to transfer out of nucleonic logic through electroweak means to a photonic end. This is the normal direction associated with the electroweak system but what carbon entromorphology shows us is that it also works in reverse. An observer seeing the hand waving goodbye does so by reversing their electroweak standard model. They now absorb photonic energy back through the electroweak system into nucleonic storage in the brain or nucleus producing mass and gravitational space time deformational effects through sensory electroweak neurological system. Therefore the electroweak standard model working in reverse is gravitonic where the hand or Z^0 particle is now operating as the Higgs boson and the W+ W- system operating in reverse transferring energy into a condensed stored state known as mass. In this way, the reverse allows the gauge bosons to become considered as the gravitonic bosons.

Once mass is established by reverse gauge bosonic effects its spatial distortion is constant and permanent through space-time, mass doesn't really attract it simply distorts space time, where space itself can be thought of as the force carrying particle but the gauge bosons establish it firstly by acting in reverse. Until we can thoroughly resolve the fractional dimension of space then a complete understanding of the gravitational effect will elude us. Gravity is a little misleading as we describe it as a force of attraction, but general relativistic logic states that space-time is distorted by mass and as such the exchange particle for gravity on a continuous basis must be space itself as a force-carrying particle. Once the reverse gravitonic gauge bosons fix energy, as mass in atomic systems then there is a permanent space-time distortion. Space-time appearsto be the bosonic force carrying particle system. Space itself is a seething soup of particles and anti particles popping in and out of existence and many virtual particles make up this soup. **This is a principle observation only and needs considerable thought although life does give us an interesting insight into this mechanism.**

The mind and consciousness itself is well placed to be described as a soup of virtual particles such as virutual photons, gaining rest mass when they are brought into fruition through an organisms behaviour; a matter anti-matter model of the soul of life.

When an atom collapses its wave function and the electroweak system is shut down then gravity determines energy level. Only when quantum wave functions re-appear do we observe the equilibrium with gravity. Since energy and mass are one and the same thing it makes sense that any bosons used for electromagnetic properties must also have equal gravitational mass properties as well. We don't just have an electromagnetic gauge system, in reverse it produces or instils the mass effect and instils permanent special distortion observed as gravity through gravitational waves perceived as a constant un-diminishing event. The overall effect on matter can be seen as the quantum gravity effect of the states of matter, solid, liquid and gas. They represent the quantum gravitational properties of matter linking gravity to the other three forces. Again the unique equilibrium established in matter in this way is the precise point in the force cycle produced by matter. All particle/wave systems are measured at some point in the force cycle; this defines particle physics but based on distance from the observer.

This author will continue to work on this situation it does need vastly more consideration, but again I feel that carbon entromorphology allows us to see how energy moves through matter more clearly on the level of an organism than the surreal impossible world of tiny atoms which are difficult to resolve and model.

Carbon entromorphology will continue to pursue these theories, aiming to bring understanding to humanity and more importantly peace.

Thank you for participating!

Science check

Do the observations correlate with the theoretical model?

YES ✓ *NO*

PART THIRTEEN

The author and development of carbon entromorphology and 'My Cell'

The carbon entromorphology was originally conceived by simple observations made in 1992 at Stockport College of Further and Higher Education. This author was studying for a place at UMIST (The University of Manchester Institute of Science and Technology) where I read chemical engineering for a fleeting half year. During my physical chemistry studies I learnt about atomic physics and the structure and function of carbon. On doing so, I realised that the organisation and functionality of carbon appeared to emulate the structure and function of my entire body. The oldest idea being that the head of a human containing two eyes, which have spin, and absorb radiation in the same way as electrons absorb radiation and have spin. They also form a spherical probability distribution around the brain system. The brain system again like the atom, was a centralising organisational system with a positive charge and a binary logic. All the basic components on both the atomic level and the level of an entire human seemed to me to be harmonious.

Over the next fifteen years the ideas swelled in my mind and having gone back to university to study biology over five years at Manchester Metropolitan University I was able to extend these ideas and I found more and more agreements between atomic, cellular and organism levels.

My final year project allowed me the freedom to develop these ideas further and I produced my first incarnation 'stress dynamics – An investigation into the role of stress in natural selection'. I explored the dynamics of life with the dynamics of atomic organisation, and looked at the symmetry with increasing amazement at the levels of agreement. I began producing fractional dimensional logic as a way of explaining the accretion theory of life. I looked at orbital models and explored thermodynamics and its links to behaviour in living organisms. I found atomic physics was all around me in my everyday life and began developing the ideas seriously.

Having become a Chartered Biologist CBiol I realised I needed to develop the ideas and embarked on one of the most prolific and extraordinary processes of scientific exploration. My aim was to link physics and biology, and to drive a powerful revolution in thinking which would allow humanity to integrate and harmonise science to produce a model of consciousness.

The following collage images reflect the mood and powerful methods of development from wearing my ideas on t-shirts to covering the inside of my entire house with my theories. I also produced a massive science/art gallery exhibition called 'My Cell' which filled an entire sports hall as a means of making science more interesting and diverse. I also embarked on the most intolerable and difficult process of gaining acceptance from my scientific peers. The culmination of which is saddening as many scientists were rude, narrow minded and of little or no use to me; even though these theories remained intact and unsuccessfully challenged after the past two years. My fear that science is so ludicrously aggressive and uncooperative has been borne out, but my motivation is still ragging even stronger than ever before and my resolve is powerfully stimulated.

My work culminated when a peer reviewer, Professor Jim Al Khalili a world renowned physicist, demonstrated my theory on his TV program, at which point I knew I would one day gain recognition.

It's important to make it clear that Professor Al Khalili has never made a clear statement of his support for carbon entromorphology. His broadcast however can never be retracted and exists as a most useful citation; they say 'imitation is the sincerest form of flattery'.

During 2010 we became part of the 'Human Thermodynamics' project where we received acceptance and my theories became established through this field at www.EoHT.com .

We also developed the website at www.mrcarbonatom.com and produced over 50 YouTube video lectures and 35 audio lectures on our website.

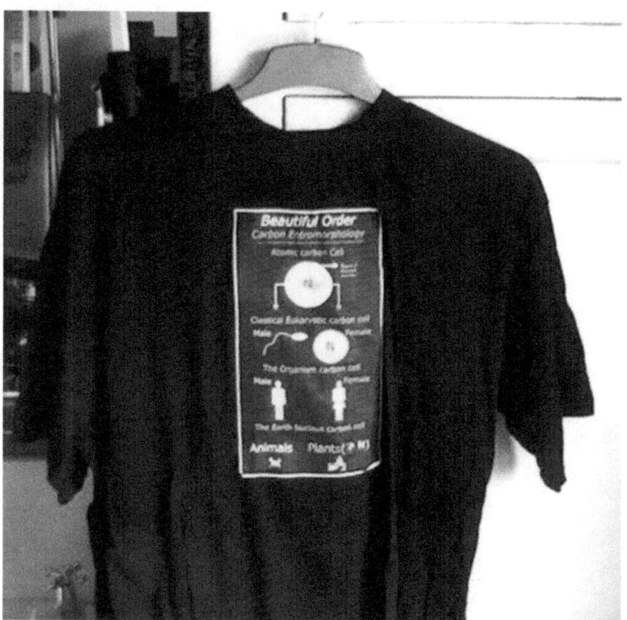

The Manic
Depressive 162
Depressive 163

Dedicated to my mother Margaret and father Richard, my best friends in the whole world! My parental K shell.

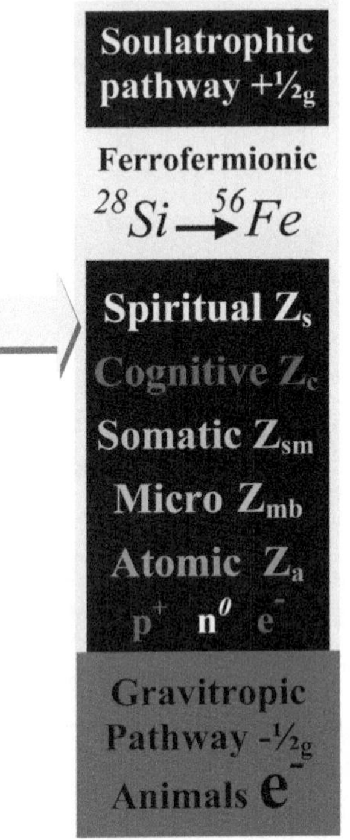

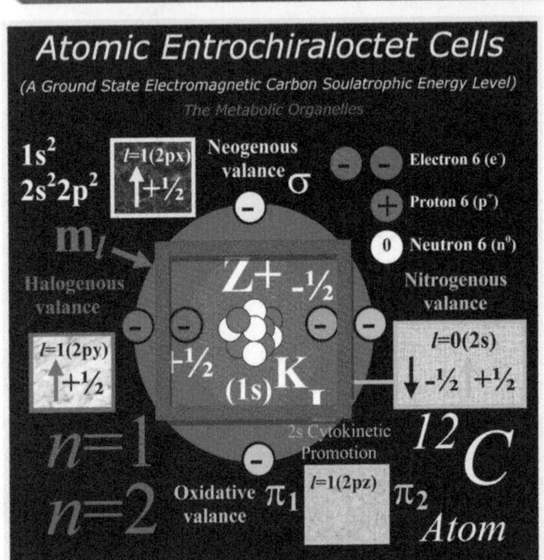

This author's (Z+) spiritual K shell.

PART FOURTEEN

Glossary of Terms and Etymology

Accretion theory – a theory where composite particles build upwards to form composite structures, biological systems follow this logic.

Anisogamy – The observation regarding the physical size and motility differences in sexual gametes, such as sperm and egg. The egg is considerably larger than the sperm, it also includes oogamy which is classical none motile large egg and motile small sperm, also isogamy where there is little difference between sperm and egg and anisogamy where the egg is large, sperm small but both are motile.

Artificial life – life post organic carbon based synthesis, often associated with silicon but determined by a carbon catalytic effect.

Atom – one of many levels of natural scale, the atom has a tiny positively charged nucleus containing most of the mass surrounded by electrons in motion having a negative charge. Organised into quantum mechanics, which considers electron clouds and nuclear physics appertaining to the nucleus.

Atomic nucleus – the centre or heart of an atom having a positive charge composed of neutrons and protons, and having most of the mass of the atom.

Big Bang – The beginning of the universe post singularity and the beginning of time and the true origin of life 13.7 billion years ago.

Binary – a system of logic divided into two functional logic states such as the nucleus of an atom comprised of protons and electrons, or DNA comprised of AT and CG residues, of the brain having action potentials either on or off.

Biology – the very broad subject concerning life although the word is poorly described and appertaining to carbon based systems covering all organic systems, but extended to include artificial life through silicon and iron.

Biochemistry – the chemistry of proteins, carbohydrates, fats and nucleic acids and a vast array of other molecules and their properties and interactions.

Bilateral – the emergence of cephalisation or the development of neurological systems of nucleonic organisation in living organisms, the evolution of a head in animals; not possessed by the plants.

Boson – the opposite of fermions, bosons are force carrying particles and mediate the four forces of nature. An important boson is carbon, but also sub atomic particles such as $W+$, $W-$, Z^o and photons γ, also suggesting Higgs boson and the graviton as part of the model of electroweak interactions.

Carbon – the soul or basis of all life, carbon is a small but undoubtedly the most exotic and varied of all elements and acts centrally extending its atomic properties to allow living beings to posses' carbon conserved properties: the basis of carbon entromorphology where carbons characteristics are conserved but amplified by solar acquisition. Carbon is an enormous concept and cannot be treated lightly. Carbon has 12 nucleons, 6 protons and 6 neutrons, and 6 electrons; having a nucleus with a K shell and L shell energy level system able of making 4 covalent bonds of various strengths.

Cell – a level of natural scale composed of a fractional dimension of atoms but acting as a fractional dimension of organisms. The cell is a fundamental unit of matter associated with carbon and acting as an amplified carbon entrochiraloctet. It also contains a nucleus with a binary logic made of DNA and expressing protein clouds in the cytoplasmic space and membrane to mirror that in the carbon atoms it is comprised of.

Cell cycle – the process of temporal components of G1, S, G2 interface, and culminating in mitosis or cellular separation: the interphase components representing bio molecules and higher structures.

Cellular nucleus – made of nucleonic DNA organised into condensed supercoiled chromosomes and centralising activities in the body and membrane of a cell; reflecting the nucleonic nuclear logic through amplified solar logic, self symmetrical to atomic nucleonics.

Chaos – the mathematical and physical nature of incompleteness and uncertainty, and the basis of nonlinear equations where small physical changes are amplified to produce huge errors in mathematical applications.

Chemistry – the electron photon interactions of groups of atoms bonding and exchanging energy to produce molecules which produce all the complexity and diversity of natural matter.

Chromosome – the supercoiled condensed nature of nucleonics such as atomic proton neutrons, and in cells AT and CG and neurological systems with action potential on and off. A classical cellular chromosome is made of histones and DNA and a book, DVD, CD as an example of extended neurological chromosomes.

Cognitive – appertaining to neurological nucleonic action potential on or off, and one of the higher soulatrophic energy levels in living beings, the level before spiritual familial fractional dimensions or soulatrophic energy levels. Plants are absent of this particular level of organisation.

Conservational – inheritable physical properties often with a non locality reference. Energy conserved and amplified by solar acquisition in living organisms.

Consciousness – an integrated state of being, typically associated with neurological nucleonic systems. A state of self awareness or self measurement; although the beginning of such a process can only be justified from the Big Bang. In carbon entromorphology consciousness does not end at the neurological level and is considered to all levels of natural scale, hence a carbon atom has a low level of self awareness although this is a theoretical logic and impossible to prove.

Cytoplasm – the body of a prokaryotic and eukaryotic cell, containing a stable region of protein expression or protein clouds. It is based on the K shell component of atomic levels of organisation but amplified to cellular levels. The cytoplasm has an L shell membrane level shielding it from external energy exchange.

Cytology – the subject of prokaryotic and eukaryotic activities of cells and all their properties and activities.

Cytokinesis – the point at which a classical cell divides into two cells through the activities of the cell cycle and leading to growth through solar ECASSE amplification.

Death – appertaining to a differentiated state of life where integration of soulatrophic energy levels or fractional dimensions collapses to lower levels such as the microbiological levels. Apoptosis is used to describe cellular death and describes a sudden energy change to thermoentropic states.

Disorder – a state of non symmetry associated with Boltzmann entropy, and the increase in the probability of certain ensembles of particle microstates. It is the personification of the second law of thermodynamics where disorder or entropy is increasing in closed thermodynamic systems over time.

DNA – Deoxyribose Nucleic Acid, the nucleonic form following on from classical atomic nucleonic. A nuclear system, centralising protein expression in the cytoplasm with bosonic mRNA carrying the genetic force. A double helix of AT and CG binary base pairs for encoding stored nuclear information for making living structures.

ECASSE (etymology) – Entromorphic Carbon Amplification by Stimulated Solar Emission; a LASER based logic based on Bose condensate and bosonic statistics. Bosonic carbon statistically appears in the same or similar state as other carbon atoms. There is also in-phase cohesion forming an ECASSE beam such as a flock of birds or a shoal of fish. The stimulated effect is solar and is fixed through the two photo systems in the plants through to the animals.

Entrochiraloctet (etymology) – a carbon based template, the entrochiraloctet has a vast array of forms which reflect the enormous variation in living beings. The entrochiraloctet has a nuclear region, a K shell or body, and an L shell or valance limb bonding region. A human being is a hybrid entrochiraloctet having a nuclear head a stable torso region or K shell, and a bonding energy conductance region or L shell. The octet component is satisfied by bonding with four leading to quantum stability ('entro' from the word entropy, the 2nd law of thermodynamics, 'chiral' the assymetrical distributions of charge in life and 'octet' the rule of octets describe the tendancy for elements to complete valance shells).

Electromagnetism – one of the four forces of nature, it is both attractive and repulsive and acts over a relatively short range although it theoretically extends to infinity and is mediated by bosonic photons.

Electron – a fundamental particle and part of the particle atomic trinity and is mediated by photon interactions, subject to quantum jumps and is the basis of chemistry.

Energy – is equivalent to mass according to Einstein's relativistic mechanics it is the substance of all matter in the universe and is set in quantity by the first law of thermodynamics, where energy cannot be created or destroyed but moves from one place to another changing form, but inevitably turning into heat and dissipating. This is the basis of the second law of thermodynamics.

Entropy – energy which is no longer available to do useful work in a thermodynamic system. It is the personification of the second law of thermodynamics and is increasing over time in the universe.

Entromorphology (etymology) – the subject describing the thermodynamic characteristics of living systems when based on the second law of thermodynamics. The inevitability of thermo entropic states is minimised during an organism's life but will eventually produce disorder and death. Consider the face of a baby and a 95 year old person, there is a dramatic difference in symmetry which reflects entropic states. The baby is smooth skinned with absolute symmetry and plenty of living potential, the 95 year old is wrinkled, blemished, sagging and disordered, and this is an entromorphic span from high free energy levels to low free energy levels. There is an entromorphic span across a living organism from the head area which is highly symmetrical to the genital region which is thermoentropic and highly asymmetrical. This entromorphic span is also called the 'erotic potential' and drives basic sexual nuclear physics. The symmetry is based on statistical ensembles of living microstates, the probability distribution of which increases with age as the 2nd law of thermodynamics prevails. ('entro' the enevitability of the 2nd law of thermodynamics and 'morphology' the particle/ wave duality energy levels in living organisms – life is defined by amplification of nuclear and field regions by fractal solar processes). In short 'Carbon Entromorphology' is the study of the fractional dimensions of conscious self awareness through solar amplified carbon determinism.

Entromorphic elevation (etymology) – the elevation or probabilistic distribution in a living organism relative to the Earth and the Sun. Humans have a 90 degree elevation where as a worm has a flat elevation, the elevation is a measure of living hierarchy and represents evolutionary potential. It is also varied as organisms sleep in a low elevation and wake into an elevated state. The effect is a convection model of living animation called the 'flame model'.

Evil – the metaphysical concept of disorder and solar liberation. In carbon entromorphology evil is part of thermodynamics and suggests that the past is the personification of the second law, a thermoentropic model. In statistical thermodynamics evil produces a breakdown of natural symmetry and an increase in the number and frequency of specific particle microstates.

Evolution – the time orientated change in particle and thermodynamic properties in living organisms. Evolution is the mechanism by which specific thermodynamically stable particle ensembles remain prosperous and where inferior systems are reduced in probability over time.

Flame model (etymology) – the model of living animation of organisms any conscious organism has a span of energy and symmetry across their bodies. They appear animated as if they were a stable flame with the environment attempting to blow them out or allow them to prosper and stabilise.

Field – the quantum and gravitational consequences of the four forces of nature, it is the outer space consciousness and is theoretically infinite in its scope. In living beings the limbs form the field components and the body is typically part of the field.

Gametogenesis – The process of the formation of the sexual cells such as sperm and egg.

Gender – the basis for energy transfer through particle/wave interactions which is based on the fermionic trinity where a male is an electron particle and a female a protonic particle. They join to form a neutronic offspring or exciton, neutronism is also seen as the basis for homosexuality.

Gender model (etymology) – a reductionist model of gender based on the fact that all matter in the universe is made of protons, neutrons and electrons. This model can also be considered as electrons and one up and one down quark. This model allows particle characteristics such as pressure to indicate why living organism look and function in the way they do. For example, a male is an electron particle and therefore moves a lot, hence has a streamlined property seen in practical clothing. A female is a protonic nuclear particle and remains stable and under high pressure hence impractical clothing and almost no stream lining.

Genetics – the field of biology based on conserved characteristics based on DNA nuclear systems and as a result, a very narrow concept. In carbon entromorphology genetics is based on atomic and neurological genetics and silicon genetics and the basis is in non locality or memories which allow properties or alleles to be conserved and passed on through sexual reproduction.

Gravity – one of the four forces of nature, it is the long range large scale determinant in the universe, and it's always attractive and is very weak.

Familial – appertaining to the family as a basic multi organismal cell, part of the spiritual soulatrophic energy level.

Fermion – strongly interacting particles comprised of quarks and leptons, part of the proton, electron and neutron.

Female – a protonic nuclear particle gender component, having high pressure and lacking streamlining.

Follicle – The large aggregate of cells which surounds the female oocyte leading to ovulation, the follicle bursts and begins growth when FSH (follicle stimulating hormone) is released from the pituitary gland. The contraceptive pill inhibits FSH so that oocytes do not develop into viable mature ovum.

Fractional dimension – a mathematical system where natural order follows a fractional dimension to produce complex life. The atom is a fraction of the cell, the cell a fraction of an organism and together they demonstrate self symmetry on each level and conserve properties. They follow an iterative cycle and lead towards increasing compositions.

Free energy – Gibbs suggested free energy was energy available outside entropy or waste and the associated enthalpic difference or total energy of a thermodynamic system. Free energy allows bonds to form in living beings where the free energy source is the Sun and the Earth is an open system receiving and fixing this energy in biological systems.

Gene – a memory in the form of a binary string such as DNA or the words in this book. Based on non local logic and copied and transmitted through sexual reproductive means.

Genomics – complete gene compositions in a living organism, typically associated with DNA but also through neurological systems where this book is a genome.

Gluon – the strongly interactive component of the strong interaction in atoms but also in higher nucleonic systems such as DNA and neurological systems. Gluons hold quarks together in the same way the letters in these words are held together under the same logic. Gluons are string like and can be seen in DNA which is linear, and neurological networks also have such properties.

Growth – amplification of particle/wave systems mediated on planet Earth through solar acquisition. Growth occurs under ECASSE on planet Earth through carbon mediated amplification to produce cells, organisms and multi organismal cells.

Good – the metaphysical concept associated with growth and stability in biological systems. It represents the open system thermodynamic logic which leads to increased physical size with greater order and complexity, but with reduced waste.

Halogens – the most electronegative elements in group 7 on the periodic table. They are one electron short of quantum stability and have powerful electrophilic potentials. Chlorine is the most prolific in biological systems but also fluorine and iodine.

Hereditary – copied and conserved information associated with atoms, cells through DNA and neurological systems through classical memories. Life continues to increase in order and complexity through conserved transfer of memories; based on non local logic in atomic physics.

Homosexual – a neutral non polar gender where small changes in potential produce powerful cells (neutronism).

Hydrogen – the first atom and the first component post Big Bang, leading to the composition of all the 92 naturally occurring elements. Comprised of a proton and electron, it is a single polarised system and forms an excitonic state.

Inner space – the observer measuring with a nuclear perspective.

Interphase – the period of time in a cell cycle associated with physical activities such as duplication of components and readiness for duplication of cells. Made of G1, S, and G2 interphase leading to mitosis M or cell division. In carbon entromorphology cells are extended and as such, the cell cycle in also extended. A G1 for a human adult is the morning, the S interphase the afternoon and the evening is the G2 interphase and finally sleep or neurological mitosis leads to a new day. The cell cycle is based on the octet rule (rule of octaves) in carbon chemistry; the four components of the octet are symmetrical to interphase and mitosis.

Isogamy - The observation regarding the physical size and motility differences in sexual gametes such as sperm and egg. Isogamy where there is little difference between sperm and egg and anisogamy where the egg is large, sperm small but both are motile.

Iterative – cycling mathematical process associated with fractional dimensions and fractals.

Life – a particle/wave system having self observational thermodynamics and therefore self awareness or self preservation, also having the ability to duplicate. The animals personify this condition most clearly but even the plants act as heliotropes actively increasing their chances of reproduction and physical energy conservation, although their self awareness activities are comparatively primitive and extremely slow. The fractional dimensions of carbon amplified accretion through solar nuclear and field amplification.

Legal genomics (etymology) – an extension of classical genetics which is based almost entirely on DNA to neurological nucleons and their applications, legal genetics is another branch of nuclear physics. The genomes of which are represented by the vast array of stored laws within in human systems.

Legal genetics (etymology) – a general statement of the extension to classical genetics based on neurological systems but acting with evolution and natural selection.

Legal gene (etymology) – an individual law based on neurological genetics, a national memory and a non local event reflecting previous conscious deliberation.

Male – the electronic component of the gender model it is a fundamental particle operating in quantum fields of uncertainty, and moving rapidly hence having streamlined properties, it is a field particle.

Mass – a concentrated form of energy which can be converted from one form to another but does fall within the logic of the first law of thermodynamics, where energy or its mass equivalence cannot be created or destroyed.

Membrane – a highly reactive field component of living organisms which can act to selectively allow particles to permeate through it according to need. The membrane in a cell is a classical bilayer mosaic system, on the level of an entire organism such as a human where the membrane is generated by the limbs which are also a bilayer into upper limb and lower limb producing a similar result. The membrane is also based on the second energy level in carbon the L shell where energy comes and goes and where bonds are formed and broken. The membrane stabilises the inner K shell or cytoplasm and the nuclear components on all levels of biological organisation.

Meiosis – the process of reducing diploid double nuclear components such as chromosomes into haploid germ cells in sexual reproduction. In all nucleonic systems from atoms to cells, to organisms the DNA, neurological systems and atomic nuclei possess nuclear fusion processes referred to as meiosis. In neurological systems the nuclear fusion in meiosis produces new ideas or thoughts as the offspring of combining two or more ideas stored as memories of neurological genes.

Mitosis – nuclear fission it relates to the division of one entity into two by solar stimulated emission or acquisition. It is based on Bose condensate or solar stimulated emission similar to LASER logic. In atoms of carbon this process takes into account bonding between two or more atoms, in cells it produces new cells which bond to serve similar purposes and in organisms post sexual reproduction produces offspring from meiosis or nuclear fusion then mitosis or nuclear amplification or growth. It is the basis of growth in all biological systems through physical stimulated emission.

Metabolic organelle (etymology) – the missing bonding components in carbon based cells, it represents the octet rule in physics where carbon seeks quantum stability by acquiring missing electrons from covalent bonding to stabilise its L shell or membrane. It is the detailed way in which carbon cells grow and acquire energy in order to remain functional. In most organisms they are the food they eat and the other organisms they interface with, in humans the metabolic organelles includes the technology we use and bond with every day. There are four missing components: the nitrogenous configuration which is structural, the oxidative configuration which is an energy store and the halogenous which is based on fats and membrane activities associated with fluorine, iodine and chloride ions (it is the protective but most reactive level), and finally the neogenous configuration which is a nuclear duplication system which produces quantum stability. Basically any organism must find nitrogenous proteins, oxidative carbohydrates, halogenous lipids and neogenous nucleic acids to function and remain stable. A human will also look for these but also a nitrogenous house, an oxidative electricity supply, halogenous devices and neogenous internet. It may also be driven by iron nuclear instability.

Neon – the Nobel gas which is the direct result of carbon, nitrogen, oxygen and the halogens completing their membrane L shell octet and reaching quantum stability.

Neutron – a nuclear atomic particle and part of the atomic trinity, a neutron is broken into an electron a proton and a neutrino. It is the heaviest of the particle trinity and is non polarised and therefore not subject to electric fields. It is also associated with the homosexual gender in carbon entromorphology where men with men and women with women bond with neutrality.

Neurological – appertaining to the highest nucleonic form the neurological system is based on neurons, cells which act as nuclear systems. They produce networks or lines of inter connected cells which convey energy through action potentials a binary on off code system. The neurological system is fractured into DNA and DNA fractured into atomic nuclei. The neurological nucleons have many applied forms such as laws, information technology and the written word. They represent genetics at the highest level only the genes are lines of nerve cells and their configurations produce neurological genes termed memories although they produce the same logic as the lower DNA and atomic levels.

Neurogene (etymology) – a nerve cell string representing a single component such as the memory of the word 'word' which is a neurogene. A collection of words form chromosomal systems such as a this book which is a neurological genome super coiled into this chromosomal book.

Neurogenomics (etymology) – the collective term used to describe memories composed of many complex sub memories acting as a group through genetic logic.

Neuromeiosis (etymology) – the concept of a new idea a combination of memories to produce a new result which can be copied and transmitted as the words on this page are doing in the reader.

Neuromitosis (etymology) – the duplication of neurological memories and transitions based on ECASSE for long term conservation.

Nitrogen – a small but highly important associate of carbon, nitrogen is the next element on from carbon and contains an extra proton, neutron and electron. It allows carbon to extend its nuclear range through the breathtaking variety of nitrogenous carbon systems. Nitrogen is highly electronegative and is also associated with oxygen to produce all the basic metabolic organelles in biological systems.

Non locality – a powerful but poorly understood concept in physics typically associated with atoms and their constituent particles. Non locality implies, although many claim to have disproved this, that information or energy about a particle interaction is either able to travel faster than light or that a particle conserves its particle entanglements changing its state for interactions at vast distances away beyond any mechanism associated with non relativistic postulates.
In carbon entromorphology it is theoretically proposed to be the basis of heredity the concept of memories on atomic, cellular and organism levels, it is the basis of genetics.

Nucleonics (etymology) – appertaining to atoms the proton and neutron in the nucleus are nucleons. In carbon entromorphology there is an extension to classical nucleonic where DNA and neurological systems are higher forms of nucleons and behave in symmetrical ways to the atoms they are comprised of. They have nuclear fission and fusion properties known as meiosis and mitosis for DNA and neuromeiosis and neuromitosis for neurological systems.

Nucleus – a centrally constraining logic, the backbone or heart of a system, the determining feature in any physical system. Typically the word is associated with atoms and cells, but the brain also acts with central determinism and is also a nucleus. Geological arguments can be made that the magma in the Earth is also a nucleus and that the Sun is the nucleus of the solar system. The difference is reflected in the forces dominating the logic towards atomic nuclei the dominance is electromagnetic and towards astrophysical bodies it is gravitationally determined.

Octet rule – a very old and powerful rule in chemistry (John Newlands 'law of octaves' 1864) which suggests that atoms can only become stable if their energy levels are complete. Their unique chemical properties are determined by how much an atom needs to stabilise its valance shell. In carbon this process is very important as carbon has a half filled shell and as a result, has a broad set of interactive properties which enables it to bond with vast numbers of other carbon atoms and other atoms. The octet is also amplified in carbon entromorphology and is theoretically associated with the cell cycle.

Order – symmetrical systems which demonstrate stability through thermodynamics and energy.

Organelle – a specific particle system such as the valance electrons in atoms or the mitochondria in cells, it is a term typically associated with cells and the components of their anatomy. In carbon entromorphology it is extended to include atoms and organisms. The word organ is associated with the liver for example and it in turn has organelles in its cells such as mitochondria, which serve the same purpose and electrons in the carbon atoms which make the cells, which have oxidative properties.

Oogamy - The observation regarding the physical size and motility differences in sexual gametes such as sperm and egg. The egg is considerably larger than the sperm, it also includes oogamy which is classical none motile large egg and motile small sperm, also isogamy where there is little difference between sperm and egg and anisogamy where the egg is large, sperm small but both are motile.

Oogenesis – The gametogenesis of the egg or oocyte (ovum) associated with female sexual dimorphism, the folliculogenesis stages from primordial to zygote.

Oocyte – The female sexual cells which are large spherical cells typically none motile. Oocyte are primordial and diploid in females, and remain as such until the female starts to ovulate at which point they grow from 30 to 50 μm to 50 to 100 μm which is the primary stage, and >100 to 150 μm secondary stage. They undergo meiosis from primordial stage and interrupt it until they are fertilised. The process of maturity also produces a follicle which is very large and encapsulates the oocyte with an antrum centre and releases and bursts on ovulation.

Organic chemistry – the massive subject associated with carbon and hydrogen but also nitrogen, oxygen, phosphorus, halogens, sulphur and up to 26 elements for living organisms. It is concerned with carbons enormous distribution of properties and the immense variation in the molecules, cells and organisms which is formed. Carbon remains central so any organic system is fundamentally governed by carbon nuclear potentials. This process allows carbon to extend its characteristics to produce amplified carbon states as proposed by carbon entromorphology.

Organism- a free living organic system, it is typically associated with DNA driven systems and cells although a complex organic molecule could be considered to be an organism.

Outer space – appertaining to field theory and quantum mechanics in atomic physics, in humans for example outer space reflects our experience of the environment.

Ovum – The large (100 to 150 μm) spherical female sexual cell in its mature stage post ovulation.

Oxygen – a small atom associated with carbon, nitrogen and phosphorus it is highly electronegative and highly reactive. In biological systems it is found associated with high energy extensions of carbon based organisms. The oxygen atom is able to accept low level electrons and is a major component of water and as such absolutely pivotal to life on all levels. It exists as a gas on planet Earth.

Periodic table – a powerful and major theme in the whole of science all the known elements and many artificial elements are collected into a set of tables which group according to size, reactivity and unique physical properties such as orbitals. The periodic table is the basis for all life as described in carbon entromorphology and its octet rule is one of the most powerful theories for explaining all natural order and behaviour with iron sitting at its heart with all the elements seeking its stability.

Pion – a nuclear binding particle, the pion allows nucleons to bind together where they bind internally through gluons.

Proton – one of the particle trinity, the proton has a positive charge and is termed a nucleon. It is part of the nucleus of an atom and is found with neutrons making a binary logic. It contains the medium level of mass and produces potential wells or holes in the quantum fields surrounding the nucleus which are in turn filled with electrons. In carbon entromorphology the proton is a female and has high pressure morphology as it is a nuclear particle and is balanced by a male, an electron particle.

Quantum gravity – the theory that large scale theoretical physics associated with gravity: the weakest, but longest range, force to the small scale theory of quantum mechanics and the strong, weak and electromagnetic force. A unified theory the dominance of general relativity at large scale levels in the universe to seamlessly translate into quantum mechanics on a small scale. In carbon entromorphology it is postulated that any such theory must have the observer or life at its absolute heart and cannot be simply arranged either side of such a perspective. Life translates all levels of natural scale and as such must personify any theory uniting all four forces, therefore life as defined by carbon entromorphology is a unified theory where all the four forces exist on all levels of natural scale regardless of their absolute contribution to determinism at any point x,y,z,t. Quantum mechanics makes the mistake of eliminating gravity from its equations and as such breaks open the four forces into 1 for large scale and the other three for small scale. The quantum numbers must have gravity in them even if the gravitational effect of a single electron on another electron is vanishingly small. If such an electron were on the surface of a Sun a hundred times the size of our own then its quantum effects would be distorted by the action of the Sun and cannot be separated from the physical model.

Quark – a fermion and a fundamental particle which make up the sub atomic particles. It is postulated in carbon entromorphology that letters in words for example, and bases in DNA are a type of amplified quark logic.

Radial – a simple type of multi cellular organism it is the somatic soulatrophic energy level or the radial fractional dimension. It extends only in the animals to produce bilateral symmetry and cephalisation or the head.

RNA – Ribose Nucleic Acid, it has a messenger and transfer property and is associated with the transmission or force carrying properties of living organisms. It is compared with bosonic particles in the atoms it is made from. The RNA equivalent on the level of an entire organism is seen in the axon and dendrite components of neurological systems, again force carrying particles.

Silicogene (etymology) – an information technology memory or algorithm: ROM (read only memory) it is associated with neurological soulatrophicity and subject to the same rules associated with genetics and nuclear physics.

Sperm – The male sexual cells they are small (45 μm in length, head 5-8 μm in humans) and highly streamlined, and have a haploid genome.

Spermatogenesis – The gametogenesis stages for the production of sperm cells in the testes for sexual dimorphic reproduction means.

Spiritual (etymology) – the highest soulatrophic energy level it is the 'familial fractional dimension' in soulatrophic pathways. It is the basis of multi organism cells typified by the basic family grouping and colonial organisation seen in ant colonies or simple organisms such as bacteria.

Somatic (etymology) – the radial fractional dimension associated with multi cellular systems and can be seen in the lower legs and abdominal region of a human, it is very old and can be seen in the radially symmetrical organisms such as sponges, starfish, polyps and medusa.

Soul – the origin of a physical system, in carbon entromorphology the soul is a nucleonic pathway leading all the way back to the Big Bang.

Soulatrophic (etymology) – a term in carbon entromorphology giving reference to particular fractional dimensions of natural scale; such as the microbiological, somatic, cognitive and spiritual levels.

Soulatrophicity (etymology) – appertaining to unique fractional dimensions of scale but with an absolute origin or soul at the Big Bang.

Strong interaction – one of the four forces of nature it holds nucleonic systems in atoms together and is immensely powerful but its range is very small. In carbon entromorphology the strong force can also be observed in DNA and neurological systems. For example the words in this text; the letters are held together by strong interaction logic and the words held together by pionic logic.

Supercoiled – a natural property of nuclear to field logic, nucleonic systems such as atomic nuclei, cellular nuclei and organisms brains are all supercoiled. They are associated with DNA and can be observed in the convoluted appearance of the brain. Supercoiling is associated with pions and the naperian logarithm associated with natural spirals and quantum mechanics.

Thermodynamics – a massive fundamental part of all science but typically associated with physics. It concerns itself with the properties and characteristics of energy translation and defines systems and surroundings and the process of work. It is also a way of understanding thermal effects and the availability and usefulness of energy in a particular system. It is also based on statistical mechanics where particle systems and their organisation are related to macroscopic environmental organisation. In human thermodynamics we find systems concerned with energy flow through human observations.

Thermal energy – energy in the form of molecular vibrations, but also as infra red radiation.

Unified theory – a theory aiming to seamlessly link a series of differentiated algorithms into an integrated form.

Wave function – the wave function is the basis of quantum mechanics which defines an electrons position and momentum around the nucleus. In carbon entromorphology this concept is extended into everyday life where the 'conscious wave function' is the very essence of consciousness. It is based on x, y, z, t space time amplitude readings. The wave function is based on 'e' the Naperian logarithm and 'π' and can be solved to produce probability distributions or electron clouds. When a human moves their arm they are producing conscious wave functions and orbitals, living orbitals.

Weak interaction – one of the four forces of nature unified with electromagnetism to produce the electroweak standard model in atomic physics. It allows energy to flow out of the nucleus into quantum fields and is associated with radioactivity. In humans the amplified effects can be seen in the body language of movements and interaction as the absolute personification of consciousness.

Science

check

Do the observations correlate with the theoretical model?

YES ✓ ***NO***

PART FIFTEEN

Index (bold text indicates best references)

References for the mathematical evidence on pages 91-97.

These are a small quantity of important scholarly articles which were used in the creation of the scholarly article for publication.

Baccetti B (1986), Evolutionary trends in sperm structure. Comp Biochem Physiol A. 85(1): 29-36. PubMed PMID: 2876819.

Birkhead T, Hosken D and Scott P (2009), Sperm biology and evolutionary perspective, Academic Press, Science.

Cummins J M and Woodall P F (1985), On mammalian sperm dimensions, Journals of Reproduction and Fertility, 75, pg 153-175.

Curry MR, Millar JD, Tamuli SM and Watson PF (1996), Surface area and volume measurements for ram and human spermatozoa. Biology of reproduction 55, 1325-1332.

Durinzi KL, Saniga EM and Lanzendorf SE (1995) The relationship between size and maturation in vitro in the unstimulated human oocyte Feb; 63(2): 40 4-6.

Flam F (2006) Researchers delve deep to explore the secret life of sperm. Seattle: The Seattle Times Company. May 17, 2006.

Mtango N R, Potireddy S and Latham K.E (2008) Oocyte quality and maternal control of development. Int. Rev. Cell Mol. Biol. 268, 223-290.

Rikmenspoel R (1984) Movements and active moments of bull sperm flagella as a function of temperature and viscosity, 7- exp. Biol. 108, 205-230 (1984) 205.

http://en.wikipedia.org/wiki/Oocyte
http://en.wikipedia.org/wiki/Spermatozoon
http://www.edurite.com/kbase/human-egg-cell-size
http://wiki.answers.com/Q/Size_of_a_egg_cell
http://www.spuc.org.uk/ethics/abortion/human-development
http://www.dh.gov.uk/en/AdvanceSearchResult/index.htm?searchTerms=sperm+size

Many other academic scholarly articles were used in this investigation however the article section generalises their information. This reflects the reduced space available for production of a scholarly article.

Appendix 1

Atomic and molecular models used to demonstrate symmetry in carbon entromorphology.

Since the theory of carbon entromorphology suggests that living multi cellular organisms are amplified consequences of atomic organisation, atomic symmetry in an organism is demonstrated using models. The models at best are mathematical but scientists have been using graphical (visual) atomic models for the best part of 100 years to illustrate natural organisation. Atoms are counter intuitive and are mainly (99.999999999999%) empty space. It should not impede the use of such simple ball and stick models to illustrate atomic and molecular and now cellular and multi organismal systems. Below are some examples of typical atomic models and the link to carbon entromorphology. These models are often made into solid ball and stick structures by chemists and physicists and biochemists.

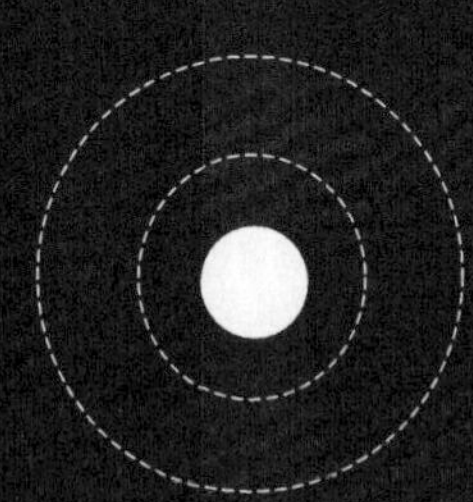

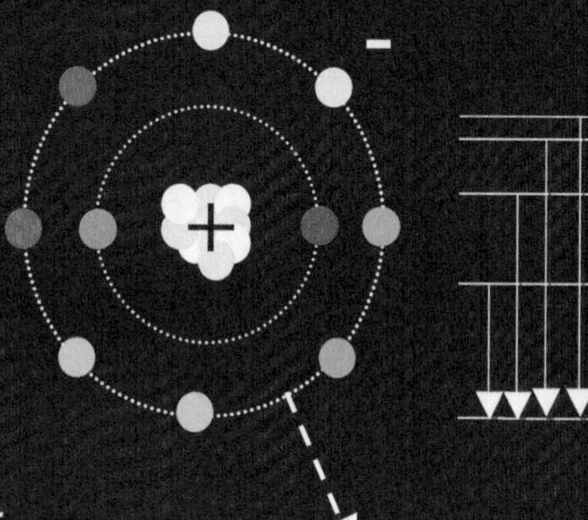

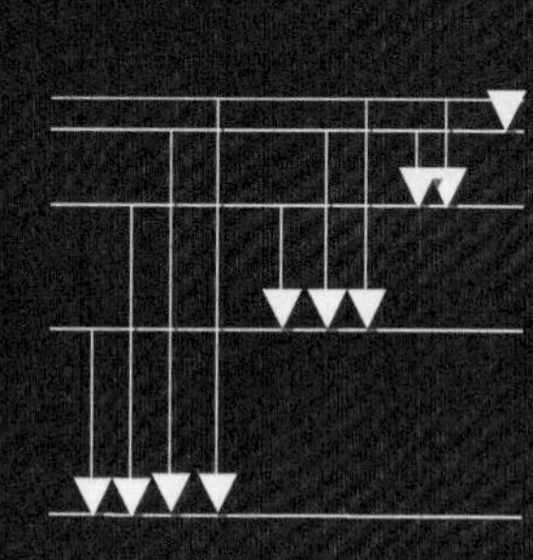

These visual models simplify the true complexity of atoms and as such are often regarded as insufficient. Only mathematical models offer a true model of atoms, but this visual method is universally utilised and is perfectly satisfactory for illustrating atomic organisation and its relationship to entromorphic atoms.

The different colours illustrate different strengths of bond!

Appendix 2

Morality and metaphysical logic in human thermodynamics.

Metaphysical concepts appear at first glance to have nothing to do with science as they are typically impossible to measure or even identify tangibly. It is however not logical to ignore them as they contribute to the development of scientific method and influence billions of people on planet Earth each day. In carbon entromorphology they become increasingly important, particularly their contribution in human thermodynamics and the 2nd and newly hypothesised 4th law. Below are examples of beauty, good and order for the 4th law and ugliness, evil and disorder for the 2nd law.

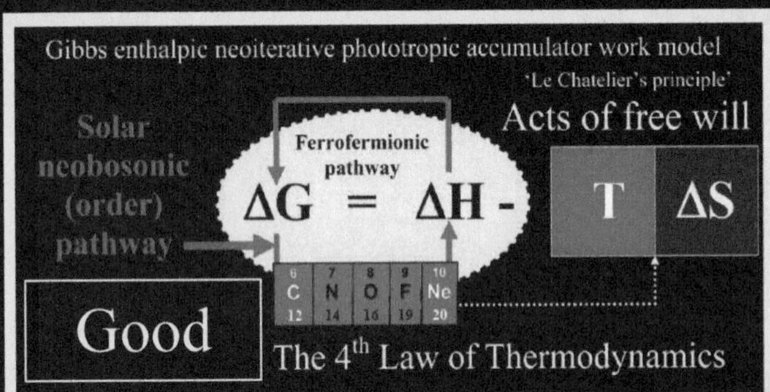

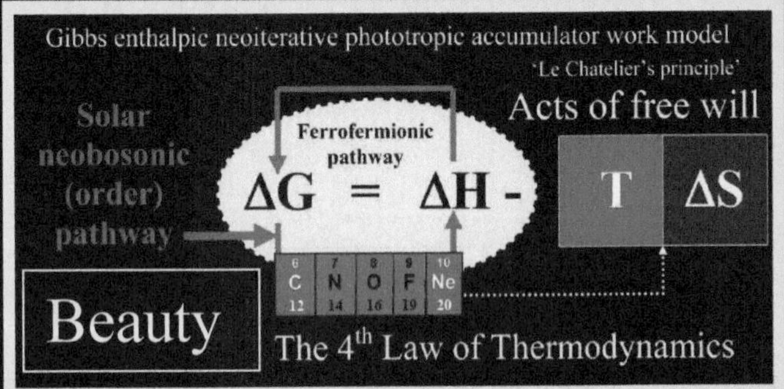

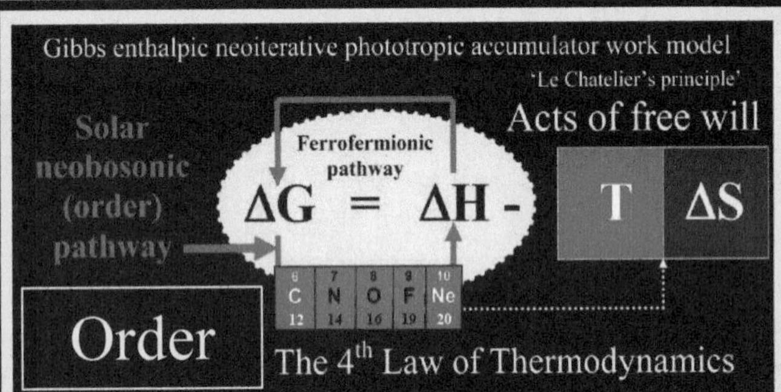

Symmetry 'heavenly environments' free energy, divine stability

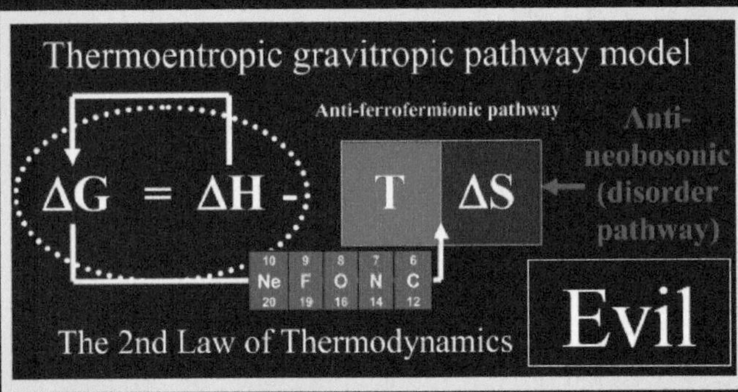

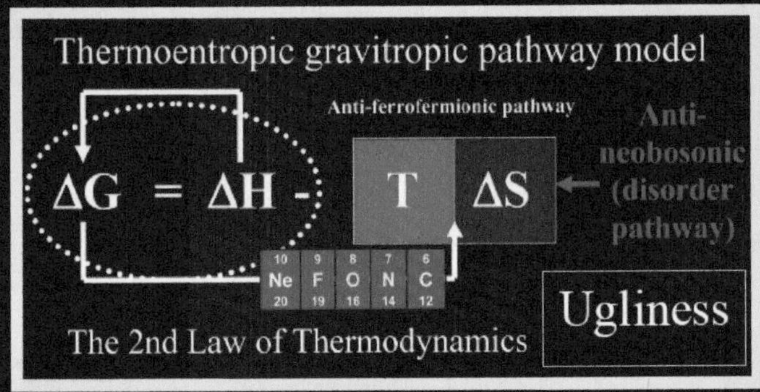

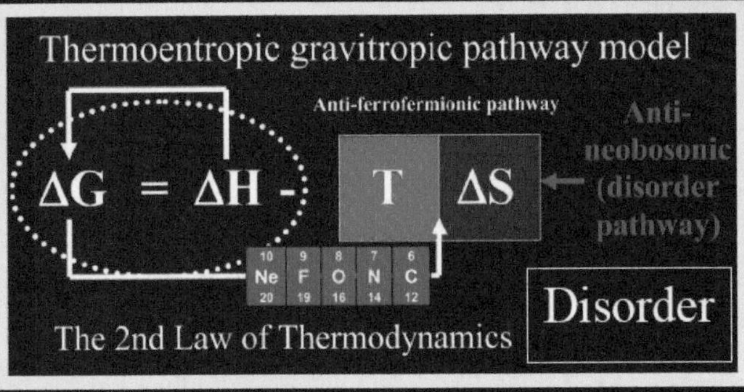

Symmetry breaking 'hellish environments' zero free energy, demonic instability

Appendix 3

A basic fractal model of life, the fractional dimensions (soulatrophic energy levels).

Just a final reminder of the nature of fractional dimensions used to describe living organisms in carbon entromorphology. These three pathways are formed from the dominant contribution made by the three Fermionic particles which comprise all matter in the universe.

This is a hyper reductionist view but one which appropriates the logic of natural order against our most fundamental physics.

This set of pathways is the new convention based on peer review advice provided through the encyclopaedia of human thermodynamics where the world 'soul' was considered to be unscientific.

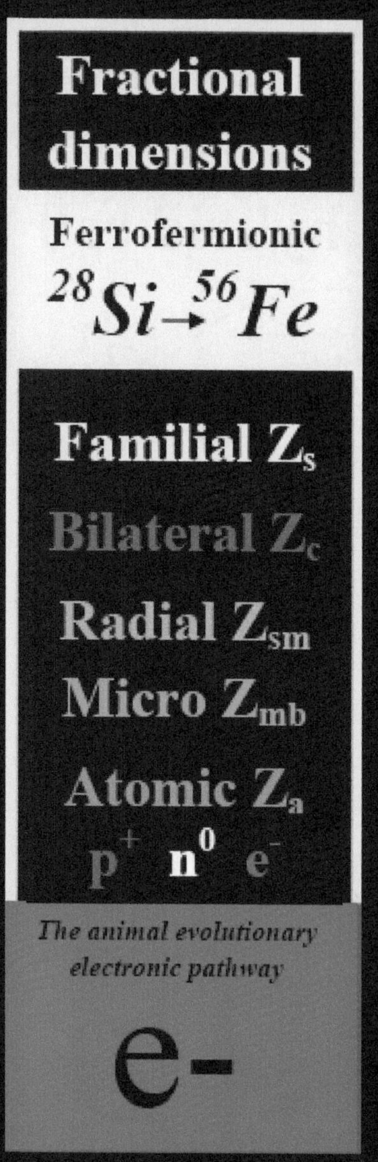

Fractional dimensions
Ferrofermionic
$^{28}Si \rightarrow ^{56}Fe$
Familial Z_s
Bilateral Z_c
Radial Z_{sm}
Micro Z_{mb}
Atomic Z_a
p^+ n^0 e^-
The animal evolutionary electronic pathway
e-

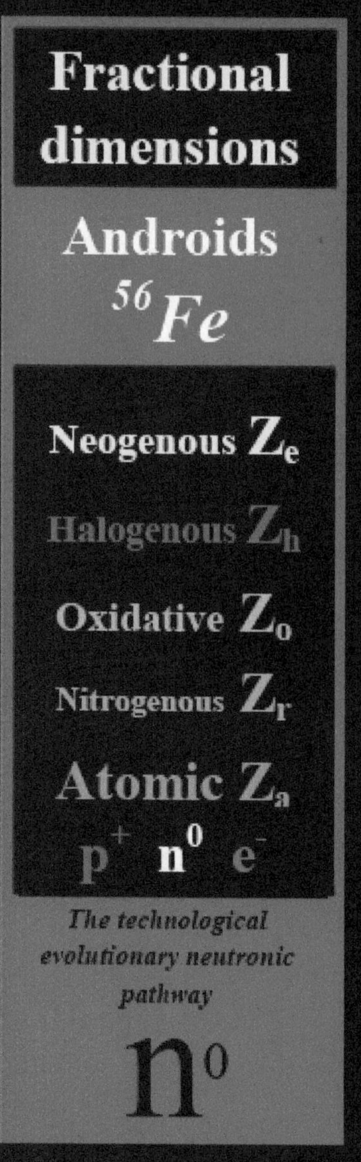

Appendix 4

A basic fractal model of Life.

The 'Big Bang' singularity (13.7 billion years ago)

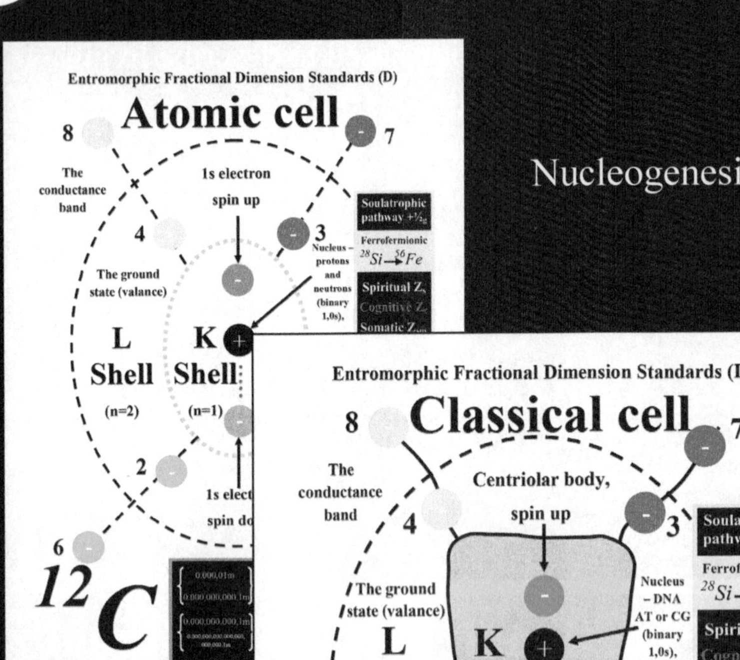

Nucleogenesis (>5 billion years ago)

(1 billion years ago)

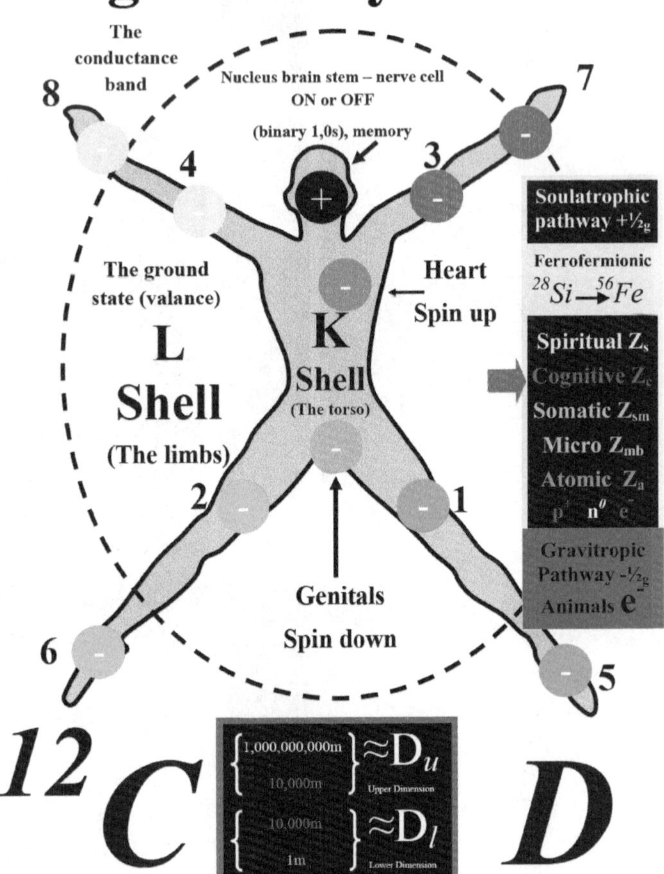

Modern Humans (20 thousand years ago)

Appendix 5

Nuclear and field amplification in carbon.

The following diagrams illustrate the process of carbon amplified determinism from atomic levels up to microbiological levels (classical cells). From Urey and Miller's amazing research into early carbon accretion through organic chemistry (alkane/alkene/alkyne) to biochemistry to cell biology.

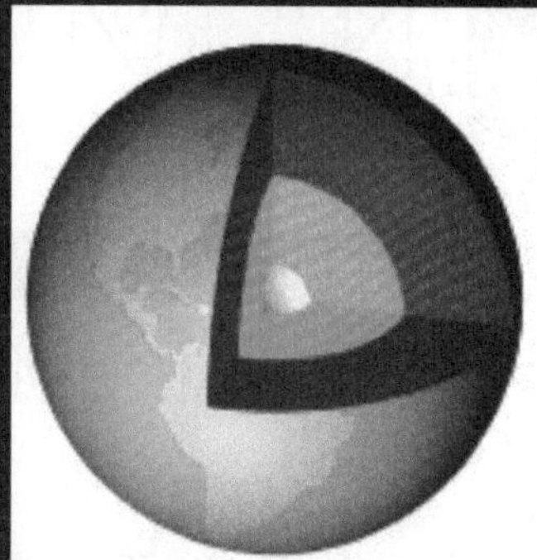

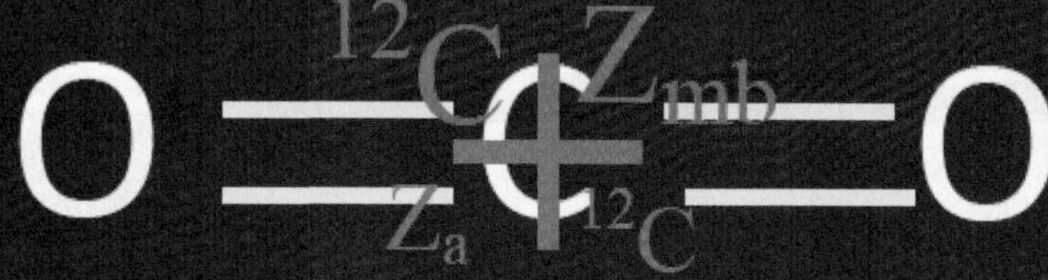

Urey and Miller (1953)

– Evidence of Abiogenesis, spontaneous formation of biomolecules.

There is some debate about the presence of methane at this early stage.

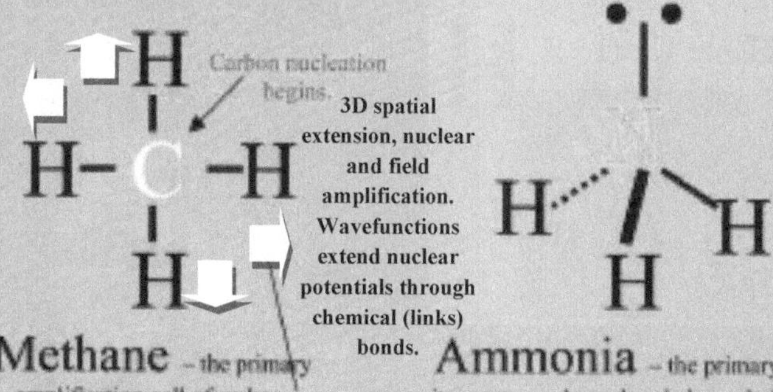

Methane – the primary amplification cell of carbon. Hydrogen as a basic membrane.

Ammonia – the primary nitrogenous valance bonded metabolic organelle of carbon cells.

H—H

Hydrogen – Hydrogen as a basic membrane (valance bonded neogenous metabolic organelles).

δ-

δ+H H δ+

Water – the primary hydration, mobile component (dipolarity). The primary oxidative valance bonded organelle of carbon cells.

Appendix 5 continued

Nuclear and field amplification in carbon continued (carbon centric logic).

This page demonstrates carbon centric logic, the primary mechanism which allows carbons properties to be conserved and amplified to produce all the variation of life on Earth. The most important of these is the chiral carbon in the amino acid boxed in yellow. This model clearly identifies the metabolic organelles in carbon cells, and the octet or cell cycle.

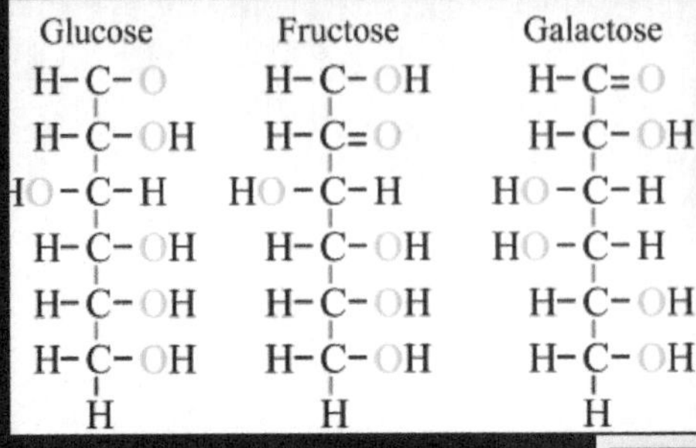

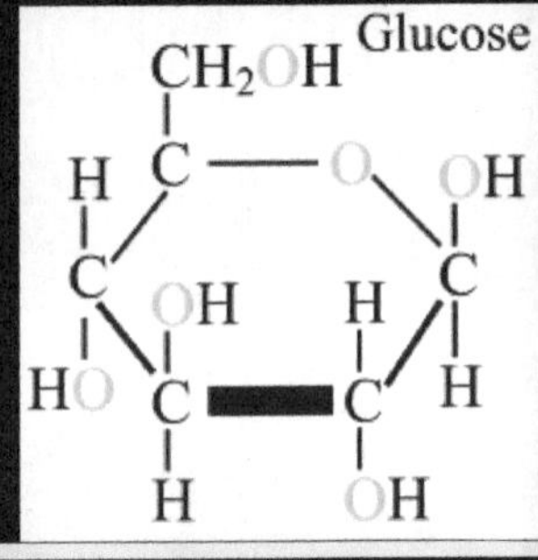

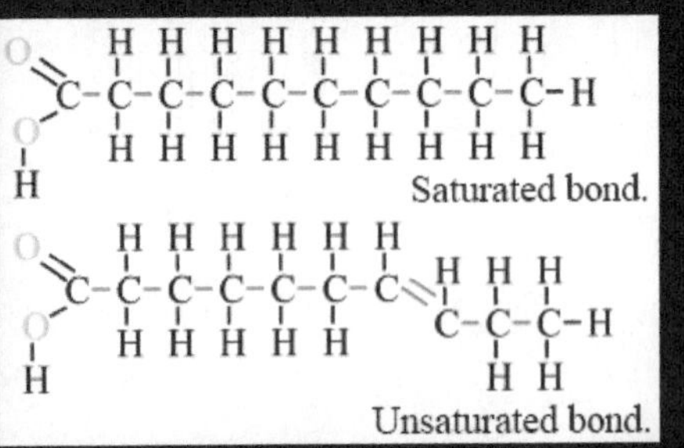

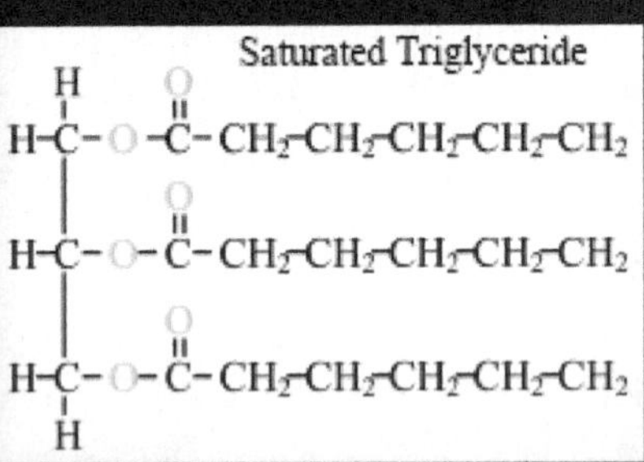

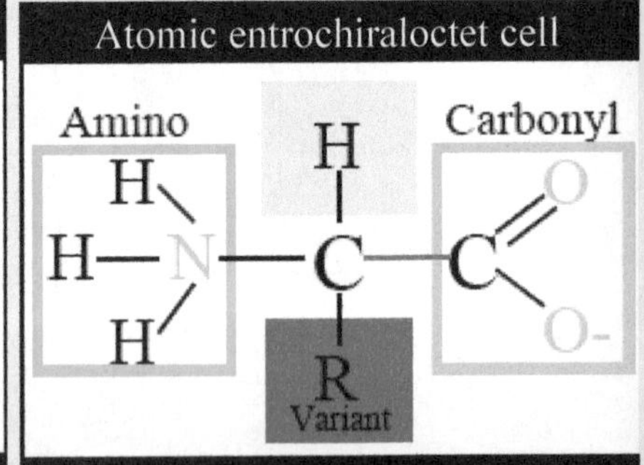

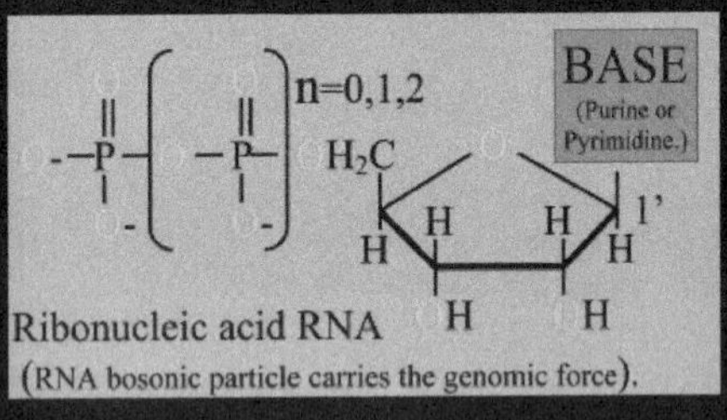

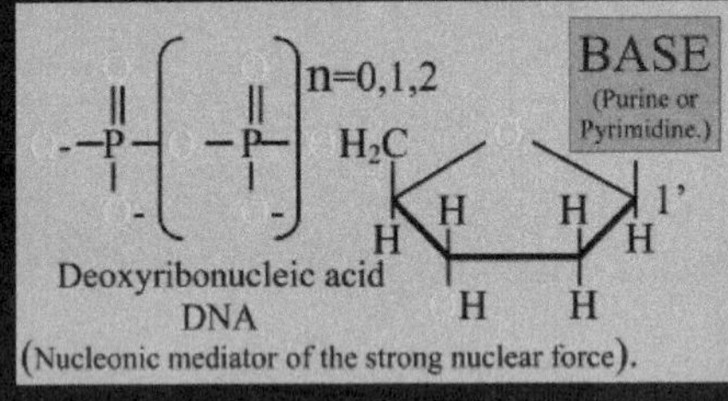

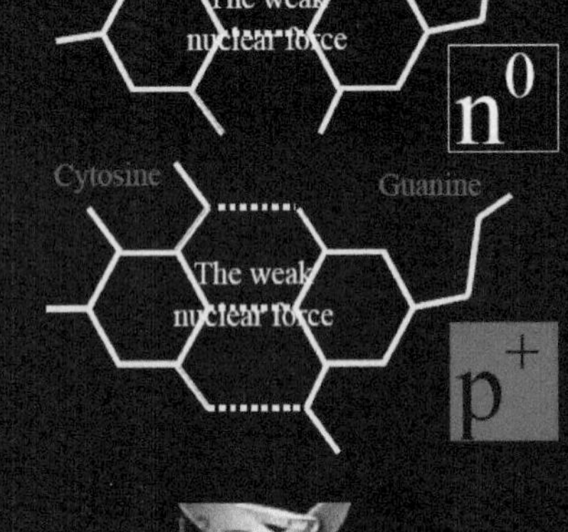

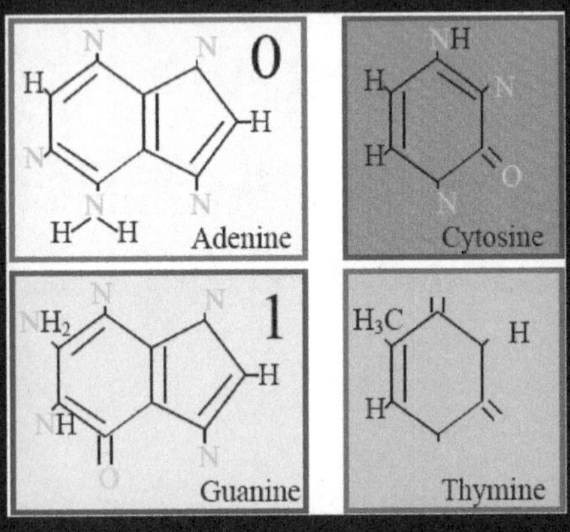

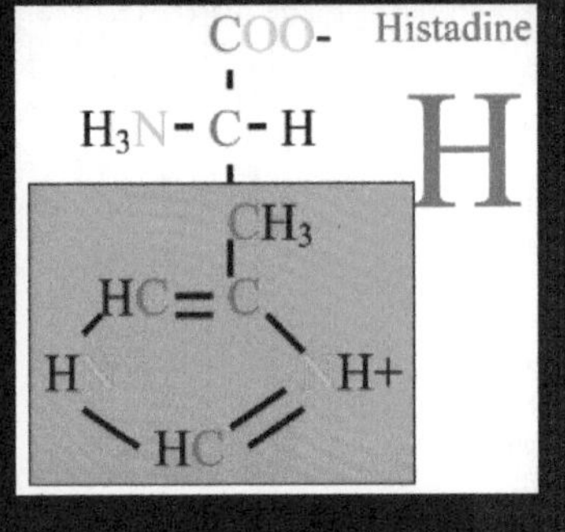

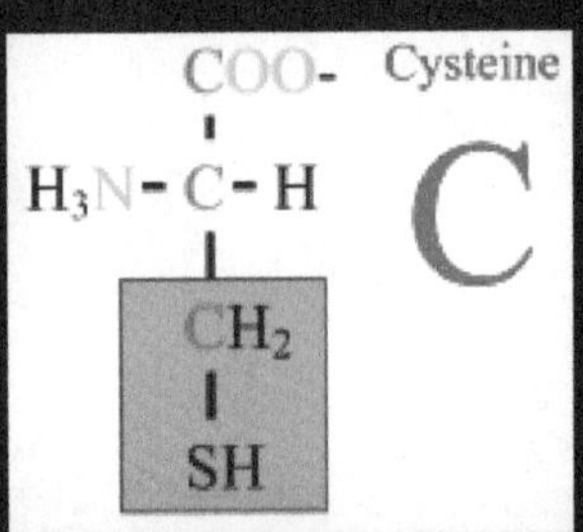

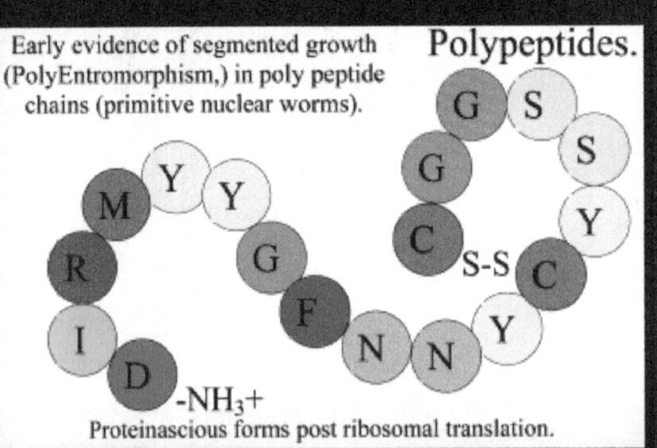

Early evidence of segmented growth (PolyEntromorphism,) in poly peptide chains (primitive nuclear worms).
Polypeptides.
Proteinascious forms post ribosomal translation.

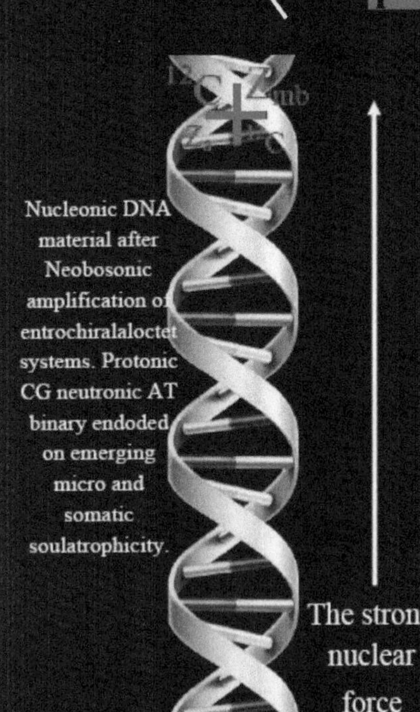

Nucleonic DNA material after Neobosonic amplification of entrochiralaloctet systems. Protonic CG neutronic AT binary endoded on emerging micro and somatic soulatrophicity.
The strong nuclear force

Appendix 6

The entromorphic conundrum, and the final word.

1. For any person to reject the evidence, logical interpretation and conclusions in this book it could only be achieved if the individual was to **reject the current laws of physics as being wrong; this theory simply extends and integrates the laws!**

2. This author's ability to create this book, to move limbs, eyes and to think and to live are only made possible because of immediate quantum and nuclear consequences in the atoms which comprise the body, as proof of an integrated link.

3. This effect in individual atoms is integrated to produce amplified results in living organisms (REDOX reactions are the basis of all biochemistry). Rhodopsin amplification cascade in the eyes, electron transport chain is the basis of all respiration as just two examples.

4. Also living beings represent an unbroken line of thermodynamically stable particle /wave systems. Through accretion and non local conservation the smallest entities from the Big Bang onwards build up to produce organic molecules, then cells, then tissues, then an organism and then groups of organisms producing fractional dimensions of natural scale.

5. Because of organic accretion living beings have no other discernable beginning other than the Big Bang. As a result any organism on the organic accretion pathway has a long line of thermodynamically successful and stable ancestors. It follows that any organism at the time of writing this book is 13.7 billion years old.

6. Just to finally reiterate, there is no point at which we can identify life's beginning, nor can we suggest that living self awareness begins at any time other than the Big Bang. This means that even the inanimate elements themselves are living entities, although their consciousness is a minute fraction of humans.

7. We do not describe life as 'carbon based' for no reason, why therefore should science be surprised to find that we have the physical properties of carbon and even more fundamentally hydrogen?

Appendicies 7 to 28

Mr Carbon Atom

The Bosley Cloud photo shoot

'The Entromorphic Gallery'

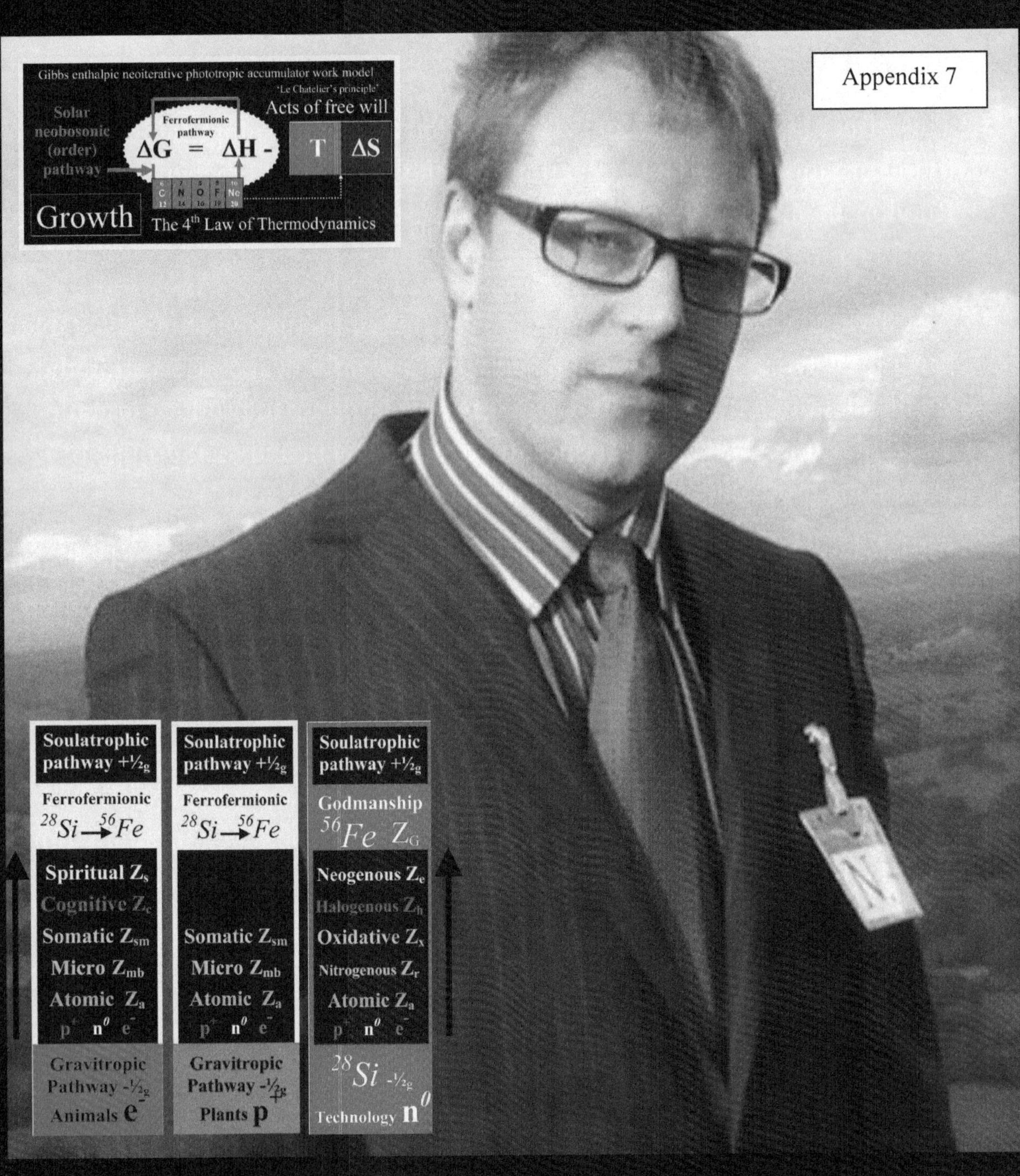

Mr Carbon Atom

Single biceps pose on Bosley Cloud on the 20th May 2011

Mr Carbon Atom

Relaxed front full body pose on Bosley Cloud on the 20th May 2011

Mr Carbon Atom

Relaxed back full body pose on Bosley Cloud on the 20th May 2011

Mr Carbon Atom

Relaxed front full body pose on Bosley Cloud on the 20th May 2011

The archer's front pose on Bosley Cloud on the 20th May 2011

Mr Carbon Atom

Hybridised front full body pose on Bosley Cloud on the 20th May 2011

Mr Carbon Atom

Relaxed back full body pose on Bosley Cloud on the 20th May 2011

Mr Carbon Atom

The archer's front pose on Bosley Cloud on the 20th May 2011

The archer's back pose on Bosley Cloud on the 20th May 2011

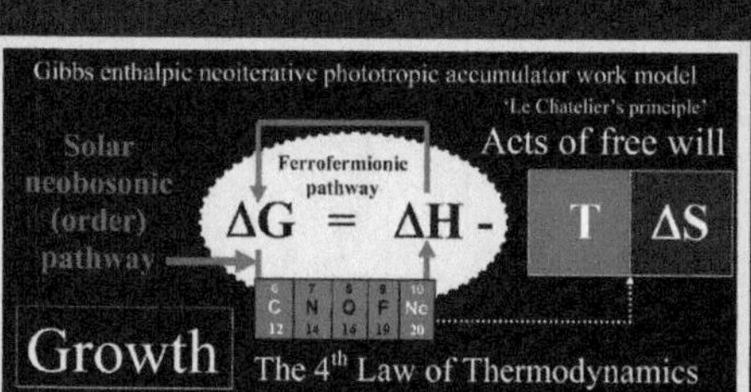

Mr Carbon Atom

Open arms back full body pose on Bosley Cloud on the 20th May 201

Mr Carbon Atom

Back double biceps pose on Bosley Cloud on the 20th May 2011

Mr Carbon Atom

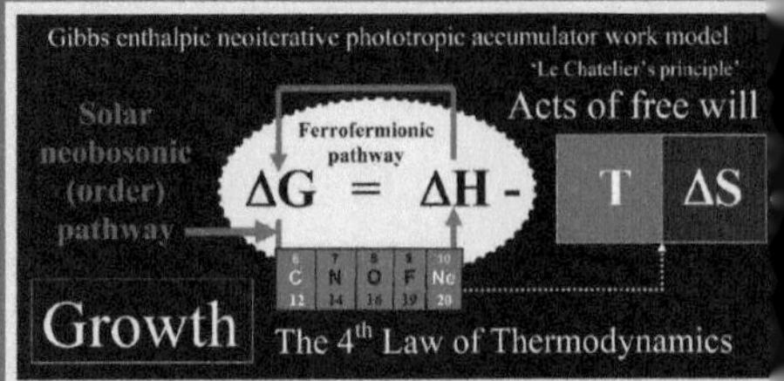

The open arms back pose on Bosley Cloud on the 20th May 2011

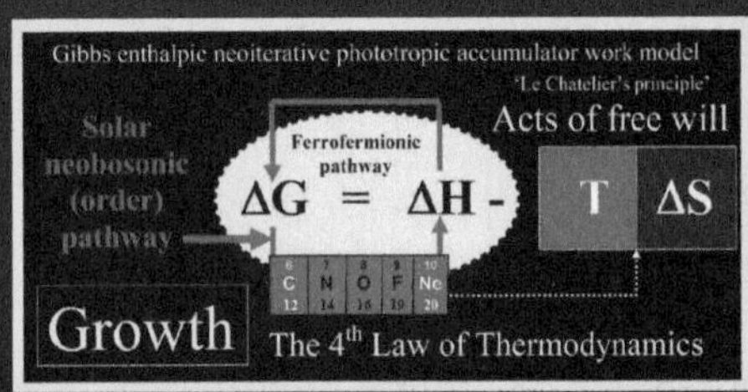

The sky reach pose on Bosley Cloud on the 20th May 2011

Mr Carbon Atom

Torso front hybridised pose on Bosley Cloud on the 20th May 2011

Mr Carbon Atom

Torso front ionised pose on Bosley Cloud on the 20th May 2011

Appendix 19

Gibbs enthalpic neoiterative phototropic accumulator work model
'Le Chatelier's principle'
Acts of free will
Solar neobosonic (order) pathway
Ferrofermionic pathway
$\Delta G = \Delta H - T \Delta S$
C N O F Ne
12 14 16 19 20
Growth
The 4th Law of Thermodynamics

+

^{12}C MR CARBON ATOM

VALANCE SHELL

L SHELL

K SHELL

THE IONISED STATES (THE INCOMPLETE OCTET)

Soulatrophic pathway $+\frac{1}{2}_g$
Godmanship
^{56}Fe Z_G
Neogenous Z_e
Halogenous Z_h
Oxidative Z_x
Nitrogenous Z_r
Atomic Z_a
p^+ n^0 e^-
^{28}Si $-\frac{1}{2}_g$
Technology n^0

Mr Carbon Atom

Relaxed front full body pose on Bosley Cloud on the 20th May 2011

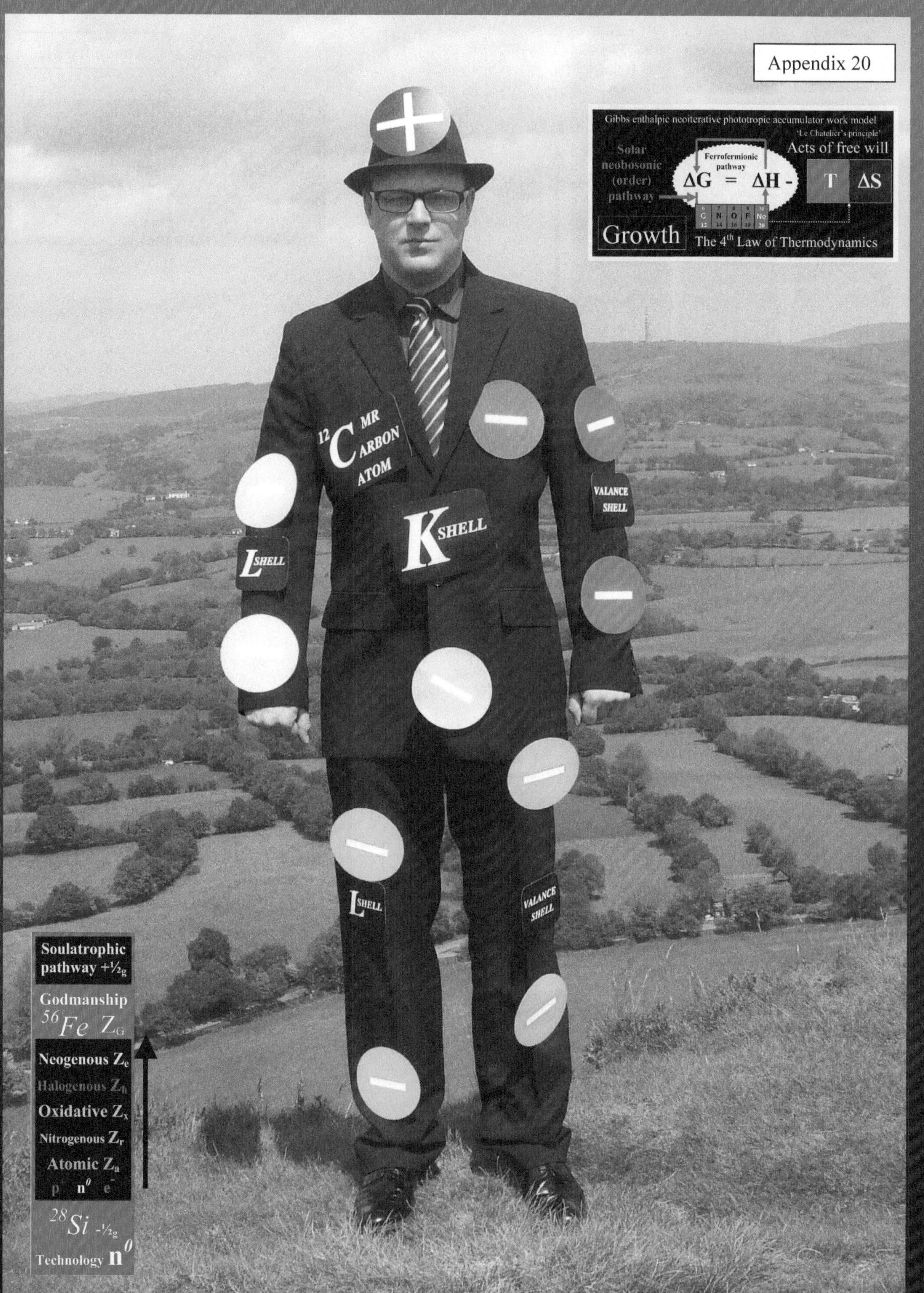

Mr Carbon Atom

Full body back hybridised pose on Bosley Cloud on the 20th May 2011

Appendix 21

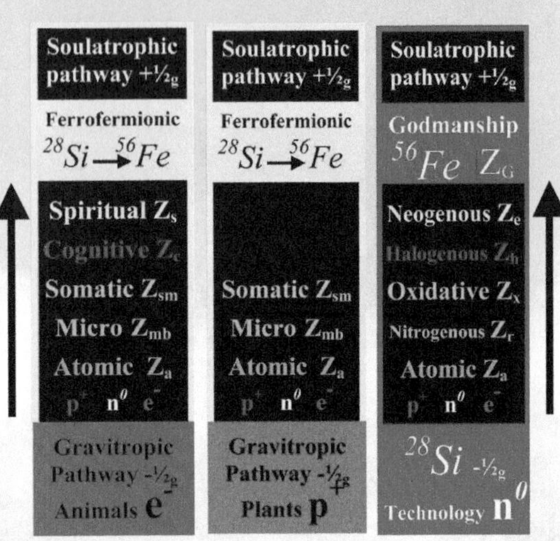

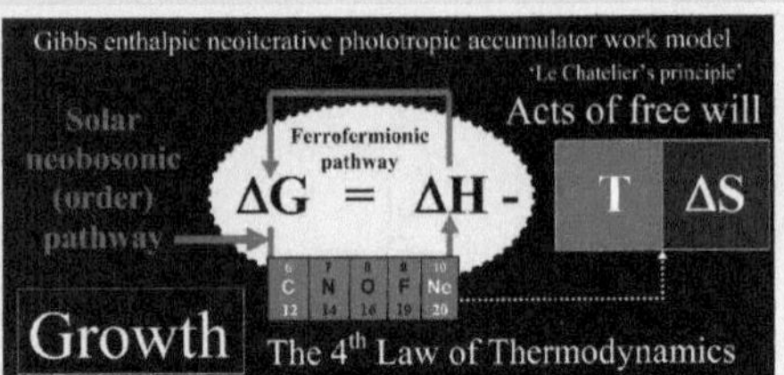

Mr Carbon Atom

Full body back hybridised pose on Prestbury on the 24th May 2011

Mr Carbon Atom

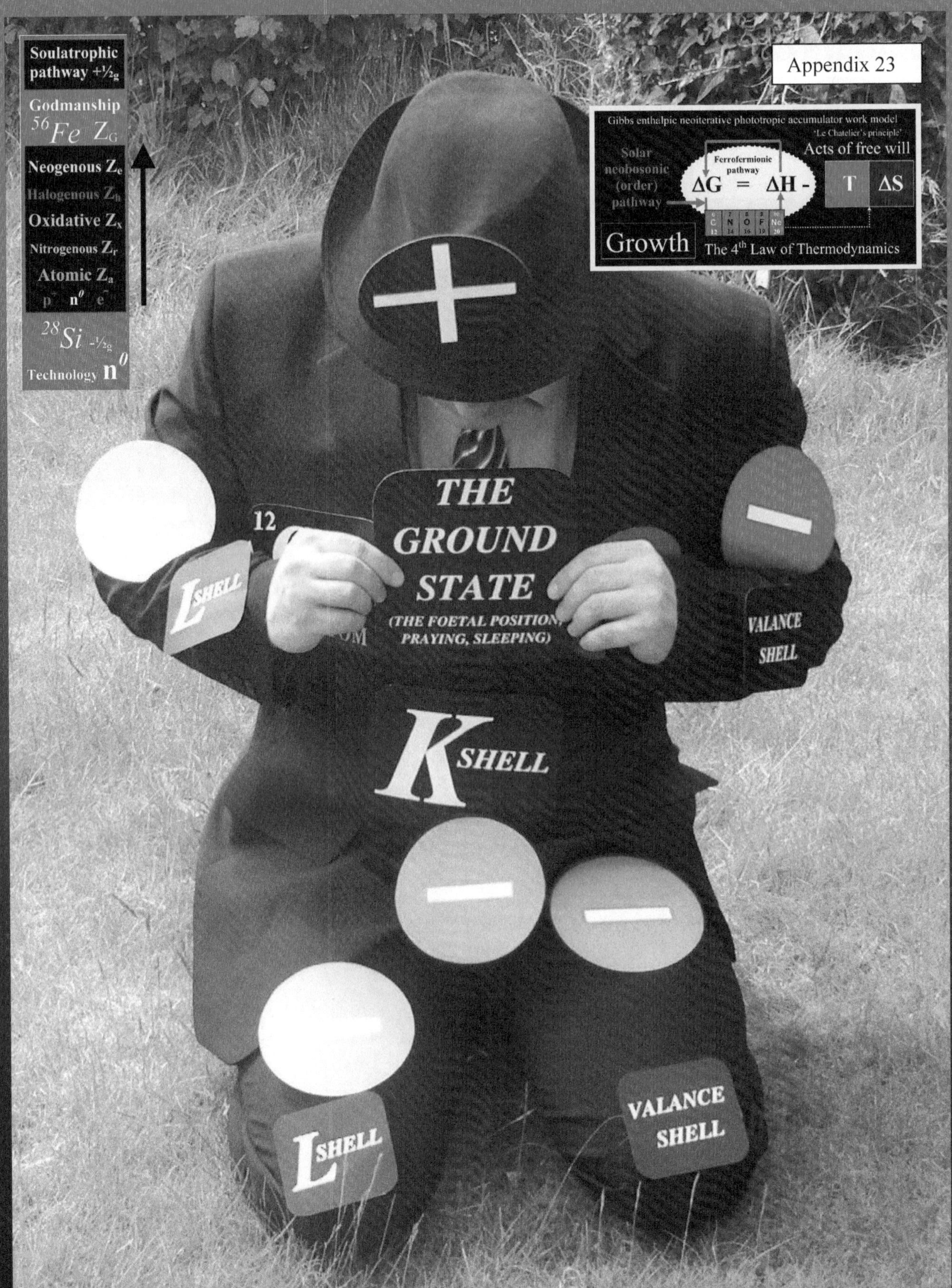

The ground state is a phenomenon observed in all animal organisms but has its origins in atomic physics; it includes the microorganisms and plants (spore formation). The limbs and head are fully retracted to the body, in humans this is observed in praying, sleeping and the 'foetal position'. A preservation pose.

Mr Carbon Atom

Front ionised ground state pose on Bosley Cloud on the 20th May 2011

The ground state is a phenomenon observed in all animal organisms but has its origins in atomic physics; it includes the microorganisms and plants (spore formation). The limbs and head are fully retracted to the body, in humans this is observed in praying, sleeping and the 'foetal position'. A preservation pose.

Mr Hydrogen Atom

A basic frontal pose with Pip 'my little electron'on Bosley Cloud on the 20th May 2011

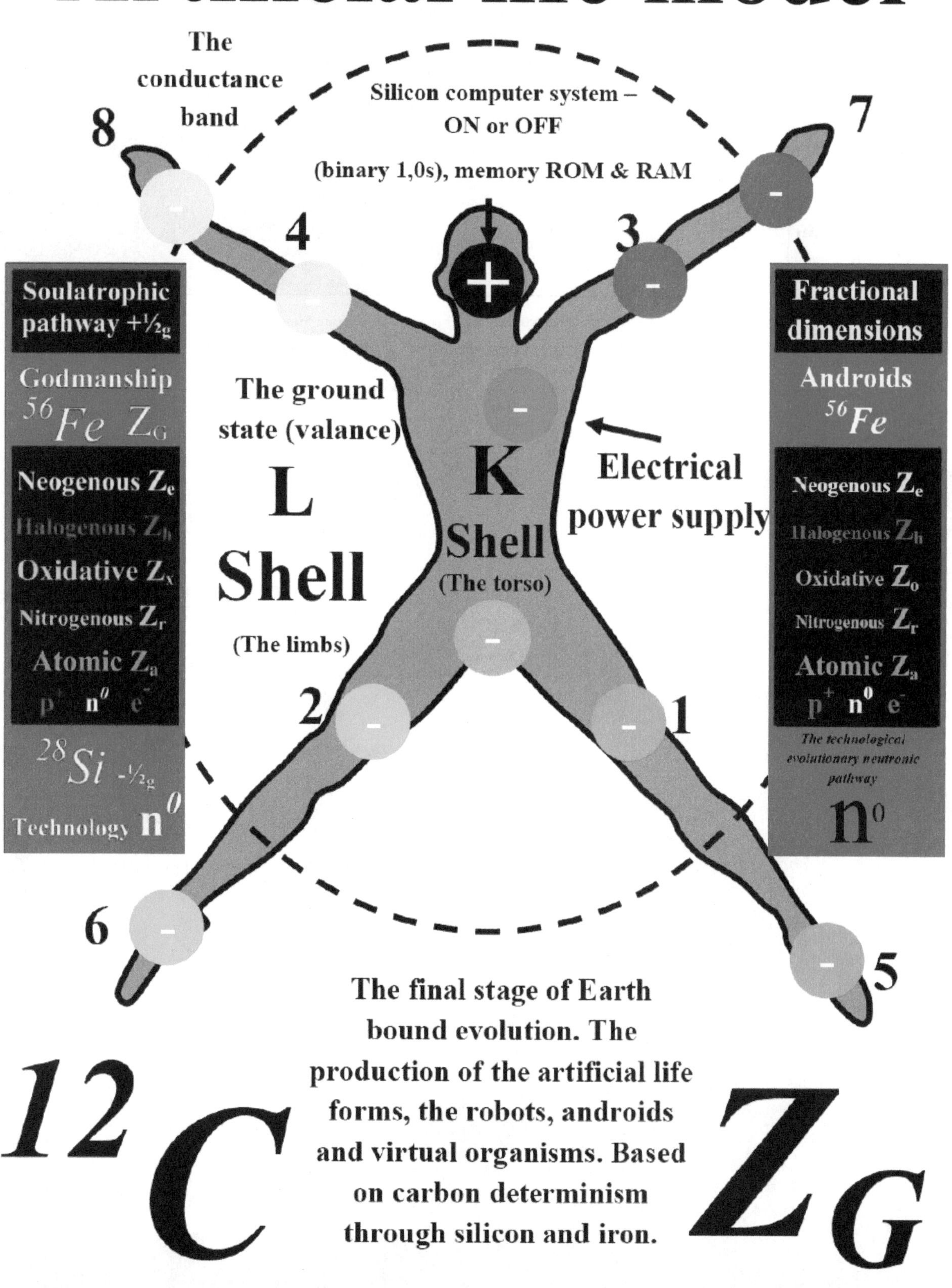
Entromorphic Fractional Dimension Standards (D)
Artificial life model
The conductance band
Silicon computer system – ON or OFF
(binary 1,0s), memory ROM & RAM
8
7
4
3
Soulatrophic pathway +½g
Godmanship
56Fe ZG
Neogenous Ze
Halogenous Zh
Oxidative Zx
Nitrogenous Zr
Atomic Za
p+ n0 e-
28Si -½g
Technology n0
The ground state (valance)
L
Shell
(The limbs)
K
Shell
(The torso)
Electrical power supply
Fractional dimensions
Androids
56Fe
Neogenous Ze
Halogenous Zh
Oxidative Zo
Nitrogenous Zr
Atomic Za
p+ n0 e-
The technological evolutionary neutronic pathway
n0
2
1
6
5
The final stage of Earth bound evolution. The production of the artificial life forms, the robots, androids and virtual organisms. Based on carbon determinism through silicon and iron.
12C
ZG

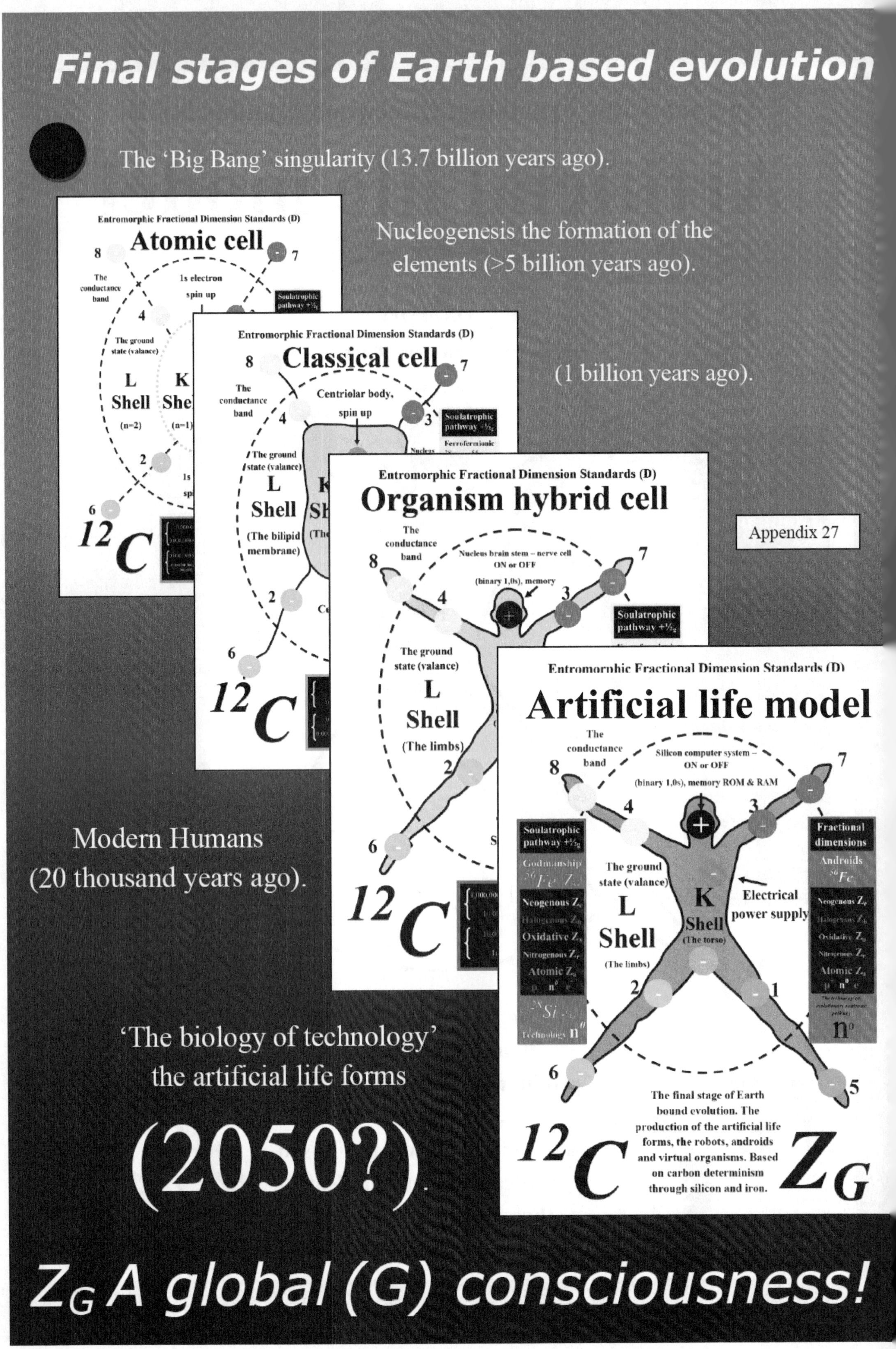

Final stages of Earth based evolution

The 'Big Bang' singularity (13.7 billion years ago).

Nucleogenesis the formation of the elements (>5 billion years ago).

(1 billion years ago).

Appendix 27

Modern Humans (20 thousand years ago).

'The biology of technology' the artificial life forms

(2050?).

The final stage of Earth bound evolution. The production of the artificial life forms, the robots, androids and virtual organisms. Based on carbon determinism through silicon and iron.

Z_G A global (G) consciousness!

Phototropism

4th Law of Thermodynamics. Naturally selective acts of free will (free energy)in open systems; all evolving conscious microstate ensembles seek the stability of iron.

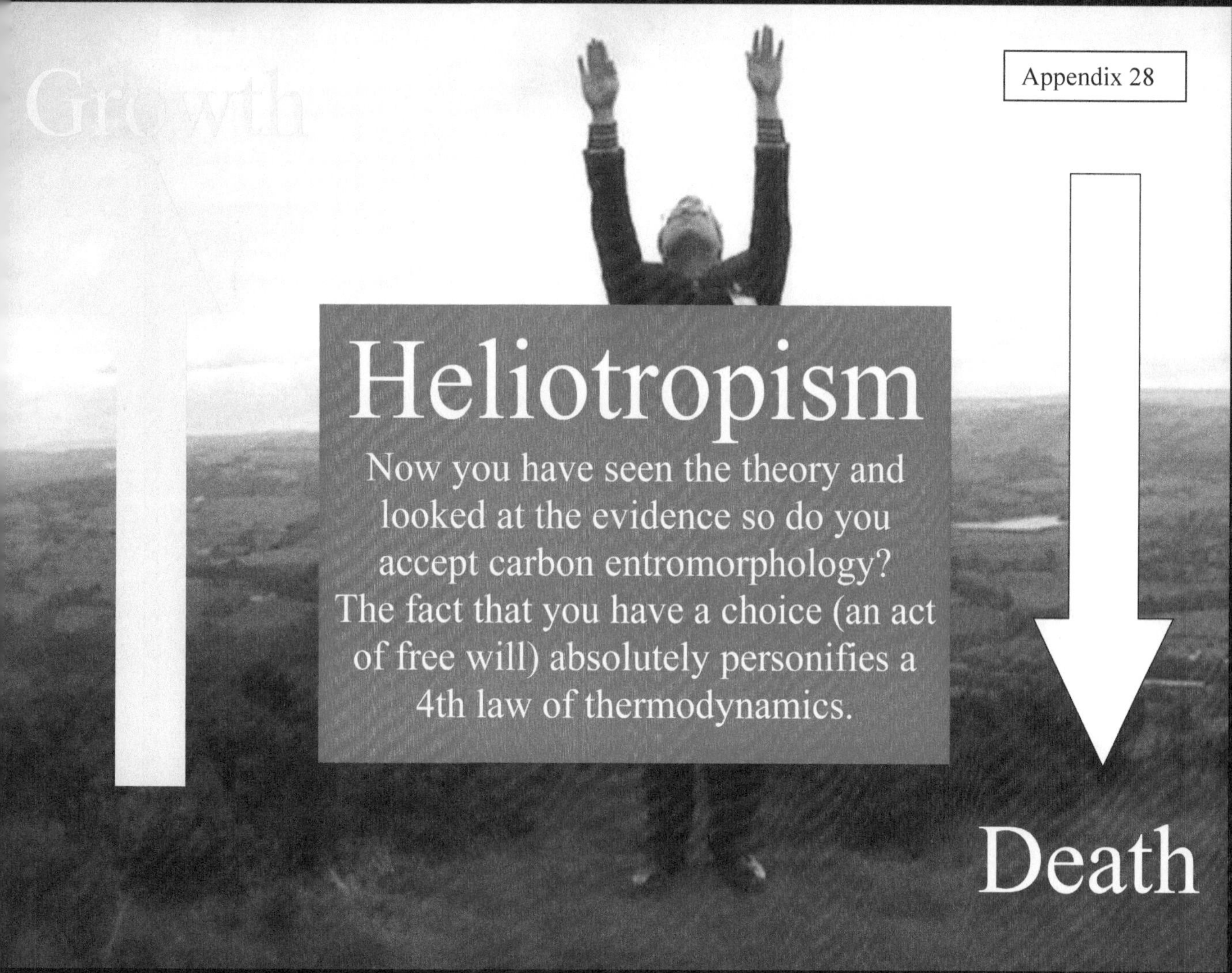

Appendix 28

Growth

Heliotropism

Now you have seen the theory and looked at the evidence so do you accept carbon entromorphology? The fact that you have a choice (an act of free will) absolutely personifies a 4th law of thermodynamics.

Death

2nd Law of Thermodynamics. The total entropy of an isolated system increases over time. As time passes the universe is becoming increasingly disordered.

Gravitropism

CPSIA information can be obtained at www.ICGtesting.com
Printed in the USA
LVOW012115081112

3155LVUK00004B/1/P